An Introduction to Primate Conse

An Introduction to Primate Conservation

EDITED BY

Serge A. Wich

School of Natural Sciences and Psychology, Liverpool John Moores University

Andrew J. Marshall

Department of Anthropology, Program in the Environment, and School of Natural Resources and Environment, University of Michigan

OXFORD
UNIVERSITY PRESS

OXFORD
UNIVERSITY PRESS

Great Clarendon Street, Oxford, OX2 6DP,
United Kingdom

Oxford University Press is a department of the University of Oxford.
It furthers the University's objective of excellence in research, scholarship,
and education by publishing worldwide. Oxford is a registered trade mark of
Oxford University Press in the UK and in certain other countries

First Edition published in 2016
Impression: 1

Published in the United States of America by Oxford University Press
198 Madison Avenue, New York, NY 10016, United States of America

British Library Cataloguing in Publication Data
Data available

Library of Congress Control Number: 2015957820

ISBN 978–0–19–870338–9 (hbk.)
ISBN 978–0–19–870339–6 (pbk.)

Printed and bound by
CPI Group (UK) Ltd, Croydon, CR0 4YY

Acknowledgements

This book never would have come to fruition without the hard work of many people.

We would like to thank Erin Vogel for initial discussions on a book about primate conservation that helped shape the current volume. We are very grateful to the five anonymous reviewers of our initial book proposal for providing thoughtful comments and valuable feedback that helped shape the content and structure of the volume. We hope that they will be pleased with the end result and see their input reflected in it.

We deeply appreciate all of the authors' willingness to contribute chapters to this volume. We realize that book chapters often are undervalued by academic institutions and are therefore frequently written in outside of normal working hours. We thank all of the authors for their hard work and excellent contributions.

We are extremely grateful to the external reviewers who provided extensive, constructive comments on one or more individual chapters: Marc Ancrenaz, Ralph Buij, Gail Campbell-Smith, Tim Caro, Colin Chapman, Guy Cowlishaw, Todd Disotell, Amy Dunham, Alejandro Estrada, Katie Feilen, Agustin Fuentes, Katie Gonder, Sandy Harcourt, Tatanya Humle, Simon Husson, Ian Singleton, Tony King, Fiona Maisels, Erik Meijaard, John Oates, Unai Pascual, Stuart Pimm, Ian Redmond, Johannes Refisch, Sadie Ryan, Chris Shepherd, Karen Strier, Matt Struebig, and Jessica Walsh. All were generous with their time and expertise, and their contributions improved this volume immensely.

We appreciate the support of our academic institutions: Liverpool John Moores University (SW) and the University of California, Davis and the University of Michigan (AJM).

We thank Oxford University Press and particularly Lucy Nash for their cheerful assistance and sound guidance through all phases of this project.

Finally, we thank our families, Tine, Amara, and Lenn (SW) and Raven (AJM) for their love, support, and encouragement.

Contents

List of contributors

Benjamin B. Beck Newark, MD, USA.

Jean P. Boubli School of Environment and Life Sciences, University of Salford, Salford, UK.

Genevieve Campbell The Biodiversity Consultancy, Cambridge, Cambridgeshire, UK.

Lounès Chikhi Instituto Gulbenkian de Ciência, Oeiras, Portugal, and Laboratoire Evolution & Diversité Biologique, UMR 5174, Centre National de la Recherche Scientifique (CNRS), Université Paul Sabatier, Toulouse, France.

Fay Clark Bristol Zoological Society, Bristol, UK.

Alison Cotton Bristol Zoological Society, Bristol, UK.

John E. Fa Division of Biology and Conservation Ecology, School of Science and the Environment, Manchester Metropolitan University, Manchester, UK, and Center for International Forestry Research (CIFOR), Bogor, Indonesia.

John Garcia-Ulloa Institute of Terrestrial Ecosystems, ETH Zürich, Zürich, Switzerland.

D.L.A. Gaveau Center for International Forestry Research, Bogor, Indonesia.

Thomas R. Gillespie Departments of Environmental Sciences and Environmental Health, Emory University & Rollins School of Public Health, Atlanta, GA, USA.

Benoit Goossens Danau Girang Field Centre, Sabah, Malaysia and Danau Girang Field Centre, School of Biosciences, Cardiff University, Cardiff, UK.

Colin Groves School of Archaeology & Anthropology, Australian National University, Canberra, Australia.

Josephine Head Chameleon Strategy, London, UK.

Aoife Healy Oxford Wildlife Trade Research Group, Oxford Brookes University, Oxford, UK.

Catherine Hill Department of Social Sciences, Faculty of Humanities and Social Sciences, Oxford Brookes University, Oxford, UK.

Alison P. Hillyer School of Applied Sciences, Bournemouth University, Poole, UK.

Tatyana Humle Durrell Institute of Conservation and Ecology (DICE), School of Anthropology and Conservation, University of Kent, Canterbury, UK.

Mitchell Irwin Department of Anthropology, Northern Illinois University, DeKalb, IL, USA.

Jessica Junker Mack Planck Institute for Evolutionary Anthropology, Leipzig, Germany.

Lian Pin Koh Environment Institute and School of Earth and Environmental Sciences, University of Adelaide, Adelaide, Australia.

Amanda H. Korstjens School of Applied Sciences, Bournemouth University, Poole, UK.

Andrew J. Marshall Department of Anthropology, Program in the Environment, and School of Natural Resources and Environment, University of Michigan, Ann Arbor, MI, USA.

Erik Meijaard Borneo Futures, Jakarta, Indonesia.

K.A.I. Nekaris Oxford Brookes University, Nocturnal Primate Research Group, Faculty of Humanities and Social Sciences, Department of Social Sciences, Oxford, UK.

Vincent Nijman Department of Anthropology, Oxford Brookes University, Oxford, UK.

Charles L. Nunn Department of Evolutionary Anthropology, Duke University, Durham, NC, USA.

Milena Salgado Lynn Danau Girang Field Centre, Sandakan, Sabah, Malaysia; Sabah Wildlife Department, Sabah, Malaysia; Wildlife Health, Genetics and Forensics Laboratory, Kota Kinabalu, Sabah, Malaysia; and School of Biosciences, and Sustainable Places Research Institute, Cardiff University, Cardiff, UK.

Christopher Schwitzer Bristol Zoological Society, Bristol, UK.

Pierfrancesco Sechi Consiglio Nazionale delle Ricerche, Istituto per lo Studio degli Ecosistemi, Sassari, Italy.

Nikki Tagg Centre for Research and Conservation, Royal Zoological Society of Antwerp, Antwerp, Belgium.

Sandra Tranquilli Centre for Biocultural History, Aarhus University, Aarhus, Denmark.

Serge A. Wich School of Natural Sciences and Psychology, Liverpool John Moores University, Liverpool, UK, and Institute for Biodiversity and Ecosystem Dynamics, University of Amsterdam, Amsterdam, The Netherlands.

CHAPTER 1

An introduction to primate conservation

Serge A. Wich and Andrew J. Marshall

The forest edge in North Sumatra, Indonesia. Copyright: conservationdrones.org.

1.1 Introduction

Primate conservation is a discipline that aims to develop the scientific understanding necessary to implement actions that will ensure long-term preservation of non-human primates and their habitats (Cowlishaw and Dunbar 2000). Interest in primate conservation has grown substantially in recent decades, reflected in substantial increases in the numbers of publications related to the topic (Figure 1.1). There are likely many reasons for this increase, including the growing threats to primates and an increased awareness of their biological, intellectual, economic, and ecological importance (Marshall and Wich, Chapter 2, this volume).

The last systematic treatment of primate conservation, Cowlishaw and Dunbar's widely read *Primate Conservation Biology*, was published a decade and a half ago (2000). At that time, 200 to 230 species of primates were recognized (Groves 1993), 31% of which were classified as threatened by the International Union for the Conservation of Nature (IUCN) (Baillie and Groombridge 1996). Since then, both the number of primate species recognized (479) and the proportion classified as threatened (48%) have risen sharply (Mittermeier *et al.* 2013). The increase in the number of threatened taxa is due to a variety of threats, especially important among these being hunting and the loss, fragmentation, and degradation of primate habitats

Wich, S.A. and Marshall, A.J., *An introduction to primate conservation*. In: *An Introduction to Primate Conservation*. Edited by: Serge A. Wich and Andrew J. Marshall, Oxford University Press (2016). © Oxford University Press.
DOI 10.1093/acprof:oso/9780198703389.003.0001

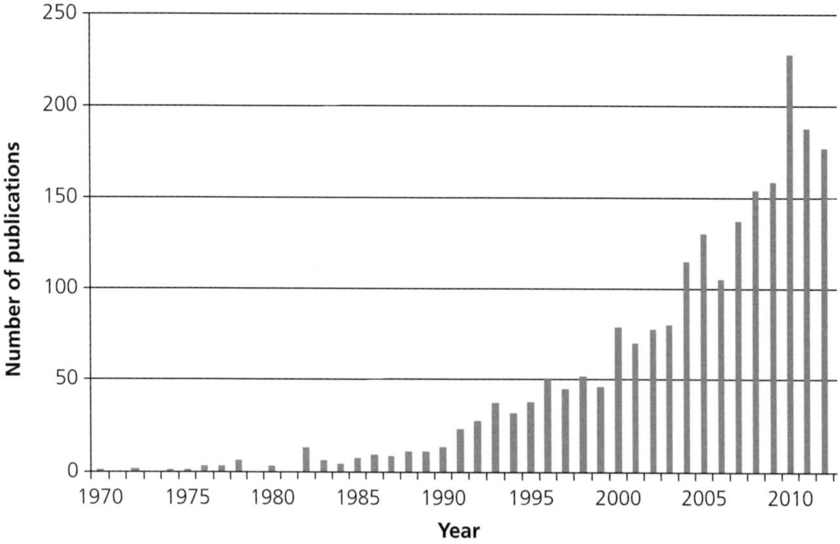

Figure 1.1 The number of scientific publications on primate conservation published annually from 1970 to 2012. Data reflect the number of publications returned using a keyword search in Scopus (http://www.scopus.com) for publications that contained both the keywords 'primate' and 'conservation' in each year. The data search was conducted on 11 November 2013.

(Bennett *et al.* 2002; Fa *et al.* 2005; Hansen *et al.* 2008; DeFries *et al.* 2010; FAO 2010; Meijaard *et al.* 2011; Wich *et al.* 2012; Hansen *et al.* 2013). While only one primate species is thought to have gone extinct over this period (*Procolobus badius waldroni*; McGraw and Oates (2007)), the threats to most primate populations are increasing and the survival of many populations and several species is severely in doubt (Mittermeier *et al.* 2013). In this edited volume we provide an overview of the current state of primate conservation, incorporating much of the information on primate behaviour, ecology, species distribution, conservation challenges, and conservation solutions that has emerged in the time since publication of Cowlishaw and Dunbar's influential book.

1.2 The primate order

Primates are an order of mammals. Recent analyses of molecular data indicate that primates diverged from other placental mammals roughly 75.8 million years ago (Steiper and Seiffert 2012). This date is considerably more recent than previous estimates (Tavaré *et al.* 2002) and comports more closely with the estimated divergence time based on the earliest known primate fossils (Smith. 2006). Although humans are primates, in this book, we use the term 'primate' to refer only to non-human primates.

The primate order is commonly divided into either Prosimians and Anthropoids (Fleagle 2013) or Strepsirhines and Haplorhines (e.g. Disotell 2008). The difference between these classifications depends on whether gradistic or phyletic information is used and as a result the position of the tarsiers differs (Figure 1.2). Traditionally, Tarsioidea were classified together with Lemuroidea (lemurs) and Lorisoidea (lorises) in the suborder Prosimii, but recent taxonomies based on phylogenetic relatedness have classified the tarsiers with the traditional members of the suborder Anthropoidea (Ceboidea, Cercopithecoidea, and Hominoidea) to form the new suborder Haplorhinii. Modern taxonomies typically follow this classification, with the lemurs and lorises grouped in Strepsirhinii and the monkeys, apes, and tarsiers forming Haplorhinii. The IUCN currently recognizes 16 primate families, 77 genera, 479 species, and 681 taxa (Mittermeier *et al.* 2013). They occur mainly in the tropical and subtropical areas of South America, Africa, and Asia (Figure 1.3).

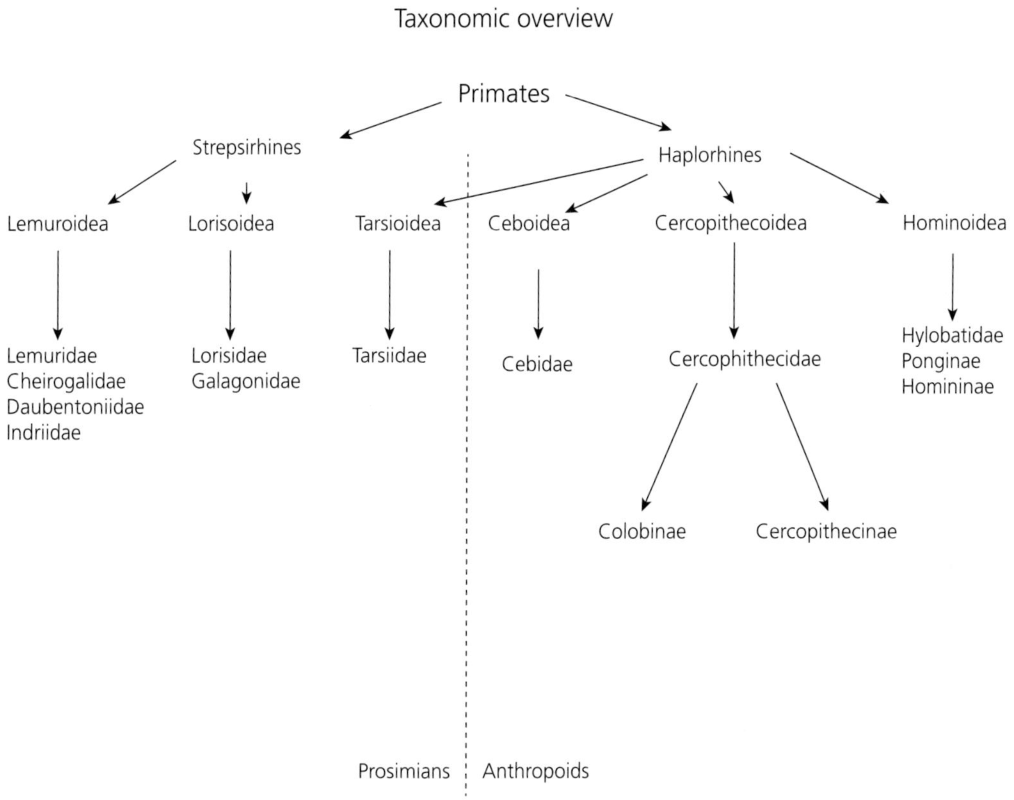

Figure 1.2 Overview of primate taxonomy illustrating the two alternative classification systems. Under one system the Tarsioidea are grouped with Ceboidea, Cercopithecoidea, and Hominoidea as Haplorhines (and Lemuroidea and Lorisoidea comprise the Strepsirhines); under the other Tarsioidea are grouped with Lemuroidea and Lorisoidea as Prosimians while the Ceboidea, Cercopithecoidea, and Hominoidea comprise the Anthropoids. After Stanford *et al.* 2013.

Primates exhibit tremendous variation in size from male gorillas (*Gorilla* sp.) that can weigh over 200 kg (Harcourt *et al.* 1981) to mouse lemurs (*Microcebus berthae*) that can weigh as little as 30 g (Dammhahn and Kappeler 2005). There is no single trait that unambiguously distinguishes primates from other mammals; instead they are generally identified by a shared suite of characteristic traits (Stanford *et al.* 2013). Primates have a fairly generalized mammalian body plan, with the exception of distinct anatomical adaptations for specialized locomotion (Fleagle 2013). Primates also generally exhibit opposable thumbs, which permit precision grasp and sensitive tactile manipulation—although in some strepsirrhines this trait is less fully developed. In contrast to humans, other primates also have an opposable big toe that allows them to use their feet as hands (Stanford *et al.* 2013). Most primates are also characterized by having flattened nails instead of claws or hooves, but in the family Callitrichidae these have evolved into claw-like nails on all digits except the big toe, presumably as an adaptation to their diet and arboreal lifestyle (Garber *et al.* 1996). Primates also have forward-facing eyes with overlapping fields of view, permitting stereoscopic vision that, often in combination with their grasping hands, is thought to have initially evolved as an adaptation to ancestral life in the trees (Elliot Smith 1924), catching small prey (Cartmill 1992), foraging on small objects in an arboreal environment (Sussman 1991), or avoiding predation by snakes (Isbell 2009).

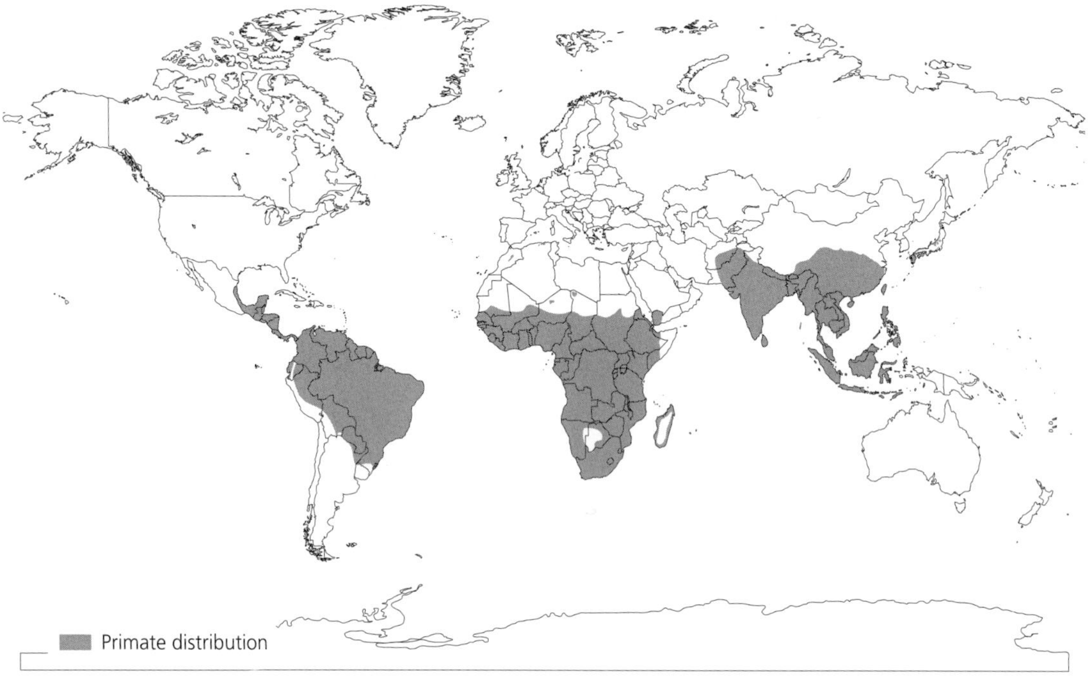

Figure 1.3 Global primate distribution. Based on the IUCN Red List Primate Distribution Layer: http://www.iucnredlist.org.

Most primate species consume a diverse diet, which is reflected in their rather unspecialized teeth compared to other mammals, although variation in dental characteristics such as incisor length, structure of molars, and enamel thickness has been linked to diet (Swindler 2002; Lucas 2004), particularly the components of the diet that are most difficult to process (Rosenberger 1992; Marshall and Wrangham 2007). In the past it was argued that all primates could be distinguished from other mammals due to their possession of a small part of the inner ear called the petrosal bulla (Szalay Frederick and Eric 1979), but it is now unclear whether the earliest primates possessed this trait (MacPhee *et al.* 1983). Another anatomical feature that has been of considerable interest is the bony post-orbital bar that completely or almost completely encloses the eyes of primates, but is absent in most other mammals (Fleagle 2013). The function of this trait is under considerable debate, but a recent hypothesis suggests that it protects the eye from the mechanical strains of chewing (Heesy 2005).

Primates also have unusual life histories. Compared to other mammals, primates in general have slower life histories, living longer, maturing more slowly, and producing smaller numbers of offspring (Charnov and Berrigan 1993). Proposed explanations for the slow life history exhibited by primates include their relatively large brains (Harvey *et al.* 1987), social complexity (Dunbar and Shultz 2007), cooperative breeding (van Schaik *et al.* 2012), arboreality (Shattuck and Williams 2010), and low levels of daily energy expenditure (Pontzer *et al.* 2014).

Most primate species live in permanent groups with complex social interactions (Mitani *et al.* 2012). These groups can be as large as a few hundred at the sleeping sites of gelada baboons or very small as in the case of one-male, multi-female groups of some langur species. There are, however, notable exceptions such as orang-utans (*Pongo* sp.) that do not live in permanent groups (Wich *et al.* 2009) and primate species that live in pairs (Palombit 1996). There are also species that exhibit fission–fusion social systems (*Pan* spp., *Ateles* spp.) where individuals reside in a large community but spend their time in smaller subgroups that frequently change in composition as individuals move among them. The evolution of group living in primates is typically considered to have been shaped by food distribution, feeding competition, predation, and infanticide (Wrangham 1980; Schaik 1983; Schaik and Hooff 1983; Sterck *et al.* 1997). Socio-ecological models aim to explain the diversity of primate social systems on the basis of those factors (Kappeler and van Schaik 2002). Although these models have shed light on many aspects of primate socio-ecology, they have yet to explain the full range of variation seen in extant primates (Kappeler and van Schaik 2002; Clutton-Brock and Janson 2012). In addition, while typical socio-ecological models imply great flexibility in sociality in response to ecological conditions, there is clearly substantial phylogenetic inertia in primate social systems as well (Schultz *et al.* 2011).

Although some primate species are found in colder areas in countries like China, Nepal, and Japan, most species live in warmer tropical climates (Figure 1.3). Within the tropics, primates occupy a large variety of habitat types that range from tropical evergreen rainforests to arid savannahs. Within forests, primates can often be found at all heights, with some species being exclusively arboreal and others almost exclusively terrestrial. In contrast to many other mammals, the majority of primate species are active during the day (diurnal), although there are a substantial number of primate species that are active at night (nocturnal) and some that are active around dawn or dusk (crepuscular).

Most primate species consume a broad diet that consists mostly of fruit, leaves, flowers, and other vegetative materials. Often protein-rich foods such as insects, and occasionally meat, are also added. Perhaps the most well-known primate meat eaters are the chimpanzees, who hunt colobus monkeys on a regular basis in many populations (Boesch 1994; Stanford 1996; Watts and Mitani 2002). Other primates, such as mandrills, have also been observed catching duikers (Kudo and Mitani 1985), while orang-utans have been witnessed to feed on a small nocturnal primate, the slow loris (*Nyctecebus coucang*) (Utami and van Hooff 1997; Hardu *et al.* 2012). Some primates also feed on gum from trees.

1.3 Threats to primates

A key goal of primate conservation is to determine and quantify the nature and magnitude of threats to primates. Below we note the most widely discussed threats to primates, most of which are considered in greater detail in subsequent chapters.

1.3.1 Habitat loss

There are three main threats to primate habitats: (1) total loss of habitat (e.g. due to clear cut logging or conversion to agriculture); (2) degradation of habitat (e.g. due to logging or mining); and (3) fragmentation of habitat (e.g. due to loss of surrounding habitat or roads). Habitat loss is the most severe of these because it reduces the total amount of suitable habitat available to a species and often results in the loss of all individuals contained in the destroyed habitat patch. Habitat loss is straightforward to quantify in the field because it is a complete replacement of natural vegetation by something else (e.g. agriculture, timber plantations, roads, houses). Despite this, assessments of habitat loss from satellites can be more complex because some mature timber plantations are difficult to distinguish from forest (Hansen *et al.* 2013). Forests in the tropics, with the exception of Brazil, are declining at faster rates in recent years (Hansen *et al.* 2013); a worrying trend because most primates disappear once their habitat is cleared (e.g. chimpanzees; Campbell *et al.* (2008)).

1.3.2 Habitat degradation

Habitat degradation most often occurs due to legal or illegal logging. The effects of such degradation on primates differ greatly, depending on the species, logging intensity, the methods used, and the time since degradation (Johns and Skorupa 1987; Thomas 1991; Peres 1993; Chapman *et al.* 2000; Cowlishaw *et al.* 2009; Husson *et al.* 2009). An important unanswered question for research on primate survival in altered habitats is whether primates can survive in areas that are under continuous, albeit well-managed logging, or whether a period of recovery between logging bouts is required (Chapman *et al.* 2000; Ancrenaz *et al.* 2010). Although the effects of habitat degradation on primates have been studied for decades and knowledge continues to accrue, we have yet to develop a predictive model of primate persistence under varying logging intensities.

1.3.3 Habitat fragmentation

The effect of habitat fragmentation on primate survival depends on a variety of factors, including fragment size, species' home range size and diet, inter-patch distance, and the nature of the matrix in which the habitat fragments are embedded (Chiarello 1999; Michalski and Peres 2005; Boyle and Smith 2010; Meijaard *et al.* 2010).

1.3.4 Drivers of habitat loss, fragmentation, and degradation

A main cause of habitat loss and fragmentation is conversion to agricultural lands (Gibbs *et al.* 2010). During the 1980s and the 1990s more than 55% of new agricultural land in the tropics was created by replacing primary forest, while another 28% came from conversion of logged forests (Gibbs *et al.* 2010). Other proximate drivers of habitat loss, fragmentation, and degradation are wood extraction and infrastructure development (Geist and Lambin 2002). Ultimate drivers of deforestation include economic factors such as the demand for timber; institutional factors related to land-use policies; and demographic factors such as population growth (Geist and Lambin 2002). A recent analysis of the drivers of tropical deforestation showed that it is driven by urban population growth and urban and international demands for agricultural products; perhaps somewhat counterintuitively, rural population growth was not associated with deforestation (DeFries *et al.* 2010).

1.3.5 Hunting

Hunting occurs both within forests, when people hunt primates for either their own consumption or for trade, and at the interface between forests and agricultural lands, when primates raid crops (Peres 2001; Fa and Brown 2009; Meijaard *et al.* 2011). For many species hunting is not sustainable and can lead to local extinctions (Fa and Brown 2009). Killing also often occurs in the context of the pet trade because collection of live primates for trade often entails killing of other individuals, especially mothers (Malone *et al.* 2002; Shepherd *et al.* 2004; Nijman *et al.* 2011; Stiles *et al.* 2013). Although the trade in live primates is difficult to quantify, estimates are that tens if not hundreds of thousands of primates are caught for the pet trade per year (Nijman *et al.* 2011).

1.3.6 Disease

The threat of disease to primates is perhaps best understood in the African great apes. It is well documented that the Ebola virus has decimated gorilla and chimpanzee populations in Gabon and Congo (Walsh *et al.* 2003; Bermejo *et al.* 2006). In addition, the increasing contact between humans and great apes through hunting, encroachment into ape habitat, research, and tourism has led to numerous cases where primates have been negatively affected by diseases acquired from humans (Wallis and Lee 1999). Recent research on chimpanzees in Ivory Coast has unequivocally established that disease can directly spread from humans to apes (Koendgen *et al.* 2008).

1.3.7 Climate change

There is increasing concern that climate change will negatively affect primates, through altering their time budgets, changing their food supplies, or reducing their habitats (Lehmann *et al.* 2010; Gregory *et al.* 2012). There has to date been little research conducted on the potential impact of climate change on primates, and the effects of climate change are unlikely to be easy to predict. This is partly due our limited understanding of the nature and rates of change in key factors such as food species distribution and phenology, primate species' dispersal abilities, and inter-specific interactions. In addition, potential interactions among these factors, and the adaptive potential of primate species to a changing climate, are virtually unknown (Marshall and Wich, Chapter 18, this volume).

1.3.8 Roads

The global road network is expanding rapidly as a result of increased demand for natural resources and arable land. Although the importance of this infrastructure for human health and development is clear, it is raising many problems of great concern to conservationists. Roads facilitate access for hunters and loggers, reduce forest cover, fragment primate habitats, and kill wildlife through collisions with vehicles (Laurance and Balmford 2013; Caro *et al.* 2014). Road construction is therefore likely to limit primate distribution and reduce primate population density (Laurance *et al.* 2008; Gaveau *et al.* 2009; Laurance *et al.* 2009). There is thus an urgent need for careful planning so that future roads will not be established in areas that are important for conservation (Laurance and Balmford 2013; Laurance *et al.* 2013). In addition, greater consultation between infrastructure developers and natural resource managers, a case-by-case examination of each proposed road, effective enforcement of traffic speed and volume, and the development of policies that make international development aid conditional upon appropriate assessments of the long-term costs of road development are also required (Caro *et al.* 2014).

1.4 Approaches to primate conservation

Given the magnitude and multitude of threats it is easy to despair for the persistence of primate populations; the challenges are great and the available resources are scant. Nevertheless, primate conservationists have successfully protected primate populations, sometimes despite apparently insurmountable hurdles (McNeilage 1996; Gray *et al.* 2013). Successes tend to be the result of a combination of multiple strategies and tactics and a careful tailoring of conservation approaches to the particular conditions and challenges on the ground. Below we briefly highlight some of the major current approaches to primate conservation. This is not meant to be an exhaustive list, rather we highlight some that are well represented in the literature or are novel and might be promising for the future.

1.4.1 Protected areas

Conservation efforts for primates and biodiversity have traditionally focused on the establishment of protected areas (Terborgh *et al.* 2002; Chape *et al.* 2005) and approximately 12% of the Earth's terrestrial surface area carries some level of protected status (Gaveau *et al.*, Chapter 12, this volume). Researchers have conducted various studies to evaluate the effectiveness of protected areas at preventing wildlife habitat loss, degradation, and fragmentation (Gaveau *et al.*, Chapter 12, this volume; DeFries *et al.* 2005; Andam *et al.* 2008; Joppa *et al.* 2008; Nagendra 2008; Newmark 2008; Joppa and Pfaff 2009). The conclusions of these studies differ, at least in part due to the use of different analytical approaches. In addition, broad conclusions are unlikely to be appropriate because conditions, threats, opportunities, resources, and implementation differ widely across protected areas. A recent large-scale study found that only half of the protected areas assessed protected biodiversity well (Laurance *et al.* 2012). There is an ongoing debate over why protected areas might not be as efficient as we might wish. Part of this debate focuses on the management of protected areas and the issue of whether strictly protected areas that exclude local community participation are more effective than methods in which local communities have a more participatory role in management (Bruner *et al.* 2001; Wilshusen *et al.* 2002; Hayes 2006; Hansen and DeFries 2007). Other aspects of this debate focus on how protected area management plans should interact with larger scale landscape conservation strategies for multi-use landscapes (Meijaard, Chapter 13, this volume).

1.4.2 Law enforcement

A recurrent structural problem for primate conservation is the lack of appropriate law enforcement in many of the countries that primates inhabit (Struhsaker *et al.* 2005; Fischer 2008; Tranquilli *et al.* 2012). A recent Africa-wide review showed that law enforcement is the best predictor of ape persistence in resource management areas (Tranquilli *et al.* 2012). Law enforcement is necessary to prevent illegal habitat conversion in protected areas or lands where alteration is prohibited (e.g. steep areas in Indonesia; Wich *et al.* (2011)). It is also required to reduce the intense, illegal hunting of many primate populations (Fa and Tagg, Chapter 9, this volume; Stiles *et al.* 2013). Although improving law enforcement is complicated, organizations that commit themselves to the endeavour—such as the Last Great Ape Organization (LAGA) and the Project for the Application of Law for Fauna (PALF)—appear to hold promise.

1.4.3 Payments for ecosystem services

There is a growing realization that natural areas provide a large number of services that are beneficial to human society. These benefits include reducing the impact of floods, pollination services, capturing carbon dioxide from the atmosphere, and maintaining a carbon stock (MEA 2005). Estimating the economic value of such services and explicitly incorporating their value (and the potential cost of their loss) into policy discussions may provide a valuable new approach to achieving conservation goals. Recognition of ecosystem services as financial assets can provide an incentive for their protection. For example, as a result of efforts to curb the emission of greenhouse gases, a considerable amount of attention has been paid to reducing carbon emissions through placing an economic value on sequestered carbon through efforts such as REDD (Reducing Emissions from Deforestation and forest Degradation) or more recently REDD +, which incorporates the biodiversity benefits of carbon sequestration. Despite their initial promise, establishing REDD + projects remains a challenge (Garcia-Ulloa and Koh, Chapter 16, this volume).

1.5 Overview of the book

The chapters in this book are grouped into three sections: (1) background and conceptual issues, (2) threats, and (3) solutions. In the first section, the authors consider why we should conserve primates (Marshall and Wich, Chapter 2), summarize the conservation status of primates (Cotton *et al.*, Chapter 3), discuss species concepts and their relevance to conservation (Groves, Chapter 4), review primate conservation genetics (Salgado-Lynn *et al.*, Chapter 5) and describe primate abundance and distributions (Campbell *et al.*, Chapter 6). The second section includes discussion of threats from habitat destruction and degradation (Irwin, Chapter 7), primate trade (Nijman and Healey, Chapter 8), hunting (Fa and Tagg, Chapter 9), infectious diseases (Nunn and Gillespie, Chapter 10), and climate change (Korstjens and Hillyer, Chapter 11). The third section considers solutions to primate conservation challenges from several perspectives: protected areas (Gaveau *et al.*, Chapter 12), landscape mosaics (Meijaard, Chapter 13), human–primate conflict (Humle and Hill, Chapter 14), reintroduction (Beck, Chapter 15), ecosystem services (Garcia-Ulloa and Koh, Chapter 16), and evidence-based conservation (Tranquilli, Chapter 17). We conclude the book with a consideration of some future directions for primate conservation research (Marshall and Wich, Chapter 18).

Acknowledgements

We thank Guy Cowlishaw for reviewing the chapter.

References

Ancrenaz, M., Ambu, L., Sunjoto, I., Ahmad, E., Manokaran, K., *et al.* (2010). 'Recent surveys in the forests of Ulu Segama Malua, Sabah, Malaysia, show that orangutans (P.p. morio) can be maintained in slightly logged forests.' *PLoS One* **5**(7): e11510.

Andam, K. S., Ferraro, P. J., Pfaff, A., Sanchez-Azofeifa, G. A., and J. A. Robalino (2008). Measuring the effectiveness of protected area networks in reducing deforestation. *Proceedings of the National Academy of Sciences* **105**(42): 16089–160944.

Baillie, J. and Groombridge, B. (1996). 1996 IUCN Red List of Threatened Animals. Gland, Switzerland: IUCN.

Bennett, E. L., Milner-Gulland, E. J., Bakarr, M., Eves, H. E., Robinson, J. G., *et al.* (2002). Hunting the world's wildlife to extinction. *Oryx* **36**(4): 328–329.

Bermejo, M., Rodriguez-Teijeiro, J. D., Illera, G., Barroso, A., Vila, C., *et al.* (2006). Ebola outbreak killed 5000 gorillas. *Science* **314**(5805): 1564.

Boesch, C. (1994). Cooperative hunting in wild chimpanzees. *Animal Behaviour* **48**: 653–667.

Boyle, S. A. and Smith, A. T. (2010). Can landscape and species characteristics predict primate presence in forest fragments in the Brazilian Amazon? *Biological Conservation* **143**(5): 1134–1143.

Bruner, A. G., Gullison, R. E., Rice, R. E., and da Fonseca, G. A. B. (2001). Effectiveness of parks in protecting topical biodiversity. *Science* **291**: 125–128.

Campbell, G., Kuehl, H., N'Goran Kouame, P., and Boesch, C. (2008). Alarming decline of West African chimpanzees in Cote d'Ivoire. *Current Biology* **18**(19): R903–R904.

Caro, T., Dobson, A., Marshall, A. J., and Peres, C. A. (2014). Compromise solutions between conservation and road building in the tropics. *Current Biology* **24**(16): R722–R725.

Cartmill, M. (1992). New views on primate origins. *Evolutionary Anthropology: Issues, News, and Reviews* **1**(3): 105–111.

Chape, S., Harrison, J., Spalding, M., and Lysenko, I. (2005). Measuring the extent and effectiveness of protected areas as an indicator for meeting global biodiversity targets. *Philosophical Transactions of the Royal Society B: Biological Sciences* **360**(1454): 443–455.

Chapman, C. A., Balcomb, S. R. Gillespie, T. R., Skorupa, J. P., and Struhsaker, T. T. (2000). Long-term effects of logging on African primate communities: a 28-year comparison from Kibale National Park, Uganda. *Conservation Biology* **14**(1): 207–217.

Charnov, E. L. and Berrigan, D. (1993). Why do female primates have such long lifespans and so few babies? Or life in the slow lane. *Evolutionary Anthropology: Issues, News, and Reviews* **1**(6): 191–194.

Chiarello, A. G. (1999). Effects of fragmentation of the Atlantic forest on mammal communities in south-eastern Brazil. *Biological Conservation* **89**(1): 71–82.

Clutton-Brock, T. and Janson, C. (2012). Primate socioecology at the crossroads: past, present, and future. *Evolutionary Anthropology: Issues, News, and Reviews* **21**(4): 136–150.

Cowlishaw, G. and Dunbar, R. I. (2000). *Primate Conservation Biology*. Chicago, Il: University of Chicago Press.

Cowlishaw, G., Pettifor, R. A., and Isaac, N. J. (2009). High variability in patterns of population decline: the importance of local processes in species extinctions. *Proceedings of the Royal Society of London B: Biological Sciences* **276**(1654): 63–69.

Dammhahn, M. and Kappeler, P. M. (2005). Social system of Microcebus berthae, the world's smallest primate. *International Journal of Primatology* **26**(2): 407–435.

DeFries, R., Hansen, A., Newton, A. C., and Hansen, M. C. (2005). Increasing isolation of protected areas in tropical forests over the past twenty years. *Ecological Applications* **15**(1): 19–26.

DeFries, R. S., Rudel, T. Uriarte, M., and Hansen, M. (2010). Deforestation driven by urban population growth and agricultural trade in the twenty-first century. *Nature Geoscience* **3**(3): 178–181.

Disotell, T. R. (2008). Primate phylogenetics. *eLS*. DOI: 10.1002/9780470015902.a0005833.pub2.

Dunbar, R. I. and Shultz, S. (2007). Evolution in the social brain. *Science* **317**(5843): 1344–1347.

Elliot Smith, G. E. (1924). *The Evolution of Man*. London and New York, NY: Oxford University Press.

Fa, J. E. and Brown, D. (2009). Impacts of hunting on mammals in African tropical moist forests: a review and synthesis. *Mammal Review* **39**: 231–264.

Fa, J. E., Ryan, S. F., and Bell, D. J. (2005). Hunting vulnerability, ecological characteristics and harvest rates of bushmeat species in afrotropical forests. *Biological Conservation* **121**(2): 167–176.

FAO (2010). Global Forest Resources Assessment. Rome: FAO.

Fischer, F. (2008). The importance of law enforcement for protected areas: don't step back! Be honest–protect!' *GAIA-Ecological Perspectives for Science and Society* **17**(Supplement 1): 101–103.

Fleagle, J. G. (2013). *Primate Adaptation and Evolution* 3 edn. San Diego, CA and London: Academic Press.

Garber, P. A., Rosenberger, A. L., and Norconk, M. A. (1996). Marmoset misconceptions. In: Norconk, M. A., Rosenberger, A. L., and Garber, P. A. (Eds), *Adaptive Radiations of Neotropical Primates*. pp. 87–95. Springer.

Gaveau, D. L., Wich, S., Epting, J., Juhn, D. Kanninen, M., *et al.* (2009). The future of forests and orangutans (Pongo abelii) in Sumatra: predicting impacts of oil palm plantations, road construction, and mechanisms for reducing carbon emissions from deforestation. *Environmental Research Letters* **4**(3): 034013.

Geist, H. J. and Lambin, E. F. (2002). Proximate causes and underlying driving forces of tropical deforestation tropical forests are disappearing as the result of many pressures, both local and regional, acting in various combinations in different geographical locations. *BioScience* **52**(2): 143–150.

Gibbs, H. K., Ruesch, A. S., Achard, F., Clayton, M. K., Holmgren, P., *et al.* (2010). Tropical forests were the primary sources of new agricultural land in the 1980s and 1990s. *Proceedings of the National Academy of Sciences* **107**(38): 16732–16737.

Gray, M., Roy, J., Vigilant, L., Fawcett, K., Basabose, A., *et al.* (2013). Genetic census reveals increased but uneven growth of a critically endangered mountain gorilla population. *Biological Conservation* **158**: 230–238.

Gregory, S. D., Brook, B. W., Goossens, B., Ancrenaz, M., Alfred, R., *et al.* (2012). Long-term field data and climate-habitat models show that orangutan persistence depends on effective forest management and greenhouse gas mitigation. *PloS One* **7**(9): e43846.

Groves, C. P. (1993). Order Primates. In: Wilson D. E. and Reeder D. M. (Eds), *Mammal Species of the World. A Taxonomic and Geographic Reference*. pp. 243–278. Washington, D.C: Smithsonian Institution Press.

Hansen, A. J. and DeFries, R. (2007). Ecological mechanisms linking protected areas to surrounding lands. *Ecological Applications* **17**(4): 974–988.

Hansen, M. C., Potapov, P. V., Moore, R., Hancher, M., Turubanova, S. A., *et al.* (2013). High-resolution global maps of 21st-century forest cover change. *Science* **342**(6160): 850–853.

Hansen, M. C., Stehman, S. V. Potapov, P. V., Loveland, T. R., Townshend, J. R., *et al.* (2008). Humid tropical forest clearing from 2000 to 2005 quantified by using multitemporal and multiresolution remotely sensed data. *Proceedings of the National Academy of Sciences* **105**(27): 9439–9444.

Harcourt, A. H., Harvey, P. H., Larson, S. G., and Short, R. (1981). Testis weight, body weight and breeding system in primates. *Nature* **293**(5827): 55–57.

Hardus, M. E., Lameira, A. R., Zulfa, A., Atmoko, S. S. U., de Vries, H., *et al.* (2012). Behavioral, ecological, and evolutionary aspects of meat-eating by Sumatran orangutans (*Pongo abelii*). *International Journal of Primatology* **33**(2): 287–304.

Harvey, P. H., Martin, R. D., and Clutton-Brock, T. H. (1987). Life histories in comparative perspective. In: Smuts, B. B., Cheney, D. L., Seyfarth, R. M., Wrangham R. W., and Struhsaker T. T. (Eds), *Primate Societies*. pp. 181–196. Chicago: University of Chicago Press.

Hayes, T. M. (2006). Parks, people, and forest protection: an institutional assessment of the effectiveness of protected areas. *World Development* **34**(12): 2064–2075.

Heesy, C. P. (2005). Function of the mammalian postorbital bar. *Journal of Morphology* **264**(3): 363–380.

Husson, S. J., Wich, S. A., Marshall, A. J., Dennis, R. D., Ancrenaz, M., *et al.* (2009). Orangutan distribution, density, abundance and impacts of disturbance. In: Wich, S. A. (Ed.), *Orangutans: Geographic Variation in Behavioral Ecology and Conservation*. pp. 77–96. New York: Oxford University Press.

Isbell, L. A. (2009). *The Fruit, the Tree, and the Serpent: Why We See So Well*. Cambridge, MA: Harvard University Press.

Johns, A. D. and Skorupa, J. P. (1987). Responses of rainforest primates to habitat disturbance: a review. *International Journal of Primatology* **8**: 157–191.

Joppa, L. N. and Pfaff, A. (2009). High and far: biases in the location of protected areas. *PLoS One* **4**(12): e8273.

Joppa, L. N., Loarie, S. R., and Pimm, S. L. (2008). On the protection of "protected areas". Proceedings of the National Academy of Sciences *Proceedings of the National Academy of Sciences* **105**(18): 6673–6678.

Kappeler, P. M. and van Schaik, C. P. (2002). Evolution of primate social systems. *International Journal of Primatology* **23**: 707–740.

Koendgen, S., Kuehl, H., N'Goran, P. K. Walsh, P. D., Schenk, S., *et al.* (2008). Pandemic human viruses cause decline of endangered great apes. *Current Biology* **18** (4): 260–264.

Kudo, H. and Mitani, M. (1985). New record of predatory behavior by the mandrill in Cameroon. *Primates* **26**(2): 161–167.

Laurance, W. F. and Balmford, A. (2013). Land use: a global map for road building. *Nature* **495**(7441): 308–309.

Laurance, W. F., Croes, B. M., Guissouegou, N., Buij, R., Dethier, M., *et al.* (2008). Impacts of roads, hunting, and habitat alteration on nocturnal mammals in African rainforests. *Conservation Biology* **22**(3): 721–732.

Laurance, W. F., Goosem, M., and Laurance, S. G. (2009). Impacts of roads and linear clearings on tropical forests. *Trends in Ecology and Evolution* **24**(12): 659–669.

Laurance, W. F., Useche, D. C., Rendeiro, J., Kalka, M., Bradshaw, C. J., *et al.* (2012). Averting biodiversity collapse in tropical forest protected areas. *Nature* **489**(7415): 290–294.

Lehmann, J., Korstjens, A. H., and Dunbar, R. I. (2010). Apes in a changing world–the effects of global warming on the behaviour and distribution of African apes. *Journal of Biogeography* **37**(12): 2217–2231.

Lucas, P. W. (2004). *Dental Functional Morphology: How Teeth Work*. Cambridge: Cambridge University Press.

MacPhee, R., Cartmill, M., and Gingerich, P. D. (1983). New Palaeogene primate basicrania and the definition of the order Primates. *Nature* **301**(5900): 509–511.

Malone, N., Fuentes, A., Purnama, A., and Wedana, I. (2002). Displaced hylobatids: biological, cultural, and economic aspects of the primate trade in Jawa and Bali, Indonesia. *Tropical Biodiversity* **8**: 41–50.

Marshall, A. J. and Wrangham, R. W. (2007). Evolutionary consequences of fallback foods. *International Journal of Primatology* **28**: 1219–1235.

McGraw, W. S. and Oates, J. F. (2007). Miss Waldron's Red Colobus, *Procolobus badius waldroni* (Hayman, 1936). Mittermeier, R. A., Wallis, J., Rylands, A. B., *et al.* (Eds), *Primates in Peril: The World's 25 Most Endangered Primates 2006–2008*. Arlington, VA: IUCN/SSC Primate Specialist Group (PSG), International Primatological Society (IPS), and Conservation International (CI).

McNeilage, A. (1996). Ecotourism and mountain gorillas in the Virunga Volcanoes. In: Taylor, V. J. and Dunstone, N. (Eds), *The Exploitation of Mammal Populations*. pp. 334–344. Springer.

Millennium Ecosystem Assessment (MEA) (2005). *Ecosystems and Human Well-being: Synthesis*. Washington, D.C: Island Press.

Meijaard, E., Albar, G., Nardiyono, Y. Rayadin, Y., Ancrenaz, M., *et al.* (2010). Unexpected ecological resilience in Bornean orangutans and implications for pulp and paper plantation management. *PLoS One* **5**(9): e12813.

Meijaard, E., Buchori, D., Hadiprakarsa, Y., Utami-Atmoko, S. S., Nurcahyo, A. A., *et al.* (2011). Quantifying killing of orangutans and human-orangutan conflict in Kalimantan, Indonesia. *PLoS One* **6**(11): e27491.

Michalski, F. and Peres, C. A. (2005). Anthropogenic determinants of primate and carnivore local extinctions in a fragmented forest landscape of southern Amazonia. *Biological Conservation* **124**(3): 383–396.

Mitani, J. C., Call, J., Kappeler, P. M., Palombit, R. A., and Silk, J. B. (2012). *The Evolution of Primate Societies*. Chicago, Il: University of Chicago Press.

Mittermeier, R. A., Rylands, A. B., and Wilson, D. E. (2013). *Handbook of the Mammals of the World*. Barcelona: Lynx Edicions.

Nagendra, H. (2008). Do parks work? Impact of protected areas on land cover clearing. *AMBIO: A Journal of the Human Environment* **37**(5): 330–337.

Newmark, W. D. (2008). Isolation of African protected areas. *Frontiers in Ecology and the Environment* **6**(6): 321–328.

Nijman, V., Nekaris, K., Donati, G., Bruford, M., and Fa, J. (2011). Primate conservation: measuring and mitigating trade in primates. *Endangered Species Research* **13**: 159–161.

Palombit, R. A. (1996). Pair bonds in monogamous apes: a comparison of the siamang, Hylobates syndactylus, and the white-handed gibbon, Hylobates lar. *Behaviour* **133**: 321–356.

Peres, C. A. (1993). Structure and organization of an Amazonian terra firme primate community. *Journal of Tropical Ecology* **9**: 259–276.

Peres, C. A. (2001). Synergistic effects of subsistence hunting and habitat fragmentation on Amazonian forest vertebrates. *Conservation Biology* **15**(6): 1490–1505.

Rosenberger, A. L. (1992). Evolution of feeding niches in new world monkeys. *American Journal of Physical Anthropology* **88**: 545–562.

Pontzer, H., Raichlen, D. A., Gordon, A. D., Schroepfer-Walker, K. K., Hare, B., *et al.* (2014). Primate energy expenditure and life history. *Proceedings of the National Academy of Sciences* **111**(4): 1433–1437.

Schaik, C. P. v. (1983). Why are diurnal primates living in groups? *Behaviour* **87**: 120–143.

Schaik, C. P. v. and Hooff, J. A. R. A. M. v. (1983). On the ultimate causes of primate social systems. *Behaviour* **85**: 91–117.

Schaik, C. P. v., Isler, K., and Burkart, J. M. (2012). Explaining brain size variation: from social to cultural brain. *Trends in Cognitive Sciences* **16**(5): 277–284.

Shattuck, M. R. and Williams, S. A. (2010). Arboreality has allowed for the evolution of increased longevity in mammals. *Proceedings of the National Academy of Sciences* **107**(10): 4635–4639.

Shepherd, C. R., Sukumaran, J., and Wich, S. A. (2004). *Open Season: An Analysis of the pet Trade in Medan, Sumatra 1997–2001*. Selangor, Malaysia: TRAFFIC Southeast Asia. Available at: <https://portals.iucn.org/library/sites/library/files/documents/Traf-088.pdf> [Accessed November 2015].

Shultz, S., Opie, C., and Atkinson, Q. D. (2011). Stepwise evolution of stable sociality in primates. *Nature* **479**(7372): 219–222.

Smith, T., Rose, K. D., and Gingerich, P. D. (2006). Rapid Asia-Europe-North America geographic dispersal of earliest Eocene primate Teilhardina during the Paleocene-Eocene thermal maximum. *Proceedings of the National Academy of Sciences* **103**(30): 11223–11227.

Stanford, C. B. (1996). The hunting ecology of wild chimpanzees: implications for the evolutionary ecology of Pliocene hominids. *American Anthropologist*. **98**: 96–113.

Stanford, C. B., Allen, J. S., and Anton, S. C. (2013). *Biological Anthropology* 3rd Edn. Boston: Pearson.

Steiper, M. E. and Seiffert, E. R. (2012). Evidence for a convergent slowdown in primate molecular rates and its implications for the timing of early primate evolution. *Proceedings of the National Academy of Sciences* **109**(16): 6006–6011.

Sterck, E. H. M., Watts, D. P., and van Schaik, C. P. (1997). The evolution of female social relationships in nonhuman primates. *Behavioral Ecology and Sociobiology* **41**(5): 291–309.

Stiles, D., Redmond, I., Cress, D., Nellemann, C., and Formo, R. (2013). *Stolen Apes—The Illicit Trade in Chimpanzees, Gorillas, Bonobos and Orangutans. A Rapid Response Assessment*. United Nations Environment Programme. Arendal, Norway: GRID-Arendal. Available at: <http://www.grida.no/publications/rr/apes/> [Accessed November 2015].

Struhsaker, T. T., Struhsaker, P. J., and Siex, K. S. (2005). Conserving Africa's rain forests: problems in protected areas and possible solutions. *Biological Conservation* **123**(1): 45–54.

Sussman, R. W. (1991). Primate origins and the evolution of angiosperms. *American Journal of Primatology* **23**: 209–223.

Swindler, D. R. (2002). Primate Dentition: An Introduction to the Teeth of Non-human Primates. Cambridge: Cambridge University Press.

Szalay F. S. and Eric, D. (1979). *Evolutionary History of the Primates*. New York, NY: Academic Press.

Tavaré, S., Marshall, C. R., Will, O., Soligo, C., and Martin. R. D. (2002). Using the fossil record to estimate the age of the last common ancestor of extant primates. *Nature* **416**: 726–729.

Terborgh, J., van Schaik, C., Davenport, L., and Rao, M. (2002). *Making Parks Work: Strategies for Preserving Tropical Nature*. Washington, D.C: Island Press: pp. **i-xix**, 1–511.

Thomas, S. C. (1991). Population densities and patterns of habitat use among anthropoid primates in the Ituri Forest, Zaire. *Biotropica* **23**(1): 68–83.

Tranquilli, S., Abedi-Lartey, M., Amsini, F. Arranz, L., Asamoah, A., *et al.* (2012). Lack of conservation effort rapidly increases African great ape extinction risk. *Conservation Letters* **5**(1): 48–55.

Utami, S. S. and Hooff, J. A. R. A. M. v. (1997). Meat-eating by adult female Sumatran orangutans (Pongo pygmaeus abelii). *American Journal of Primatology* **43**: 159–165.

Wallis, J. and Lee, D. R. (1999). Primate conservation: the prevention of disease transmission. *International Journal of Primatology* **20**(6): 803–826.

Walsh, P. D., Abernethy, K. A., Bermejo, M., Beyers, R., De Wachter, P., *et al.* (2003). Catastrophic ape decline in western equatorial Africa. *Nature* **422**(6932): 611–614.

Watts, D. P. and Mitani, J. C. (2002). Hunting behavior of chimpanzees at Ngogo, Kibale National Park, Uganda. *International Journal of Primatology* **23**: 1–28.

Wich, S. A., Fredriksson, G. M., Usher, G. Peters, G., Priatna, D., *et al.* (2012). Hunting of Sumatran orang-utans and its importance in determining distribution and density. *Biological Conservation* **146**(1): 163–169.

Wich, S. A., Riswan, J., Refisch, J., and Nellemann, C. (2011). Orangutans and the Economics of Sustainable Forest Management in Sumatra, United Nations Environment Programme. Available at: < http://www.unep.org/pdf/orangutan_report_scr.pdf > [Accessed November 2015].

Wich, S. A., Utami Atmoko, S. S., Mitra Setia, T., and van Schaik, C. P. (2009). *Orangutans: Geographic Variation in Behavioral Ecology and Conservation*. New York, NY: Oxford University Press.

Wilshusen, P. R., Brechin, S. R., Fortwangler, C. L., and West, P. C. (2002). Reinventing a square wheel: critique of a resurgent "protection paradigm" in international biodiversity conservation. *Society and Natural Resources* **15**(1): 17–40.

Wrangham, R. W. (1980). An ecological model of female bonded primate groups. *Behaviour* **75**: 262–300.

CHAPTER 2

Why conserve primates?

Andrew J. Marshall and Serge A. Wich

Bonnet Macaque feeding on fruit at Thekkady, Western Ghats, India
Photograph copyright: Swapna Nelaballi.

2.1 A basic question

Most primate populations are declining in numbers and many primate species are under threat of extinction for a variety of reasons, including hunting, disease, climate change, and the loss, degradation, and fragmentation of their habitats (Cowlishaw and Dunbar 2000; Schwitzer *et al.* 2014). For some, this knowledge alone is sufficient reason to conserve primates—it both provides a clear justification for conservation and implies a moral obligation to do so. This view is not universally held, however, and it is therefore important to consider explicitly various answers to the very basic question: why should we conserve primates?

Readers of this book likely require little convincing that non-human primates (hereafter 'primates') deserve targeted conservation attention. Many of us first became involved in primate research because of a deep concern for wild primate populations and a desire to contribute positively to their conservation. Others have become more involved in primate conservation over time, perhaps due to threats to their own study populations or in response to accumulating knowledge of the increasingly dire status of many primate species. Still others may

Marshall, A.J. and Wich, S.A., *Why conserve primates?* In: *An Introduction to Primate Conservation*.
Edited by: Serge A. Wich and Andrew J. Marshall, Oxford University Press (2016). © Oxford University Press.
DOI 10.1093/acprof:oso/9780198703389.003.0002

be relatively new to the topic, but have strong convictions about the importance of primate conservation. Whatever our personal motivations may be, we will encounter individuals, organizations, companies, and governments that do not share our values. We may be challenged by fellow conservationists who disagree with us about the importance of protecting primates over other taxa, activists that remind us that our proposed conservation actions may have negative consequences on local people, or government officials who argue that economics and development trump all other concerns. In such situations, inability to provide a convincing answer to the simple question of why we should conserve primates will likely doom our efforts to failure before they begin.

In this chapter we summarize several justifications for conserving primates. Our goal is to compile general information that will help primate conservationists make strong cases for the need to engage in specific conservation actions aimed at protecting particular primate populations in particular places. Not all arguments will work in all instances, of course, and there is no substitute for a well-considered, creative, and site-specific justification to support a particular policy. Nevertheless, some of the general points considered here may bolster specific arguments. We present eight broad justifications for conserving primates, starting with those that are most anthropocentric and progressing to more biocentric ones. After considering these justifications, we discuss some factors that complicate attempts to make convincing arguments in favour of primate conservation.

2.2 Primates promote human health

Primates have long been considered crucial to research that improves human health. Although primates comprise a small proportion of animals used in biomedical work, their close genetic and physiological similarity to humans makes them uniquely valuable in developing treatments for and vaccines against human illnesses (Bontrop 2001; Carlsson *et al.* 2004; Sibal and Samson 2001). Indeed, researchers studying a wide range of diseases and disorders consider primates to be irreplaceable to research that is ultimately aimed at enhancing human health and wellbeing (Bennett 2015; Capitanio and Emborg 2008; Evans and Silvestri 2013; Joyner *et al.* 2015; VandeBerg and Zola 2005). Most primates used in biomedical research are bred in captivity for this purpose (California Biomedical Research Association 2015). Nevertheless, wild populations are still occasionally used as source populations in exceptional cases where captive-bred primates are inappropriate (e.g. Home Office 2004; United States Department of Agriculture 2013). The escalating threat from emerging infectious diseases and the rapidly changing environmental conditions resulting from global climate change may increase the importance of wild primate populations as sources of research subjects. For example, primate populations that harbour natural immunity to novel pathogens may provide unique insights that help fight future human diseases. The extinction of wild primate populations could mean the loss of information vital to human survival in a future of emerging infectious diseases and global climate change.

2.3 Primates provide benefits to local communities

Wild primate populations can provide important benefits to people living in proximity to them. In some areas, primates and other sources of wild meat can serve as important food resources for communities living inside or adjacent to tropical forests (Brashares *et al.* 2011; Millner-Gulland *et al.* 2003). For example, consumption of meat from wild animals, including primates, was associated with substantially reduced incidence of anaemia in children living in villages around the Makira Protected Area in northeastern Madagascar (Golden *et al.* 2011). Such hunting is, however, usually unsustainable (Cowlishaw and Dunbar 2000; Fa *et al.* 2002, 2005; Fa and Tagg, Chapter 9, this volume; Golden 2009) and can lead to local extinction of species (Nunez-Iturri *et al.* 2008). Nevertheless, truly sustainable management of primate populations for food would, by definition, ensure their long-term persistence and therefore could conceivably be used as a justification for primate conservation under certain, special circumstances (Cowlishaw and Dunbar 2000; Crockett *et al.* 1996; de Thoisy *et al.* 2009; Ramirez

1984). This argument is, of course, incompatible with some alternative justifications for their protection (e.g. those that invoke the intrinsic value of life, Section 2.9), highlighting the complexities inherent to most conservation and reminding us that groups that share a common goal may do so for very different reasons.

In some areas, conservation of a particular primate population might provide economic benefits to local communities (Siex and Struhsaker 1999; Davenport *et al.* 2002). For example, substantial revenue is generated in some communities from primate and rainforest tourism (Adams and Infield 2003; Archabald and Naughton-Treves 2001; Kirby *et al.* 2010); such tourism may in turn promote conservation under certain circumstances (Pusey *et al.* 2008; Kirby *et al.* 2010; Savage *et al.* 2010).

In addition to tangible benefits they may provide, primates may have cultural or religious significance for people living in nearby areas (Fuentes and Wolfe 2002; Riley 2010; Humle and Hill, Chapter 14, this volume). For instance, the Hanuman langur in India is considered holy in the Hindu religion and the Iyaelima people in the Democratic Republic of Congo have taboos that prevent them from eating bonobos (Fuentes and Wolfe 2002). In such instances local people have likely been living in close proximity to wild primates for millenia (e.g. Tutin and Oslisly 1995), and in some cases can be powerful advocates for primate conservation.

Thus, for local people, extinction of nearby primate populations could reduce sources of wild meat, decrease economic opportunities, or erode deeply held cultural beliefs.

2.4 Primates serve key ecological functions

Primates often perform critical ecological functions in the ecosystems they inhabit. First, primates provide pollination services in some ecosystems (Carthew and Goldingay 1997; Gautier-Hion and Maisels 1994; Janson *et al.* 1981). For instance, Kress *et al.* (1984) conducted a detailed study of the relationship between the traveller's tree (*Ravenala madagascarensis*) and ruffed lemurs (*Varecia variegata*) and concluded that the system showed features of a co-evolved plant–pollinator relationship. Second, primates are widely acknowledged to be important seed dispersers (Chapman 1995; Lambert and Garber 1998; Norconk *et al.* 1998; Sato 2102; Tutin *et al.* 1991). In some plants, seed germination rates are positively influenced by passage through the primate gut; other plant species depend solely on primates for dispersal (Chapman and Onderdonk 1998; Wrangham *et al.* 1994). There is mounting evidence that local extinction of primates substantially alters plant species composition (Effiom *et al.* 2013a; Nunez-Iturri *et al.* 2008; Vanthomme *et al.* 2010). Third, primates are important seed predators in some ecosystems (Peres 1991; Peters 1993; Norconk and Veres 2011), and while it has to date received little attention, it is possible that primate seed predators may help maintain plant species diversity by disproportionately preying on seeds of common plant taxa (cf. Paine and Beck 2007; Power *et al.* 1996; Terborgh 2012). Fourth, folivory by primates can affect the mortality, fecundity, and growth rates of tree species (Chapman *et al.* 2013). Fifth, the presence of primates can influence community structure across multiple trophic levels. For example, the loss of primates due to hunting in Nigerian tropical forests has resulted in changes in the relative abundances of other mammals, with cascading effects on plant communities (Effiom *et al.* 2013b). Sixth, primates are important prey species in some ecological communities (Isbell 1994; Hart 2007); some species, most notably chimpanzees, can also have considerable impacts as predators on primates and other animals (Stanford 1995; Teelen 2008). Finally, primates may play a role in buffering against the detrimental effects of global climate change. Primates are typically the key dispersers of larger-seeded plant species (Howe 1986), and large-seeded tree species often have higher carbon densities than trees with small seeds (Queenborough *et al.* 2009; Wright *et al.* 2007). Thus, the presence of primates promotes the sequestration of additional carbon in tropical forests, which serve as key buffers against global climate change (Van der Werf *et al.* 2009).

These examples demonstrate that primates play an important role in maintaining well-functioning ecosystems. It has generally been difficult to determine whether primates serve keystone functions in ecological systems, in part because it is unclear to what extent the ecological roles of primates would be filled by other taxa were primates absent (e.g. Chapman and Onderdonk 1998; Gautier-Hion *et al.* 1985; Poulsen *et al.* 2002; Russo and Chapman 2011; Chapman *et al.* 2013). Mounting evidence suggests, however, that at least in some systems primates serve uniquely important roles, and that their loss has large effects that are not offset by other taxa (Effiom *et al.* 2013a, b; Muller-Landau 2007; Nunez-Iturri *et al.* 2008). Primate conservation is therefore crucially important to maintain intact ecosystems and the many services these ecosystems provide to people, including clean and stable water supplies, prevention from floods and landslide, pollination, stable micro-climates, and buffering of global warming (Wich *et al.* 2011).

2.5 Primates provide unique insights into human evolution

Humans are primates and therefore the protection of wild primate populations preserves our ability to study the ecology, sociality, and behaviour of our close relatives (Boyd and Silk 2012; Fleagle 2013). A deep understanding of humans is impossible without placing our evolution, biology, and culture in broad phylogenetic context. Extinction of a primate species would diminish our capacity to understand ourselves, our evolution, and our place in nature. For example, consideration of humans in the context of non-human primates has enhanced our understanding of human cognition (Matsuzawa 2001; Tomasello 2009), genetics (The Chimpanzee Sequencing and Analysis Consortium 2005; Patterson *et al.* 2006), communication (Savage-Rumbaugh *et al.* 1998; Tomasello 2008), aggression (Smuts 1992; Wrangham and Peterson 1996), reconciliation (Aureli *et al.* 2002; de Waal 2000), ecology (Hill 1982; Ulijaszek 2002), and much more. Studies of extant primate tool use, hunting, cultural traditions, and diet importantly inform reconstructions of human evolution (e.g. McGrew 1992; van Schaik *et al.* 1999; Matsuzawa 2001; Boyd and Silk 2012). For this reason, most primatologists in the United States, and many in Europe and Japan, are affiliated with academic departments or institutes primarily dedicated to the study of anthropology and human evolution.

Study of great apes has been of particular interest, given their close phylogenetic relatedness to humans (de Waal 2005; Knott 2001; Semendeferi *et al.* 2002; Wrangham 1987; Wrangham and Pilbeam 2001), but other taxa have also been argued to provide valuable insights into the evolution of human behaviour (DeVore and Washburn 1963; Kinzey 1987). Loss of any primate species, but especially an ape taxon, would both hamper our ability to distinguish homologies (characters shared based on common descent) from homoplasies (characters evolved independently through convergence) in hominoid evolution and limit our understanding of the range of variation possible in some traits. For instance, consider how different our understanding of ape social relationships, aggression, and dominance would have been had bonobos gone extinct before they were studied in the wild. Bonobos exhibit several features that contrast starkly with general patterns seen in other great apes: bonobo females are more social, form stronger bonds with one another, and are subject to greatly reduced threats of sexual aggression or infanticide compared to other apes (Hare *et al.* 2012; Stumpf 2006; Surbeck *et al.* 2012). These observations helped spark investigation of and appreciation for the importance of female social relationships and the social function of sexuality in apes, thereby broadening conceptions of the range of variation possible in the lineage producing chimpanzees, bonobos, and humans (Kano 1992; Parish 1994; de Waal 2005). Conserving primates preserves precious information about ourselves and our past.

2.6 Primates are of immense biological interest and importance

The primate order is a diverse group and exhibits substantial variation in ecology, social system, and behaviour (Smuts *et al.* 1987; Kappeler 1999; Mitani *et al.* 2012; Rylands and Mittermeier 2014). Primate species span at least four orders of magnitude in body size, consume a wide variety of diets, exhibit the most diverse set of locomotor adaptations of any animal order, live in many types of social system, and inhabit a range of environments (Clutton Brock 1989; Fleagle 2013; Rowe and

Myers 2015; Wright 1999). This variation presents a treasure trove of raw material that biologists can explore to promote our general understanding of how morphology, sociality, and behaviour evolve under a range of ecological conditions. Of particular interest are questions regarding how different species adapt to the same conditions (e.g. studies of primate communities, examination of different responses to environmental change and habitat degradation) and how the same species adapts to different conditions (e.g. documentation of variation in behaviour, sociality, and life history across environmental gradients). Extinction of primate species and loss of populations will hamper our ability to make sense of the natural world and elucidate general biological principles that apply to many other taxa.

The crucial role of primates in furthering biological understanding is especially evident when one considers our general ignorance of the tropics. The tropics house the majority of the world's biodiversity (Ceballos and Ehrlich 2006; Kreft and Jetz 2007) and yet for many groups we lack even the most basic understanding of their diversity, biology, or conservation status. For example, in amphibians a much larger proportion of tropical species than temperate species are classified as data deficient by the IUCN (Collen *et al.* 2008; Stuart *et al.* 2004), a pattern that appears to be true of other taxa as well (e.g. mammals: Schipper *et al.* (2008); birds: Butchart and Bird (2010)). Against this backdrop of general ignorance about the tropics, primates stand out as a relatively well-studied group, in part because many species are gregarious, diurnal, and relatively easy to study in the wild (Emmons 1999; Harcourt 2000, 2006; Beaudrot *et al.* 2013). In a recent study of research conducted in tropical protected areas, Marshall *et al.* (2016) found that 47.5% of all works returned by a Google Scholar search of the names of all terrestrial protected areas in great ape range countries concerned primates, compared to 23.6% for other mammals, 5.9% for birds, 11.3% for plants, and 11.7% for other taxa. This suggests that, at least in the paleotropics, much more scientific research is published on primates than any other taxon.

Research on primates often sheds light on other taxa inhabiting the same forests, thereby raising our general understanding of the tropics. For instance, *Cola lizae*, a tree endemic to Gabon that was long known locally as an important timber species, was only recognized as a species in 1987 when discovered by primatologist Liz Williamson, for whom the tree is named (Hallé 1987). Research by primatologists demonstrated that the tree is only dispersed by gorillas, despite being fed on by many primate species (Tutin *et al.* 1991). Were gorillas to be lost from these forests, an important resource (as food for frugivores and timber for people) would be lost—an insight gained only through the work of primatologists. Primatologists have discovered other ecologically important relationships between primates and other forest species (see Section 2.4). Protection and study of primates is therefore vitally important to promote our biological understanding of some of the most diverse and least understood communities on Earth.

Merely protecting primate species from extinction is inadequate to preserve their value as subjects of biological investigation. Alteration and degradation of primate environments and loss of populations permanently reduces our ability to understand basic aspects of their behaviour, ecology, and adaptability (Caro and Sherman 2011). Studies of primate taxa in distinct environments have demonstrated considerable within-taxon variation in diet, life history, ecology, sociality, and behaviour (e.g. baboons: Kamilar (2006); orang-utans: van Schaik *et al.* (2009); red colobus: Struhsaker (2010)). Such variability is likely quite common in primates (see Groves, Chapter 4, this volume), suggesting that extinction of local populations will result in permanent losses of diversity (Caro and Sherman 2012). This reduction of diversity will not only reduce our ability to understand the biology and ecology of wild primates; it may also remove from a species' behavioural repertoire the ability to adapt to climate change, or eliminate from a species' gene pool resistance to emerging infectious diseases. Therefore, preserving populations across the full range of environments that a primate species occupies and protecting at least a portion of each habitat type from degradation is necessary to capitalize fully on the scientific value of primates as subjects of biological study.

2.7 Primates may promote conservation of other taxa

Primates can serve as important surrogate species (*sensu* Caro 2010) that contribute positively to the conservation of other taxa by acting as flagship, umbrella, or indicator species. Many primate species are charismatic, emotionally evocative, and interesting to people (e.g. Nishida *et al.* 2001; Wrangham *et al.* 2008; Meijaard *et al.* 2012) and can therefore serve as effective flagships that raise awareness, funds, and support for conservation actions that protect multiple species (Alexander 2000; Clucas *et al.* 2008). Some primates may also serve as classic umbrella species, meaning that the protection of sufficient habitat to secure long-term viability of the primate taxon also ensures persistence of other threatened species (Caro 2003, 2010). This is most likely to be true for large-bodied species that live at relatively low population densities and therefore need large blocks of habitat to ensure the demographic and genetic health necessary for long-term persistence (e.g. orang-utans: Marshall *et al.* (2009)). Finally, primates have been argued to be valuable indicator species (Hill 2002), because their species richness serves as a surrogate for diversity in other taxa or because the health of their populations reflects the general health of an ecosystem.

In addition to the potential value of primates as surrogates, their presence at particular sites can promote conservation of sympatric taxa. It has become increasingly appreciated that researchers provide direct conservation benefits at the sites where they work by promoting awareness of the value of the natural world, training the next generation of scientists and managers, building capacity, facilitating law enforcement, and providing alternative sources of income to people who may otherwise engage in activities detrimental to biodiversity (Paaby *et al.* 1991; Wrangham 2008; Campbell *et al.* 2011; Sekercioglu 2012; Laurance 2013). The presence of charismatic taxa, such as apes, attracts researchers and the positive conservation effects of their attention (Magin *et al.* 1994; Sitas *et al.* 2009; Marshall *et al.* 2016). In essence, then, primates attract researchers, and researcher presence provides a protective umbrella that promotes conservation of primate habitats and other taxa inhabiting them.

2.8 Some primates are particularly susceptible to extinction

Many primates exhibit traits that have been shown to increase extinction probability in other taxa. The most important factor predicting extinction risk is small population size because small populations are at high risk of extinction due to demographic, genetic, and environmental stochasticity (Soulé and Wilcox 1980; Soulé 1987; Caughley 1994). Primate populations across the globe are shrinking due to habitat loss and degradation, hunting, and disease, and are becoming divided into smaller units by habitat fragmentation. These small primate populations continue to be affected by the deterministic processes that led to their declines, but once small are subject to the additional stochastic effects that magnify extinction risk. Small population sizes are also often linked to small geographic ranges and low population densities, which both significantly increase extinction risk in primates and other species (Purvis *et al.* 2000; Johnson 1998; Harcourt and Schwartz 2001; Harcourt *et al.* 2005; Harcourt 2006). Species with slow life histories are also identified as extinction prone in many analyses (Terborgh 1974; Cox 1997; but see Purvis *et al.* 2000), and primates have famously slow life histories compared to other mammals (Charnov and Berrigan 1993). Large-bodied primates are even more vulnerable to extinction, both because large body size is a strong independent predictor of vulnerability (Purvis *et al.* 2000; Cardillo *et al.* 2005) and because large species, such as great apes, have slow life histories (Wich *et al.* 2004, 2009; Marshall *et al.* 2009). More work is needed to fully understand the factors that predict extinction risk, in part because interpretation of broad comparative analyses of extinction risk is complicated by biases due to missing data (González-Suárez *et al.* 2012). Nevertheless, many primate taxa exhibit multiple traits that consistently predict extinction risk in comparative analyses, suggesting that primates warrant conservation under models that allocate conservation effort based on vulnerability.

2.9 Ethical arguments

For many people there are ethical reasons to protect primates (e.g. Cavalieri and Singer 1993). Although this sentiment might be more frequently (although certainly not solely) expressed in developed nations (Hill 2002), it is perhaps one of the most fundamental justifications to protect any species. Ethics were an important impetus for the creation of the first National Parks (Callicott 1990) and the founding of conservation NGOs (e.g. WWF: Schwarzenbach (2011)). Primatologists often cite ethical arguments as their personal reason for becoming involved in conservation. Such arguments are often rooted in the belief that all life has equal inherent value and the loss of any species due to human actions represents a failing of our moral obligation to protect species from human-induced extinction (Naess 1986; Hargrove 1989). In addition to their intrinsic value, the fact that primates are our closest genetic relatives and share many other characteristics with humans has been used to bolster ethical arguments for specific conservation efforts for primates, and in particular great apes (Nishida *et al.* 2001; Wrangham *et al.* 2008).

2.10 Complications

Taken together, the arguments reviewed here comprise both compelling justification for primate conservation and imply that we have an obligation to do so. While there are many reasons to protect primates, to be truly effective advocates for their conservation we must be aware of some complications attendant to the justifications discussed above. We consider four of these below. We begin by discussing the basic question of whether primates are uniquely deserving of conservation attention. We then note that some alternative justifications for primate conservation are contradictory, that cultural factors often complicate primate conservation, and that most justifications are unlikely to be successful in every context. We next discuss potential risks to some justifications for primate conservation, and end with a consideration of opportunity costs. These complications do not undermine all justifications for primate conservation, but they do highlight the need to be strategic when applying them.

2.10.1 Are primates special?

Many arguments made in support of primate conservation begin with the tacit assumption that primates are more special, and more deserving of protection, than other taxa. Some of the reasons given as justification are demonstrably true. For example, non-human primates are undoubtedly our closest phylogenetic relatives, and if one accepts the premise that studying other taxa is important to better understand ourselves, then it is difficult to take issue with the contention that primates are special because they provide unique insights into human evolution. Similarly, it is hard to argue that the primates are not among the most well studied of tropical animals and as such are special because they provide a valuable insight into otherwise often poorly known ecosystems.

The contention that primates are special is not, however, always so easy to justify (Lovett and Marshall 2006). One reason for this is that many of the justifications given for primate conservation are not unique to primates. Primates may not be the only, or even the most important, provider of a particular ecological function in some systems (e.g. seed dispersal: Corlett (1998); Stevens *et al.* (2014)). Many primate taxa are threatened with extinction, but it is not always true that they are the most threatened in a particular region or country (e.g. in many places amphibians are more severely threatened than primates: Baillie *et al.* (2004)). And while primates can be important flagship species, they are not the only such taxa (Caro and O'Doherty 1999; Clucas *et al.* 2008) nor necessarily the most effective (Bowen-Jones and Entwistle 2002; Smith *et al.* 2012). In other words, primates may not always be *especially* important seed dispersers, *especially* severely threatened, or *especially* effective flagship species. In such cases, basing justification for conservation investment on the contention that primates are special may not be wise or effective.

A second complication of invoking the 'primates are special' argument in support of primate conservation is that individuals may raise principled objections to the focus on any particular taxon. There is a strain of thought in conservation suggesting that all species have the same inherent value and are therefore equally deserving of conservation funds (Hargrove 1989; Naess 1986). There is also merit in the argument that conservation should not be principally organized around preservation of particular taxonomic groups, and that we should instead focus on, for example, provision of ecosystem services (Tallis *et al.* 2008), optimal allocation of limited resources (Wilson *et al.* 2006), maximizing preserved phylogenetic diversity (Faith 1992), areas of high endemism and threat (Myers *et al.* 2000), or regions that are otherwise especially vulnerable or irreplaceable (Brooks *et al.* 2006).

We do not wish to undermine the many defensible arguments that can be made in support of the contention that primates are special. Primates *are* special in important ways, and pointing this out can be quite effective in arguing for primate conservation in some contexts. Primates are not, however, special in *all* ways. Uncritical application of the 'primates are special' argument is unlikely to be successful. We should be careful to limit our use of this justification to situations where it is demonstrably true or empirically defensible.

2.10.2 Contradictions, complexities, and limitations

As with many areas of conservation science and practice, contradictions and complexities abound in debates of whether and why to protect primates. Some of the preceding arguments in favour of primate conservation are at odds with one another; others are complex and difficult to apply in specific situations. For instance, ethical arguments invoking our moral obligation not to harm individual primates cannot be easily squared with the perspective that primate populations should be valued for the wild meat that they provide some communities (Hill 2002) or justifications for primate protection rooted in their value for biomedical research. In addition, although not necessarily inherently contradictory (Guy *et al.* 2014), actions to help primates taken in the name of animal welfare (e.g. rehabilitation, release) are often not the most cost-effective or beneficial tactics to promote conservation of wild primate populations or their habitats (Wilson *et al.* 2014; Yeager 1997). Indeed, under some circumstances, actions undertaken to promote individual welfare, such as release of sick individuals into the wild, may endanger wild populations (Harcourt 1987; Bennett 1992).

Complexities arise when the attitudes or cultural beliefs of distinct stakeholder groups clash. For instance, the perspectives of people in high-biodiversity, developing countries are often sharply at odds with the views of conservationists largely based in developed countries that have already substantially degraded their own wildlife (Meijaard and Sheil 2011). Even people living side by side can have very different cultural values and attitudes towards wild primates; some may view them as sacred while others consider them agricultural pests or sources of food (Hill 2002; Humle and Hill, Chapter 14, this volume). This is particularly true in instances of migration, where immigrant communities often lack long-term ownership over the land and consequently have little incentive to utilize it sustainably (Cowlishaw and Dunbar 2000; Ekadinata *et al.* 2013; Levang *et al.* 2007; López *et al.* 1988).

Finally, most justifications used in support of primate conservation will not work in all contexts. For example, tourism is not a panacea. Although tourism can generate income for local communities, complexities and conflicts surrounding such arrangements (e.g. Adams and Infield 2003; Archabald and Naughton-Treves 2001) highlight more general concerns with tourism (Kiss 2004; Weaver and Lawton 2007). It is likely that the conditions necessary to promote successful primate tourism exist at only a limited number of sites. In addition, even when the economics of a tourism operation are effectively designed, it may be inadvisable to develop tourism everywhere because of the risk of transmission of diseases from humans to primates (Goldberg *et al.* 2007; Pusey *et al.* 2008) and the stress that tourism may impose on individuals being observed by tourists (Maréchal *et al.* 2011). Similarly, as discussed elsewhere in this chapter, justifications invoking ethics, spirituality, extinction risk, or biological interest will have different probabilities of success depending on context and the relevant stakeholder groups.

These contradictions, complexities, and limitations highlight the need for tactical, situation-specific justifications for primate conservation. We cannot uncritically apply justifications or approaches that were successful in one context and assume they will work elsewhere. Similarly, failure of a tactic or strategy in one context does not necessarily mean that it would not work somewhere else. Primate conservation requires creative, open minds and informed understanding of the cultural, social, economic, and ecological particulars of a given conservation context.

2.10.3 Risky justifications

Some justifications for conserving primates have risks of backfiring: they may be used to support primate conservation in some instances but could be used to argue against it in others. For example, while economic arguments for conservation have the potential to substantively influence policy in ways other justifications cannot (Balmford *et al.* 2002; Pearce *et al.* 2008), they are risky because conservation will not always be the most economically rational choice. Often the deck is stacked against conservation because it is difficult to assess the value of biodiversity benefits and the costs are often ignored (e.g. comparisons of the cost effectiveness of alternative fuel sources typically exclude environmental costs associated with global climate change). In addition, the economic benefits of environmental degradation are usually immediate and reaped by a relatively small set of (typically powerful) individuals, whereas the costs are not fully felt until much later and are often largely born by those who do not share in the benefits (e.g. Balmford and Whitten 2003; Barber and Schweithelm 2000). Even in situations where conditions are conducive to sustainable management, the most economically rational decision may well be to clear cut a forest and invest the funds wisely, rather than protect the forest for the ecosystem services it provides or extract timber in a sustainable way (Alvard 1988; Harcourt 2001). Thus, relying solely on economic arguments in favour of primate conservation, such as touting the potential tourism benefits to local people, runs the risk that an alternative, more lucrative proposal entailing destruction of primate habitat could win out.

Justifying conservation of primates on the basis of their phylogenetic relatedness to humans can likewise backfire. If one argues that protection of primates is important because they are closely related to humans, then a logical retort is that humans must be the most important of all, and therefore steps taken to conserve primates at the expense of people are unjustified. In Indonesia, we frequently encounter people who struggle to understand why so much international funding and attention is devoted to orang-utans when the majority of the people who live in close proximity to them exist on less than USD2 per day (Meijaard *et al.* 2012). In such circumstances it is easy to understand why politicians find it expedient to campaign on platforms that promote helping people, not orang-utans, as a candidate for governor of East Kalimantan, Indonesian Borneo, did in 2008 (Meijaard and Sheil 2008).

Finally, the fact that local communities may have beliefs, attitudes, or practices that appear to be consistent with the preservation of nearby primate communities should not be a cause for complacency. In part, this is because veneration does not necessarily prohibit utilization (Hill 2002) or conflict (Fuentes *et al.* 2005; Fuentes 2012). For example, in the Mentawai Islands, Indonesia, while primates are sacred cultural symbols in art, music, and folklore among traditional communities, they are also frequently hunted and consumed (Mitchell and Tilson 1986). Similarly, although some individual primates are kept as pets and incorporated into fictive kinship systems by the Guaja Indians of Brazil, other individuals of the same species are hunted for food (Cormier 2002; Hill 2002). Also, like other elements of culture, peoples' beliefs and attitudes about primates are not fixed. Species that were once considered sacred may come to be viewed as less so in the face of basic economic needs (Hill 2002; Leach 1994 in Hill 2002; Humle and Hill, Chapter 14, this volume). For instance, the Hindu beliefs that once protected monkeys in rural India have not protected them from persecution when raiding crops in recent decades (Mukherjee *et al.* 1986; Southwick *et al.* 1983). Even when values do not shift, improvements in hunting technology, transportation, and access, or changes in human population density can render unsustainable practices that once were far less damaging (Alvard 1993, 1988; Hames 1979; Harcourt 2001). Thus, the protections afforded by traditional beliefs will not necessarily persist in the face of changing economic conditions or shifting social customs.

Conservationists, like people more generally, are notoriously reticent to explicitly incorporate the risk of failure into their decisions (Plous 1993; Redford and Taber 2000; Game *et al.* 2013). Nevertheless, selection of justifications for primate conservation must include consideration of the possibility that their invocation may have unwanted effects.

2.10.4 Opportunity costs

Because the resources available for conservation are insufficient to meet all needs, a major focus of conservation over the last several decades has been determining effective, efficient ways to allocate limited resources (Brooks *et al.* 2006; Carwardine *et al.* 2008; Wilson *et al.* 2007). A range of different conservation prioritization strategies have been proposed and implemented, and while they differ in important ways, they all seek to explicitly integrate opportunity costs into conservation decision-making (Game *et al.* 2013; Kirkpatrick 1983; Wilson *et al.* 2006, 2009). Opportunity costs formalize the intuition that investment to address one conservation problem reduces or precludes investment in a different problem; as Game *et al.* (2013: 480) succinctly state: 'every good thing we do is another good thing we do not'. Thus, when we advocate expenditure of funds to conserve primates, we must recognize that resources allocated to primate conservation will often therefore be unavailable to address other conservation goals. It is possible that the conservation funding we seek to help a threatened primate species could be better spent to protect a critically endangered bird, or perhaps the chances of success at protecting our target primate population are so low that a wiser use of funds would be to invest them on a taxon with a more reasonable chance of persistence (Bottrill *et al.* 2008).

Choices among competing conservation demands are not easy to make, but we make them, whether we choose to acknowledge them or not (see Marshall and Wich, Chapter 18, this volume). There are occasionally win–win situations, where investment in a primate species may provide ancillary benefits to other taxa (e.g. when primates are umbrella species), but such instances are probably rarer than we imagine. It is also sometimes true that funding sources are earmarked for a particular taxon, due to interests of private donors, targeted fundraising campaigns, or legislated government policies, and in such cases use of resources to conserve primates may not present opportunity costs for conservation of other taxa. But even in these more targeted instances, it is generally the case that there are not sufficient funds to support all worthwhile primate projects, so consideration of opportunity costs will still be necessary. In such instances, use of formal, quantitative methods provide defensible, rigorous, and transparent algorithms to allocate limited conservation funds (Wilson *et al.* 2007; Gregory *et al.* 2012; Game *et al.* 2013).

Acknowledgements

We thank Tim Caro, Katie Feilen, and Sandy Harcourt for thoughtful reviews that substantially improved this chapter, and Swapna Nelaballi for providing the photograph for the chapter front page. A. J. M. also thanks the participants in his Fall 2014 Primate Conservation Biology seminar at the University of Michigan for reading an early draft of this chapter and for stimulating discussions of some of the topics considered here.

References

Adams, W. M. and Infield, M. (2003). Who is on the gorilla's payroll? Claims on tourist revenue from a Ugandan National Park. *World Development* **31**: 177–190.

Alexander, S. E. (2000). Resident attitudes towards conservation and black howler monkeys in Belize: the community baboon sanctuary. *Environmental Conservation* **27**: 341–350.

Alvard, M. S. (1993). Testing the 'ecologically noble savage' hypothesis: inter-specific prey choice by Piro hunters of Amazonian Peru. *Human Ecology* **21**: 355–387

Alvard, M. S. (1998). Evolutionary ecology and resource conservation. *Evolutionary Anthropology* **7**: 62–74.

Archabald, K. and Naughton-Treves, L. (2001). Tourism revenue-sharing around national parks in Western Uganda: early efforts to identify and reward local communities. *Environmental Conservation* **28**: 135–149.

Aureli, F., Cords, M., and Van Schaik, C. P. (2002). Conflict resolution following aggression in gregarious animals: a predictive framework. *Animal Behaviour* **64**(3): 325–343.

Baillie, J. E. M., Hilton-Taylor, C., and Stuart, S. N. (Eds) (2004). 2004 IUCN Red List of Threatened Species. A Global Species Assessment. Gland, Switzerland and Cambridge: IUCN.

Balmford, A., Bruner, A., Cooper, P., Costanza, R., Farber, S., *et al.* (2002). Economic reasons for conserving wild nature. *Science* **397**: 950–953.

Balmford, A. and Whitten, T. (2003). Who should pay for tropical conservation, and how could the costs be met? *Oryx* **37**: 238–250.

Barber, C. V. and Schweithelm, J. (2000). *Trial by Fire: Forest Fires and Forestry Policy in Indonesia's Era of Crisis and Reform*. Washington, DC: World Resources Institute.

Beaudrot, L., Struebig, M. J., Meijaard, E., van Balen, S., Husson, S., *et al.* (2013). Co-occurrence patterns of Bornean vertebrates suggest competitive exclusion is strongest among distantly related species. *Oecologia* **173**: 1053–1062.

Bennett, J. (1992). A glut of gibbons in Sarawak–is rehabilitation the answer? *Oryx* **26**: 157–164.

Bennett, A. J. (2015). New era for chimpanzee research: broad implications of chimpanzee research decisions. *Developmental Psychobiology* **57:** 279–288.

Bontrop, R. E. (2001). Non-human primates: essential partners in biomedical research. *Immunological Reviews* **183**: 5–9.

Bottrill, M. C., Joseph, L. N., Carwardine, J., Bode, M., Cook, C., *et al.* (2008). Is conservation triage just smart decision making? *Trends in Ecology & Evolution* **23**(12): 649–654.

Bowen-Jones, E. and Entwistle, A. (2002). Identifying appropriate flagship species: the importance of culture and local contexts. *Oryx* **36**(02): 189–195.

Boyd, R. and Silk, J. (2012). *How Humans Evolved*, 6th edn. New York NY: W. W. Norton & Co.

Brashares, J. S, Golden, C. D., Weinbaum, K. Z., Barrett, C. B., and Okello, G. V. (2011). Economic and geographic drivers of wildlife consumption in rural Africa. *Proceedings of the National Academy of Sciences* **108**: 13931–13936.

Brooks, T. M., Mittermeier, R. A., da Fonseca, G. A., Gerlach, J., Hoffmann, M., *et al.* (2006). Global biodiversity conservation priorities. *Science* **313**: 58–61.

Butchart, S. H. and Bird J. P. (2010). Data deficient birds on the IUCN Red List: what don't we know and why does it matter? *Biological Conservation* **143:** 239–247.

California Biomedical Research Association (2015). Fact sheet: primates in biomedical research. Available at: <http://www.ca-biomed.org/pdf/media-kit/fact-sheets/FS-Primate.pdf.> [Accessed November 2015].

Callicott, J. B. (1990). Whither conservation ethics? *Conservation Biology* **4**(1): 15–20.

Campbell, G., Kuehl, H., Diarrassouba, A., N'Goran, P. K., and Boesch, C. (2011). Long-term research sites as refugia for threatened and over-harvested species. *Biology Letters* **7**: 723–726.

Capitanio, J. P. and Emborg, M. E. (2008). Contributions of non-human primates to neuroscience research. *The Lancet* **371**: 1126–1135.

Cardillo M., Mace, G. M., Jones, K. E., Bielby, J., Bininda-Emonds, O. R. P., *et al.* (2005). Multiple causes of high extinction risk in large mammal species. *Science* **309**: 1239–1241.

Carlsson, H.-E., Schapiro, S. J., Farah, I., and Hau, J. (2004). Use of primates in research: a global review. *American Journal of Primatology* **63**: 225–237

Caro, T. (2003). Umbrella species: critique and lessons from East Africa. *Animal Conservation* **6**: 171–181.

Caro, T. (2010). *Conservation by Proxy: Indicator, Umbrella, Keystone, Flagship, and Other Surrogate Species*. Washington, DC: Island Press.

Caro, T. and O'Doherty, G. (1999). On the use of surrogate species in conservation biology. *Conservation Biology* **13**(4): 805–814.

Caro, T. and Sherman, P. W. (2011). Endangered species and a threatened discipline: behavioural ecology. *Trends in Ecology and Evolution* **26**: 111–118.

Caro, T. and Sherman, P. W. (2012). Vanishing behaviors. *Conservation Letters* **5**: 159–166.

Carthew, S. M. and Goldingay, R. L. (1997). Non-flying mammals as pollinators. *Trends in Ecology and Evolution* **12**: 104–108.

Carwardine, J., Wilson, K. A., Ceballos, G., Ehrlich, P. R., Naidoo, R., *et al.* (2008). Cost-effective priorities for global mammal conservation. *Proceedings of the National Academy of Sciences* **105**(32): 11446–11450.

Caughley G. (1994). Directions in conservation biology. *Journal of Animal Ecology* **63:** 215–244.

Cavalieri, P. and Singer, P. (1993). A declaration of great apes. In: Cavalierai P. and Singer P. (Eds), *The Great Ape Project: Equality Beyond Humanity*. pp. 4–7. London: Fourth Estate.

Ceballos, G. and Ehrlich, P. R. (2006). Global mammal distributions, biodiversity hotspots, and conservation. *Proceedings of the National Academy of Sciences* **103**: 19374–19379.

Chapman, A. P. (1995). Primate seed dispersal: coevolution and conservation implications. *Evolutionary. Anthropology* **4**: 74–82.

Chapman, C. A. and Onderdonk, D. A. (1998). Forests without primates: primate/plant codependency. *American Journal of Primatology* **45**: 127–141.

Chapman, C. A., Bonnell, T. R., Gogarten, J. F., Lambert, J. E., Omeja, P. A., *et al.* (2013). Are primates ecosystem engineers? *International Journal of Primatology* **34**: 1–14.

Charnov E. L. and Berrigan, D. (1993). Why do female primates have such long lifespans and so few babies? Or life in the slow lane. *Evolutionary Anthropology* **1**: 191–194.

Clucas, B., McHugh, K., and Caro, T. (2008). Flagship species on covers of US conservation and nature magazines. *Biodiversity and Conservation* **17**(6): 1517–1528.

Clutton-Brock, T. H. (1989). Review lecture: mammalian mating systems. *Proceedings of the Royal Society of London. B. Biological Sciences* **236**(1285): 339–372.

Collen, B., Ram, M., Zamin, T., and McRae, L. (2008). The tropical biodiversity data gap: addressing disparity in global monitoring. *Tropical Conservation Science* **1**: 75–88.

Corlett, R. T. (1998). Frugivory and seed dispersal by vertebrates in the Oriental (Indomalayan) Region. *Biological Reviews of the Cambridge Philosophical Society* **73**(04): 413–448.

Cormier, L. A. (2002). Monkey as food, monkey as child: Guaja symbolic cannibalism. In: Fuentes, A. and Wolfe, L. (Eds), *Primates Face to Face*. pp. 63–84. Cambridge: Cambridge University Press.

Cowlishaw, G. and Dunbar, R. I. M. (2000). *Primate Conservation Biology*. Chicago, IL: University of Chicago Press.

Cox, G. W. (1997). *Conservation Biology*. Boston, MA: McGraw-Hill.

Crockett, C. M., Kyes, R. C., and Sajuthi, D. S. (1996). Modeling managed monkey populations: sustainable harvest of longtailed macaques on a natural habitat island. *American Journal of Primatology* **40**: 343–360.

Davenport, L., Brockelman, W. Y., Wright, P. C., Ruf, K., and Rubio del Valle, F. B. (2002). Ecotourism tools for parks. In: Terborgh, J., Van Schaik, C., Davenport, L., and Rao M. (Eds), *Making Parks Work*. pp. 279–306. Washington, DC: Island Press.

de Thoisy, B., Richard-Hansen, C., and Peres, C. A. (2009). Impacts of subsistence game hunting on Amazonian primates. In: Garber, P. A., Estrada, A., Bicca-Marques, J. C., Heymann E. W., and Strier K. B. (Eds), *South American Primates: Comparative Perspectives in the Study of Behavior, Ecology, and Conservation*. pp. 389–412. New York, NY: Springer.

DeVore, I. and Washburn, S. L. (1963). Baboon ecology and human evolution. Reprinted in: Bourlière, F. and Howell, C. F. (Eds), *African Ecology and Human Evolution* (2013). pp. 335–367. London: Routledge.

de Waal, F. M. B. (2000). Primates—A natural heritage of conflict resolution. *Science* **289**: 586–590.

de Waal, F. M. B. (2005). A century of getting to know the chimpanzee. *Nature* **437**: 56–59.

Effiom, E. O., Nuñez-Iturri, G., Smith, H. G., Ottosson, U., and Olsson, O. (2013a). Bushmeat hunting changes regeneration of African rainforests. *Proceedings of the Royal Society B: Biological Sciences* **280**: DOI: 10.1098/rspb.2013.0246.

Effiom, E. O., Birkhofer, K., Smith, H. G., and Olsson, O. (2013b). Changes of community composition at multiple trophic levels due to hunting in Nigerian tropical forests. *Ecography* **37**: 367–377.

Ekadinata, S., van Noordwijk, M., Budidarsono, S., and Dewi, S. (2013). Hotspots in Riau, haze in Singapore: the June 2013 event analyzed. *ASB Policy Brief No 33*. Nairobi: ASB Partnership for the Tropical Forest Margins: http://www.worldagroforestry.org/downloads/Publications/PDFS/BR13072.pdf [Accessed November 2015].

Emmons, L. H. (1999). Of mice and monkeys: primates as predictors of mammal community richness. In: Fleagle, J. G., Janson C., and Reed, K. E. (Eds), *Primate Communities*. pp. 171–188. Cambridge: Cambridge University Press.

Evans, D. T. and Silvestri, G. (2013). Non-human primate models in AIDS research. *Current Opinion in HIV and AIDS* **8**: 255.

Fa, J. E., Peres, C. A., and Meeuwig, J. (2002). Bushmeat exploitation in tropical forests: an intercontinental comparison. *Conservation Biology* **16**: 232–237.

Fa, J. E., Ryan, S. F., and Bell, D. J. 2005. Hunting vulnerability, ecological characteristics and harvest rates of bushmeat species in afrotropical forests. *Biological Conservation* **121**: 167–176.

Faith, D. P. (1992). Conservation evaluation and phylogenetic diversity. *Biological Conservation* **61**: 1–10.

Fleagle, J. G. (2013). *Primate Adaptation and Evolution*, 3rd edn. San Diego, CA and London: Academic Press.

Fuentes, A. (2012). Ethnoprimatology and the anthropology of the human-primate interface. *Annual Review of Anthropology* **41**: 101–117.

Fuentes, A. and Wolfe, L. D. (Eds) (2002). *Primates Face to Face. The Conservation Implications of Human-Nonhuman Primate Interconnections*. Cambridge: Cambridge University Press.

Fuentes, A., Southern, M., and Suaryana, K. G. (2005). Monkey forests and human landscapes: is extensive sympatry sustainable for *homo sapiens* and *macaca fascicularis* in Bali? In: Patterson, J. and Wallis, J. (Eds), *Commensalism and Conflict: The Primate-Human Interface*. Norman, OK: American Society of Primatology Publications.

Game, E. T., Kareiva, P., and Possingham, H. P. (2013). Six common mistakes in conservation priority setting. *Conservation Biology* **27**(3): 480–485.

Gautier-Hion, A. and Maisels, F. (1994). Mutualism between a leguminous tree and large African monkeys as pollinators. *Behavioral Ecology and Sociobiology* **34**: 203–210.

Gautier-Hion, A., Duplantier, J.-M., Quris, R., Feer, F., Sourd, C., *et al.* (1985). Fruit characters as a basis of fruit choice and seed dispersal in a tropical forest vertebrate community. *Oecologia* **65**: 324–337.

Goldberg, T. L., Gillespie, T. R., Rwego, I. B., Wheeler, E., Estoff, E. L., *et al.* (2007). Patterns of gastrointestinal bacterial exchange between chimpanzees and humans involved in research and tourism in western Uganda. *Biological Conservation* **135**: 511–517.

Golden, C. D. (2009). Bushmeat hunting and use in the Makira Forest north-eastern Madagascar: a conservation and livelihoods issue. *Oryx* **43**: 386–392.

Golden, C. D., Fernald, L. C. H., Brasheres, J. S., Rasolofoniaina, B. J. R., and Kremen, C. (2011). Benefits of wildlife consumption to child nutrition in a biodiversity hotspot. *Proceedings of the National Academy of Sciences USA* **108**: 19653–19656.

González-Suárez, M., Lucas, P. M., and Revilla, E. (2012). Biases in comparative analyses of extinction risk: mind the gap. *Journal of Animal Ecology* **81**: 1211–1222.

Gregory, R., Failing, L., Harstone, M., Long, G., McDaniels, T., *et al.* (2012). *Structured Decision Making: A Practical Guide to Environmental Management Choices*. Oxford: Wiley-Blackwell.

Guy, A. J., Curnoe, D., and Banks, P. B. (2014) Welfare based primate rehabilitation as a potential conservation strategy: does it measure up? *Primates* **55**: 139–147.

Hallé, N. (1987). *Cola lizae* N. Hallé (Sterculiaecea) Nouvelle espece du Moyen Ogooue (Gabon). *Adansonia* **3**: 229–237.

Hames, R. B. (1979). A comparison of the efficiencies of the shotgun and the bow in Neotropical forest hunting. *Human Ecology* **7**: 219–252.

Harcourt, A. H. (1987). Options for unwanted or confiscated primates. *Primate Conservation* **8**: 111–113

Harcourt, A. H. (2000). Coincidence and mismatch of biodiversity hotspots: a global survey for the order, primates. *Biological Conservation* **93**: 163–175.

Harcourt, A. H. (2001). Conservation in practice. *Evolutionary Anthropology* **9**: 258–265.

Harcourt, A. H. (2006). Rarity in the tropics: biogeography and macroecology of the primates. *Journal of Biogeography* **33**: 2077–2087.

Harcourt A. H. and Schwartz M. W. (2001). Primate evolution: a biology of Holocene extinction and survival on the southeast Asian Sunda Shelf islands. *American Journal of Physical Anthropology* **114**: 4–17.

Harcourt, A.H., Coppetto, S. A., and Parks, S. A. (2005). The distribution-abundance (density) relationship: its form and causes in a tropical mammal order, Primates. *Journal of Biogeography* **32**: 565–579.

Hare, B., Wobber, V., and Wrangham, R. (2012). The self-domestication hypothesis: evolution of bonobo psychology is due to selection against aggression. *Animal Behaviour* **83**(3): 573–585.

Hargrove, E. C. (1989). An overview of conservation and human values: are conservation goals merely cultural attitudes? In: Western, D. and Pearl, M. C. (Eds), *Conservation in the Twenty-First Century*. pp. 227–231. New York: Oxford University Press.

Hart, D. (2007). Predation on primates: a biogeographical analysis. In: Gursky-Doyen, S. and Nekaris, K. A. I. (Eds), *Primate Anti-predator Strategies*. pp. 27–59. New York, NY: Springer,

Hill, C. M. (2002). Primate conservation and local communities: ethical issues and debates. *American Anthropologist* **104**: 1184–1194.

Hill, K. (1982). Hunting and human evolution. *Journal of Human Evolution* **11**(6): 521–544.

Home Office (2004). Statistics of scientific procedures on living animals. UK: Home Office. Available at: <https://www.gov.uk/government/uploads/system/uploads/attachment_data/file/272232/6713.pdf> [Accessed November 2015].

Howe, H. F. (1986). Seed dispersal by fruit-eating birds and mammals. In: Murray, D. R. (Ed.), *Seed Dispersal*. pp. 123–189. New York, NY: Academic Press.

Isbell, L. A. (1994). Predation on primates: ecological patterns and evolutionary consequences. *Evolutionary Anthropology* **3**(2): 61–71.

Janson, C. H., Terborgh, J., and Emmons, L. H. (1981). Non-flying mammals as pollinating agents in the Amazonian forest. *Biotropica* **13**: 1–6.

Johnson C. N. (1998). Species extinction and the relation- ship between distribution and abundance. *Nature* **394**: 272–274.

Joyner, C., Barnwell, J. W., and Galinski, M. R. (2015). No more monkeying around: primate malaria model systems are key to understanding Plasmodium vivax liver-stage biology, hypnozoites, and relapses. *Frontiers in Microbiology* **6**: 1–8.

Kamilar J. M. (2006). Geographic variation in savanna baboon (*Papio*) ecology and its taxonomic and evolutionary implications. In: Lehman, S. M. and Fleagle, J. G. (Eds), *Primate Biogeography: Progress and Prospects*. pp. 169–200. New York, NY: Springer.

Kano, T. (1992). *The Last Ape: Pygmy Chimpanzee Behavior and Ecology*. Stanford, CT: Stanford University Press.

Kappeler, P. M. (1999). Convergence and divergence in primate social systems. In: Fleagle, J. G., Janson, C., and Reed, K. E. (Eds), *Primate Communities*. pp. 158–170. Cambridge: Cambridge University Press.

Kinzey, W. G. (Ed.) (1987). *Evolution of Human Behavior: Primate Models*. New York, NY: SUNY Press.

Kirkby C. A., Giudice-Granados, R., Day, B., Turner, K., Velarde-Andrade, L. M., Dueñas-Dueñas, A., *et al.* (2010). The market triumph of ecotourism: an economic investigation of the private and social benefits of competing land uses in the Peruvian Amazon. *PLoS One* **5**(9): e13015. doi:10.1371/journal.pone.0013015.

Kirkpatrick, J. B. (1983). An iterative method for establishing priorities for the selection of nature reserves: an example from Tasmania. *Biological Conservation* **25**: 127–134.

Kiss, A. (2004). Is community-based ecotourism a good use of biodiversity conservation funds? *Trends in Ecology and Evolution* **19**: 232–237.

Knott, C. D. (2001). Female reproductive ecology of the apes: implications for human evolution. In: Ellison, P. T. (Ed.), *Reproductive Ecology and Human Evolution*. pp. 429–463. New Brunswick, NJ: Transaction Publishers.

Kreft, H. and Jetz, W. (2007). Global patterns and determinants of vascular plant diversity. *Proceedings of the National Academy of Sciences* **104**(14): 5925–5930.

Kress, W. J., Schatz, G. E., Andrianifahanana, M., and Morland, H. S. (1994). Pollination of Ravenala madagascariensis (Strelitziaceae) by lemurs in Madagascar: evidence for an archaic coevolutionary system? *American Journal of Botany* **81**: 542–551.

Lambert, J. E. and Garber, P. A. (1998). Evolutionary and ecological implications of primate seed dispersal. *American Journal of Primatology* **45**: 9–28.

Laurance, W. F. (2013). Does research help to safeguard protected areas? *Trends in Ecological Evolution* **28**: 261–266.

Levang, P., Sitorus, S., Gaveau, D. L. A., and Abidin, Z. (2007). Elites' perceptions about the Bukit Barisan Selatan National Park. Centre for International Forestry Research Bogor Bar, Indonesia.

López, G. S., Orduña, F. G., and Luna, E. R. (1988). The status of *Ateles geoffroyi* and *Alouatta palliata* in disturbed forest areas in Sierra de Santa Marta, Mexico. *Primate Conservation* **9**: 53–61.

Lovett, J. C. and Marshall A. R. (2006). Why should we conserve primates? *African Journal of Ecology* **44**: 113–115.

Magin, C. D., Johnson, T. H., Groombridge, B., Jenkins, M., and Smith, H., *et al.* (1994). Species extinctions, endangerment and captive breeding. In: Olney, P. J. S., Mace, G. M., and Feistner, A. T. C. (Eds), *Creative Conservation: Interactive Management of Wild and Captive Animals*. London: Chapman and Hall.

Maréchal, L., Semple, S., Majolo, B., Qarro, M., Heistermann, M., *et al.* (2011). Impacts of tourism on anxiety and physiological stress levels in wild male Barbary macaques. *Biological Conservation* **144**: 2188–2193.

Marshall, A. J., Lacy, R., Ancrenaz, M., Byers, O., Husson, S., Leighton, M., *et al.* (2009). Orangutan population biology, life history, and conservation: perspectives from PVA models. In: Wich, S. A., Utami, S., Mitra Setia T., and van Schaik C. P. (Eds), *Orangutans: Geographic Variation in Behavioral Ecology and Conservation*. pp.311–326. Oxford: Oxford University Press.

Marshall, A. J., Meijaard, E., Van Cleave, E., and Sheil, D. (2016). Charisma counts: the presence of great apes affects the allocation of research effort in the paleotropics. *Frontiers in Ecology and the Environment* **14**: 13–19.

Matsuzawa, T. (2001). *Primate foundations of human intelligence: a view of tool use in nonhuman primates and fossil hominids*. In: Matsuzawa, T. (Ed.), *Primate Origins of Human Cognition and Behavior*. pp. 3–25. Tokyo, Japan: Springer.

McGrew, W. C. (1992). *Chimpanzee Material Culture: Implications for Human Evolution*. Cambridge: Cambridge University Press.

Meijaard, E. and Sheil, D. (2008). Cuddly animals don't persuade poor people to back conservation. *Nature* **454**: 159.

Meijaard, E. and Sheil, D. (2011). A modest proposal for wealthy countries to reforest their land for the common good. *Biotropica* **43**: 524–528.

Meijaard, E., Wich, S. A., Ancrenaz, M., and Marshall, A.J. (2012). Not by science alone: why orangutan conservationists must think outside the box. *Annals of the New York Academy of Sciences* **1249**: 29–44.

Milner-Gulland, E. J., Bennett, E. L., and the SCB 2002 Annual Meeting Wild Meat Group (2003). Wild meat: the bigger picture. *Trends in Ecology & Evolution* **18**: 351–7.

Mitani, J. C., Call, J., Kappeler, P. M., Palombit, R. A., and Silk, J. B. (Eds) (2012). *The Evolution of Primate Societies*. Chicago, Il: University of Chicago Press.

Mitchell, A. H. and Tilson R. L. (1986). Restoring the balance: traditional hunting and primate conservation in the Mentawai Islands, Indonesia. In: Else, J. and Lee, P. (Eds), *Primate Ecology and Conservation*. pp. 249–260. Cambridge: Cambridge University Press.

Mukherjee, R. P., Mukherjee G. D., and Bhuinya, S. (1986) Population trends of Hanuman langurs in agricultural areas of Midnapur District, West Bengal, India. *Primate Conservation* **7**: 53–54.

Muller-Landau, H. C. (2007). Predicting the long-term effects of hunting on plant species composition and diversity in tropical forests. *Biotropica* **39**: 372–384.

Myers, N., Mittermeier, R. A., Mittermeier, C. G., Da Fonseca, G. A., and Kent, J. (2000). Biodiversity hotspots for conservation priorities. *Nature* **403**: 853–858.

Naess, A. (1986). Intrinsic value: will the defenders of nature please rise? In: Soulé, M. E. (Ed.), *Conservation Biology: The Science of Scarcity and Diversity*. pp. 504–516. Sunderland, MA: Sinauer.

Nishida, T., Wrangham R. W., Jones, J. H., Marshall, A. J., and Wakibara, J. (2001). Do chimpanzees survive the 21st century? In: *The Apes: Challenges for the 21st Century*. Conference Proceedings. pp. 43–51. Brookfield, Il: Brookfield Zoo.

Norconk, M. A. and Veres, M. (2011). Physical properties of fruit and seeds ingested by primate seed predators with emphasis on sakis and bearded sakis. *The Anatomical Record* **294**(12): 2092–2111.

Norconk, M. A., Grafton, B. W., and Conklin-Brittain, N. L. (1998). Seed dispersal by neotropical seed predators. *American Journal of Primatology* **45**(1): 103–126.

Nunez-Iturri, G., Olsson, O., and Howe, H. F. (2008). Hunting reduces recruitment of primate-dispersed trees in Amazonian Peru. *Biological Conservation* **141**: 1536–1546.

Paaby, P., Clark, D. B., and Gonzalez, H. (1991). Training rural residents as naturalist guides: evaluation of a pilot project in Costa-Rica. *Conservation Biology* **5**: 542–546.

Paine, C. T. and Beck, H. (2007). Seed predation by neotropical rain forest mammals increases diversity in seedling recruitment. *Ecology* **88**: 3076–3087.

Parish, A. R. (1994). Sex and food control in the 'uncommon chimpanzee': how bonobo females overcome a phylogenetic legacy of male dominance. *Ethology and Sociobiology* **15**(3): 157–179.

Patterson, N., Richter, D. J., Gnerre, S., Lander, E. S., and Reich, D. (2006). Genetic evidence for complex speciation of humans and chimpanzees. *Nature* **441**: 1103–1108.

Pearce, D., Hecht, S., and Vorhies, F. (2008). What is biodiversity worth? Economics as a problem and a solution. In: Macdonald, D. W. and Service, K. (Eds), *Key Topics in Conservation Biology*. pp. 35–45. Oxford: Blackwell Publishing.

Peres, C. A. (1991). Seed predation of Cariniana micrantha (Lecythidaceae) by brown capuchin monkeys in Central Amazonia. *Biotropica* **23**(3): 262–270.

Peters, C. R. (1993). Shell strength and primate seed predation of nontoxic species in eastern and southern Africa. *International Journal of Primatology* **14**(2): 315–344.

Plous, S. (1993). *The Psychology of Judgment and Decision Making*. New York, NY: McGraw-Hill.

Poulsen, J. R., Clark, C. J., Connor, E. F., and Smith, T. B. (2002). Differential resource use by primates and hornbills: implications for seed dispersal. *Ecology* **83**: 228–240.

Power, M. E., Tilman, D., Estes, J. A., Menge, B. A., Bond, W. J., *et al.* (1996). Challenges in the quest for keystones. *BioScience* **46**: 609–620.

Purvis, A., Gittleman, J. L., Cowlishaw, G. C., and Mace, G. M. (2000). Predicting extinction risk in declining species. *Proceedings of the Royal Society of London, Series B* **267**: 1947–1952.

Pusey, A. E., Wilson, M. L., and Anthony Collins, D. (2008). Human impacts, disease risk, and population dynamics in the chimpanzees of Gombe National Park, Tanzania. *American Journal of Primatology* **70**: 738–744.

Queenborough, S. A., Mazer, S. J., Vamosi, S. M., Garwood, N. C., Valencia, R., *et al.* (2009). Seed mass, abundance and breeding system among tropical forest species: do dioecious species exhibit compensatory reproduction or abundances? *Journal of Ecology* **97**: 555–566.

Ramirez, M. (1984). Population recovery in the moustached tamarin (*Saguinus mystax*): management strategies and mechanisms of recovery. *American Journal of Primatology* **7**: 245–259.

Redford, K. H. and Taber, S. (2000). Writing the wrongs: developing a safe-fail culture in conservation. *Conservation Biology* **14**: 1567–1568.

Riley, E. P. (2010). The importance of human-macaque folklore for conservation in Lore Lindu National Park, Sulawesi, Indonesia. *Oryx* **44**: 235–240.

Rowe, N. and Myers, M. (2015) *All the World's Primates* website. Available at: <http://alltheworldsprimates.org/Home.aspx> [Accessed November 2015].

Russo, S. S. and Chapman, C. A. (2011). Primate seed dispersal: linking behavioural ecology and forest community structure. In: Campbell, C. J., Fuentes, A. F., MacKinnon, J. C., Panger, M., and Bearder S. (Eds), *Primates in Perspective*. pp. 523–524. Oxford: Oxford University Press.

Rylands, A. B. and Mittermeier R. A. (2014). Primate taxonomy: species and conservation. *Evolutionary Anthropology* **23**: 8–10.

Savage, A., Guillen, R., Lamilla, I., and Soto, L. (2010). Developing an effective community conservation program for cotton-top tamarins (*saguinus oedipus*) in Colombia. *American Journal of Primatology* **72**: 379–390.

Savage-Rumbaugh, S., Shanker, S. G., and Taylor, T. J. (1998). *Apes, Language, and the Human Mind*. Oxford: Oxford University Press.

Schipper, J., Chanson, J. S., Chiozza, F., Cox, N. A., Hoffmann, M., *et al.* (2008). The status of the world's land and marine mammals: diversity, threat, and knowledge. *Science* **322**(5899): 225–230.

Schwarzenbach, A. (2011). *Saving the World's Wildlife*. London: Profile Books Limited.

Schwitzer, C., Mittermeier, R. A., Rylands, A. B., Taylor, L. A., Chiozza, F., *et al.* (Eds) (2014). *Primates in Peril: The World's 25 Most Endangered Primates 2012–2014*. IUCN SSC Primate Specialist Group (PSG), International Primatological Society (IPS), Conservation International (CI), and Bristol Zoological Society, Arlington, VA.

Sekercioglu, C. H. (2012). Promoting community-based bird monitoring in the tropics: conservation, research, environmental education, capacity-building, and local incomes. *Biological Conservation* **151**: 69–73.

Semendeferi, K., Lu, A., Schenker, N., and Damásio, H. (2002). Humans and great apes share a large frontal cortex. *Nature Neuroscience* **5**(3): 272–276.

Sibal, L. R. and Samson, K. J. (2001). Nonhuman primates: a critical role in current disease research. *ILAR Journal* **42**: 74–84.

Siex, K. S. and Struhsaker, T. T. (1999). Colobus monkeys and coconuts: a study of perceived human-wildlife conflicts. *Journal of Applied Ecology* **36**: 1009–1020.

Sitas, N., Baillie, J. E. M., and Isaac, N. J. B. (2009). What are we saving? Developing a standardized approach for conservation action. *Animal Conservation* **12**: 231–237.

Smith, R. J., Veríssimo, D., Isaac, N. J., and Jones, K. E. (2012). Identifying Cinderella species: uncovering mammals with conservation flagship appeal. *Conservation Letter* **5**(3): 205–212.

Smuts, B. B. (1992). Male aggression against women. *Human Nature* **3**(1): 1–44.

Smuts, B. B., Cheney, D. L., Seyfarth, R. M., Wrangham, R. W., and Struhsaker, T. T. (1987). *Primate Societies*. Chicago, Il: University of Chicago Press.

Soulé, M. E. (1987). *Viable Populations for Conservation*. Cambridge: Cambridge University Press.

Soulé, M. E. and Wilcox, B. A. (1980). *Conservation Biology*. Sunderland, MA: Sinauer Associates, Inc.

Southwick, C. H., Siddiqi, M. F., and Oppenheimer, J. R. (1983). Twenty-year changes in rhesus monkey populations in agricultural areas of Northern India. *Ecology* **64**: 434–439.

Stanford, C. B. (1995). The influence of chimpanzee predation on group size and anti-predator behaviour in red colobus monkeys. *Animal Behaviour* **49**(3): 577–587.

Stevens, V. M., Whitmee, S. Le Galliard, J.-F., Clobert, J., Böhning-Gaese, K., *et al.* (2014). A comparative analysis of dispersal syndromes in terrestrial and semi-terrestrial animals. *Ecology Letters* **17**: 1039–1052.

Struhsaker, T. T. (2010). *The Red Colobus Monkeys: Variation in Demography, Behaviour, and Ecology of Endangered Species*. Oxford: Oxford University Press.

Stuart, S. N., Chanson, J. S., Cox, N. A., Young, B. E., Rodrigues, A. S. L., *et al.* (2004). Status and trends of amphibian declines and extinctions worldwide. *Science* **306**: 1783–1786.

Stumpf, R. (2006). Chimpanzees and bonobos: diversity within and between species. In: Campbell, C. J., Fuentes, A. F., MacKinnon, J. C., Panger, M., and Bearder S. (Eds), *Primates in Perspective*. pp. 321–344. Oxford: Oxford University Press.

Surbeck, M., Deschner, T., Schubert, G., Weltring, A., and Hohmann, G. (2012). Mate competition, testosterone and intersexual relationships in bonobos, *Pan paniscus*. *Animal Behaviour* **83**(3): 659–669.

Tallis, H., Kareiva, P., Marvier, M., and Chang, A. (2008). An ecosystem services framework to support both practical conservation and economic development. *Proceedings of the National Academy of Sciences* **105**(28): 9457–9464.

Teelen, S. (2008). Influence of chimpanzee predation on the red colobus population at Ngogo, Kibale National Park, Uganda. *Primates* **49**: 41–49.

Terborgh, J. (1974). Preservation of natural diversity: the problem of extinction-prone species. *BioScience* **24**: 715–722.

Terborgh, J. (2012). Enemies maintain hyperdiverse tropical forests. *The American Naturalist* **179**(3): 303–314.

The Chimpanzee Sequencing and Analysis Consortium. (2005). Initial sequence of the chimpanzee genome and comparison with the human genome. *Nature* **437**: 69–87.

Tomasello, M. (2008). *Origins of Human Communication*. Cambridge, MA: MIT Press.

Tomasello, M. (2009). *The Cultural Origins of Human Cognition*. Cambridge, MA: Harvard University Press.

Tutin, C. E. G. and Oslisly, R. (1995). *Homo, Pan,* and *Gorilla*: co-existence over 60,000 years at Lopé in central Gabon. *Journal of Human Evolution* **28**: 597–602.

Tutin, C. E. G., Williamson, E. A., Rogers, M. E., and Fernandez, M. (1991). A case study of a plant-animal relationship: *Cola lizae* and lowland gorillas in the Lopé Reserve, Gabon. *Journal of Tropical Ecology* **7**: 181–199.

Ulijaszek, S. J. (2002). Human eating behaviour in an evolutionary ecological context. *Proceedings of the Nutrition Society* **61**(04): 517–526.

United States Department of Agriculture (2013). Use of animals in research and education. Available at: <http://www.aphis.usda.gov/import_export/animals/oie/downloads/tahc_feb13/tahc_use_animals_research_and_education_82_feb13_rpt.pdf> [Accessed November 2015].

VandeBerg, J. L. and Zola, S. M. (2005). A unique biomedical resource at risk. *Nature* **437**: 30–32.

Van der Werf, G. R., Morton, D. C., DeFries, R. S., Olivier, J. G., Kasibhatla, P. S., *et al.* (2009). CO2 emissions from forest loss. *Nature Geoscience* **2**: 737–738.

van Schaik, C. P., Deaner, R. O., and Merrill, M. Y. (1999). The conditions for tool use in primates: implications for the evolution of material culture. *Journal of Human Evolution* **36**(6): 719–741.

van Schaik, C. P., Marshall, A. J., and Wich, S. A. (2009). Geographic variation in orangutan behavior and biology: its functional interpretation and its mechanistic basis. In: Wich, S. A., Utami, S., Mitra Setia, T., and van Schaik, C. P. (Eds), *Orangutans: Geographic Variation in Behavioral Ecology and Conservation*. pp. 351–361. Oxford: Oxford University Press.

Vanthomme, H., Belle, B., and Forget, P. M. (2010). Bushmeat hunting alters recruitment of large-seeded plant species in central Africa. *Biotropica* **42**: 672–679.

Weaver, D. B. and Lawton, L. J.(2007). Twenty years on: the state of contemporary ecotourism research. *Tourism Management* **28**: 1168–1179.

Wich, S. A., De Vries, H., Ancrenaz, M., Perkins, L., Shumaker, R. W., *et al.* (2009). Orangutan life history variation. In Wich, S. A. Utami, S. Mitra Setia, T. and van Schaik C. P. (Eds), *Orangutans: Geographic Variation in Behavioral Ecology and Conservation*. pp. 65–75. New York: Oxford University Press.

Wich, S., Riswan, Jenson, J. Refish J., and Nelleman, C. (2011). *Orangutans and the Economics of Sustainable Forest Management in Sumatra*. UNEP/GRASP/PanEco/YEL/ICRAF/GRID-Arendal. Norway: Birkeland Trykkeri AS.

Wilson, K. A., McBride, M. F., Bode, M., and Possingham, H. P. (2006). Prioritizing global conservation efforts. *Nature* **440**(7082): 337–340.

Wilson, K. A., Underwood, E. C., Morrison, S. A., Klausmeyer, K. R., Murdoch, W. W., *et al.* (2007). Conserving biodiversity efficiently: what to do, where, and when. *PLoS Biology* **5**(9): e223.

Wilson, K. A., Carwardine, J., and Possingham, H. P. (2009). Setting conservation priorities. *Annals of the New York Academy of Sciences* **1162**(1): 237–264.

Wilson, H. B., Meijaard, E., Venter, O., Ancrenaz, M., and Possingham, H. P. (2014). Conservation strategies for orangutans: reintroduction versus habitat preservation and the benefits of sustainably logged forest. *PLoS One* **9**(7): e102174. DOI: 10.1371/journal.pone.0102174.

Wrangham, R. and Pilbeam, D. (2001). African apes as time machines. In: Briggs, N. E., Sheeran, L. K., Shapiro G. L., and Goodall J. (Eds), *All Apes Great and Small, Volume 1: African Apes*. pp. 5–17. New York, NY: Springer.

Wrangham, R. W. (1987). The significance of African apes for reconstructing human social evolution. In: Kinzey, W. G. (Ed.), *The Evolution of Human Behavior: Primate Models*. pp. 51–71. Albany, NY: SUNY Press.

Wrangham, R. W. (2008). Why the link between long-term research and conservation is a case worth making. In: Wrangham, R. W. and Ross, E. (Eds), *Science and Conservation in African Forests*. Cambridge: Cambridge University Press.

Wrangham, R. W., Chapman, C. A., and Chapman, L. J. (1994). Seed dispersal by forest chimpanzees in Uganda. *Journal of Tropical Ecology* **10**: 355–368.

Wrangham, R. W. and Peterson, D. (1996). *Demonic Males*. Boston, MA: Houghton Mifflin Company.

Wrangham, R. W., Hagel, G., Leighton, M., Marshall, A. J., *et al.* (2008). The Great Ape World Heritage Species Project. In: Mehlman, P., Steklis D., and Stoinski T. (Eds), *Conservation in the 21st Century: Gorillas as a Case Study*. pp. 282–295. New York, NY: Kluwer Academic/Plenum Publishers.

Wright, I. J., Ackerly, D. D., Bongers, F. Harms, K. E., Ibarra-Manriquez, G., *et al.* (2007). Relationships among ecologically important dimensions of plant trait variation in seven Neotropical forests. *Annals of Botany* **99**: 1003–1015.

Wright, P. C. (1999). Lemur traits and Madagascar ecology: coping with an island environment. *Yearbook of Physical Anthropology* **42**: 31–72.

Yeager, C. P. (1997). Orangutan rehabilitation in Tanjung Puting National Park, Indonesia. *Conservation Biology* **11**: 802–805.

CHAPTER 3

IUCN Red List of Threatened Primate Species

Alison Cotton, Fay Clark, Jean P. Boubli, and Christopher Schwitzer

Juvenile black uakari monkey (Cacajao hosomi) eating a Mauritia flexuosa palm fruit in Pico da Neblina, Brazil Photo copyright: J. P. Boubli.

3.1 Introduction

The IUCN Red List of Threatened Species is the most comprehensive source of data on the global conservation status of animal, fungi, and plant taxa. Each species or subspecies receives its own assessment; an assessment entails the compilation of all relevant data on the species and application of Red List Criteria in order to identify the species' conservation status. The assessment of the conservation status of primates is undertaken by the Primate Specialist Group (PSG), one of approximately 120 taxonomic and disciplinary specialist groups of the Species Survival Commission (SSC) of the International Union for Conservation of Nature (IUCN). Assessments often take place during specialist workshops for each taxonomic group.

The latest IUCN assessment of primates reveals that the Primate order is facing a global extinction crisis: 63% of all species are currently classified as threatened with extinction. Of these, 14% are Critically Endangered (IUCN 2015). The main threats to primates are habitat loss, primarily through the burning and clearing of tropical forests, and hunting for bushmeat and the pet trade (Mittermeier *et al.* 2013; Fa and Tagg, Chapter 9, this volume; Irwin, Chapter 7, this volume). Emerging infectious diseases are also an overlooked yet significant threat (e.g. ebola, Smith *et al.* (2014); Nunn and Gillespie, Chapter 10, this volume). Primates serve important ecological functions, and their

Cotton, A.J., Clark, F.E., Boubli, J.P. and Schwitzer, C., *IUCN Red List of Threatened Primate Species.* In: *An Introduction to Primate Conservation.* Edited by: Serge A. Wich and Andrew J. Marshall, Oxford University Press (2016). © Oxford University Press.
DOI 10.1093/acprof:oso/9780198703389.003.0003

extirpation can have severe cascading ecological effects (Wright *et al.* 2007; Dirzo *et al.* 2014; Marshall and Wich, Chapter 2, this volume).

This chapter analyses trends in primate conservation status and patterns of extinction risk across regions and taxonomic groups, using data from the 423 primate species assessed on the IUCN Red List through 2014. Red List assessments are dynamic, as new data on species are acquired and species are re-assessed over time. In this chapter, we primarily refer to data collated from the Red List in June 2014, unless otherwise stated. We do not consider humans or extinct primate species.

3.2 IUCN Red List of Threatened Species

The IUCN Red List of Threatened Species, now 50 years old, is the most objective, scientifically based source of information on the conservation status of animals, fungi, and plants (Rodrigues *et al.* 2006; Hoffmann *et al.* 2008; Schipper *et al.* 2008). It has been described as 'The Barometer of Life' (Stuart *et al.* 2010) because it not only does list species and their conservation status, but also serves to inform and encourage conservation activities and policy.

Over the past 50 years the IUCN Global Species Programme and the IUCN Species Survival Commission (SSC) have been working together to improve the global conservation efforts of an ever-increasing number of species and subspecies. This has been achieved through extensive and increasingly comprehensive extinction risk assessments.

The IUCN Red List is far more than simply a list of threatened species. It is a comprehensive document containing detailed information on all assessed species providing conservation planners, funding agencies, campaigners, and policy officials with essential information for these decision-makers to set their conservation priorities and design their strategies. The information that is provided can include, but is not limited to, trends and fluctuations in population size as well as absolute numbers of individuals, geographical range and specific areas of occupancy, as well as habitat requirements and current threats.

As an example of the wide acceptance of the IUCN Red List as an authoritative summary of the status of global biodiversity, the United Nations Convention on Biological Diversity (CBD) has adopted IUCN's Red List to compose its Red List Index (see McLellan 2014), an indicator for Aichi Target 12. Target 12 is one of 20 targets proposed by the CBD in 2010 in the city of Nagoya, Aichi Prefecture, Japan, to achieve its goals for their 2011–20 Global Biodiversity Strategic Plan. It states that: 'By 2020 the extinction of known threatened species has been prevented and their conservation status, particularly of those most in decline, has been improved and sustained' (CBD Strategic Plan 2010).

The Red List Index records temporal changes in the extinction risk of the assessed species, thus tracking the Parties' (signatory countries) progress on Target 12. The World Wide Fund for Nature also incorporates the Red List Index in their Living Planet Report, a biannual assessment of global biodiversity (McLellan 2014).

3.2.1 Categories and criteria

The IUCN Red List classifies the relative risk of extinction that is faced by each species. The primary objective of the Red List is to focus attention on those plants, animals, and fungi that are at the greatest risk of extinction due to increasing human threats. Those at the greatest risk are categorized under the umbrella term of 'Threatened'. There are three categories that fall within this group; these are Critically Endangered, Endangered, and Vulnerable (Table 3.1). In addition to those taxa that are Threatened, the Red List also provides information on those that are categorized as Extinct and Extinct in the Wild as well as those that cannot be evaluated due to lack of information (Data Deficient). Those species that are not currently threatened but are either close to the set thresholds for this classification, or only stable due to ongoing, specific conservation work, are listed as being Near Threatened. Prior to 2003, Red Lists did not consider species that were not under threat. This has changed since and all those that are at a low extinction risk are grouped as Least Concern. This was done to increase the transparency of the system and avoid bias, putting all species into a more meaningful global context.

While the number of species that have been assessed is growing, overall there remains a strong taxonomic bias, with mammals, birds, and

Table 3.1 Definitions of the IUCN Threat categories.

Critically Endangered (CR)	The taxon (species or subspecies) faces an extremely high risk of extinction in the wild
Endangered (EN)	The taxon faces a very high risk of going extinct in the wild
Vulnerable (VU)	The taxon faces a high risk of extinction in the wild
Near Threatened (NT)	Although not currently threatened, the taxon is very close to qualifying for threatened status
Least Concern (LC)	The taxon does not qualify for threatened status; it may be declining, but is still widespread and abundant
Data Deficient (DD)	There is not enough information to make a decision about the taxon's status. Please note this is not an assessment of its status per se, but rather of the inadequacy of the available data, and a species or subspecies listed as DD may still be in serious danger
Not Evaluated (NE)	Usually applied to recently described forms, this marks a taxon that has not yet been assessed

Critically Endangered, Endangered and Vulnerable categories all form the broader 'Threatened' classification.

amphibians all comprehensively assessed, in stark contrast to invertebrates, fungi, and plants, which remain largely unassessed. While there were a large number of plants assessed using old criteria in 1997, they have yet to be re-assessed using current methods and therefore do not appear on the Red List. In addition, freshwater crabs, warm-water reef building corals, conifers, and cycads have all been broadly evaluated.

3.3 IUCN Species Survival Commission and Primate Specialist Group

The IUCN Species Survival Commission is a network of volunteer experts organized into over 120 taxonomic and disciplinary specialist groups. The Primate Specialist Group (PSG) is one such taxonomic specialist group that functions as the Red List Authority for all of the world's primate species.

The PSG was founded in the early 1960s. Russell A. Mittermeier took over as Chair in 1977 and has remained in this position since. The Deputy Chair is Anthony B. Rylands. The group currently has 585 members organized by expertise and geography. The PSG is subdivided into regional sections coordinated by regional Vice Chairs representing Mesoamerica, Andean Countries, Brazil and Guianas, Africa, Madagascar, China, Southeast Asia, and South Asia. It also has specific sections on Great Apes and Lesser Apes. Several journals and regional newsletters are periodically published by the PSG. These include *Primate Conservation* published by Conservation International and the Margot Marsh Biodiversity Foundation, and four regional Newsletters: *African Primates, Asian Primates, Neotropical Primates*, and *Lemur News*.

3.4 The dynamism of Red List assessments

With threats increasing or decreasing in scope and severity, the conservation status of a species may change over time. Hence, as a general rule, all species on the IUCN Red List should be re-assessed at least once every 10 years. For mammals, this is being done under the heading of the Global Mammal Assessment initiative, which published the first comprehensive mammal assessment on the IUCN Red List in 2008. The next global assessment of mammals will be published in 2016.

Of the 423 extant non-human primate species assessed on the Red List, 24% (100/423) were last assessed, or assessed for the first time, in 2014. This does not necessarily mean that the actual expert discussions for estimating extinction risk of these species all happened in 2014; in fact, 99 of the 100 aforementioned species, namely all lemurs, were assessed at a workshop in Madagascar in July 2012. However, it takes the Red List Authority Coordinator a considerable amount of time to review and organize all the information produced during the assessments, followed by two rounds of consistency checks by the Global Mammal Assessment team in Rome and the IUCN Red List Unit in Cambridge. The 'assessment date' on the IUCN Red List thus refers to the publication date of the assessment rather than the date of the assessment workshop itself. Just above 1% (6/423) of all assessed primate species were last assessed, or assessed for the first time, in 2013, just under 1% (2/423) in 2012, and 74% (315/423) in 2008.

3.5 Trends in primate diversity

Over the last 15–20 years there has been a considerable leap forward in our understanding of the level of global primate diversity (Groves, Chapter 4, this volume). At the time of writing, the IUCN Red List recognized a total of 423 extant non-human primate species comprising 695 taxa in 17 families and 76 genera. According to Mittermeier *et al.* (2013), 96 primate taxa (species and subspecies) have been discovered and described since 1990; 66 taxa have been described since 2000.

The number of recognized primate species has significantly increased since the 1970s for two main reasons: (1) increasing use of molecular genetics to test previous taxonomic hypotheses, and (2) an increase in the number of field expeditions to survey new areas. Molecular genetics has revolutionized the field of systematics and phylogenetics by uncovering unknown lineages and species relationships with greater statistical power than previous morphological comparisons. As a consequence, a large number of new taxa have been identified, requiring taxonomists to scrutinize old taxonomic arrangements and to name new taxa (Groves, Chapter 4, this volume; Groves 2014; Rylands and Mittermeier 2014).

Although molecular genetics has been an important tool contributing to the flurry of new species descriptions, the most surprising discoveries have come from an increased number of scientific expeditions to remote regions of the world. According to Ceballos and Erlich (2009), nearly 40% of the new mammal species described from 1993 to 2007 resulted from the exploration of new areas that until very recently were not easily accessible to collecting expeditions. These authors listed more than 400 species of mammals described in this period, including 55 new primates. Only bats and rodents had more newly described species than primates (174 and 94 respectively: Ceballos and Erlich (2009)).

In some cases, the discovery and description of new species in the field lead to further taxonomic revisions in order to accommodate the newly discovered animal. As an example, after discovering the Ayres black uakari monkey (*Cacajao ayresi*) along the Aracá River in northern Amazonia, Boubli *et al.* (2008) had to revise the taxonomy of all black uakaris, resulting in the description of a second species, the Neblina uakari (*Cacajao hosomi*), a known animal but one thought previously to be the type specimen for *Cacajao melanocephalus* collected by Alexander von Humboldt in 1801. In this case, then, the discovery of one new species resulted in an increase of recognized black uakari species richness from one to three species.

In addition to the ongoing discovery of new primate species and subspecies, there have also been changes in taxonomy at the genus level. For example, the African guenon genus *Cercopithecus* was recently revised, separating out the genera *Allochrocebus* and *Chlorocebus* (Perelman *et al.* 2011); even more recently, it was proposed to divide the New World genus *Saguinus* into *Leontocebus* and *Saguinus* (Buckner *et al.* 2015). Boubli *et al.* (2014) have also proposed the division of the genus *Callicebus* into three new genera, based on the time of split between the different species groups, which is much older than the split of well-recognized genera such as *Lagothrix* and *Ateles.*

Paradoxically, another reason for an increase in recognized primate species is deforestation. Dramatic habitat destruction has inspired researchers to discover the existence and distribution of new species of primates in their range countries, and to form urgent conservation action plans, before it is too late. It is ironic that forest clearance, increased transport, and infrastructure have allowed researchers to access previously impenetrable forest. A good example is the recent discovery of Milton's titi monkey, *Callicebus miltoni*, found in the Amazonian Arch of Deforestation, that is the area most impacted by the advance of the Brazilian agricultural frontier (Dalponte *et al.* 2014).

3.6 Trends in primate conservation status

For 96% (405/423) of all primate species assessed so far, there is sufficient information available about their distribution, population sizes, and threats to assign them to a conservation category (defined in Table 3.1), with the remainder assigned to the Data Deficient category. Of those species that have been assigned a conservation status, 63% (257/405) are currently classified as being threatened (Critically Endangered, Endangered, or Vulnerable), and the remaining are classified as Near Threatened (21/405, 5%) or Least Concern (127/405, 31%).

Overall, there has been a general increase over time, from 1996 to 2014 (i.e. over successive Red List assessments) in the percentage of assessed primate species classified as threatened. In 1996, 41% of primate species were threatened with extinction (96/233 assessed species), while in 2004 this decreased slightly to 39% (114/296). In 2008, the Red List revealed that 48% of primates were threatened. The 2012 Red List classified 57% (206/362) of assessed primate species as threatened; with 42 species (71 taxa) Critically Endangered, 85 (137 taxa) Endangered, and 79 (103 taxa) Vulnerable. In 2014, the percentage of threatened species increased once again to 63% (259/408). According to the most current data available, 74% (314/423) of primate species have a decreasing population trend; the remaining 26% have stable (9%), increasing (1%), or unknown population trends (16%). When only those species with known population trends are examined, 90% of primates are currently experiencing a decline in population size.

3.6.1 Trends by region and taxonomic group

The most recent IUCN assessments have shown that 93.9% of lemur species are classified as threatened, thus rendering Madagascar as the place with the highest number and percentage of threatened primate species (Schwitzer *et al.* 2013). This region is followed by Asia, where 78.9% of primates are threatened, the Neotropics (44.4%), and Africa (43.3%) (Table 3.2).

Apes (greater and lesser) are the most threatened group of primates, with 100% of taxa classified as threatened. Prosimians are second, with 67.2% of species threatened, followed by monkeys (51.9%). The primate families most at risk, with 100% of classified species categorized as threatened, are the gibbons (Hylobatidae), great apes (Hominidae), sportive lemurs (Lepilemuridae), aye-ayes (Daubentoniidae), and the indriids (Indriidae), with the tarsiers following closely behind with six out of seven species classified as threatened (Tarsiidae, 86%).

3.7 How Red List data can be utilized in primate conservation

3.7.1 Case study: lemur conservation

Madagascar is arguably the world's most important biodiversity hotspot (Schwitzer *et al.* 2013). It is the world's fourth largest island and has been isolated

Table 3.2 Summary of regional and taxonomical distribution of threat (data from the Primate Specialist Group website, current as of 1 September 2014). Column headings indicate IUCN threat status, using abbreviations in Table 3.1. Cells indicate numbers of taxa (species and subspecies).

	CR	EN	VU	NT	LC	DD	NE	TOTAL
All primates	64	141	98	37	195	83	16	634
Prosimians	10	30	21	9	38	25	2	135
Monkeys	43	87	75	28	157	38	14	442
Apes	11	24	2	0	0	1	0	38
Madagascar	8	18	14	5	11	40	0	96
Africa	12	30	21	9	70	10	12	164
Asia	23	65	33	14	19	17	4	175
Neotropics	21	28	30	9	95	16	0	199

from other landmasses for at least 88 million years. As a result, levels of primate diversity are unrivalled by any other country. This results in conservation on Madagascar not just being about saving species, but preserving entire lineages and a substantial segment of evolutionary history (Schwitzer *et al.* 2013). Madagascar faces an extensive array of conservation threats, where a large percentage of the original vegetation has been destroyed and the remaining patches are subject to severe fragmentation. In addition, Madagascar is one of the world's poorest countries, with more than 92% of people living on less than USD2 a day (World Bank 2013). Since 2009, the country has seen a period of heightened political instability (Schwitzer *et al.* 2014), during which hunting of lemurs has increased dramatically (Jenkins *et al.* 2011). The long-term wellbeing of the Malagasy people is heavily dependent on a healthy, functioning ecosystem (Schwitzer *et al.* 2013). This is represented by the continued survival of not only lemurs and other endemic species, but also the forests and other habitats that harbour them. It remains vital, therefore, that conservation efforts are directed towards ensuring the survival of Madagascar's endemic flora and fauna.

As a part of the Global Mammal Assessment, in July 2012 a workshop took place to re-assess 103 lemur taxa using the IUCN Red List criteria. This took place in Antananarivo, Madagascar, and involved a group of more than 60 Malagasy and international experts. This workshop highlighted the dire conservation status of many of Madagascar's lemurs, with almost 94% of taxa (of those that had sufficient data to be categorized) classified as threatened with extinction. Of these threatened species, 24 were assessed as Critically Endangered, 49 as Endangered, and 20 as Vulnerable. Only three were assessed outside these categories, with three assigned to Near Threatened, three to Least Concern, and four to the Data Deficient category. These data prompted an immediate response, and during the latter half of the Red Listing workshop, experts prepared an action plan designed to promote and guide lemur conservation (Schwitzer *et al.* 2013).

This action plan used the IUCN Red List data to highlight the predicament of Madagascar's lemurs and propose constructive solutions to the current crisis, such as community involvement and lemur tourism. In addition to international and national scale approaches, 30 specific site-based action plans were included, with a total proposed budget of USD7.6 million. A companion scientific paper was published in the high-profile journal *Science*, outlining the findings of the Red List workshop and highlighting a number of key areas that the authors believed were crucial to the long-term survival of lemurs (Schwitzer *et al.* 2014).

Both publications disseminated key information about lemur conservation to a wide audience. They highlighted the following points. First, they noted that the creation of protected areas to slow habitat loss is critical to the long-term survival of lemurs. In 2010, 29 new protected areas were created and formally protected areas now cover 8% of Madagascar's land area, up from 3% in 2003. Political instability and the associated lack of enforcement have historically resulted in deforestation in key protected sites, and so successful management and enforcement of these new areas will prove crucial moving into the future (Schwitzer *et al.* 2014). In addition to habitat protection, the authors emphasized the importance of working closely with local communities. There is a strong tradition in Madagascar of local groups implementing conservation in areas surrounding protected locations (Schwitzer *et al.* 2013). In some areas where training and support is given, such as Andasibe, local guide and community groups have even formed community reserves neighbouring national protected areas (Schwitzer *et al.* 2013). In addition, the authors championed the development of lemur ecotourism. Despite the unstable political situation, lemurs are already Madagascar's number one tourist attraction. With suitable funding, the scope for this could increase in the future, with economic benefits at both the local and national level. Schwitzer *et al.* (2013: 10) claim '. . . there is no reason why tourism should not become the number one foreign exchange earner in the next five years'. The final key element highlighted for lemur conservation was the creation and maintenance of a long-term research presence in important sites, such as the western dry forest in Kirindy and the southern spiny desert of Beza-Mahafaly. Scientific activity encourages community engagement and education, provides employment opportunities, and functions as a watchdog against prohibited activities (Laurance 2013; Schwitzer *et al.* 2013).

3.8 Summary

Primates are one of the most recognizable and celebrated groups of animals on the planet. Recently there have been numerous phylogenetic changes to the Primate order due to an increase in the use of molecular techniques as well as an increase in field expeditions to remote locations. In addition to changes in taxonomy, primates are currently undergoing a very serious population decline, with 90% of species with known population trends currently experiencing a decrease in population size. The IUCN Red List has been instrumental in highlighting the scale of this decline during the recent Global Mammal Assessment and this has resulted in a wealth of media coverage as well as action plans to tackle the crisis. There is no doubt that the IUCN Red List's impact on conservation and species information will continue to grow and will continue to be the most authoritative document for conservation practitioners and policy-makers worldwide.

Acknowledgements

The authors thank the editors for inviting us to contribute with this chapter.

References

Boubli, J. P., da Silva, M. N. F., Amado, M. V., Hrbek, T., Pontual, F. B., *et al.* (2008). A taxonomic reassessment of Cacajao melanocephalus Humboldt (1811), with the description of two new species. *International Journal of Primatology* **29**(3):723–741.

Boubli, J. P., Sampaio, I., and Schneider, H. (2014). New discoveries in taxonomy and phylogeny of new world primates. In: Symposium in the XXV Congress of the International Primatological Society, Hanoi, Vietnam.

Buckner, J. C., Alfaro, J. L., Rylands, A. B., and Alfaro, M. E. (2015). Biogeography of the marmosets and tamarins (Callitrichidae). *Molecular Phylogenetics and Evolution* **82**: 413–425.

Ceballos, G. and Ehrlich, P. R. (2009). Discoveries of new mammal species and their implications for conservation and ecosystem services. *Proceedings of the National Academy of Sciences* **106**(10): 3841–3846.

Convention on Biological Diversity (CBD) (2010). Strategic Plan for Biodiversity 2011–2020, including Aichi Biodiversity Targets. Available at: <https://www.cbd.int/sp/> [Accessed November 2015].

Dalponte, J. C., Silva, F. E., and Silva-Júnior, S. (2014). New species of titi monkey, genus Callicebus Thomas, 1903 (Primates, Pitheciidae), from Southern Amazonia, Brazil. *Papéis Avulsos de Zoologia (São Paulo)* **54**(32): 457–472.

Dirzo, R., Young, H. S., Galetti, M., Ceballos, G., Isaac, N. J. B., *et al.* (2014). Defaunation in the Anthropocene. *Science* **345**(6195): 401–406.

Groves, C. P. (2014). Primate taxonomy: inflation or real? *Annual Review of Anthropology* **43**(1): 27–36.

Hoffmann, M., Brooks, T. M., da Fonseca, G. A. B., Gascon, C., Hawkins, A. F. A., *et al.* (2008). Conservation planning and the IUCN Red List. *Endangered Species Research* **6**(2): 113–125.

IUCN (2015). Red list category summary for all animal classes and orders, IUCN Red List. Available at: <http://cmsdocs.s3.amazonaws.com/summarystats/2015–4_Summary_Stats_Page_Documents/2015_4_RL_Stats_Table_4a.pdf> [Accessed November 2015].

Jenkins, R. K. B., Keane, A., Rakotoarivelo, A. R., Rakotomboavonjy, V., Randrianandrianina, F. H., *et al.* (2011). Analysis of patterns of bushmeat consumption reveals extensive exploitation of protected species in eastern Madagascar. *PLoS One* **6**(12): p. e 27570.

Laurance, W. (2013). The scientist as guardian: a tool for protecting the wild. *Yale Environment 360*: Available at: <http://e360.yale.edu/feature/william_laurance_scientist_as_guardian_protecting_flora_fauna/2632/> [Accessed November 2015].

McLellan, R. (Ed.) (2014). *Living Planet Report 2014: Species and Spaces, People and Places.* Gland, Switzerland: WWF International. Available at: <https://www.wwf.or.jp/activities/lib/lpr/WWF_LPR_2014.pdf> [Accessed November 2015].

Mittermeier, R. A., Rylands, A. B., and Wilson, D. E. (Eds) (2013). *Handbook of the Mammals of the World. Vol. 3. Primates.* Barcelona: Lynx Edicions.

Perelman, P., Johnson, W. E., Roos, C., Seuánez, H. N., Horvath, J. E., *et al.* (2011). A molecular phylogeny of living primates. *PLoS Genetics* **7**(3): p. e 1001342.

Rodrigues, A. S. L., Pilgrim, J. D., Lamoreux, J. F., Michael Hoffmann, M., and Brooks, T. M. (2006). The value of the IUCN Red List for conservation. *Trends in Ecology & Evolution* **21**(2): 71–76.

Rylands, A. B. and Mittermeier, R. A. (2014). Primate taxonomy: species and conservation. *Evolutionary Anthropology* **23**(1): 8–10.

Schipper, J., Chanson, J. S., Chiozza, F., Cox, N. A., Hoffmann, M., *et al.* (2008). The status of the world's land and marine mammals: diversity, threat, and knowledge. *Science* **322**(5899): 225–230.

Schwitzer, C., Russell, A., Mittermeier, R. A., Davies, N., Johnson, S., *et al.* (2013). Lemurs of Madagascar: a Strategy for their Conservation 2013–2016. IUCN SSC Primate Specialist Group, Bristol Conservation and Science Foundation, and Conservation International, Bristol, UK. Available at <https://portals.iucn.org/library/efiles/documents/2013-020.pdf> [Accessed November 2015].

Schwitzer, C., Mittermeier, R. A., Johnson, S. E., Donati, G., Irwin, M., *et al.* (2014). Averting lemur extinctions amid Madagascar's political crisis. *Science* **343**(6173): 842–843.

Smith, K. F., Goldberg, M., Rosenthal, S., Carlson, L., Chen, J., *et al.* (2014). Global rise in human infectious disease outbreaks. *Journal of the Royal Society Interface* **11**(101): 20140950.

Stuart, S. N., Wilson, E. O., McNeely, J. A., Mittermeier, R. A., and Rodríguez, J. P. (2010). The barometer of life. *Science* **328**(5975): 177.

World Bank (2013). Madagascar: Measuring the Impact of the Political Crisis. World Bank. Available at: <https://www.worldbank.org/en/news/feature/2013/06/05/madagascar-measuring-the-impact-of-the-political-crisis> [Accessed November 2015].

Wright, S. J., Stoner, K. E., Beckman, N., Corlett, R. T., Dirzo, R., *et al.* (2007). The plight of large animals in tropical forests and the consequences for plant regeneration. *Biotropica* **39**(3): 289–291.

CHAPTER 4

Species concepts and conservation

Colin Groves

Tephrosceles red colobus in Kibale, Uganda. Photo copyright: A. H. Korstjens.

4.1 Introduction: what exactly are species?

What, precisely, are species? Since the Evolutionary Synthesis of the 1930s, species have been thought of as populations (or groups of populations). The late 1930s added to this: species are populations (or groups of populations) that do not interbreed under natural conditions (Mayr (1963): known as the Biological Species Concept), but since about 1970 biologists have been gradually finding the problems with such a concept, and proposing alternatives (for a brief history of this, see Groves (2001)). Genetic analysis has been gradually uncovering the ubiquity of interbreeding between 'good' species, making Mayr's concept untenable, and the question of how to judge whether allopatric populations are, or are not, 'good' species is ultimately insoluble under Mayr's concept.

Species constitute a major component of the pattern of biodiversity; the processes by which they may have come into being, or be maintained, are something to be studied only after the pattern has been elucidated (Groves 2001, 2004). Tattersall (1992) put it this way: '. . . the evolutionary history of a clade is the history of the innovations which arise within it. We risk losing sight of most of those innovations, and of their significance, if, in time or space, we disguise varying morphologies under the rubric of a single species.' In other words, we must be very careful not to carelessly lump populations that are different, and call the result 'species'—we must carefully examine these different populations in case there are several species involved.

What Tattersall was promoting, and what I urged, was use of the Phylogenetic Species Concept (PSC), in which species are the smallest diagnosable units: 'diagnosable' means that there are features by

Groves, C.P., *Species concepts and conservation.* In: *An Introduction to Primate Conservation.* Edited by: Serge A. Wich and Andrew J. Marshall, Oxford University Press (2016). © Oxford University Press.
DOI 10.1093/acprof:oso/9780198703389.003.0004

which each one may be uniquely recognized, and the meaning of 'the smallest' is that there are no subordinate diagnosable subunits within them. And they are, ipso facto, the terminals on a cladogram. In this way, species are evolutionary lineages; the PSC corresponds to the 'general lineage concept' of species of de Queiroz (2007), which emphasizes that a species is an evolutionary lineage, and the task of the taxonomist is to find whether there is actual evidence, in the form of unique heritable features (be they DNA, morphology, behaviour), that a given population actually is an evolutionary lineage.

The PSC is now widely adopted in vertebrate taxonomy in general, although it is still making only slow headway in ornithology (perhaps because the founder of the traditional view, Ernst Mayr, was an ornithologist?). Among ornithologists, Zink (2006) has argued most strongly for the objectivity of the PSC, stressing problems such as the impossibility, under the traditional view (i.e. the BSC), of deciding upon the taxonomic status of allopatric populations, and the amount of interbreeding now known to occur between species that have always been acknowledged as being distinct. Within species, in the traditional view, we have the allopatric (by definition) entities called subspecies; not only these are frequently not themselves historical entities, but also there is a significant difference between 'subspecies' which are 100% different (which are in reality distinct species) and those which differ only by having greater or lesser frequencies of some character (and are to a greater or lesser extent arbitrary).

Under the PSC, species are diagnosable units; there is no necessity that they have been separate for a particular length of time, that they must have a particular level of genetic distance between them, or that they are not reproductively compatible—the ability to hybridize is an ancestral condition which may be retained for longer or shorter time since two species became separate (Zink and Davis 1999).

The only stipulation about the features by which species may be recognized is that they must be heritable. They may be phenotypic—morphological, behavioural, physiological—as long as there is a reasonable (in theory testable) presumption that they have a genetic basis, even though we may well not know what this basis actually is. Or, of course, these features may include unique DNA sequences, and in such a case we do have direct evidence that the differences are heritable; but species whose phenotypic differences are evident, even though they are not supported by *known* DNA sequence differences, are not less distinct for that.

The PSC is testable, as a scientific hypothesis must be. No other view of what species are—certainly not the Biological Species Concept—is testable: that is to say, no other view of the nature of species is genuinely scientific. Nonetheless, there is still an 'old school' which, apparently appalled by the increase in the number of species recognized under the PSC, continues to resist it (see, e.g., Rosenberger (2014)). The adoption of the PSC in primatology at first came fairly slowly, then, since the early 2000s, all in a rush, and resistance to it has largely (but not entirely!) fallen away.

Glossary

Allopatric. When two populations live in different areas, and their ranges do not meet, they are said to be allopatric. Obviously, such populations do not have a chance to 'interbreed under natural conditions', or not to interbreed, so, under the Biological Species Concept, it is impossible to determine whether they are distinct species or not.

Biological Species Concept. This conception of species, proposed by the great ornithologist Ernst Mayr in the 1930s, defined species based on whether they interbreed under natural conditions with other species. Many biologists still accept this definition, but if we do so, we can never test whether allopatric (*quod vide (q.v)*) populations are different species or not. More recently, it has been found that many species have incorporated DNA, especially mtDNA (*q.v.*), from related species, in other words they have been interbreeding on the quiet all the time, so flouting the Biological Species Concept.

Glossary (Continued)

Caecotrophy. The caecum is a blind pouch at the junction of the small intestine and large intestine in the gut; it contains bacteria which ferment cellulose and hemicellulose, breaking them down into short-chain fatty acids. But most absorption of nutrients takes place in the small intestine, so in animals with a very large caecum the fermentation products must be taken through the gut a second time in order to be absorbed. Rabbits are one such species. *Lepilemur leucopus* is said to be similar; Charles-Dominique and Hladik (1971) reported that, in the late afternoon, sportive lemurs of this species wake up (they are nocturnal) and ingest a first (rather liquid) set of faeces, which therefore can now go through the small intestine so that the fermentation products can be absorbed; the second, 'standard', set of faeces are then extruded during the animal's nocturnal ramblings.

Cladogram. A depiction of evolutionary relationships, showing which species is related to which, but not what the ancestors are/were, or when they lived, that is not a 'family tree'.

Diagnosable. Refers to traits that differ completely between taxa, and are (in genetic terms) fixed. The basic criterion of the Phylogenetic Species Concept is that species are identified by diagnosable/diagnostic characters, meaning characters that enable a species to be infallibly recognized.

Evolutionary synthesis. The coming together in the 1930s of the different schools of thought on exactly how evolution occurred. The result was 'population thinking': evolution occurs in populations, basically by natural selection acting on genetic mutations. The evolutionary synthesis is regarded as the genesis of modern evolutionary theory.

General lineage concept of species. The idea that every lineage is a species; conversely, every species is an evolutionary lineage.

mtDNA. A form of DNA that is inherited entirely through the female line (because, not being in the cell nucleus, none is supplied to the zygote by sperm). Because it remains intact through the generations (albeit accumulating mutations), even if males of species A dominate those of species B and mate with species B females, then with the hybrid females, then with the backcrosses, and so on, until a population more or less indistinguishable from species A results, the mtDNA of species B remains in that population, documenting the hybridization that went on generations ago.

Phylogenetic Species Concept. The principle that species are defined by diagnosability (*q.v.*). Essentially, this is the way in which species may be recognized in the General Lineage Concept (*q.v.*).

Subspecies. Geographic segments of a species that differ at some level (different gene frequencies, average size, or colour differences) from each other. They are often dignified by trinomial scientific names, but what proportion of individuals should be distinguishable before one can recognize a subspecies is subjective. They cannot be reified as if they were species.

Synonymy. In biological nomenclature, different names that refer to the same species (or other taxonomic entity). Synonymy arises because, in the early years of taxonomy (late eighteenth century, most of the nineteenth century), communications were poor and the same species might be described, under different names, by two or more different people; this occasionally still happens. Sometimes, the describer of a species underestimates the range of variation in that group, and inadvertently describes a species which is actually just an unusual specimen of an already-described species.

Taxonomy. The science of biological classification into phyla, classes, orders, families, genera, and species.

Testability. Able to be falsified; strictly speaking, a scientific proposition can never be proved to be true, but it can be proved to be untrue. The great twentieth-century philosopher of science, Karl Popper, argued that falsifiability (testability) is a requirement of any proposition if it is to be considered scientific.

4.2 Taxonomic inflation: what does it mean?

If a date can be put on the decisive swing towards the PSC in primatology, I would choose 1991, when de Vivo published his revision of the genus *Callithrix* (as the marmosets, exclusive of the pygmy marmoset, were still being called). The simple expedient of finding the diagnosable units, and rating them all as species, cut through decades of disputes over whether there was interbreeding, whether two were more closely related to each other than they were to a third, and so on. This seemed to establish the principle that one should identify the units (i.e. the pattern) first, and then start investigating their interrelationships, their ecology, their biogeography, and other significant questions. This initiative slowly gathered force; other species of marmosets were described through the 1990s, and new species of platyrrhines in general are now being described quite comfortably at almost yearly intervals, mainly by re-examining what used to be thought of as single species, and finding that there are actually several diagnosable populations, that is species, within them.

The spread of the PSC to other groups of primates began in 2000, with the publication of Rasoloarison *et al.*'s revision of the mouse lemurs of western Madagascar. In its way, this was more startling than de Vivo (1991), because there were not just subspecies being raised to species rank, but in this case species were actually being resurrected from synonymy, and some totally new species were being described. The lemur taxonomy revolution again took a while to catch on, but when it did so it introduced a new concept to primatology—new species being described largely on the basis of DNA sequences.

The first of these 'DNA taxonomy' papers (Louis *et al.* 2006a) began to do for the mouse lemurs of eastern Madagascar what Rasoloarison *et al.* (2000) had done for those of the west, but from a different starting point: mtDNA sequences were primary, and morphological differences essentially subsidiary. Three new species of *Microcebus* were described in the Louis *et al.* (2006a) paper. This was as nothing, however, compared to a monograph later in the same year, when Louis *et al.* (2006b) again primarily (but not entirely) used mtDNA to describe 11 new species of *Lepilemur*.

As these changes were startling, however, they seemed to correspond to a felt need; and other groups descended on Madagascar to collect samples, to the extent that at least one new species of *Lepilemur* was described twice, by two independent teams (see Zinner *et al.* (2007)). Subsequently, more new species have been described, and 'subspecies' confirmed as distinct species, using mtDNA as the leading (but not the sole) evidence: further new species described in *Microcebus* and *Lepilemur*, while *Mirza*, *Avahi*, *Propithecus*, and *Eulemur* have all experienced DNA-assisted taxonomic clarification, with *Cheirogaleus* in progress (Mittermeier *et al.* 2010; Thiele *et al.* 2013; Lei *et al.* 2014).

I will pass over the further revisions undertaken in the past 15 years or so in the Platyrrhini, as well as the revisions to the Tarsiidae, to mention a group of Old World monkeys that have also undergone what has been rather insultingly dubbed 'taxonomic inflation': the red colobus (genus *Piliocolobus*). Formerly all red colobus were placed (with the occasional exception of the Zanzibar red colobus) into a single species, *Piliocolobus* (or *Procolobus*, or even *Colobus*) *badius*; Groves (2007) recognized 15 species, plus a 'muddle in the middle' that has so far resisted clarification, while the following year Ting (2008) published a DNA study which showed that the divisions between (some of) them were very ancient, well pre-Pleistocene.

Among the Hominoidea, not only the gibbons but even the great apes have undergone 'taxonomic inflation'. Prominent among these is the orang-utan; over the past 15 years or so, the acceptance that Bornean and Sumatran orang-utans belong to two different species (*Pongo pygmaeus*, *Pongo abelii*) has become almost universal, quite contrary to the mid twentieth century when it was not even accepted that they were different subspecies.

4.3 The newly recognized species: what is their significance? The lemur case

It is still the case that some primatologists are sceptical of the value of these newly recognized species, and uncomprehending questions like 'how much difference do you need before you can recognize it as a species?' are bandied about. (The answer to this is that a species is not a quantity, it is a quality.) It was unexpected that one of the most prominent objectors to the multiplication of species, with reference to Malagasy lemurs, was Tattersall (2007), whose espousal of the PSC in palaeoanthropology was quoted earlier. In this section, I hope to show that the species thrown up by taxonomic inflation are not simply a bunch of DNA substitutions or colour variants, but have real meaning: where we have evidence, it tends to indicate that other aspects of biology have been dragged along with the diagnostic features, whether by natural selection, or by genetic drift or genetic hitchhiking.

The paper by Boinski and Cropp (1999) was perhaps one of the first to show that previously recognized species differences have biological meaning: that three rather similar-looking species of squirrel monkey (*Saimiri*) actually differ strongly in social organization, rates of aggression, reproductive parameters, and other features. This may not be an ideal example with which to commence, because it can be argued that multiple species of *Saimiri* had begun to be recognized even before the PSC revolution, but it does illustrate the vital importance of species differences for understanding physiology, reproductive biology, and behaviour: a squirrel monkey is not just a squirrel monkey.

In the case of the newly described species of Malagasy lemur, we are hamstrung to some extent by the lack of in-depth field studies, but there are some very interesting pointers. In *Lepilemur*, there are differences in such features as diet, resting places, and general activity patterns at least between *Lepilemur mustelinus, L. microdon, L. ankaranensis, L. tymerlachsoni, L. sahamalazensis, L. edwardsi, L. ruficaudatus*, and *L. leucopus* (see summaries in Mittermeier *et al.* (2010)). For example: *Lepilemur leucopus*, a species that admittedly was resurrected prior to the coming of the PSC to primatology, is the only one known to practice caecotrophy (Charles-Dominique and Hladik 1971). Another example is *Lepilemur ankaranensis*: Lowin (2010) reported that a large proportion (nearly one third) of dietary observations were of fruit eating, quite contrary to any other *Lepilemur* species, which are overwhelmingly folivores; he also observed them snapping leaf stems from trees and consuming the latex that flowed, another unprecedented dietary source for *Lepilemur* taxa.

Let us pause for a moment: instead of 7 species, there are now 26 species of sportive lemurs (*Lepilemur*: Figure 4.1). Much of this increase has been because each of the original 7 has been shown to consist of several previously unrecognized species; instead of 7 widespread species, we now have 26 species of very restricted distribution, many of them Endangered or even Critically Endangered (see Cotton *et al.*, Chapter 3, this volume). Yet, as we have seen, of these 26, those that have been investigated in detail turn out to be very real entities, and the loss if any of them should become extinct would be very grave. Protecting sportive lemurs as if they were still only seven species will just not suffice.

Field studies on different species of *Microcebus* have also uncovered noticeable differences between them. Thus Randrianambanina *et al.* (2003) detailed differences between two *Microcebus* species—*M. ravelobensis* and *M. rufus*—according to the occurrence and pattern of seasonal torpor, body weight fluctuations, tissue storage in the tail, and the length of the period during which females went into oestrus. There are also differences in populations of *M. murinus* in western and northwestern Madagascar, and these need to be reconsidered because the concept of a widespread *M. murinus*, which survived the Rasoloarison *et al.* (2000) divisions, has begun to be questioned more recently. *M. griseorufus* is, perhaps uniquely, a specialist gummivore, especially in the dry season (Génin 2008). *M. berthae* differs in many aspects of its diet, social organization, periodicity of torpor, and so on from sympatric *M. murinus* and, perhaps, from all other species (Dammhahn and Kappeler 2008a). Finally *M. lehilahytsara* may be a highland specialist, the only one so far identified among mouse lemurs (Radespiel *et al.* 2012). Again, when we can investigate properly, these 'new' species are not just fantasies of a taxonomic splitter.

It seems there are likewise remarkable differences between the two thus-far described species of *Mirza*. Initially, *M. zaza* lives at much higher population densities than *M. coquereli*, and seems to be more social (Markolf *et al.* 2008a). In the genus *Hapalemur, H. alaotrensis* and (what is presumed to be) *H. griseus* live in very different types of habitat, and their social organization and diet are different (see, e.g., Mutschler (2002)). There is of course a possibility that these instances are simply phenotypic plasticity; but, as it is clear that in so many other cases cited in this chapter there is no question of such plasticity, the precautionary principle dictates that we must first and foremost look to a more basic, genetic explanation.

It is not unexpected that there have been relatively few field studies comparing different species of the

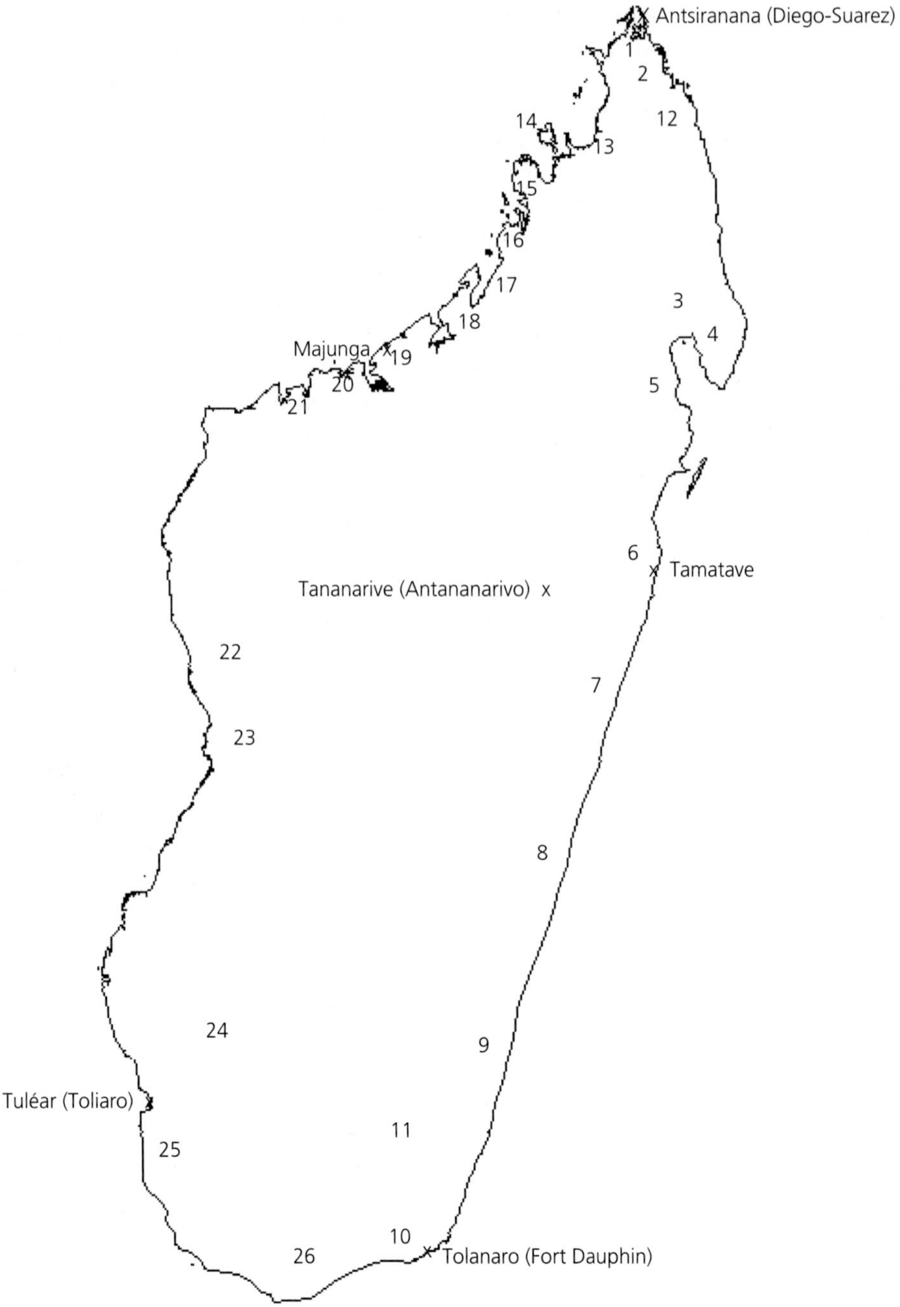

Figure 4.1 Map showing the approximate locations of 26 *Lepilemur* species. 1: *Lepilemur septentrionalis*, 2: *L. ankaranensis*, 3: *L. seali*, 4: *L. scottorum*, 5: *L. hubbardorum*, 6: *L. mustelinus*, 7: *L. betsieo*, 8: *L. microdon*, 9: *L. jamesorum*, 10: *L. fleuretae*, 11: *L. wrightae*, 12: *L. milanoii*, 13: *L. mittermeieri*, *14: L. dorsalis*, *15: L. tylermachsoni*, 16: *L. sahamalazensis*, 17: *L. grewcockorum*, 18: *L. otto*, 19: *L. edwardsi*, 20: *L. aeeclis*, 21: *L. ahmansonorum*, 22: *L. randrianasoli*, 23: *L. ruficaudatus*, 24: *L. hubbardorum*, 25: *L. petteri*, 26: *L. leucopus*

Eulemur fulvus group, maybe because they were all classed as one species, an all-purpose *E. fulvus*. This lemur group was specifically mentioned by Tattersall (2007) in his critique as surely not deserving to be recognized as being more than one species. Because of the lack of field studies, we cannot yet be sure that there are behavioural/ecological differences between the members of the group; yet they differ strongly in the presence and/or expression of sexual dichromatism, and this surely means something, presumably in the area of sexual selection. A study of intersexual relations should prove very enlightening.

4.4 Case study: the red colobus

Tom Struhsaker has long taken a special interest in variations among red colobus over different parts of their wide distribution, and, although he himself has not accepted the PSC, his massive recent compilation of all available data has in effect highlighted that red colobus species are not just convenience categories—variations on a theme—but have ecological and behavioural meaning (Struhsaker 2010: Figure 4.2). Oates (2011) has additionally highlighted these differences, specifically among the West African species.

P. temminckii, the westernmost species, characteristic of dry, open forest and gallery forest, has small groups of generally fewer than 30 individuals, and both males and females transfer between groups. Different groups share their ranges, but there seems to be dominance ranking between them. The diet, which in most red colobus is predominantly young leaves and leaf buds, has at least in one study slightly higher proportions of fruits and seeds. At the other end of the range of the genus is the easternmost (likewise open forest and gallery forest) species, *P. rufomitratus*, which shares the characteristics of small group size and of both males and females transferring between groups, but there is also a high proportion of solitary individuals and of groups of males, and half of the normal bisexual groups have only a single male, the rest having only two males. Much less (about one third) of the home range seems to be shared with other groups.

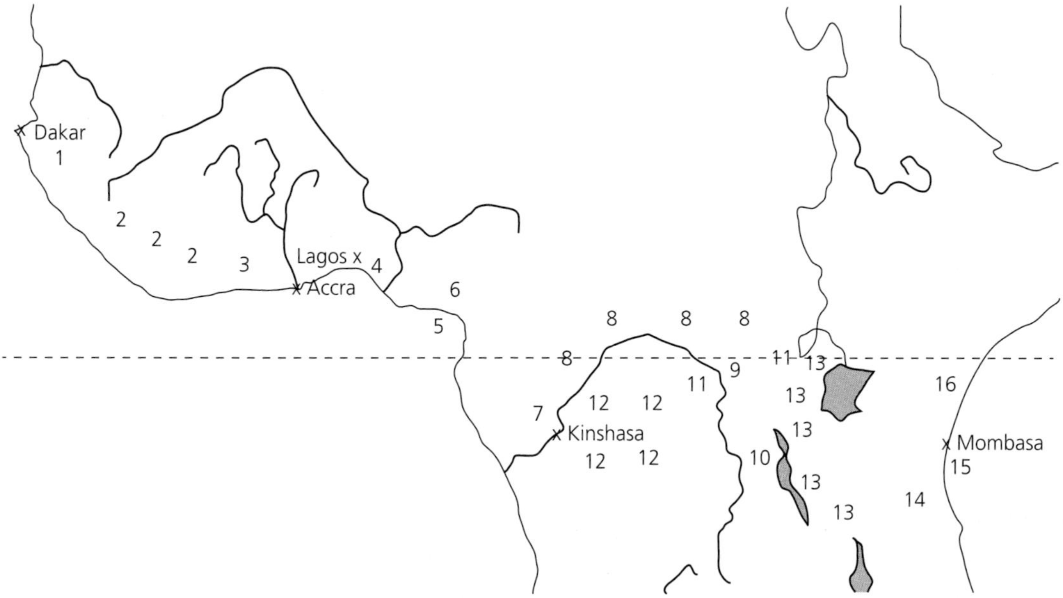

Figure 4.2 Idealized range map of red colobus (*Piliocolobus*) species. 1: *P. temminckii*, 2: *P. badius*, 3: *P. waldronae* (probably extinct), 4: *P. epieni*, 5: *P. pennantii* (Bioko I.), 6: *P. preussi*, 7: *P. bouvieri*, 8: *P. oustaleti*, 9: *P. langi*, 10: *P. foai*, 11: *P. parmentieri*, 12: *P. tholloni*, 13: *P. tephrosceles*, 14: *P. gordonorum*, 15: *P. kirkii* (Zanzibar I.), 16: *P. rufomitratus*.

Another difference is that the diet is slightly more than half young leaves, although there is again a fairly high proportion (about one quarter) of fruit pulp, but very few seeds. The message from this is that the fact of living in similar habitats does not give them similar social behaviours, social organization, or even diet.

The closest relative of *P. temminckii* (and even regarded as conspecific by Groves (2007)) is the rainforest-living *P. badius*. This species lives in much larger social groups, ranging from 33 to 90. Quite contrary to the two dry-forest species, the males seem to remain in their natal groups, and only the females disperse. Again contrary to *P. temminckii*, the ranges of different groups overlap very little. The diet is about one third young leaf parts and (in one study) one quarter seeds, and 20% mature leaves. The probably extinct *P. waldronae*, whose distribution abutted that of *P. badius* to the east, seemed as far as brief observations are concerned to have large social groups and to have a broadly similar diet, but we just don't know—it's gone. As long as it was 'just a subspecies', not much attention was paid to it.

Further to the east, but in isolated patches, occur *P. epieni* in the swamp forests of the Niger Delta, and *P. preussi* from the leguminous forests of the Nigeria–Cameroon border. Both seem to have group sizes, group interrelationships, and diets much like *P. badius*, except that seeds are not recorded in the diet (but more fieldwork needs to be done to determine whether these distinctions are robust). On the island of Bioko lives *P. pennantii*, predominantly (nowadays) in montane forest, an unusual habitat for red colobus, and this by contrast has small group sizes, under 30, many of them unimale, and solitary females are often encountered.

In the Ngotto Forest of the Central African Republic lives *P. oustaleti*, where it has been observed in very small groups of up to 18, all with a single male, in marshy forest, walking along branches up to their bellies in water and feeding on water plants (Galat-Luong and Galat 1979). This same species is found all along the northern banks of the Congo and its tributaries, and into dry gallery forest; further field studies would be of great interest, to compare with the brief study in the Ngotto Forest. South of the Congo lives *P. tholloni*, in leguminous forest like *P. preussi* and in similarly large troops, but feeding for nearly one third of its diet on seeds (Maisels *et al.* 1994).

P. tephrosceles, from the forests east of the Western Rift lakes in Uganda and Tanzania, seems to have social organization rather like that of *P. badius*, with very large social groups between which it is the females that transfer, though these groups may be fission–fusion; but the diet is overwhelmingly leaves, both young leaves and mature leaves making up nearly half of the diet, and very few seeds are consumed. The related *P. gordonorum*, from the Udzungwa mountains and adjoining lowland forests, has much smaller groups, about 25–33, though with more than one adult male, and was seen eating leaves, mainly young leaves (Struhsaker and Leland 1980). The Zanzibar red colobus, *P. kirkii*, the only one that has 'traditionally' often been ranked as a separate species from the others, again has relatively small groups, whose ranges overlap, and these have multiple males and it is just the females that transfer between them; the diet is mainly leaf buds and young leaves, but up to one third fruit and seeds—and, curiously, charcoal from the dying fires outside people's huts (Mturi 1993).

This brief survey shows that, even on such gross comparisons as group size, which sex transfers, whether group ranges overlap, and what proportion of seeds are in the diet, the different species of red colobus differ from one another, though admittedly in some cases there is variation between different locales and even between troops in different parts of the same forest (Chapman *et al.* 2002), and there is a possibility that some of these differences result from differences in sampling period or duration (Andy Marshall, pers. comm.).

The message from all this is that species differences are real; even between vicariant species one cannot predict the life ways of one from knowing all about its neighbour. The flexibility characteristic of so many mammals, including probably all primates, certainly could not explain anything like all these differences. In the case of red colobus, for example, while the type of forest a species inhabits, or the resources available in a given area, seems to predict some features, like small group size, it does not do so for others, such as how many males there are in a group and which sex disperses.

4.5 The case of the orang-utan

As recounted by Groves (1986), until the advent of two key papers published at the end of the 1960s, it was doubted whether Bornean and Sumatran orang-utans constituted even different subspecies; thereafter, as the case for their separation grew, the question began to arise of whether they might be more than subspecies (Groves 1986). The recognition of two distinct species seems now well established, and this has freed up the mind and stimulated a great deal of research with the aim of discovering what it means that they are distinct.

The search for Borneo–Sumatra differences began with Delgado and van Schaik (2000). Following on this, Wich *et al.* (2004) argued that, because of adaptation to greater productivity and stability of food supply (especially fruits) in Sumatra, whose forests are based on richer volcanic soils, life history is slower in *P. abeli* than in *P. pygmaeus*, especially as concerns interbirth intervals (111.5 months at Ketambe and 98.4 months at Suaq Balimbing, both Sumatran sites, versus 92.6 months at Tanjung Puting and 86.4 months at Gunung Palung, both Bornean sites). Anderson *et al.* (2008) countered that other aspects of life-history traits are not consistent with this, and that interbirth intervals in captivity are actually less in Sumatran orang-utans; although they did admit that possibly 'each species has evolved a different norm of reaction in response to the ecological change'. Wich *et al.* (2009) acknowledged some of their argument, but re-emphasized the predictability of the percentage of fruit in the diet of Sumatran versus Bornean orang-utans, which in addition have a higher percentage of inner bark in the diet. Marshall *et al.* (2009) tested whether indeed Sumatran forests are more productive, and concluded that they truly are, and that this correlates with a lower density of dipterocarps in Sumatra compared to Borneo.

Meanwhile other aspects of life history attracted attention, in particular the discovery that a great many of the supposedly adolescent (unflanged) males in Sumatra are actually well into their adult years—'arrested adolescents'. Do these 'arrested adolescents' also occur in Borneo? Maggioncalda *et al.* (2002) studied the hormonal status of males of different degrees of development in captivity, and listed the status of the individuals that they had studied: of the 'arrested adolescents' in their sample, four were known or probable Sumatran, two were known or probable hybrids, and only one was a 'probable Bornean'. Chance, or real? (Work at present under way by U. Welsch and myself indicates that 'arrested adolescence' does occur in Borneo, but we cannot say how common it is.) Van Schaik *et al.* (2009), however, say that whereas in Sumatra there are two arrested adolescents for every flanged male, in Borneo it is the other way about—approximately two flanged males for every unflanged male, and this has more recently been corroborated, and its implications argued, by Dunkel *et al.* (2013).

Most recently, van Schaik *et al.* (2009) have listed all the dietary, behavioural, and life-history differences which they could detect. Sumatran orang-utans live in forests with higher productivity, and with tigers roaming the forest floor it is little surprise that they do not come to the ground very much, whereas Borneans frequently come to the ground. Sumatran orang-utans have more gracile mandibles with thinner tooth enamel, they do not rely on non-fruit fallbacks as do Bornean orang-utans, females have larger home ranges and travel a greater daily distance, nestbuilding is more frequent, population density is greater, they are more sociable, male developmental arrest is much more marked (as noted above), courtships are much longer, there is only a low frequency of forced matings, their cultural repertoire is larger, and of course their interbirth intervals are longer.

Now, orang-utans are among the most intelligent and behaviourally flexible of non-human primates, and the question asked above in the context of lemurs and red colobus—how much is genetic and how much is behavioural flexibility—is writ large. Yet among the characters that are in theory subject to great individual flexibility, like home range size or fear of tigers, are others that must have at least some genetic predisposition, like maturity in males—and the dietary differences correlate very well with their dental and mandibular differences (although we cannot be sure about the direction of causation). It is difficult to avoid the conclusion that, as in lemurs, squirrel monkeys, and red colobus, the species of orangutans are more than just a bunch of random DNA sequences and pelage colour. It has recently been argued (Krützen *et al.* 2011) that there is much more plasticity, as well as effects of culture, in orang-utan variation, but that there is biological differentiation is not denied.

4.6 Species and the biodiversity crisis

The documentation of biodiversity requires 'an unambiguous measurement unit' (Chaitra *et al.* 2004), and I have argued above that this absolutely must be the Phylogenetic (evolutionary) concept of species, but it is still the case that different species concepts are in vogue in different groups of organisms (concerning ornithology, see Section 4.1). Agapow *et al.* (2004) contemplated that the search for a common approach to the species concept 'may have to be abandoned for a flexible spectrum of methodologies that either employ a range of species concepts or dispense with species altogether', but this is simply a counsel of despair. The use of different species concepts actually undermines species conservation, because species will mean different things in different groups (if one can say that they mean anything objective at all in a system which does not use the PSC), and dispensing with testable species dispenses with the very units of biology.

There is in addition a danger that morphologically based taxonomy will be made to take second place (if it is allowed to exist at all) to genetically based systems. We simply may not know at present where in the DNA a particular morphological character is determined. It may turn out to be entirely a product of phenotypic plasticity, but this should under no circumstances be assumed; rather, if there is some doubt in the matter, it should be tested. It is better to list a species that ultimately turns out to be spurious than to miss one that is genuine.

National and international funding for biodiversity and its conservation has never been a priority as far as politicians are concerned (Amato and DeSalle 2012). How best to use limited conservation funding? There is no solution as long as conservation funding is given such a low priority. But consider: biodiversity underlies all human endeavours. Why should we expect biologists be coy about this?

Species are the objects of study of biologists. Before the PSC became widespread in primatology, it was assumed that there were just three species of *Callithrix*, one or two species of *Microcebus*, six or seven of *Lepilemur*, one of *Mirza*, and at the most two species of *Piliocolobus*. Research since the beginning of the 1990s has shown that, in each case, the number of species has been drastically underestimated, and 'accepted' species have consequently been split up and shown actually to consist of more than one species. It is no good planning conservation action for, say, 'the Gray Mouse Lemur *Microcebus murinus* of western Madagascar', because this 'species', a relic of the old polytypic species concept, actually consists of (at last count) 12 distinct species, and unless we recognize this and take conservation action we are going to lose a great deal of biodiversity. These species are not just idle 'taxonomic inflation'—they have a real existence, and cannot stand in for one another (see also Marshall and Wich, Chapter 2, this volume). If one goes extinct, nothing can replace it. Conservation funding must take biodiversity into account, not the other way around.

Scientists have come to expect that their research should be underfunded, though, unexpectedly, some extremely large and costly projects do get funding (think of space exploration and the Large Hadron Collider); but no other field of science would accept that its objects of study should actually be destroyed, as species and ecosystems are daily.

Moreover, primatologists—probably all biologists—would agree that the species we work on have enormous aesthetic appeal. Artists, architects, and others working in aesthetic fields have come to expect, like scientists, that their endeavours should be underfunded, although again some extremely large and costly projects do get funding (think of the dismantling and reconstruction of Borobudur between 1975 and 1985, costing USD25 million—corresponding to about USD80 million today), but no one, practising artist or no, would accept the destruction of the Prado and all its contents.

It has recently been calculated (in the case of endangered bird species) how small are the costs of species conservation—'1 to 4% of the estimated net value of ecosystem services that are lost per year [estimated at $2 to $6.6 trillion]. More prosaically, the total required is less than 20% of annual global consumer spending on soft drinks' (McCarthy *et al.* 2012).

4.7 Summary

Since the early 1990s, the number of species of primates recognized by primatologists has risen enormously, from about 200 to 400 or so. This is largely due to an intense restudy of formerly accepted species, showing that many of them are actually composed of a number of distinct species. This restudy has been stimulated by the adoption of the Phylogenetic Species Concept, under which species are 'diagnosable'. I have shown that these newly unearthed species—in such diverse primate groups as mouse lemurs, marmosets, red colobus, and orangutans—tend to be distinct in unpredicted ways, for example in their behaviour. Conservationists and conservation funding must come to terms with this: the biological world is an even richer place than we had imagined, and we stand to lose so much if we ignore it.

References

Agapow, P-M., Bininda-Emonds, O. R. P., Crandall, K. A., Gittleman, J. L., Mace, G. M., *et al.* (2004). The impact of species concept on biodiversity studies. *Quarterly Review of Biology* **79**: 161–179.

Amato, G. and DeSalle, R. (2012). Assessing biodiversity funding during the sixth extinction. *Bioessays* **34**: 658–660.

Anderson, H. B., Emery Thompson, M., Knott, C. D., and Perkins, L. (2008). Fertility and mortality patterns of captive Bornean and Sumatran orangutans: is there a species difference in life history? *Journal of Human Evolution* **54**: 34–42.

Boinski, S. and Cropp, S. J. (1999). Disparate data sets resolve squirrel monkey *(Saimiri)* taxonomy: implications for behavioral ecology and biomedical usage. *International Journal of Primatology* **20**: 237–256.

Chaitra, M. S., Vasudevan, K., and Shanker, K. (2004). The biodiversity bandwagon: the splitters have it. *Current Science* **86**: 897–899.

Chapman, C. A., Chapman, L. J., and Gillespie, T. R. (2002). Scale issues in the study of primate foraging: red colobus of Kibale National Park. *American Journal of Physical Anthropology* **117**: 349–363.

Charles-Dominique, P. and Hladik, C. M. (1971). Le *Lepilemur* du sud de Madagascar: écologie, alimentation et vie sociale. *Terre et Vie* **25**: 3–66.

Dammhahn, M. and Kappeler, P. (2008). Small-scale coexistence of two mouse lemur species (*Microcebus berthae* and *M.murinus*) within a homogeneous competitive environment. *Oecologia* **157**: 473–483.

Delgado, R. and van Schaik, C. P. (2000). The behavioural ecology and conservation of the orangutan (*Pongo pygmaeus*): a tale of two islands. *Evolutionary Anthropology* **9**: 201–218.

De Queiroz, K. (2007). Species concepts and species delimitation. *Systematic Biology* **56**: 879–886.

De Vivo, M. (1991). *Taxonomia de Callithrix Erxleben, 1777 (Callitrichidae, Primates)*. Belo Horizonte: Fundação Biodiversitas.

Dunkel, L. P., Arora, N., van Noordwijk, M. A., Atmoko, S., Putra, A., *et al.* (2013). Variation in developmental arrest among male orangutans: a comparison between a Sumatran and a Bornean population. *Frontiers in Zoology* **10**: 12.

Galat-Luong, A. and Galat, G. (1979). Quelques observations sur l'écologie de *Colobus pennant oustaleti* en Empire Centrafricain. *Mammalia* **43**: 309–312.

Génin, F. (2008). Life in unpredictable environments: first investigation of the natural history of *Microcebus griseorufus*. *International Journal of Primatology* **29**: 303–334.

Groves, C. P. (1986). Systematics of the great apes. In: Swindler, D. R. and Erwin, J. (Eds), *Comparative Primate Biology. Volume 1: Systematics, Evolution and Anatomy*, pp. 187–217.

Groves, C. P. (2001). *Primate Taxonomy*. Washington: Smithsonian Institution Press.

Groves, C. P. (2004). The what, why and how of primate taxonomy. *International Journal of Primatology* **25**: 1105–1126.

Groves, C. P. (2007). The taxonomic diversity of the Colobinae of Africa. *Journal of Anthropological Sciences* **85**: 7–34.

Krützen, M., Willems, E. P., and van Schaik, C. P. (2011.) Culture and geographic variation in orangutan behavior. *Current Biology* **21**: 1808–1812.

Lei, R., Frasier, C., McLain, A. T., Taylor, A., Bailey, C., *et al.* (2014). Revision of Madagascar's Dwarf Lemurs (Cheirogaleidae: *Cheirogaleus*): designation of species, candidate species status and geographic boundaries based on molecular morphological data. *Primate Conservation* **28**: 9–35.

Louis, E. E., Coles, M. S., Andriantompohavana, R, Sommer, J. A., Engberg, S. E., *et al.* (2006a). Revision of the Mouse Lemurs (*Microcebus*) of Eastern Madagascar. *International Journal of Primatology* **27**: 347–389.

Louis, E. E., Engberg, S. E., Lei, R., Geng, H., Sommer, J. A., *et al.*. (2006b). Molecular and morphological analyses of the Sportive Lemurs (Family Megaladapidae: genus *Lepilemur*) reveals 11 previously unrecognised species. Special Publications, Museum of Texas Tech University, number 49.

Lowin, A. J. (2010). *Lepilemur* feeding observations from Northern Madagascar. *Lemur News* **15**: 20–21.

Maggioncalda, A. N., Czekala, N. M., and Sapolsky, R. M. (2002). Male orangutan subadulthood: a new twist on the relationship between chronic stress and developmental arrest. *American Journal of Physical Anthropology* **11**: 825–832.

Maisels, F., Gautier-Hion, A., and Gautier, J-P. (1994). Diets of two sympatric colobines in Zaire: more evidence on seed-eating in forests on poor soils. *International Journal of Primatology* **15**: 681–701.

Markolf, M., Kappeler, P. M., and Rasoloarison, R. (2008). Distribution and conservation status of *Mirza zaza*. *Lemur News* **13**: 37–40.

Marshall, A. J., Ancrenaz, M., Brearley, F. Q., Fredriksson, G. M., Gaffar, N, *et al.* (2009). The effects of forest phenology and floristics on populations of Bornean and Sumatran orangutans. In: Wich, S. A., Utami Atmoko, S. S., Mitra Setia, T., and Schaik, C. P. (Eds), *Orangutans: Geographic Variation in Behavioural Ecology and Conservation*. pp. 97–117. Oxford University Press.

Mayr, E. (1963). *Animal Species in Evolution*. Harvard: Belknap Press.

McCarthy, D. P., Donald, P. F., Scharlemann, J. P. W., Buchanan, G. M., Balmford, A., *et al.* (2012). Financial costs of meeting global biodiversity conservation targets: current spending and unmet needs. *Science* **338**: 946–949.

Mittermeier, R. A., Louis, E. E., Richardson, M., Schwitzer, C., Langrand, O., *et al.* (2010). *Lemurs of Madagascar*, 3rd edn. Conservation International.

Mturi, F. A. (1993). Ecology of the Zanzibar red colobus monkey, *Colobus badius kirkii* (Gray, 1868), in comparison with other red colobines. In: Lovett, J. C. and Wasser, S. K. (Eds), *Biogeography and Ecology of the Rain Forests of Eastern Africa*. pp. 243–266. Cambridge University Press.

Mutschler, T. (2002). Alaotran gentle lemur: some aspects of its behavioural ecology. *Evolutionary Anthropology* **11** (supplement): 101–104.

Oates, J. F. (2011). *Primates of West Africa*. Conservation International, Tropical Field Guide Series.

Radespiel, U., Ratsimbazafy, J., Rasoloharijaona, S., Raveloson, H., Andriaholinirina, N., *et al.* (2012). First indications of a Highland specialist among mouse lemurs (*Microcebus* spp.) and evidence for a new mouse lemur species from eastern Madagascar. *Primates* **53**: 157–170.

Randrianambinina, B., Rakotondravony, D., Radespiel, U., and Zimmermann, E. (2003). Seasonal changes in general activity, body mass and reproduction of two small nocturnal primates: a comparison of the golden brown mouse lemur (*Microcebus ravelobensis*) in Northwestern Madagascar and the brown mouse lemur (*Microcebus rufus*) in Eastern Madagascar. *Primates* **44**: 321–331.

Rasoloarison, R. M., Goodman, S. M., and Ganzhorn, J. U. (2000). Taxonomic revision of mouse lemurs (*Microcebus*) in the western portions of Madagascar. *International Journal of Primatology* **21**: 963–1019.

Rosenberger, A. L. (2014). Species: beasts of burden. *Evolutionary Anthropology* **23**: 27–29.

Struhsaker, T. T. (2010). *The Red Colobus monkeys: Variation in Demography, Behaviour, and Ecology of Endangered Species*. Oxford: Oxford University Press.

Struhsaker, T. T. and Leland, L. (1980). Observations on to rare and endangered populations of red colobus monkeys in East Africa: *Colobus badius gordonorum* and *Colobus badius kirkii*. *African Journal of Ecology* **18**: 191–216.

Tattersall, I. (1992). Species concepts and species identification in human evolution. *Journal of Human Evolution* **22**: 341–349.

Tattersall, I. (2007). Madagascar's Lemurs: cryptic diversity or taxonomic inflation? *Evolutionary Anthropology* **16**: 12–23.

Thiele, D., Razafimahatratra, E., and Hapke, A. (2013). Discrepant partitioning of genetic diversity in mouse lemurs and dwarf lemurs—Biological reality or taxonomic bias? *Molecular Phylogenetics and Evolution* **69**: 593–609.

Ting, N. (2008). Mitochondrial relationships and divergence dates of the African colobines: evidence of Miocene origins for the living colobus monkeys. *Journal of Human Evolution* **55**: 312–325.

van Schaik, C. P., Marshall, A. J., and Wich, S. A. (2009). Geographic variation in orangutans behaviour and biology. In: Wich, S. A., Utami Atmoko, S. S., Mitra Setia, T., and van Schaik C. P. (Eds), *Orangutans: Geographic Variation in Behavioural Ecology and Conservation*. pp. 351–361. Oxford University Press.

Wich, S. A., de Vries, H., Ancrenaz, M., Perkins, L., Shumaker, R. W., *et al.* (2009). Orangutan life history variation. In: Wich, S. A., Utami Atmoko, S. S., Mitra Setia, T., and van Schaik C. P. (Eds), *Orangutans: Geographic Variation in Behavioural Ecology and Conservation*. pp. 65–75. Oxford University Press.

Wich, S. A., Utami-Atmoko, S. S., Mitra Setia, T., Rijksen, H. D., Schurmann, C., *et al.* (2004). Life history of wild Sumatran orangutans (*Pongo abelii*). *Journal of Human Evolution* **47**: 385–398.

Zink, R. M. (2006). Rigor and species concepts. *The Auk* **123**: 887–891.

Zink, R. M. and Davis, J. I. (1999). New perspectives on the nature of species. In: Adams, N. J. and Slotow, R. H. (Eds), *Proceedings of the 22 International Ornithological Congress, Durban*. pp. 1505–1518. Johannesburg: BirdLife South Africa.

Zinner, D., Roos, C., Fausser, J. L., Groves, C., and Rumpler, Y. (2007). Disputed taxonomy classification of sportive lemurs (*Lepilemur*) in NW Madagascar. *Lemur News* **12**: 53–56.

CHAPTER 5

Primate conservation genetics at the dawn of conservation genomics

Milena Salgado Lynn, Pierfrancesco Sechi, Lounès Chikhi, and Benoit Goossens

Wild species must have available a pool of genetic diversity if they are to survive environmental pressures exceeding the limits of developmental plasticity. If this is not the case, extinction would appear inevitable.
O. H. Frankel, 1983

Non-invasive (faecal) sampling of wild long-tailed macaques (Macaca fascicularis). Photo copyright: Danau Girang Field Centre.

5.1 Introduction

With almost 60% of the primate taxa identified as threatened with extinction (IUCN 2015; Cotton *et al.*, Chapter 5, this volume), primatologists must take advantage of all the tools available to guarantee their survival. Molecular genetics is one of these tools that is frequently incorporated into the conservation biologist's toolbox. Molecular studies permit investigation of aspects of the

Salgado-Lynn, M., Sechi, P., Chikhi, L. and Goossens, B., *Primate conservation genetics at the dawn of conservation genomics.* In: *An Introduction to Primate Conservation.* Edited by: Serge A. Wich and Andrew J. Marshall, Oxford University Press (2016).

DOI 10.1093/acprof:oso/9780198703389.003.0005

biology and ecology (feeding and social behaviour, parental relationships, natural selection) of threatened species that are sometimes impossible to reliably assess with observational field studies alone, particularly in elusive species. In addition, it is the only way to determine if deleterious genetic effects are likely to threaten a species' survival in the future through loss of genetic diversity and inbreeding depression. Conservation genetics is a young, but maturing, discipline that has undergone rapid transformations as genetic and genomic datasets have become increasingly and rapidly available for **non-model organisms** (note: all terms in bold are defined in the glossary or in the accompanying boxes) (Ouborg *et al.* 2010; Shafer *et al.* 2015) including non-human primates (McManus *et al.* 2015; Prado-Martinez *et al.* 2013; Zhou *et al.* 2014). Yet, it has been noted that wildlife managers (and primatologists) often lack adequate genetic training to apply genetic information to the management of wild populations (Frankham 2009; Shafer *et al.* 2015). In this chapter, the concepts of conservation and non-invasive genetics are reviewed along with their challenges and applications to primate conservation. The potential of incorporating population genetics modelling and genomic approaches into primate conservation is also discussed.

Glossary

Allele: an allele defines one or more variants of a gene. In diploid organisms, each organism can carry two identical alleles (homozygous) or two different alleles (heterozygous) for each gene. A **deleterious allele** is a version of a gene that, on average, decreases the fitness of the organism carrying it.

Bottleneck: A marked reduction in population size that often results in the loss of genetic variation and more frequent matings among closely related individuals.

Genetic drift: the process by which allele frequencies increase and decrease by chance over time. The chance of fixation or loss of a genetic variant by genetic drift is higher in smaller populations.

Genome: the complete set of genetic material/genetic information (nuclear, mitochondrial, chloroplastic) present in an organism.

Genotyping: is the process of determining differences in the genetic make-up (**genotype**) of an individual by examining the individual's DNA sequence using biological assays and comparing it to another individual's sequence or a reference sequence. It reveals the alleles individuals have inherited from their parents.

Model and non-model organism: a model organism is a non-human species selected by the scientific community in order to be extensively studied to understand its biology, with the aim that the discovery of the biological functioning of this organism will give insights into the functioning of other species. Model organisms have some common characteristics: for example, they are easy to rear in a laboratory, have typically short life spans and high fecundity. Thus, a non-model organism is the opposite; they have not been selected by the research community for extensive studies, either for historical reasons or because they lack of the characteristics mentioned earlier.

Restriction sites: are locations characterized by more or less specific DNA sequences, which are recognized by enzymes that cut the DNA in that particular site. Restriction sites are usually between 4 and 8 base pairs in length.

Sanger sequencing: a method of DNA sequencing based on the incorporation of chain-terminating dideoxynucleotides by PCR. It was developed by Frederick Sanger in 1977, and it was the most widely used sequencing method for approximately 25 years. As it is implemented in most available sequencing machines, it allows the sequencing of a maximum of 96 templates per run, each having a maximum length of ~1,000 base pairs (bp).

5.2 Conservation/population genetics

Conservation genetics may be considered as the scientific discipline that attempts to quantify, and study the genetic diversity present on the planet (Allendorf and Luikart 2007; Frankham *et al.* 2010). In an era characterized by large-scale habitat destruction, global warming, increasing toxic wastes, and the spread of exotic species, it may not necessarily be obvious why studying or safeguarding genetic diversity should become a priority for conservation. While demographic and environmental factors are likely to be the most important aspects influencing population persistence on a short timescale (Lande 1988), genetic factors will likely matter on longer timescales, particularly on an evolutionary timescale (Lande 1988; Lynch and Gabriel 1990; Lynch *et al.* 1990). Genetic diversity is indeed the likely basis for a species' adaptability to environmental change; therefore its loss should be of serious concern to conservationists. The main threats to genetic diversity are considered to be stochastic events, whose probability and effects are more important in small and isolated populations (Frankham *et al.* 2010). For instance, mutations across the **genome** generate **deleterious alleles** that, in turn, accumulate and become randomly fixed through **genetic drift**. This can lead to serious genetic defects when populations are isolated or experience long-duration **bottlenecks**, as has been shown in several vertebrates species such as the Florida panther (*Felis concolor coryi*; Roelke *et al.* (1993)) and in the lions of the Ngorongoro Crater (*Panthera leo*; Munson *et al.* (1996)). In the worst cases, the loss of genetic diversity and accumulation of deleterious alleles by genetic drift can contribute to the extinction of such groups or populations (see Frankham (2015) for a review). In recent decades, there has been a gradual increase among conservationists in the application of population genetics methods to the management of natural and captive populations (Bertorelle *et al.* 2009). For instance, several studies have shown that carefully planned introduction of individuals into a closed population can lead to genetic enrichment and viability improvement of populations (Hostetler *et al.* 2012; Madsen *et al.* 1999). However, it must also be acknowledged that the application of genetics to management of populations is never easy for various reasons, some of which will be outlined in this chapter.

Population geneticists typically infer past demographic events or natural selection by analysing present-day populations. Data used by population genetics include the distribution of genetic diversity within and among populations, the level of genetic variation in a sampled population, and the differences (if any) between **alleles** or their frequencies and the geographical patterns they might follow. This discipline also explores whether those patterns (or lack thereof) reflect a particular past demographic or selective scenario, and attempts to quantify the relative importance of the main evolutionary factors that led to the current patterns. To answer these questions, DNA sequence variation has to be measured. The tools of choice are molecular markers, which are small sections of the genome that can be used as indicators of genome-wide variation (Box 5.1). Some of the most popular or commonly

Box 5.1 Molecular markers frequently used in conservation genetics studies

Mitochondrial DNA (mtDNA)

The DNA inside the mitochondria is a haploid molecule present in many copies per cell. It is almost exclusively inherited clonally through the maternal line. It is often considered as selectively neutral, but because it is non-recombining any selection affecting one region of the molecule should also affect the rest of the molecule (see Ballard and Rand (2005) and Ballard and Whitlock (2004)). mtDNA is composed of various genes, some of which are sufficiently conserved to provide comparisons across species and enable markers to be developed across closely related species. Its applications include phylogenetics, phylogeography, identification of genetically distinct units for conservation, inference of patterns

continued

Box 5.1 *Continued*

of maternal relatedness, and sex-specific population structure (Avise 1995; Avise *et al.* 1987; Di Fiore 2003; Hurst and Jiggins 2005; Moritz 1994; Wang 2010). In humans it has been used jointly with Y-chromosome data to compare female- and male-related patterns of diversity and differentiation (Rasteiro and Chikhi 2013; Rasteiro *et al.* 2012). Nonetheless it has been suggested that any information obtained with this marker is likely to be associated with a large variance and all inferences made should be corroborated with data from nuclear genes (Ballard and Rand 2005; Ballard and Whitlock 2004; Hurst and Jiggins 2005). This means that any inference and interpretation based on mtDNA alone should be made with caution (see Drummond *et al.* (2005)). In addition, population structure can generate false signals of population size change (Heller *et al.* 2013). Such spurious signals can be minimized to some extent by analysing the data with various sampling schemes (Heller *et al.* 2013). But again, the main message is that mtDNA behaves as a single highly polymorphic locus in which all loci are physically linked.

Microsatellites

Also known as simple sequence length polymorphisms (SSLPs), short tandem repeats (STRs), or simple sequence repeats (SSRs). They are present in the genome as single copy nuclear DNA consisting of tandemly (i.e. consecutively) repeated short sequence motifs, each between one and ten base pairs (bp) in length. Widely scattered throughout the genome and highly polymorphic, these markers are inherited in a Mendelian fashion and are believed or assumed to be selectively neutral with high mutation rates (Bhargava and Fuentes 2010; Buschiazzo and Gemmell 2006). The combination of many, highly variable, independent microsatellite loci, creates a unique multilocus 'fingerprint' for every individual in a population in a process called genotyping. In general, their characteristics make them useful for studying paternity and kinship, genetic variation, population genetic structure and gene flow, migration rates (Goossens *et al.* 2005), and intra- and inter-locus disequilibrium. Moreover, it is also possible to identify a recently bottlenecked population when no information exists on the current or historical population size (i.e. Goossens *et al.* (2006)—for orang-utans; Quéméré *et al.* (2012)—for sifakas) and combined with other methods it is possible to simultaneously estimate the approximate date and rate of a recent reduction (or increase) in the effective population size (Allendorf and Luikart 2007; Chikhi and Bruford 2005; Ellegren 2004; Jarne and Lagoda 1996; Selkoe and Toonen 2006; Zhang and Hewitt 2003). Their distribution across the genome may have them linked with regions under selection. This may either bias estimates of demographic parameters or be used to detect regions under selection. Finally, microsatellites are highly variable and are thus less prone to ascertainment bias (a systematic distortion when measuring the true frequency of a phenomenon, due to the method by which data are collected) than less variable markers such as SNPs (Chikhi 2008; Romero *et al.* 2009) and are thus expected to be less biased for demographic inference.

Single nucleotide polymorphisms (SNPs)

Known also as single base substitution polymorphisms, SNPs are changes in the nucleotide composition of a DNA sequence at a single site. These changes are typically found every 300–1,000 bp in most vertebrate non-coding regions in the genome (Kim and Misra (2007); Lercher and Hurst (2002); but see Morin *et al.* (2004); Vignal *et al.* (2002)). However, a fraction of the substitutions have functional significance and are the basis for the diversity found among humans (and other primates) (Kwok and Chen 2003). SNPs are the most abundant and widespread polymorphic markers across genomes, and the possibility of genotyping a great quantity of them has been recognized to have great potential for ecological and evolutionary studies (Morin *et al.* 2004). SNPs have the advantage of providing information on both neutral markers and those under selection (Goossens and Bruford 2009). Compared to microsatellites, SNPs allow the amplification of extremely small fragments, which makes them very useful for population and conservation genetics using non-invasive samples, and are much easier to automate, for example on microarrays (Vernesi and Bruford 2009). However, many more SNPs (relative to microsatellites) must be used in population analyses to compensate for the limited information contained in each and to obtain statistical power (Aitken *et al.* 2004). Therefore, one challenge will be to weigh the greater discriminatory power of microsatellites against the lower cost of SNPs (Kwok and Chen 2003; Morin *et al.* 2004; Weir *et al.* 2006), and another is the (perhaps not widely recognized) bioinformatics work required to analyse and exploit SNPs. As an advantage, SNPs can provide data with absolute scores (i.e. sequence data that is not subject to differences among laboratories and sequencing platforms; see Gonen *et al.* (2015)), thus facilitating international collaboration between researchers studying the same species. Moreover, databases such as MonkeySNP are constantly growing, making the comparison of genetic variability among species and subpopulations easier (Khouangsathiene *et al.* 2008).

used ones for conservation genetics will be briefly described and discussed in this chapter. For a long time, inferences regarding genome-wide evolutionary processes were mostly based on a limited number of markers and were therefore necessarily constrained (Narum *et al.* 2013). This problem can now potentially be solved by advances in sequencing technologies that have taken place within the last two decades. They allow movement towards the analysis of genome-wide variation, permitting the study of the full range of evolutionary processes that act across a genome. There are currently different types of genomic techniques that can be applied to primates, but in this chapter we will focus on those that are currently most commonly used in the field of conservation genomics, that is next-generation sequencing (NGS) techniques (note that better terms would probably be 'second' and 'third' generation sequencing techniques since new methods are already being developed, e.g. Schadt *et al.* (2010)). In recent years, the number of articles concerning population and conservation genomics has been consistently increasing (although, to date, just a few studies concerning primates have been published).

As a closing remark for this section, it must be noted that conservation genetics, population genetics, molecular markers, and genomics are broad topics that have been addressed in various books, chapters, and reviews that should be consulted for further information (to name a few: Allendorf and Luikart (2007); Allendorf *et al.* (2010a, 2009a, 2004); Bertorelle *et al.* (2009); Chikhi and Bruford (2005); Frankham (2009); Frankham *et al.* (2010); Groom *et al.* (2006); Hedrick (2001, 2004); Kohn *et al.* (2006); Luikart *et al.* (2003); Ouborg *et al.* (2010); Shafer *et al.* (2015); Wayne and Morin (2004); Zhang and Hewitt (2003)).

5.3 To invade or not to invade . . . that is the question

Many primate species are threatened, elusive, rare, occupy inaccessible areas, and occur at low densities (Harcourt 2006; Mittermeier *et al.* 2009). For the primatologist concerned with genetic studies, all of these characteristics present challenges for the collection of samples that can yield sufficient amounts of good-quality DNA (i.e. blood and tissue) to permit analysis. Trapping or darting primates poses a series of complicated logistic and ethical problems. Furthermore, authorizations from CITES and health/agriculture departments and Universities or Research Institutes will likely be needed to collect, store, transport, and transfer blood and tissue samples within or between countries. A key component in conservation genetics has been the development of non-invasive molecular methods to assess and monitor wild primate populations (Goossens and Bruford 2009; Goossens *et al.* 2011; Taberlet *et al.* 1996). The most common non-invasive sources of DNA that have been used to study wild populations of primates include shed (and plucked) hairs and faeces (Goossens and Bruford 2009). Other DNA sources are urine, saliva, menstrual blood, and male ejaculates (Goossens *et al.* 2011). Although these may appear to represent a large number of DNA sources, in reality these samples are very uncommon because they and their associated DNA degrade rapidly in the tropical forests that characterize most primates' habitats (90% of taxa are found in equatorial tropical rainforest: Cowlishaw and Dunbar (2000); Mittermeier and Cheney (1987)). For example, faecal samples may not persist for more than a few days under tropical conditions or in places where dung beetles and other scavengers will rapidly decompose the faeces. Given that UV radiation degrades DNA, sunlight together with degradation by enzymes (whose activity also increases with temperature) and microorganisms will likely influence the quality of the samples and further reduce the quality of the DNA that is available for extraction. Therefore, obtaining suitable samples such as hair and faeces in a tropical forest is a real challenge. Below are some considerations when using these two DNA sources; for additional technical details on sample collection, see Box 5.2 and the review by Goossens *et al.* (2011).

5.3.1 Hair

Hair samples can be obtained under various circumstances, although plucked hairs with root material are the best source of hair DNA. With roots packed with mitotic cells, a single plucked hair

Box 5.2 Sample storage and DNA extraction techniques for non-invasive samples

Hair

Storage

In paper envelopes, dried with self-indicating silica granules either at room temperature, or frozen at -80 °C (for more details see Goossens *et al.* (2011)).

DNA extraction

Normally done with Chelex® 100 and Proteinase K (Walsh *et al.* 1991), but Vigilant (1999) obtained better results using Taq polymerase PCR buffer as the extraction buffer (see Allen *et al.* (1998)). In our experience with orangutan hair, the method described by Jeffery *et al.* (2007) (PCR buffer, Proteinase K, and H_2O in a small extraction volume) is very reliable for shed and plucked hairs (Goossens *et al.* 2005).

Faeces

Storage

Popular methods include desiccation with silica gel beads, freezing, ethanol and DETs (DMSO-EDTA-Tris-salt), and RNAlater® (Bradley and Vigilant 2002; Frantzen *et al.* 1998; Goossens *et al.* 2005; Nsubuga *et al.* 2004). In our experience, the two-step approach (soaking with ethanol followed by dessication with silica gel) described by Roeder *et al.* (2004) is a reliable method to store faeces since it has yielded significantly more high-quality DNA than in those stored solely in silica gel.

DNA extraction

Use a method that involves fewer steps and sample transfers, although the removal of substances that may inhibit PCRs usually requires repeated purification processes involving several centrifugation steps. The QIAamp Stool Mini Kit (QIAGEN) has given reliable results in apes and monkeys (Bayes *et al.* 2000; Bradley *et al.* 2000; Eriksson *et al.* 2004; Goossens *et al.* 2005; Liu *et al.* 2009; Salgado-Lynn *et al.* 2010) as has the CTAB/2PCI method (Vallet *et al.* 2007) in Lemuriformes (Quéméré *et al.* 2009, 2010). Other methods described in the literature include: silica-based (Boom *et al.* 1990), diatomaceous earth (Gerloff *et al.* 1999), Chelex-100® (Walsh *et al.* 1991), CTAB (Lathuillière *et al.* 2001; Launhardt *et al.* 1998), and a standard phenol/chloroform extraction (Radespiel *et al.* 2008).

Contamination

Contamination is probably the most important factor to consider during sample collection and processing. It can occur in the field or laboratory and can be a major concern for non-invasive sampling studies. To limit contact with material that can contaminate the target sample, it is highly recommended to use latex gloves and sterile mechanical devices (e.g. tweezers) for handling samples in the field. Gloves should be changed between the handling of different samples, and mechanical devices should also be replaced or sterilized with ethanol and a lighter between samples. These precautions can be difficult to follow in a tropical (humid) environment (especially wearing gloves). Wearing a mask that will limit human contamination may sometimes be more important than wearing gloves. Also, using small sticks or leaves to collect faecal samples may actually be better than using poorly cleaned tweezers, provided that they are only used once and discarded. The success of any non-invasive genetic study can depend on adhering to these basic rules. See Goossens *et al.* (2011) for more suggestions on proper lab technique when working with non-invasive samples.

should yield enough DNA for genetic analyses provided adequate storing conditions are used (but see following text). Shed hairs provide enough mitochondrial DNA (mtDNA), but with the roots having usually undergone apoptosis before shedding; nuclear DNA is often degraded and thus unreliable for nuclear DNA analyses (Gagneux *et al.* 1997; Jeffery *et al.* 2007). For species that utilize nests, shed hairs are easier to collect in the field from empty nests (Goossens *et al.* 2002, 2005; Jeffery *et al.* 2007; Morin *et al.* 1994). Plucking hair from free-ranging primates is rarely attempted, but Valderrama *et al.*

(1999) and Améndola-Pimenta *et al.* (2009) both offered novel methods (using sticky tape or glue) for doing so with capuchins (*Cebus olivaceus*), baboons (*Papio hamadryas*), and howler monkeys (*Alouatta pigra*).

One advantage of hair over faeces is that hair contains fewer chemical inhibitors. Another advantage is that contamination from other DNA sources (e.g. dietary items' DNA found in scat) is minimized with hair. However, some studies in great apes indicate that there are fewer cells, and therefore less DNA, generally available in a shed-hair sample than in a faecal sample (Broquet *et al.* 2007; Jeffery *et al.* 2007; Morin *et al.* 2001). Prior to launching a large-scale survey, it is highly advisable to conduct a pilot study to determine the rate of success of obtaining DNA from the hair of the target species under field conditions. For example, it is strongly recommended to collect more than 10 hairs per individual (Goossens *et al.* 1998, 2002) because of the large variation in DNA amplification success due to factors such as the morphological characteristics of the species' hair, the social characteristics of the species, the environmental conditions under which the sample is collected, storage, and laboratory methods.

5.3.2 Faeces

On the surface of and inside faecal pellets are cells shed from the intestinal lining, which are a useful source of host DNA. However, these are accompanied by a complex mixture of other compounds (microorganisms, undigested food, digestive enzymes, mucus, bile salts, and bilirubin), some of which could function as chemical inhibitors that could restrict the extraction and later the amplification of DNA (Vallet *et al.* 2007). Additionally, the amount and quality of faecal DNA are known to vary significantly across species, but also with many environmental or ecological factors, such as temperature and humidity between defecation and time of collection, season (dry or raining seasons), preservation method, species diet, storage time, and extraction protocol (Nsubuga *et al.* 2004; Piggott and Taylor 2003; Piggott 2004). Finding faecal samples of non-nesting primates in a tropical forest is probably the most challenging part of a genetic study based on non-invasive sampling, especially when the ground vegetation is dense. Opportunistic findings (along transects) are very likely to be time consuming; therefore identifying feeding or resting trees may be important to increase success rates. Several authors have suggested the use of scat detection dogs (see Vynne *et al.* (2011) and Wasser *et al.* (2004) for examples of carnivores), but this has yet to be tested on primate scats. There are, however, practical issues such as determining whether canines are allowed into ranging areas of the species of interest. Nevertheless, depending on the species of interest, faecal samples could be large enough to allow multiple extractions, which presents a considerable advantage compared to the number of extractions and the amount of DNA that can be extracted from hairs of a single individual. Another advantage of faecal samples is that they can be used to study various important ecological questions beyond the genetics of the species of interest. For instance, they have been used to monitor stress responses and the reproductive status of wild females (Heistermann *et al.* 2008; Muehlenbein *et al.* 2012), and more recently and thanks to the development of next-generation sequencing (**NGS**) technologies, to also monitor parasites and feeding ecology of the species of interest (Quéméré *et al.* 2013; Schaumburg *et al.* 2013).

There is no consensus on the best technique to store or extract DNA from faecal material. The goal of a storage method is to minimize DNA degradation by nucleases. Optimal storage methods vary depending on the species, the diet, and the environment in which the samples were collected (Piggott and Taylor 2003; Waits and Paetkau 2005). Piggott and Taylor (2003) noted an interaction between storage method and extraction technique in the laboratory (i.e. certain extraction techniques performed better with certain storage methods). In addition to the lack of consensus on an optimal storage method for storing faeces, optimal DNA extraction methods also depend on the species and diet. For example, DNA extracts from vegetarian species might include secondary metabolites, which are known inhibitors of DNA amplification (Vallet *et al.* 2007). Another thing to consider is that most of the shed cells of the target organism are found at the 'front end' and the outside of the faecal material. Hence, it is important to scrape as much as possible from the outside of the material in order to maximize DNA yield. Yet, cells containing DNA are not uniformly spread throughout faeces, and two or three extracts should be made per sample (Goossens *et al.* 2000). Extraction methods will depend on the field conditions, location, season, size, and age of the samples (see Taberlet *et al.* (1999) and Piggott (2004)). These results strongly support the value of conducting a pilot study to explore the performance of various storage and extraction techniques for the species of interest.

An additional factor influencing selection of storage and extraction techniques is that costs can vary by orders of magnitude. At an average of USD10.00 per sample, extracting DNA from thousands of samples using commercial kits may become prohibitively expensive. At the same time, different extraction protocols may damage DNA to different extents and therefore preclude future analyses. In some cases it may be important to determine whether the samples and the corresponding DNA will be kept for long periods or not. It is also advisable to divide the material sampled in the field so as to keep part of it at very low temperatures without ever thawing it. Cycles of freezing and thawing will typically lead to a deterioration of often unique biological material. Therefore it is important to discuss these crucial technical issues with molecular biologists prior to collecting samples.

5.3.3 Non-invasive samples and genomics

Recent advances in sequencing technologies and the constantly decreasing costs per nucleotide in the past decades have allowed the development of massive parallel sequencing platforms, also called NGS, second-generation or high-throughput sequencing techniques (Box 5.3). All the studies previously cited are based on the amplification of a single template by PCR. A single sequence, then, is usually obtained using the classic (and still widely used) sequencing technology (**Sanger sequencing**). In contrast, NGS techniques allow the massive parallel sequencing of millions or billions of very short sequences that can be assembled to reconstruct a significant part (if not most) of the genome of an organism.

The main challenge in applying NGS and population genomics approaches to threatened and elusive species comes from the limited quality and quantity of non-invasive DNA. This makes the use of non-invasive samples not economically feasible for the still elevated high costs of high-throughput sequencing. Technically, the amount of DNA from the target species that can be extracted may be too low for many second-generation sequencing techniques. For instance, the amount of the whole genomic DNA represented in shed hairs may be too limited due to degradation by environmental exposure prior to its collection (Jeffery *et al.* 2007). For faecal samples, besides the already mentioned content in inhibitors and contaminant agents that could affect downstream analyses (Kohn and Wayne 1997; Nechvatal *et al.* 2008), a serious technical problem comes from the presence of exogenous DNA (fungi, plants, bacteria), which may be more abundant than the target species DNA. Given that second-generation sequencing techniques will sequence all the DNA present in the sample without discrimination, most of the sequencing effort may result in a waste of parallel sequencing resources and very poor coverage of the target species (Perry *et al.* 2010). These reasons make the use of non-invasive samples not (yet) economically feasible. The first NGS approaches on faecal samples were indeed to analyses of guts' microbial diversity (Qin *et al.* 2010).

Given the limited amount of DNA from the target species usually present in DNA extracted from faecal samples, the most promising advances for the use of non-invasive samples for population genomics come from NGS techniques that can reliably target and enrich only a specific fraction of the genome. This can be done by selecting or targeting genomic regions associated with **restriction sites** (genotyping-by-sequencing or **GBS**: Baird *et al.* (2008); Hohenlohe *et al.* (2010); Narum *et al.* (2013)). This allows the analysis of populations at relatively lower prices since several populations can be analysed in one run. Another promising technique, particularly in faecal samples, concerns the capture of targeted regions of genomic DNA (DNA capture: Gnirke *et al.* (2009); Mamanova *et al.* (2010)). This technique has been already tested on chimpanzees (Perry *et al.* 2010) using the chimpanzee **reference genome**. However, in order to implement this technique, a

Box 5.3 Genomics and the next-generation sequencing technologies

NGS

Next-generation sequencing. This acronym describes a set of recently developed techniques. It is also known as 'massively parallel sequencing' or 'high-throughput sequencing'. The most remarkable characteristic of such techniques is the vast increase in genetic information obtained in sequencing runs when compared to 'first-generation' techniques, allowing the study of genetics at the genomic level. NGS techniques allow the parallel sequencing of millions or billions of short sequences. There are several machines available that can produce various amounts of DNA, but for instance a single run of an Illumina HiSeq 2000 machine, one of the massively parallel sequencing machines most used currently may generate 240 Gb or more (1 Gb = 1,000 million base pairs) of nucleotide sequence data (Perry 2013).

Coverage

In second-generation sequencing techniques, the number of times a nucleotide position in a genome is 'covered' from independently sequenced reads. This is an important parameter when assembling a genome with NGS data. In fact, these kind of data are extremely error prone. As a consequence, it is necessary to read the same DNA fragment several times (thus increasing the coverage) to distinguish a mutation from a sequencing error. This is achieved by aligning and 'stacking' the reads corresponding to the same region; if a polymorphism in a determinate position appears rarely in the stacked reads, it is probably an error; however, if it appears consistently, it is probably a real polymorphism.

De-novo assembly

The reconstruction of a genome from raw next-generation sequencing data. This process generally requires consistent informatic resources. Today, it is usually achieved using powerful linux workstations and command-line software.

DNA-capture

The targeting and enrichment of the study species' DNA by 'capturing' it via RNA-biotynilated baits that are built on a reference genome and are complementary to the regions of interest.

Referenced assembly

Also called genome mapping. The reconstruction of a genome using a previously assembled genome (of the same species, or of a closely related species) as a 'backbone' sequence, called a reference. The procedure involves the alignment of second-generation reads to the reference. Usually, model species have already a reference genome.

Variant calling

SNP detection across different genomes. This is usually achieved through genomic reads mapping to a genome reference. The variants detected are then validated according to coverage. The detection of genome-wide SNP variation is the foundation of population genomics.

RAD-sequencing and GBS

RAD-sequencing focuses on the NGS of particular markers, called restriction associated DNA tags (RAD-tags). These markers were first developed for microarrays and subsequently implemented in massively parallel sequencing platforms. RAD tags are basically the flanking sequences (~100 bp, but longer fragments can be obtained depending on the sequencing chemistry) of restriction sites, interspersed across a genome. The sites are cut, tagged, and then sequenced via NGS. Once RAD tags are identified, the information they provide is mainly in the form of single nucleotide SNPs interspersed across a genome. The advantage of this technique, when compared to whole-genome sequencing, is its capacity to provide whole-genome markers with less economical and technical resources. This feature, along with the possibility of tagging RAD markers with different labels, makes it more feasible and affordable to extend this kind of whole-genome analysis from a single individual to multiple individuals, allowing the 'multiplexing' of the analysis. See

continued

Box 5.3 *Continued*

Baird *et al.* (2008) and visit the RAD-seq wiki[1] for more information on protocols and scientific works using RAD-seq.

Genotyping-by-sequencing is similar to RAD sequencing, involving the massively parallel sequencing of the flanking regions of restriction sites. However, the restriction enzymes used are methylation sensitive; that is, they target only particular restriction sites with an important 'epigenetic' modification involved in gene expression. This strategy, impossible for normal restriction enzymes used in RAD-seq, allows avoiding of repetitive regions of the genome, where this modification is not common and variability is low, thus theoretically increasing the amount of significant information while decreasing the quantity of data and the computational needs that would be involved in normal RAD-seq. For a detailed explanation of the differences between these techniques, see Cronn *et al.* (2012).

The significance of these techniques for population and conservation genomics is discussed in Narum *et al.* (2013).

reference genome is needed, which is unavailable for most primate non-model species. In this case, as suggested by Perry (2014), the issue could be resolved in two ways: (1) assembling **de-novo genomes** from non-model species of interest, using good DNA from fresh blood and tissue as a starting point; and (2) using the reference genome of closely related species, if available. Promising results were obtained in a recent study by Miller *et al.* (2012), who used the dog (*Canis lupus familiaris*) genome as a reference to study ancient admixture events between polar (*Ursus maritimus*) and brown (*U. arctos*) bears.

To conclude, non-invasive samples are the more readily available source of primates' DNA in the field, and their use for genomics was recognized early as a potential breakthrough in conservation and ecology studies (Kohn *et al.* 2006; Ouborg *et al.* 2010; Shafer *et al.* 2015). However, invasively collected samples, such as blood and tissue, are, at the moment, the best source of material for genomic studies. If anything, a limited number of individuals for which tissue samples are available may be used to provide one or several reference genomes for further studies using non-invasive samples (see for instance Sharma *et al.* (2012b) for Bornean elephants—*Elephas maximus borneensis*), or to address specific questions related to evolution and adaptation (Zhou *et al.* 2014). However, we should keep in mind that things will likely change in the next few years and new methods will likely allow us to increasingly use genomic approaches to faecal samples. The fact that ancient DNA samples (sometimes tens of thousands of years old) have recently been used to carry out genomic studies (Orlando *et al.* 2013) clearly suggests that there is potential for genomic studies from non-invasive samples.

5.4 Molecular markers: what can they tell us and how can we use them?

At the population level and below, the use of genetic markers combined with the use of non-invasive samples can be applied in a myriad of contexts (Table 5.1). They can be used to describe present-day patterns of genetic diversity and differentiation and correlate them with environmental variable or landscape features, but also to identify the evolutionary forces that led to them (Liu *et al.* 2009; Quéméré *et al.* 2010). They can be used to address small-scale processes by estimating relatedness, uncovering kin structure, dispersal, and mating systems; or large-scale processes by using simulation-based approaches (Minhós *et al.* 2013; Vigilant *et al.* 2001; Wikberg *et al.* 2014). Inferences about the evolutionary history of organisms can thus be made by measuring genetic and genotypic variation at various spatial scales and by applying population genetic models that can either incorporate or ignore temporal phenomena such as expansions or contractions. To address these questions,

[1] https://www.wiki.ed.ac.uk/display/RADSequencing/Home.

Table 5.1 Some applications of (non-)invasive genetics and genomics in primatology, with a few examples from the literature.

Applications	Examples and references	Molecular marker(s)	Sample type
Census	Mountain gorilla (*Gorilla beringei beringei*; Guschanski *et al.* 2009)	Microsatellites	Faeces
Identification of management units and/or evolutionarily significant units	Yunnan snub-nosed monkey (*Rhinopithecus bieti*; Liu *et al.* 2009)	mtDNA & Microsatellites	Faeces
	Muriqui (*Brachyteles hypoxanthus*; Fagundes *et al.* 2008)	mtDNA	Faeces
	Squirrel monkey (*Saimiri oerstedii*; Blair *et al.* 2012)	mtDNA & microsatellites	Faeces
Individual and species identification, and general taxonomy	Chimpanzee (*Pan troglodytes verus*; Mcgrew *et al.* 2004)	Microsatellites & amelogenin	Faeces
	Leaf monkeys (genus *Presbytis*; Meyer *et al.* 2011)	mtDNA	Faeces
	Kipunji (*Rungwecebus kipunji*; Davenport *et al.* 2006; Olson *et al.* 2008)	mtDNA; LPA, CD4, X Chromosome & TSPY	Tissue
Exclusion and assignment of parentage	Mountain gorilla (Nsubuga *et al.* 2008)	Microsatellites	Faeces
	Chimpanzee (*Pan troglodytes schweinfurthii*; Constable *et al.* 2001)	Microsatellites	Hair & faeces
Kinship and relatedness patterns	Sumatran orang-utan (*Pongo abelii*; Utami *et al.* 2002)	Microsatellites	Faeces
	Black howler monkey (*Alouatta pigra*; Van Belle *et al.* 2012)	Microsatellites	Faeces
	Black-and-white colobus (*Colobus vellerosus*; Wikberg *et al.* 2012)	Microsatellites	Faeces
Dispersal patterns and individual movements	Cross River gorilla (*Gorilla gorilla diehli*; Bergl and Vigilant 2007)	Microsatellites	Faeces
	Western lowland gorilla (*Gorilla gorilla gorilla*; Douadi *et al.* 2007)	mtDNA, microsatellites, & Y-chromosome	Faeces
	Orang-utan (*Pongo* spp.; Goossens *et al.* 2006b; Nietlisbach *et al.* 2012)	Microsatellites; Y Chromosome & mtDNA	Faeces
Inferring population structure	Cross River gorilla (Bergl and Vigilant 2007)	Microsatellites	Faeces
	Golden-crowned sifaka (*Propithecus tattersalli*; Quéméré *et al.* 2009)	Microsatellites	Faeces
	Bornean orang-utan (*Pongo pygmaeus*; Goossens *et al.* 2005)	Microsatellites	Faeces
Phylogeography	Orang-utan (*Pongo* spp.; Nater *et al.* 2011)	mtDNA & microsatellites	Faeces & hair
	Bonobo (*Pan paniscus*; Kawamoto *et al.* 2013)	mtDNA	Faeces
	Yunnan snub-nosed monkey (Liu *et al.* 2007)	mtDNA	Faeces, blood, & tissue
Determination of effective population size and population size estimation	Bornean orang-utan (Sharma *et al.* 2012a)	Microsatellites	Faeces
	Savannah baboon (*Papio cynocephalus*; Storz *et al.* 2002)	Microsatellites	Blood
Detection of hybridization events	Macaques (*Macaca* spp.; Evans *et al.* 2001)	mtDNA & microsatellites	Blood & tissue
	Howler monkey (*Alouatta* spp.; Cortés-Ortiz *et al.* 2007)	mtDNA, microsatellites & Y Chromosome	Blood & hair

continued

Table 5.1 *Continued*

Applications	Examples and references	Molecular marker(s)	Sample type
Evaluation of impact of habitat fragmentation, reduced gene flow and demographic history	Golden-crowned sifaka (*Propithecus tattersalli*; Quéméré *et al.* 2010)	Microsatellites	Faeces
	Bornean orang-utan (Goossens *et al.* 2006)	Microsatellites	Faeces
	Mouse lemur (*Microcebus* spp.; Olivieri *et al.* 2008; Schneider *et al.* 2010)	Microsatellites	Tissue
Sex determination	Mountain gorilla, chimpanzee (*Pan troglodytes verus*), and gibbon (*Hylobates lar*; Bradley *et al.* 2001)	Amelogenin	Faeces
	Multi genera (Villesen and Fredsted 2006)	UTX/UTY	Hair, tissue, & blood
Disease status	Chimpanzee (*Pan troglodytes*; Kaur *et al.* 2008)	Viral DNA	Faeces & tissue
	Japanese macaque (*Macaca fuscata*; Kawai *et al.* 2014)	*Plasmodium* spp. mtDNA	Urine & faeces
	Western lowland gorilla (Hamad *et al.* 2014)	Eukaryotic 18S rRNA, ITS, and other genes	Faeces
Evolutionary study of pathogen genomes	Chimpanzees, gorillas, & bonobos (*Pan paniscus*; Liu *et al.* 2010)	*Plasmodium* spp. mtDNA	Faeces
	Chimpanzees (Keele *et al.* 2006)	SIV/HIV nucleic acids	Faeces
Forensic and legal actions	Multi genera (Rönn *et al.* 2009)	DNA microarray	Tissue & blood
	Multi genera (Minhós *et al.* 2013)	DNA barcoding	Bushmeat
	Douc langur (*Pygathrix* spp.; Liu *et al.* 2008)	mtDNA & amelogenin	Hair, bone, & tissue
Dietary analysis	Wild western gorilla (*Gorilla gorilla*) and black-and-white colobus monkey (*Colobus guereza*; Bradley *et al.* 2007)	Chloroplast DNA and plant nuclear DNA	Faeces
	Golden-crowned sifaka (Quéméré *et al.* 2013)	Metabarcoding	Faeces
	Multi genera (Pickett *et al.* 2012)	Arthropod mtDNA	Faeces
	Leaf-feeding monkey (*Pygathrix nemaeus*; Srivathsan *et al.* 2014)	Metabarcoding & metagenomics	Faeces

LPA: autosomal lipoprotein gene.
TSPY: Y-linked testis specific protein.
UTX/UTY: sex-chromosomal isoforms of the ubiquitously transcribed tetratricopeptide repeat protein gene.

a wide array of molecular markers is available nowadays: from amplified length polymorphisms (AFLP), to minisatellites and microsatellites, they can either randomly sample the genome or focus on specific regions such as the major histocompatibility complex (MHC), or mitochondrial DNA (mtDNA) genes, to name two. With the exception of MHC and mtDNA, they are all distributed along the genome, and have therefore been advocated as markers of choice for genomic studies. In the last two decades, the two most commonly used markers for non-invasive genetics were mtDNA and nuclear microsatellites (see Di Fiore (2003) for a review on the first primate studies that used non-invasive sampling). However, it is likely that in the future genome-wide single nucleotide polymorphisms (SNPs) will become one of the genetic markers of choice, if not the main genetic marker to study the ecology and conservation of wild populations (Box 5.1).

Since the 1980s, software packages, many of which are publicly available for downloading from the Internet, have been evolving to analyse various kinds of molecular data. These include packages for multiple sequence alignment, phylogeny inferences, population- and individual-based genetic analyses (Table 5.2). The models behind each software are topics for stand-alone chapters and articles. Descriptions of a few of these models can be find in Bertorelle *et al.* (2009), in Chikhi and Bruford (2005), in Gutenkunst *et al.* (2009), and in Robinson *et al.* (2014), to name a few. Additionally, some journals (e.g. *Molecular Ecology Notes/Resources*, *Bioinformatics*, and *Conservation Genetic Resources*) routinely publish notes on new software suitable for particular types of analysis.

By combining the right molecular marker with appropriate statistical analyses, many questions in primate conservation can be answered. For instance, it can elucidate primate population dynamics in a local (and broader) scale in order to evaluate threats and suggest appropriate conservation measures to wildlife managers and policy-makers (Vigilant and Guschanski 2009). The case of the orang-utans (*Pongo pygmaeus*) of the Lower Kinabatangan region in Sabah, Malaysia, is a concrete example of how genetic information has been integrated into conservation policies (for the detailed case study, see Salgado-Lynn *et al.* in press). Microsatellite data were obtained from faecal material (Goossens *et al.* 2005, 2006a, 2006b) and combined with demographic information in a stochastic

Table 5.2 List of some of the most cited software used in conservation and population genetic studies.

Name	Authors	Link	Licence
Primer design			
Primer3 web	Untergrasser *et al.*	http://primer3.ut.ee/	OS
OligoAnalyzer	IDT	https://www.idtdna.com/calc/analyzer	OS
Primer premier	Lalitha *et al.*	http://www.premierbiosoft.com/primerdesign/	C
Identity and parentage analysis			
Cervus	Kalinowski *et al.*	http://www.fieldgenetics.com/pages/aboutCervus_Overview.jsp	OS/C
Colony	Wang	http://www.zsl.org/science/software/colony	OS
Kingroup	Konovalov *et al.*	https://code.google.com/p/kingroup/	OS
Famoz	Gerber *et al.*	http://www.pierroton.inra.fr/genetics/labo/Software/Famoz/	OS
Matesoft	Moilanen *et al.*	http://www.bi.ku.dk/staff/jspedersen/matesoft/	OS
Population genetics and phylogenetics packages			
Arlequin	Excoffier *et al.*	http://cmpg.unibe.ch/software/arlequin35/	OS
GenAlEx	Peakall & Smouse	http://biology-assets.anu.edu.au/GenAlEx/Welcome.html	OS
Genepop	Raymond & Rousset	http://genepop.curtin.edu.au/	OS
Genetix	Belkhir *et al.*	http://kimura.univ-montp2.fr/genetix/	OS
DNAsp	Rozas & Rozas	http://www.ub.edu/dnasp/	OS
DAMBE	Xia	http://dambe.bio.uottawa.ca/dambe.asp	OS
MEGA	Tamura *et al.*	http://www.megasoftware.net/	OS
MrBayes	Ronquist *et al.*	http://mrbayes.sourceforge.net/	OS
Network	Fluxus Technology	http://www.fluxus-engineering.com/sharenet.htm	OS
Genetic clustering packages			
STRUCTURE	Pritchard *et al.*	http://pritchardlab.stanford.edu/structure.html	OS
NewHybrids	Anderson & Thompson	http://ib.berkeley.edu/labs/slatkin/eriq/software/software.htm#NewHybs	OS

continued

Table 5.2 *Continued*

Name	Authors	Link	Licence
Migrate-N	Beerli & Palczewski	http://popgen.sc.fsu.edu/Migrate/Migrate-n.html	OS
BAPS	Corander *et al.*	http://www.helsinki.fi/bsg/software/BAPS/baps_citation.shtl	OS
geneland	Guillot *et al.*	http://www2.imm.dtu.dk/~gigu/Geneland/	OS
Geneclass2	Piry *et al.*	http://www1.montpellier.inra.fr/CBGP/software/GeneClass/	OS
TESS	François *et al.*	http://membres-timc.imag.fr/Olivier.Francois/tess.html	OS
Demographic history and simulation software			
msvar	Storz and Beaumont	http://cbsuapps.tc.cornell.edu/msvar.aspx	OS
Bottleneck	Piry *et al.*	http://www1.montpellier.inra.fr/CBGP/software/Bottleneck/	OS
DIYABC	Cornuet *et al.*	http://www1.montpellier.inra.fr/CBGP/diyabc/	OS
File conversion packages and other general utilities			
PGDSpider	Lischer & Excoffier	http://www.cmpg.unibe.ch/software/PGDSpider/	OS
Alter	Glez-Peña *et al.*	http://sing.ei.uvigo.es/ALTER/	OS
Sequence conversion	Bugaco.com	http://sequenceconversion.bugaco.com/converter/biology/sequences/	OS
Sequencher	GeneCodes	http://www.genecodes.com/	C
Microchecker	van Oosterhout *et al.*	http://www.microchecker.hull.ac.uk/	OS

C: commercial.
OS: open software.

population modelling program (VORTEX v 9.61) (Miller and Lacy 2005) to predict the viability of various orang-utan populations within the Lower Kinabatangan region under several scenarios (Bruford *et al.* 2010). The parameters of the model were based on previous Population and Habitat Viability Assessments (PHVA) (Singleton *et al.* 2004), and research and observation in the Lower Kinabatangan (Ancrenaz *et al.* 2004). The simulation showed that a combination of modest translocation rates (one individual every 20 years) and corridor establishment would permit even the most isolated subpopulations to retain demographic stability and limit localized inbreeding. These results were incorporated into Sabah's Orang-utan State Action Plan (Sabah Wildlife Department 2011) and lead (among other things) to the construction of orang-utan bridges in strategic areas of the Lower Kinabatangan, with documented success in all of them (Ancrenaz 2010). While the incorporation of the genetic data is an important achievement for orang-utan conservation in Sabah, it is important to emphasize that the whole process (from collecting samples until the incorporation of data into the management plan) took at least 10 years. This is a timeframe that may not be useful for species that are on the brink of extinction. Unless conservation geneticists get involved in policy-making, the results of their research will seldom affect their species of interest.

5.5 Conservation genomics and the future

Conservation genomics is a new field that can be broadly defined as the use of genomic techniques to address problems in conservation biology (Allendorf *et al.* 2010). But what are the real advantages of this new field, compared to what can be obtained by traditional conservation genetics?

All conservation genetic studies based on classic genetic techniques obtain important information for conservation from the study of a limited number of neutral loci (i.e. loci that are not under selection).

These few markers can give direct estimates of parameters such as population size, inbreeding, genetic drift, population structure, hybridization, migration rates, population growth and viability, and demographic vital rates. In addition, they can give estimates on kinship relationships and pedigrees. Genomics can contribute to this by hugely increasing the density of sampling across genomes. The result is the rapid increase in the number of markers, and therefore the accuracy and precision of parameter estimates.

In addition, genomics can also address factors hitherto impossible to study with classical approaches. The most important one is the possibility to investigate adaptive variation in detail. It is now affordable to identify the genetic loci responsible for adaptive evolution in non-model organisms (Dalziel *et al.* 2009; Stinchcombe and Hoekstra 2008; Ungerer *et al.* 2008). NGS techniques allow investigation of the roles of genotype-by-environment interactions, the genetic basis of inbreeding depression, and adaptive variation (Stapley *et al.* 2010). These techniques also shed light on the effects that genetic and environmental factors have on fitness (Ouborg *et al.* 2010), with a great potential to improve conservation programmes. Detailed reviews of this topic can be found in Allendorf *et al.* (2010), Ouburg *et al.* (2010), Stapley *et al.* (2010), and Shafer *et al.* (2015).

There are already several studies that have set up genomic approaches that might be useful for conservation genomic studies of primate species. Perry *et al.* (2010) successfully tested a DNA-capture protocol on western chimpanzees (*Pan troglodytes verus*), thereby investigating the species' genetic diversity and setting up a methodology to obtain genome-wide information from primate faecal samples. Using RNA sequencing, the same team studied human and non-human primates, finding patterns of variation consistent with positive or directional selection (Perry *et al.* 2012a). This team also developed important genomic resources and population studies of the aye-aye (*Daubentonia madascariensis*; Perry *et al.* (2013, 2012b)).

There are, of course, a few pitfalls. We have already mentioned the problem of the lack of a reference genome in many primate non-model species, and the difficulty of using non-invasive sources of DNA. As close species might be used for reference (Miller *et al.* 2012), genomic resources developed for humans and other model primates might be powerful tools for non-model primate studies in conservation genomics. Depending on the species of interest and the sequencing machine used, one may be able to obtain an adequate coverage. Alternatively, techniques such as genotyping-by-sequencing and RAD-seq (Baird *et al.* 2008; Narum *et al.* 2013) allow the detection of thousands of SNPs without the need of a reference genome. In addition, when planning an experiment, researchers must allocate a considerable amount of time and resources to bioinformatics training, data analysis, and data storage (Allendorf *et al.* 2010).

As discussed in the previous section, the population genetics community has witnessed major changes in the last few years with the emergence of genomic markers applied to population studies. Table 5.3 presents links to some frequently cited software for the analysis of genomic data. Thousands if not hundreds of thousands of genetic markers may soon become available to describe genetic diversity with a precision never achieved before. While this revolution is expected to generate an improvement in the precision of parameter estimation, there is also a risk associated with that increased power. Indeed, if the demographic model assumed (for instance a model with a single population size change) is a poor representation of the real population history, we may date with high precision an event that never took place. For instance Chikhi *et al.* (2010) used simulations to demonstrate that structured populations that remain of constant size over time can produce false signals of population size reduction if one ignores their genetic structure. While models will always be imperfect representations of reality, they do facilitate understanding of major demographic events. New inferential methods necessary to accurately analyse genomic data in an efficient way are being developed (Gutenkunst *et al.* 2009; Li and Durbin 2012; Liu and Fu 2015; Mazet *et al.* 2015; Schiffels and Durbin 2014; Sheehan *et al.* 2013). It is important to understand the principle behind these methods and their limits. For instance, many methods aim at inferring changes in population size, but do so under the assumption that populations are random mating units (where

Table 5.3 List of frequently cited software in the literature for de-novo assembly, referenced assembly, and SNP calling for genomic data.

Name	Authors	Link	Licence
De-novo assembly			
ABySS	Simpson *et al.*	http://www.bcgsc.ca/platform/bioinfo/software/abyss	C
Newbler	454/Roche	http://www.454.com/	C
SOAPdenovo	Li *et al.*	http://soap.genomics.org.cn/soapdenovo.html	OS
Velvet	Zerbino *et al.*	(a) http://www.ebi.ac.uk/~zerbino/velvet/	(b) OS
(c) SSAKE	(d) Warren *et al.*	(e) http://www.bcgsc.ca/platform/bioinfo/software/ssake	(f) OS
(g) Velvet	(h) Zerbino *et al.*	(i) http://www.ebi.ac.uk/~zerbino/velvet/	(j) OS
(k) ALLPATHS-LG	(l) Gnerre *et al.*	(m) https://www.broadinstitute.org/scientific-community/science/programs/genome-sequencing-and-analysis/	(n) OS
(o) AMOS	(p) Salzberg *et al.*	(q) http://sourceforge.net/projects/amos/	(r) OS
(s) Referenced assembly	(t)	(u)	(v)
(w) Bowtie	(x) Langmead & Salzberg	(y) http://bowtie-bio.sourceforge.net/index.shtml	(z) OS
(aa) BWA	(bb) Li & Durbin	(cc) http://bio-bwa.sourceforge.net/	(dd) OS
(ee) CLC Genomic Workbench	(ff) CLC	(gg) http://www.clcbio.com/products/clc-genomics-workbench/	(hh) C
(ii) MAQ	(jj) Li *et al.*	(kk) http://maq.sourceforge.net/	(ll) OS
(mm) MUMmer	(nn) Kurtz *et al.*	(oo) http://mummer.sourceforge.net/	(pp) OS
(qq) SOAP	(rr) Li *et al.*	(ss) http://soap.genomics.org.cn/	(tt) OS
(uu) Variant calling	(vv)	(ww)	(xx)
(yy) GATK	(zz) McKenna *et al.*	(aaa) https://www.broadinstitute.org/gatk/	(bbb) OS
(ccc) SAM Tools	(ddd) Li *et al.*	(eee) http://samtools.sourceforge.net/	(fff) OS
(ggg) SOAPsnp	(hhh) Li *et al.*	(iii) http://soap.genomics.org.cn/soapsnp.html	(jjj) OS

De-novo assembly: genome assembly without a reference genome.
Referenced assembly: genome assembly using a previously assembled referenced genome.
Variant calling: software for SNP and indel detection in assembled genomes.
C: commercial.
OS: open software.

every individual has the same probability of mating with another individual of the opposite sex). This is unlikely to be true for most species. Also, many primate species live in social groups, which represent a unit that is typically ignored by population genetics (but see Chesser (1991); Di Fiore and Valencia (2014); Parreira and Chikhi (2015)). Overall, genomic data represent major promises but also challenges since many inferential methods developed over the last 20 years cannot be applied directly to genomic data. New inferential methods necessary to accurately analyse genomic data in an efficient way are being developed. Therefore, keeping pace with them could be extremely useful.

5.6 Conclusion

Primate conservation requires an integrated approach—combining ecology, demography, morphology, behaviour, and genetics—to allow the preservation of significant amounts of evolutionary diversity. While the application of genetics (and genomics: Sharma *et al.* (2012a)) in the management of threatened species is increasing (DeSalle and

Amato 2004), there seems to be a general failure to incorporate this type of data into concrete conservation actions. This failure may be due to two main factors: first the difficulty for non-geneticists to interpret the results of genetic data and, second, the difficulty for conservation geneticists to become involved in policy and practical conservation decisions. Also, there is a need to develop tools that will help conservation biologists to use and master population genetics concepts (i.e. Conservation Genetic Resources for Effective Species Survival [CONGress]).[2] The incorporation of genetic data into species action plans has recently been advocated, but will require the above-mentioned difficulties to be overcome (Frankham 2009; Laikre 2010). Genetic data must be integrated with an understanding of landscape dynamics and area-based conservation actions to achieve successful decisions concerning areas, landscapes, and species.

If interpreted correctly, genetic data may provide unique information for conservation initiatives that would be difficult to obtain solely through field studies. It thus becomes important for conservation biologists and policy-makers to have an understanding of genetic data. However, any productive application of genetic techniques lies in their effective combination with field studies, so such studies must be carefully and scientifically framed to be worth the investment. After all, 'sequencing a critically endangered species' genome does not save it from extinction in the wild' (Kohn 2010).

The various techniques briefly described here may appear as a robust set of tools for examining, analysing, and interpreting molecular variations in primate taxa, yet they do not represent a complete description of what is currently available. For instance, the area of landscape genetics and the integration of spatial data, which are increasingly becoming important, are not covered in this chapter. We point to the recent reviews of Thomassen *et al.* (2010) and the recent studies by K. Gonder's group on chimpanzees (Mitchell *et al.* 2015a, b). These studies show how the use of geo-spatial data together with spatial modelling techniques and genetic data may help identify regions of greatest importance for conservation. They also stress the importance of GPS data to give a precise geographic reference to samples in the field. Similarly, we did not cover species distribution modelling and how it could be integrated with genetic data to predict regions where the species of interest may occur under various climate change scenarios. We also did not discuss approximate Bayesian computation, even though it is providing a very powerful modelling framework (Beaumont *et al.* 2002).

Considerable time, energy, and resources are often needed to become proficient with both the theory and application of molecular techniques. In addition, it is essential to recognize the limitations of new techniques, in particular those concerning data handling and data analysis. Nonetheless, we are optimistic and believe that, together with cautious application of inferential methods from population genetics, genomic data will provide us with the possibility of: (1) reconstructing (date and quantify) major events of the demographic history of populations, (2) identifying genomic regions under selection, or (3) assessing the level of inbreeding, a crucial issue for endangered primate species.

In conclusion, we emphasize that genetic research on its own is insufficient to achieve conservation and there is a need for effective dialogue between researchers and policy-makers. Genetic information has yet to be incorporated into national biodiversity policies and therefore has contributed relatively little to the management and conservation of primate populations (Di Fiore 2003; Laikre 2010; Laikre *et al.* 2010). Meanwhile, wild primate populations often become reduced and isolated due to habitat conversion and fragmentation (Marsh and Chapman 2013), thus increasing their risk for inbreeding depression. If we really want to preserve primate populations, primate conservation geneticists worldwide must expand their work in universities and research institutions to include policy and practical conservation work.

Acknowledgements

The authors would like to thank Serge A. Wich and Andrew Marshall for their invitation to contribute to this book, for their guidance, and for their

[2] http://www.congressgenetics.eu.

patience in receiving it. We appreciate the support of Luis Cunha for his suggestions when building a software list for genomics. We are also grateful to Mary K. Gonder, Lauren Gilhooly, Inês Carvalho, and Tania Minhós Rodrigues for their useful comments and suggestions on earlier versions of the manuscript. L. C. was funded by the Fundação para a Ciência e Tecnologia (Ref. PTDC/BIA-BIC/4476/2012) and the LABEX entitled TULIP (ANR-10-LABX-41).

References

Aitken, N., Smith, S., Schwarz, C., and Morin, P. A. (2004). Single nucleotide polymorphism (SNP) discovery in mammals: a targeted-gene approach. *Molecular Ecology* **13**: 1423–1431.

Allen, M., Engström, A. S., Meyers, S., Handt, O., Saldeen, T., *et al.* (1998). Mitochondrial DNA sequencing of shed hairs and saliva on robbery caps: sensitivity and matching probabilities. *Journal of Forensic Science* **43**: 453–464.

Allendorf, F. W. and Luikart, G. (2007). *Conservation and the Genetics of Populations*, 1st edn. Oxford: Blackwell Publishing.

Allendorf, F. W., Hohenlohe, P. A., and Luikart, G. (2010). Genomics and the future of conservation genetics. *National Review of Genetics* **11**: 697–709.

Améndola-Pimenta, M., García-Feria, L., Serio-Silva, J. C., and Rico-Gray, V. (2009). Noninvasive collection of fresh hairs from free-ranging howler monkeys for DNA extraction. *American Journal of Primatology* **71**: 359–363.

Ancrenaz, M. (2010). Orang-utan Bridges in Lower Kinabatangan. Available at: <http://beta.arcus-foundation.org.s3.amazonaws.com/wp-content/uploads/2010/01/Kinabatangan-Orangutan-Rope-Bridges-Ancrenaz-2010.pdf> [Accessed November 2015].

Ancrenaz, M., Goossens, B., Gimenez, O., Sawang, A., and Lackman-Ancrenaz, I. (2004). Determination of ape distribution and population size using ground and aerial surveys: a case study with orang-utans in lower Kinabatangan, Sabah, Malaysia. *Animal Conservation* **7**: 375–385.

Avise, J. C. (1995). Mitochondrial DNA polymorphism and a connection between genetics and demography of relevance to conservation. *Conservation Biology* **9**: 686–690.

Avise, J. C. (2004). *Molecular Markers, Natural History and Evolution*, 2nd edn. Sunderland, MA: Sinauer Associates.

Avise, J. C. (2009a). Perspective: conservation genetics enters the genomics era. *Conservation Genetics* **11**: 665–669.

Avise, J. C. (2009b). Phylogeography: retrospect and prospect. *Journal of Biogeography* **36**: 3–15.

Avise, J. C., Arnold, J., Ball, R. M., Bermingham, E., Lamb, T., *et al.* (1987). Intraspecific phylogeography: the mitochondrial DNA bridge between population genetics and systematics. *Annual Review of Ecology, Evolution, and Systematics* **18**: 489–522.

Baird, N. A., Etter, P. D., Atwood, T. S., Currey, M. C., Shiver, A. L., *et al.* (2008). Rapid SNP discovery and genetic mapping using sequenced RAD markers. *PLoS One* **3**: e3376.

Ballard, J. W. O. and Rand, D. M. (2005). The population biology of mitochondrial DNA and its phylogenetic implications. *Annual Review of Ecology, Evolution, and Systematics* **36**: 621–642.

Ballard, J. W. O. and Whitlock, M. C. (2004). The incomplete natural history of mitochondria. *Molecular Ecology* **13**: 729–744.

Bayes, M. K., Smith, K. L., Alberts, S. C., Altmann, J., and Bruford, M. W. (2000). Testing the reliability of microsatellite typing from faecal DNA in the savannah baboon. *Conservation Genetics* **1**: 173–176.

Beaumont, M. A., Zhang, W., and Balding, D. J. (2002). Approximate Bayesian computation in population genetics. *Genetics* **162**: 2025–2035.

Bergl, R. A. and Vigilant, L. (2007). Genetic analysis reveals population structure and recent migration within the highly fragmented range of the Cross River gorilla (*Gorilla gorilla diehli*). *Molecular Ecology* **16**: 501–516.

Bertorelle, G., Bruford, M. W., Hauffe, H. C., Rizzoli, A., and Vernesi, C. (Eds) (2009). *Population Genetics for Animal Conservation*, 1st edn. Cambridge: Cambridge University Press.

Bhargava, A. and Fuentes, F. F. (2010). Mutational dynamics of microsatellites. *Molecular Biotechnology* **44**: 250–266.

Blair, M. E., Gutierrez-Espeleta, G. A., and Melnick, D. J. (2012). Subspecies of the Central American squirrel monkey (*Saimiri oerstedii*) as units for conservation. *International Journal of Primatology* **34**: 86–98.

Boom, R., Sol, C. J. A., Salimans, M. M. M., Jansen, C. L., Wertheim-van Dillen, P. M. E., *et al.* (1990). Rapid and simple method for purification of nucleic acids. *Journal of Clinical Microbiology* **28**: 495–503.

Bradley, B. J. and Vigilant, L. (2002). False alleles derived from microbial DNA pose a potential source of error in microsatellite genotyping of DNA from faeces. *Molecular Ecology Notes* **2**: 602–605.

Bradley, B. J., Boesch, C., and Vigilant, L. (2000). Identification and redesign of human microsatellite markers for genotyping wild chimpanzee (*Pan troglodytes verus*) and gorilla (*Gorilla gorilla gorilla*) DNA from faeces. *Conservation Genetics* **1**: 289–292.

Bradley, B. J., Chambers, K. E., and Vigilant, L. (2001). Accurate DNA-based sex identification of apes using noninvasive samples. *Conservation Genetics* **2**: 179–181.

Bradley, B. J., Stiller, M., Doran-Sheehy, D. M., Harris, T., Chapman, C. A., *et al.* (2007). Plant DNA sequences from feces: potential means for assessing diets of wild primates. *American Journal of Primatology* **69**: 1–7.

Broquet, T., Ménard, N., and Petit, E. (2007). Noninvasive population genetics: a review of sample source, diet, fragment length and microsatellite motif effects on amplification success and genotyping error rates. *Conservation Genetics* **8**: 249–260.

Bruford, M. W., Ancrenaz, M., Chikhi, L., Lackman-Ancrenaz, I., Andau, P. M., *et al.* (2010). Projecting genetic diversity and population viability for the fragmented orang-utan population in the Kinabatangan floodplain, Sabah, Malaysia. *Endangered Species Research* **12**: 249–262.

Buschiazzo, E. and Gemmell, N. J. (2006). The rise, fall and renaissance of microsatellites in eukaryotic genomes. *BioEssays* **28**: 1040–1050.

Chesser, R. K. (1991). Influence of gene flow and breeding tactics on gene diversity within populations. *Genetics* **129**: 573.

Chikhi, L. (2008). Genetic markers: how accurate can genetic data be? *Heredity* **101**: 471–472.

Chikhi, L. and Bruford, M. W. (2005). Mammalian population genetics and genomics. In: Ruvinsky, A. and Graves, J. M. (Eds), *Mammalian Genomics*. pp. 539–584. Wallingford, UK: CAB International Publishing.

Chikhi, L., Sousa, V. C., Luisi, P., Goossens, B., and Beaumont, M. A. (2010). The confounding effects of population structure, genetic diversity and the sampling scheme on the detection and quantification of population size changes. *Genetics* **186**: 983–995.

Constable, J. L., Ashley, M. V, Goodall, J., and Pusey, A. E. (2001). Noninvasive paternity assignment in Gombe chimpanzees. *Molecular Ecology* **10**: 1279–1300.

Cortés-Ortiz, L., Duda, T.F., Canales-Espinosa, D., García-Orduña, F., Rodríguez-Luna, E., *et al.* (2007). Hybridization in large-bodied New World primates. *Genetics* **176**: 2421–2425.

Cowlishaw, G. and Dunbar, R. (2000). *Primate Conservation Biology*. Chicago, IL: University of Chicago Press.

Cronn, R., Knaus, B. J., Liston, A., Maughan, P. J., Parks, M., *et al.* (2012). Targeted enrichment strategies for next-generation plant biology. *American Journal of Botany* **99**: 291–311.

Dalziel, A. C., Rogers, S. M., and Schulte, P. M. (2009). Linking genotypes to phenotypes and fitness: how mechanistic biology can inform molecular ecology. *Molecular Ecology* **18**: 4997–5017.

Davenport, T. R. B., Stanley, W. T., Sargis, E. J., De Luca, D. W., Mpunga, N. E., *et al.* (2006). A new genus of African monkey, Rungwecebus: morphology, ecology, and molecular phylogenetics. *Science* **312**: 1378–1381.

DeSalle, R. and Amato, G. (2004). The expansion of conservation genetics. *Nature Reviews Genetics*. **5**: 702–712.

Di Fiore, A. (2003). Molecular genetic approaches to the study of primate behavior, social organization, and reproduction. *American Journal of Physical Anthropology Supplement* **37**: 62–99.

Di Fiore, A. and Valencia, L. M. (2014). The interplay of landscape features and social system on the genetic structure of a primate population: an agent-based simulation study using 'tamarins.' *International Journal of Primatology* **35**: 226–257.

Douadi, M. I., Gatti, S., Levrero, F., Duhamel, G., Bermejo, M., *et al.* (2007). Sex-biased dispersal in western lowland gorillas (*Gorilla gorilla gorilla*). *Molecular Ecology* **16**: 2247–2259.

Drummond, A. J., Rambaut, A., Shapiro, B., and Pybus, O. G. (2005). Bayesian coalescent inference of past population dynamics from molecular sequences. *Molecular Biology and Evolution* **22**: 1185–1192.

Ellegren, H. (2004). Microsatellites: simple sequences with complex evolution. *Nature Reviews Genetics* **5**: 435–445.

Eriksson, J., Hohmann, G., Boesch, C., and Vigilant, L. (2004). Rivers influence the population genetic structure of bonobos (*Pan paniscus*). *Molecular Ecology* **13**: 3425–3435.

Evans, B. J., Supriatna, J., and Melnick, D. J. (2001). Hybridization and population genetics of two macaque species in Sulawesi, Indonesia. *Evolution* **55**: 1686–1702.

Fagundes, V., Paes, M. F., Chaves, P. B., Mendes, S. L., Possamai, C. B., *et al.* (2008). Genetic structure in two northern muriqui populations (Brachyteles hypoxanthus, Primates, Atelidae) as inferred from fecal DNA. *Genetics and Molecular Biology* **171**: 166–171.

Frankel, O. H. (1983). The place of management in conservation. In: Schonewald-Cox, C. M., Chambers, S. M., MacByrde, B., and Thomas, L. (Eds), *Genetics and Conservation: A Reference for Managing Wild Animal and Plant Populations*. pp. 1–14. Menlo Park, CA: Benjamin/Cummings.

Frankham, R. (2009). Where are we in conservation genetics and where do we need to go? *Conservation Genetics* **11**: 661–663.

Frankham, R. (2015). Genetics and extinction. *Biological Conservation* **126**: 131–140.

Frankham, R., Ballou, J. D., and Briscoe, D. A. (2010). *Introduction to Conservation Genetics*, 2nd edn. Cambridge: Cambridge University Press.

Frantzen, M. A. J., Silk, J. B., Ferguson, J. W. H., Wayne, R. K., and Kohn, M. H. (1998). Empirical evaluation of preservation methods for faecal DNA. *Molecular Ecology* **7**: 1423–1428.

Gagneux, P., Boesch, C., and Woodruff, D. S. (1997). Microsatellite scoring errors associated with noninvasive genotyping based on nuclear DNA amplified from shed hair. *Molecular Ecology* **6**: 861–868.

Gerloff, U., Hartung, B., Fruth, B., Hohmann, G., and Tautz, D. (1999). Intracommunity relationships, dispersal pattern and paternity success in a wild living community of Bonobos (*Pan paniscus*) determined from DNA analysis of faecal samples. *Proceedings of the Royal Society of London Series B* **266**: 1189–1195.

Gnirke, A., Melnikov, A., Maguire, J., Rogov, P., LeProust, E. M., *et al.* (2009). Solution hybrid selection with ultra-long oligonucleotides for massively parallel targeted sequencing. *Nature Biotechnology* **27**: 182–189.

Gonen, S., Bishop, S. C., and Houston, R. D. (2015). Exploring the utility of cross-laboratory RAD-sequencing datasets for phylogenetic analysis. *BMC Research Notes* **8**: 299.

Goossens, B., Waits, L. P., and Taberlet, P. (1998). Plucked hair samples as a source of DNA: reliability of dinucleotide microsatellite genotyping. *Molecular Ecology* **7**: 1237–1241.

Goossens, B., Chikhi, L., Utami, S., de Ruiter, J., and Bruford, M. (2000). A multi-samples, multi-extracts approach for microsatellite analysis of faecal samples in an arboreal ape. *Conservation Genetics* **1**: 157–162.

Goossens, B., Funk, S. M., Vidal, C., Latour, S., Jamart, A., *et al.* (2002). Measuring genetic diversity in translocation programmes: principles and application to a chimpanzee release project. *Animal Conservation* **5**: 225–236.

Goossens, B., Chikhi, L., Jalil, M. F., Ancrenaz, M., Lackman-Ancrenaz, I., *et al.* (2005). Patterns of genetic diversity and migration in increasingly fragmented and declining orang-utan (*Pongo pygmaeus*) populations from Sabah, Malaysia. *Molecular Ecology* **14**: 441–456.

Goossens, B., Chikhi, L., Ancrenaz, M., Lackman-Ancrenaz, I., Andau, P. M., *et al.* (2006a). Genetic signature of anthropogenic population collapse in orang-utans. *PLoS Biol*ogy **4**: 285–291.

Goossens, B., Setchell, J. M., James, S. S., Funk, S. M., Chikhi, L., *et al.* (2006b). Philopatry and reproductive success in Bornean orang-utans (*Pongo pygmaeus*). *Molecular Ecology* **15**: 2577–2588.

Goossens, B. and Bruford, M. W. (2009). Non-invasive genetic analysis in conservation. In: Bartorelle, G., Bruford, M. W., Hauffe, H. C., Rizzoli, A., and Vernesi, C. (Eds), *Population Genetics for Animal Conservation*. pp. 167–201. Cambridge: Cambridge University Press.

Goossens, B., Anthony, N., Jeffery, K., Johnson-Bawe, M., and Bruford, M. W. (2011). Collection, storage and analysis of non-invasive genetic material in primate biology. In: Setchell, J. M. and Curtis, D. J. (Eds), *Field and Laboratory Methods in Primatology: A Practical Guide*. pp. 371–386. Cambridge: Cambridge University Press.

Groom, M. J., Meffe, G. K., and Carroll, C. R. (2006). *Principles of Conservation Biology*, 3rd edn. Sunderland, MA: Sinauer Associates.

Guschanski, K., Vigilant, L., McNeilage, A., Gray, M., Kagoda, E., *et al.* (2009). Counting elusive animals: comparing field and genetic census of the entire mountain gorilla population of Bwindi Impenetrable National Park, Uganda. *Biological Conservation* **142**: 290–300.

Gutenkunst, R. N., Hernandez, R. D., Williamson, S. H., and Bustamante, C. D. (2009). Inferring the joint demographic history of multiple populations from multidimensional SNP frequency data. *PLoS Genetics* **5**: e1000695.

Hamad, I., Keita, M. B., Peeters, M., Delaporte, E., Raoult, D., *et al.* (2014). Pathogenic eukaryotes in gut microbiota of Western lowland gorillas as revealed by molecular survey. *Scientific Reports* **4**: 6417.

Harcourt, A. H. (2006). Rarity in the tropics: biogeography and macroecology of the primates. *Journal of Biogeography* **33**: 2077–2087.

Hedrick, P. (2001). Conservation genetics: where are we now? *Trends in Ecology and Evolution* **16**: 629–636.

Hedrick, P. (2004). Recent developments in conservation genetics. *Forest Ecology and Management* **197**: 3–19.

Heistermann, M., Brauch, K., Möhle, U., Pfefferle, D., Dittami, J., *et al.* (2008). Female ovarian cycle phase affects the timing of male sexual activity in free-ranging Barbary macaques (*Macaca sylvanus*) of Gibraltar. *American Journal of Primatology* **70**: 44–53.

Heller, R., Chikhi, L., and Siegismund, H. R. (2013). The confounding effect of population structure on Bayesian skyline plot inferences of demographic history. *PLoS One* **8**: e62992.

Hohenlohe, P. A., Bassham, S., Etter, P. D., Stiffler, N., Johnson, E.A., *et al.* (2010). Population genomics of parallel adaptation in threespine stickleback using sequenced RAD tags. *PLoS Genetics* **6**: e1000862.

Hostetler, J. A., Onorato, D. P., Bolker, B. M., Johnson, W. E., O'Brien, S. J., *et al.* (2012). Does genetic introgression improve female reproductive performance? A test on the endangered Florida panther. *Oecologia* **168**: 289–300.

Hurst, G. D. D. and Jiggins, F. M. (2005). Problems with mitochondrial DNA as a marker in population, phylogeographic and phylogenetic studies: the effects of inherited symbionts. *Proceedings of the Royal Society of London Series B* **272**: 1525–1534.

IUCN 2015. The IUCN Red List of Threatened Species. Version 2015.1. <http://www.iucnredlist.org>. Downloaded on 01 June 2015.

Jeffery, K. J., Abernethy, K. A., Tutin, C. E. G., and Bruford, M. W. (2007). Biological and environmental degradation of gorilla hair and microsatellite amplification success. *Biological Journal of the Linnean Society* **91**: 281–294.

Jarne, P. and Lagoda, P. J. L. (1996). Microsatellites, from molecules to populations and back. *Trends in Ecology and Evolution* **11**: 424–429.

Kaur, T., Singh, J., Tong, S., Humphrey, C., Clevenger, D., *et al.* (2008). Descriptive epidemiology of fatal respiratory outbreaks and detection of a human-related metapneumovirus in wild chimpanzees (*Pan troglodytes*) at Mahale Mountains National Park, Western Tanzania. *American Journal of Primatology* **70**: 755–765.

Kawai, S., Sato, M., Kato-Hayashi, N., Kishi, H., Huffman, M. A., *et al.* (2014). Detection of Plasmodium knowlesi DNA in the urine and faeces of a Japanese macaque (Macaca fuscata) over the course of an experimentally induced infection. *Malaria Journal* **13**: 373.

Kawamoto, Y., Takemoto, H., Higuchi, S., Sakamaki, T., Hart, J. A., *et al.* (2013). Genetic structure of wild bonobo populations: diversity of mitochondrial DNA and geographical distribution. *PLoS One* **8**: e59660.

Keele, B. F., Van Heuverswyn, F., Li, Y., Bailes, E., Takehisa, J., *et al.* (2006). Chimpanzee reservoirs of pandemic and nonpandemic HIV-1. *Science* **313**: 523–526.

Khouangsathiene, S., Pearson, C., Street, S., Ferguson, B., and Dubay, C. (2008). MonkeySNP: a web portal for non-human primate single nucleotide polymorphisms. *Bioinformatics* **24**: 2645–2646.

Kim, S. and Misra, A. (2007). SNP genotyping: technologies and biomedical applications. *Annual Review of Biomedical Engineering* **9**: 289–320.

Kohn, M. H. (2010). Noninvasive genome sampling in chimpanzees. *Molecular Ecology* **19**: 5328–5331.

Kohn, M. H. and Wayne, R. K. (1997). Facts from feces revisited. *Trends in Ecology and Evolution* **12**: 223–227.

Kohn, M. H., Murphy, W. J., Ostrander, E. A., and Wayne, R. K. (2006). Genomics and conservation genetics. *Trends in Ecology and Evolution* **21**: 629–637.

Kwok, P.-Y. and Chen, X. (2003). Detection of single nucleotide polymorphisms. *Current Issues in Molecular Biology* **5**: 43–60.

Laikre, L. (2010). Genetic diversity is overlooked in international conservation policy implementation. *Conservation Genetics* **11**: 349–354.

Laikre, L., Allendorf, F. W., Aroner, L. C., Baker, C. S., Gregovich, D. P., *et al.* (2010). Neglect of genetic diversity in implementation of the convention on biological diversity. *Trends in Ecology and Evolution* **24**: 86–88.

Lande, R. (1988). Genetics and demography in biological conservation. *Science* **241**: 1455–1460.

Lathuillière, M., Ménard, N., Gautier-Hion, A., and Crouau-Roy, B. (2001). Testing the reliability of noninvasive genetic sampling by comparing analyses of blood and fecal samples in Barbary macaques (*Macaca sylvanus*). *American Journal of Physical Anthropology* **55**: 151–158.

Launhardt, K., Epplen, C., Epplen, J. T., and Winkler, P. (1998). Amplification of microsatellites adapted from human systems in faecal DNA of wild Hanuman langurs (*Presbytis entellus*). *Electrophoresis* **19**: 1356–1361.

Lercher, M. J. and Hurst, L. D. (2002). Human SNP variability and mutation rate are higher in regions of high recombination. *Trends in Genetics* **18**: 337–340.

Li, H. and Durbin, R. (2012). Inference of human population history from whole genome sequence of a single individual. *Nature* **475**: 493–496.

Liu, W., Huang, C., Roos, C., Zhou, Q., Li, Y., *et al.* (2008). Identification of the species, origin and sex of smuggled douc langur (*Pygathrix* sp.) remains. *Primates* **2**: 63–69.

Liu, W., Li, Y., Learn, G. H., Rudicell, R. S., Robertson, J. D., *et al.* (2010). Origin of the human malaria parasite *Plasmodium falciparum* in gorillas. *Nature* **467**: 420–425.

Liu, X. and Fu, Y.-X. (2015). Exploring population size changes using SNP frequency spectra. *Nature Genetics* **47**: 555–559.

Liu, Z., Ren, B., Wei, F., Long, Y., Hao, Y., *et al.* (2007). Phylogeography and population structure of the Yunnan snub-nosed monkey (*Rhinopithecus bieti*) inferred from mitochondrial control region DNA sequence analysis. *Molecular Ecology* **16**: 3334–3349.

Liu, Z., Ren, B., Wu, R., Zhao, L., Hao, Y., *et al.* (2009). The effect of landscape features on population genetic structure in Yunnan snub-nosed monkeys (*Rhinopithecus bieti*) implies an anthropogenic genetic discontinuity. *Molecular Ecology* **18**: 3831–3846.

Luikart, G., England, P. R., Tallmon, D., Jordan, S., and Taberlet, P. (2003). The power and promise of population genomics: from genotyping to genome typing. *Nature Reviews Genetics* **4**: 981–994.

Lynch, M., Bürger, R., Butcher, D., and Gabriel, W. (1990). The mutational meltdown in asexual populations. *Journal of Heredity* **84**: 339–344.

Lynch, M. and Gabriel, W. (1990). Mutation load and the survival of small populations. *Evolution* **44**: 1725–1737.

Madsen, T., Shine, R., Olsson, M., and Wittzell, H. (1999). Restoration of an inbred adder population. *Nature* **402**: 34–35.

Mamanova, L., Coffey, A. J., Scott, C. E., Kozarewa, I., Turner, E. H., *et al.* (2010). Target-enrichment strategies for next- generation sequencing. *Nature Methods* **7**: 111–118.

Marsh, L. K. and Chapman, C. A. (Eds) (2013). *Primates in Fragments: Complexity and Resilience*. New York, NY: Springer.

Mazet, O., Rodríguez, W., and Chikhi, L. (2015). Demographic inference using genetic data from a single individual: separating population size variation from population structure. *Theoretical Population Biology* **104**: 46–58.

Mcgrew, W. C., Ensminger, A. L., Marchant, L. F., Pruetz, J. D., and Vigilant, L. (2004). Genotyping aids field study of unhabituated wild chimpanzees. *American Journal of Primatology* **63**: 87–93.

McManus, K. F., Kelley, J. L., Song, S., Veeramah, K. R., Woerner, A. E., *et al.* (2015). Inference of gorilla demographic and selective history from whole-genome sequence data. *Molecular Biology and Evolution* **32**: 600–612.

Meyer, D., Rinaldi, I. D., Ramlee, H., Perwitasari-Farajallah, D., Hodges, J. K., *et al.* (2011). Mitochondrial phylogeny of leaf monkeys (genus Presbytis, Eschscholtz, 1821) with implications for taxonomy and conservation. *Molecular Phylogenetics and Evolution* **59**: 311–319.

Miller, P. S. and Lacy, R. C. (2005). VORTEX: A Stochastic Simulation of the Extinction Process. Version 9.50 User's Manual. Apple Valley, MN: Conservation Breeding Specialist Group (SSC/IUCN).

Miller, W., Schuster, S. C., Welch, A. J., Ratan, A., Bedoya-Reina, O. C., *et al.* (2012). Polar and brown bear genomes reveal ancient admixture and demographic footprints of past climate change. *Proceedings of the National Academy of Sciences* **109**: E2382–E2390.

Minhõs, T., Nixon, E., Sousa, C., Vicente, L. M., Da Silva, M. F., *et al.* (2013). Genetic evidence for spatio-temporal changes in the dispersal patterns of two sympatric African colobine monkeys. *The American Journal of Physical Anthropology* **150**: 464–474.

Minhós, T., Wallace, E., Ferreira da Silva, M. J., Sá, R. M., Carmo, M., *et al.* (2013). DNA identification of primate bushmeat from urban markets in Guinea-Bissau and its implications for conservation. *Biological Conservation* **167**: 43–49.

Mitchell, M. W., Locatelli, S., Ghobrial, L., Pokempner, A. A., Sesink Clee, P. R., *et al.* (2015a). The population genetics of wild chimpanzees in Cameroon and Nigeria suggests a positive role for selection in the evolution of chimpanzee subspecies. *BMC Evolutionary Biology* **15**: 3.

Mitchell, M. W., Locatelli, S., Sesink Clee, P. R., Thomassen, H. A., and Gonder, M. K. (2015b). Environmental variation and rivers govern the structure of chimpanzee genetic diversity in a biodiversity hotspot. *BMC Evolutionary Biology* **15**: 1.

Mittermeier, R. A. and Cheney, D. L. (1987). Conservation of primates and their habitats. In: Smuts, B. B., Cheney, D. L., Seyfarth, R. M., Wrangham, R. W., and Struhsaker, T. T. (Eds), *Primate Societies*. pp. 477–490. Chicago, IL: University of Chicago Press.

Mittermeier, R. A., Wallis, J., Rylands, A. B., Ganzhorn, J. U., Oates, J. F., *et al.* (2009). Primates in peril: the world's 25 most endangered primates 2008–2010. *Primate Conservation* **24**: 1–57.

Morin, P. A., Chambers, K. E., Boesch, C., and Vigilant, L. (2001). Quantitative polymerase chain reaction analysis of DNA from noninvasive samples for accurate microsatellite genotyping of wild chimpanzees (*Pan troglodytes verus*). *Molecular Ecology* **10**: 1835–1844.

Morin, P. A., Moore, J. J., Chakraborty, R., Jin, L., Goodall, J., *et al.* (1994). Kin selection, social structure, gene flow, and the evolution of chimpanzees. *Science* **265**: 1193–1201.

Morin, P. A., Luikart, G., and Wayne, R. K. (2004). SNPs in ecology, evolution and conservation. *Trends in Ecology and Evolution* **19**: 208–216.

Moritz, C. (1994). Defining 'Evolutionarily Significant Units' for conservation. *Trends in Ecology and Evolution* **9**: 373–375.

Muehlenbein, M. P., Ancrenaz, M., Sakong, R., Ambu, L. N., Prall, S., *et al.* (2012). Ape conservation physiology: fecal glucocorticoid responses in wild *Pongo pygmaeus morio* following human visitation. *PLoS One* **7**: e33357.

Munson, L., Brown, J. L., Bush, M., Packer, C., Janssen, D., *et al.* (1996). Genetic diversity affects testicular morphology in free-ranging lions (*Panthera leo*) of the Serengeti Plains and Ngorongoro Crater. *Journal of Reproduction and Fertility* **108**: 11–15.

Narum, S. R., Buerkle, C. A., Davey, J. W., Miller, M. R., and Hohenlohe, P. A. (2013). Genotyping-by-sequencing in ecological and conservation genomics. *Molecular Ecology* **22**: 2841–2847.

Nater, A., Nietlisbach, P., Arora, N., van Schaik, C. P., van Noordwijk, M. A., *et al.* (2011). Sex-biased dispersal and volcanic activities shaped phylogeographic patterns of extant Orangutans (genus: Pongo). *Molecular Biology and Evolution* **28**: 2275–2288.

Nechvatal, J. M., Ram, J. L., Basson, M. D., Namprachan, P., Niec, S. R., *et al.* (2008). Fecal collection, ambient preservation, and DNA extraction for PCR amplification of bacterial and human markers from human feces. *Journal of Microbiological Methods* **72**: 124–132.

Nietlisbach, P., Arora, N., Nater, A., Goossens, B., Van Schaik, C. P., *et al.* (2012). Heavily male-biased long-distance dispersal of orang-utans (genus: *Pongo*), as revealed by Y-chromosomal and mitochondrial genetic markers. *Molecular Ecology* **21**: 3173–3186.

Nsubuga, A. M., Robbins, M. M., Boesch, C., and Vigilant, L. (2008). Patterns of paternity and group fission in wild multimale mountain gorilla groups. *American Journal of Physical Anthropology* **135**: 263–274.

Nsubuga, A. M., Robbins, M. M., Roeder, A. D., Morin, P. A., Boesch, C., *et al.* (2004). Factors affecting the amount of genomic DNA extracted from ape faeces and the identification of an improved sample storage method. *Molecular Ecology* **13**: 2089–2094.

Olivieri, G., Sousa, V., Chikhi, L., and Radespiel, U. (2008). From genetic diversity and structure to conservation: Genetic signature of recent population declines in three mouse lemur species (Microcebus spp.). *Biological Conservation* **141**: 1257–1271.

Olson, L. E., Sargis, E. J., Stanley, W. T., Hildebrandt, K. B. P., and Davenport, T. R. B. (2008). Additional molecular evidence strongly supports the distinction between the recently described African primate *Rungwecebus kipunji* (Cercopithecidae, Papionini) and Lophocebus. *Molecular Phylogenetics and Evolution* **48**: 789–794.

Orlando, L., Ginolhac, A., Zhang, G., Froese, D., Albrechtsen, A., *et al.* (2013). Recalibrating Equus evolution using the genome sequence of an early Middle Pleistocene horse. *Nature* **499**: 74–78.

Ouborg, N. J., Pertoldi, C., Loeschcke, V., Bijlsma, R. K., and Hedrick, P. W. (2010). Conservation genetics in transition to conservation genomics. *Trends in Genetics* **26**: 177–187.

Parreira, B. R. and Chikhi, L. (2015). On some genetic consequences of social structure, mating systems, dispersal, and sampling. *Proceedings of the National Academy of Sciences* **112**: E3318–E3326, doi: 10.1073/pnas.1414463112.

Perry, G. H. (2014). The promise and practicality of population genomics research with endangered species. *International Journal of Primatology* **35**: 55–70.

Perry, G. H., Louis, E. E., Ratan, A., Bedoya-Reina, O. C., Burhans, R. C., *et al.* (2013). Aye-aye population genomic analyses highlight an important center of endemism in northern Madagascar. *Proceedings of the National Academy of Sciences* **110**: 5823–5828.

Perry, G. H., Marioni, J. C., Melsted, P., and Gilad, Y. (2010). Genomic-scale capture and sequencing of endogenous DNA from feces. *Molecular Ecology* **19**: 5332–5344.

Perry, G. H., Melsted, P., Marioni, J. C., Wang, Y., Bainer, R., *et al.* (2012a). Comparative RNA sequencing reveals substantial genetic variation in endangered primates. *Genome Research* **22**: 602–610.

Perry, G. H., Reeves, D., Melsted, P., Ratan, A., Miller, W., *et al.* (2012b). A genome sequence resource for the aye-aye (*Daubentonia madagascariensis*), a nocturnal lemur from Madagascar. *Genome Biology and Evolution* **4**: 126–135.

Pickett, S. B., Bergey, C. M., and Di Fiore, A. (2012). A metagenomic study of primate insect diet diversity. *American Journal of Primatology* **74**: 622–631.

Piggott, M. P. (2004). Effect of sample age and season of collection on the reliability of microsatellite genotyping of faecal DNA. *Wildlife Research* **31**: 485.

Piggott, M. P. and Taylor, A. C. (2003). Remote collection of animal DNA and its applications in conservation management and understanding the population biology of rare and cryptic species. *Wildlife Research* **30**: 1.

Prado-Martinez, J., Sudmant, P. H., Kidd, J. M., Li, H., Kelley, J. L., *et al.* (2013). Great ape genetic diversity and population history. *Nature* **499**: 471–475.

Qin, J., Li, R., Raes, J., Arumugam, M., Burgdorf, K. S., *et al.* (2010). A human gut microbial gene catalogue established by metagenomic sequencing. *Nature* **464**: 59–65.

Quéméré, E., Louis, E. E., Ribéron, A., Chikhi, L., and Crouau-Roy, B. (2009). Non-invasive conservation genetics of the critically endangered golden-crowned sifaka (*Propithecus tattersalli*): high diversity and significant genetic differentiation over a small range. *Conservation Genetics* **11**: 675–687.

Quéméré, E., Crouau-Roy, B., Rabarivola, C., Louis, E. E., and Chikhi, L. (2010). Landscape genetics of an endangered lemur (*Propithecus tattersalli*) within its entire fragmented range. *Molecular Ecology* **19**: 1606–1621.

Quéméré, E., Amelot, X., Pierson, J., Crouau-Roy, B., and Chikhi, L. (2012). Genetic data suggest a natural prehuman origin of open habitats in northern Madagascar and question the deforestation narrative in this region. *Proceedings of the National Academy of Sciences* **109**: 13028–13033.

Quéméré, E., Hibert, F., Miquel, C., Lhuillier, E., Rasolondraibe, E., *et al.* (2013). A DNA metabarcoding study of a primate dietary diversity and plasticity across its entire fragmented range. *PLoS One* **8**: e58971.

Radespiel, U., Rakotondravony, R., and Chikhi, L. (2008). Natural and anthropogenic determinants of genetic structure in the largest remaining population of the endangered golden-brown mouse lemur, Microcebus ravelobensis. *American Journal of Primatology* **70**: 860–870.

Rasteiro, R., Bouttier, P.-A., Sousa, V. C., and Chikhi, L. (2012). Investigating sex-biased migration during the Neolithic transition in Europe, using an explicit spatial simulation framework. *Proceedings of the Royal Society of London Series B* **279**: 2409–2416.

Rasteiro, R. and Chikhi, L. (2013). Female and male perspectives on the Neolithic Transition in Europe: clues from ancient and modern genetic data. *PLoS One* **8**: e60944.

Robinson, J. D., Bunnefeld, L., Hearn, J., Stone, G. N., and Hickerson, M. J. (2014). ABC inference of multi-population divergence with admixture from un-phased population genomic data. *Molecular Ecology* **23**: 4458–4471.

Roelke, M. E., Martenson, J. S., and O'Brien, S. J. (1993). The consequences of demographic reduction and genetic depletion in the endangered Florida panther. *Current Biology* **3**: 340–350.

Roeder, A. D., Archer, F. I., Poinar, H. N., and Morin, P. A. (2004). A novel method for collection and preservation of faeces for genetic studies. *Molecular Ecology Notes* **4**: 761–764.

Romero, I. G., Manica, A., Goudet, J., Handley, L. L., and Balloux, F. (2009). How accurate is the current picture of human genetic variation? *Heredity* **102**: 120–126.

Rönn, A. C., Andrés, O., López-Giráldez, F., Johnsson-Glans, C., Verschoor, E. J., *et al.* (2009). First generation microarray-system for identification of primate species subject to bushmeat trade. *Endangered Species Research* **9**: 133–142.

Sabah Wildlife Department (2011). Orangutan Action Plan. Available at: <http://static1.1.sqspcdn.com/static/f/1200343/24377541/1392391814550/Sabah_Orangutan_Action_Plan_2012-2016.pdf?token=Ns%2FWK4V6mkOhaQVz93UWbOsaMjI%3D> [Accessed November 2015].

Salgado-Lynn, M., Stanton, D. W. G., Sakong, R., Cable, J., Goossens, B., *et al.* (2010). Microsatellite markers for the proboscis monkey (*Nasalis larvatus*). *Conservation Genetics Resources* **2**: 159–163.

Schadt, E. E., Turner, S., and Kasarskis, A. (2010). A window into third-generation sequencing. *Human Molecular Genetics* **19**: 227–240.

Schaumburg, F., Mugisha, L., Kappeller, P., Fichtel, C., Köck, R., *et al.* (2013). Evaluation of non-invasive biological samples to monitor *Staphylococcus aureus* colonization in great apes and lemurs. *PLoS One* **8**: 6–11.

Schiffels, S. and Durbin, R. (2014). Inferring human population size and separation history from multiple genome sequences. *Nature Genetics* **46**: 919–925.

Schneider, N., Chikhi, L., Currat, M., and Radespiel, U. (2010). Signals of recent spatial expansions in the grey mouse lemur (Microcebus murinus). *BMC Evolutionary Biology* **10**: 105.

Selkoe, K. A. and Toonen, R. J. (2006). Microsatellites for ecologists: a practical guide to using and evaluating microsatellite markers. *Ecology Letters* **9**: 615–629.

Shafer, A. B. A., Wolf, J. B. W., Alves, P. C., Bergström, L., Bruford, M. W., *et al.* (2015). Genomics and the challenging translation into conservation practice. *Trends in Ecology and Evolution* **30**: 78–87.

Sharma, R., Arora, N., Goossens, B., Nater, A., Morf, N., *et al.* (2012a). Effective population size dynamics and the demographic collapse of Bornean orang-utans. *PLoS One* **7**: e49429.

Sharma, R., Goossens, B., Kun-Rodrigues, C., Teixeira, T., Othman, N., *et al.* (2012b). Two different high throughput sequencing approaches identify thousands of *de novo* genomic markers for the genetically depleted Bornean elephant. *PLoS One* **7**(11): e49533.

Sheehan, S., Harris, K., and Song, Y. S. (2013). Estimating variable effective population sizes from multiple genomes: a sequentially markov conditional sampling distribution approach. *Genetics* **194**: 647–661.

Singleton, I., Wich, S., Husson, S., Stephens, S., Utami Atmoko, S., *et al.* (2004). Orangutan Population and Habitat Viability Assessment. Available at: <http://www.cbsg.org/sites/cbsg.org/files/documents/OrangutanPHVA04_Final%20Report.pdf > [Accessed November 2015].

Srivathsan, A., Sha, J. C. M., Vogler, A. P., and Meier, R. (2014). Comparing the effectiveness of metagenomics and metabarcoding for diet analysis of a leaf-feeding monkey (Pygathrix nemaeus). *Molecular Ecology Resources* **15**(2): 250–261.

Stapley, J., Reger, J., Feulner, P. G. D., Smadja, C., Galindo, J., *et al.* (2010). Adaptation genomics: the next generation. *Trends in Ecology and Evolution* **25**: 705–712.

Stinchcombe, J. R. and Hoekstra, H. E. (2008). Combining population genomics and quantitative genetics: finding the genes underlying ecologically important traits. *Heredity* **100**: 158–170.

Storz, J. F., Ramakrishnan, U., and Alberts, S. C. (2002). Genetic effective size of a wild primate population: influence of current and historical demography. *Evolution* **56**: 817–829.

Taberlet, P., Griffin, S., Goossens, B., Questiau, S., Manceau, V., *et al.* (1996). Reliable genotyping of samples with very low DNA quantities using PCR. *Nucleic Acids Research* **24**: 3189–3194.

Taberlet, P., Waits, L. P., and Luikart, G. (1999). Noninvasive genetic sampling: look before you leap. *Trends in Ecology and Evolution* **14**: 323–327.

Thomassen, H. A., Cheviron, Z. A., Freedman, A. H., Harrigan, R. J., Wayne, R. K., *et al.* (2010). Spatial modelling and landscape-level approaches for visualizing intraspecific variation. *Molecular Ecology* **19**: 3532–3548.

Ungerer, M. C., Johnson, L. C., and Herman, M. A. (2008). Ecological genomics: understanding gene and genome function in the natural environment. *Heredity* **100**: 178–183.

Utami, S. S., Goossens, B., Bruford, M. W., de Ruiter, J. R., and van Hooff, J. A. R. A. M. (2002). Male bimaturism and reproductive success in Sumatran orang-utans. *Behavioral Ecology* **13**: 643–652.

Valderrama, X., Karesh, W. B., Wildman, D. E., and Melnick, D. J. (1999). Noninvasive methods for collecting fresh hair tissue. *Molecular Ecology* **8**: 1749–1752.

Vallet, D., Petit, E. J., Gatti, S., Levréro, F., and Ménard, N. (2007). A new 2CTAB/PCI method improves DNA amplification success from faeces of Mediterranean (Barbary macaques) and tropical (lowland gorillas) primates. *Conservation Genetics* **9**: 677–680.

Van Belle, S., Estrada, A., Strier, K. B., and Di Fiore, A. (2012). Genetic structure and kinship patterns in a population of black howler monkeys, *Alouatta pigra*, at Palenque National Park, Mexico. *American Journal of Primatology* **74**: 948–957.

Vernesi, C. and Bruford, M. W. (2009). Recent developments in molecular tools for conservation. In: Bertorelle, G., Bruford, M. W., Hauffe, H. C., Rizzoli, A., and Vernesi, C. (Eds), *Population Genetics for*

Animal Conservation. pp. 321–344. Cambridge: Cambridge University Press.

Vigilant, L. (1999). An evaluation of techniques for the extraction and amplification of DNA from naturally shed hairs. *Biological Chemistry* **380**: 1329–1331.

Vigilant, L. and Guschanski, K. (2009). Using genetics to understand the dynamics of wild primate populations. *Primates* **50**: 105–120.

Vigilant, L., Hofreiter, M., Siedel, H., and Boesch, C. (2001). Paternity and relatedness in wild chimpanzee communities. *Proceedings of the National Academy of Sciences* **98**: 12890–12895.

Vignal, A., Milan, D., SanCristobal, M., and Eggen, A. (2002). A review on SNP and other types of molecular markers and their use in animal genetics. *Genetics Selection Evolution* **34**: 275–305.

Villesen, P. and Fredsted, T. (2006). Fast and non-invasive PCR sexing of primates: apes, Old World monkeys, New World monkeys and Strepsirrhines. *BMC Ecology* **6**: 8.

Vynne, C., Skalski, J. R., Machado, R. B., Groom, M. J., Jácomo, A. T. A., *et al.* (2011). Effectiveness of scat-detection dogs in determining species presence in a tropical savanna landscape. *Conservation Biology* **25**: 154–162.

Waits, L. and Paetkau, D. (2005). Noninvasive genetic sampling tools for wildlife biologists: a review of applications and recommendations for accurate data collection. *Journal of Wildlife Management* **69**: 1419–1433.

Walsh, P. S., Metzger, D. A., and Higuchi, R. (1991). Chelex 100 as a medium for simple extraction of DNA for PCR-based typing from forensic material. *Biotechniques* **10**: 506–513.

Wang, I. J. (2010). Recognizing the temporal distinctions between landscape genetics and phylogeography. *Molecular Ecology* **19**(13): 2605–2608.

Wasser, S. K., Davenport, B., Ramage, E. R., Hunt, K. E., Parker, M., *et al.* (2004). Scat detection dogs in wildlife research and management: application to grizzly and black bears in the Yellowhead Ecosystem, Alberta, Canada. *Canadian Journal of Zoology* **82**: 475–492.

Wayne, R. K. and Morin, P. A. (2004). Conservation genetics in the new molecular age. *Frontiers in Ecology and the Environment* **2**: 89–97.

Weir, B. S., Anderson, A. D., and Hepler, A. B. (2006). Genetic relatedness analysis: modern data and new challenges. *Nature Reviews Genetics* **7**: 771–780.

Wikberg, E. C., Sicotte, P., Campos, F. A., and Ting, N. (2012). Between-group variation in female dispersal, kin composition of groups, and proximity patterns in a black-and-white colobus monkey (*Colobus vellerosus*). *PLoS One* **7**: 1–14.

Wikberg, E. C., Ting, N., and Sicotte, P. (2014). Kinship and similarity in residency status structure female social networks in black-and-white colobus monkeys (*Colobus vellerosus*). *American Journal of Physical Anthropology* **153**: 365–376.

Zhang, D. X. and Hewitt, G. M. (2003). Nuclear DNA analyses in genetic studies of populations: practice, problems and prospects. *Molecular Ecology* **12**: 563–584.

Zhou, X., Wang, B., Pan, Q., Zhang, J., Kumar, S., *et al.* (2014). Whole-genome sequencing of the snub-nosed monkey provides insights into folivory and evolutionary history. *Nature Genetics* **46**: 1303–1310.

CHAPTER 6

Primate abundance and distribution: background concepts and methods

Genevieve Campbell, Josephine Head, Jessica Junker, and K.A.I. Nekaris

Using drone technology to survey habitats of Javan slow lorises Photo copyright: Faye Vogely

6.1 Introduction

For over a century, ecologists have been interested in surveying wildlife populations to describe their distribution, abundance, dispersion, and temporal trends. Primates are no exception and much effort and many resources have been devoted to monitoring their populations over the past 40 years. Population parameters such as species distribution and abundance are not only important for examining ecosystem functioning, but also frequently form the basis for management decisions and provide the means to evaluate effectiveness of different conservation strategies (e.g. Nichols and Williams 2006; Stokes *et al.* 2010).

There has been a catastrophic decrease in global biodiversity in recent years (Koh *et al.* 2004), and the world's primate populations are declining across large parts of their range (IUCN 2014; Cotton *et al.*, Chapter 3, this volume). Threats to their survival include habitat loss and fragmentation resulting from agricultural conversion or extractive industry (Chapman *et al.* 2000; Rabanal *et al.* 2010; Wich *et al.* 2014), growing human populations that exert pressure on remaining primate habitat (Estrada and Coates-Estrada 1996; Campbell *et al.* 2008), hunting of primates for food and the pet trade (Bowen Jones and Pendry 1999; Fa *et al.* 1995, 2002; Rosen and Smith 2010), civil unrest (Hart and Mwinyihali 2001;

Campbell, G., Head, J., Junker, J. and Nekaris, K.A.I., *Primate abundance and distribution: background concepts and methods.*
In: *An Introduction to Primate Conservation.* Edited by: Serge A. Wich and Andrew J. Marshall, Oxford University Press (2016).

DOI 10.1093/acprof:oso/9780198703389.003.0006

Kalpers *et al.* 2003), and disease (Bermejo *et al.* 2006; Williams *et al.* 2008). Often these threats are not mutually exclusive. Logging and mining operations open up remote forests to hunters through the creation of road networks, and the resulting influx of people increases hunting pressure and causes further habitat loss through agriculture and habitation. Humans settling in new areas also increase the risk of disease transmission, which may kill thousands of primates in a short period of time (e.g. Bermejo *et al.* 2006). Similarly, civil unrest may result in increased hunting pressure (Plumptre *et al.* 2003), with automatic weapons and displaced individuals flooding into remote areas, disrupting traditional hunting and trading practices, as well as weakening taboos against hunting (Jones *et al.* 2008; Jimoh *et al.* 2012).

Immediate action is required if remaining primate populations are to be protected effectively. Collecting baseline information on the distribution, abundance, and trends of primate populations is a vital first step in the fight to protect them, because once established, population monitoring enables direct measurement of the effect of local threats and assessment of the effectiveness of conservation measures. Surveying primate populations is also important for identifying priority areas for their protection, developing conservation management strategies, mitigating threats, and balancing economic and conservation priorities. However, it has proven notoriously difficult to monitor primates precisely and accurately because many species live in remote, rough, and/or densely forested habitats, and some species are shy, elusive, or fast-moving, and thus difficult to detect and count.

Furthermore, estimates of species distribution, population size, and trends may vary considerably in accuracy and precision depending on the choice of survey method and the sampling objectives and design (Kuehl *et al.* 2008), among other things (e.g. sampling intensity, frequency, team capacity).

Non-invasive monitoring is an ideal solution for estimating primate population abundance and/or trends, as estimates may be obtained without the need for intrusive practices such as trapping or collaring. Furthermore, aside from the ethical concerns of live trapping primates, diseases may be transmitted from humans to primates (Wallis and Lee 1999; Köndgen *et al.* 2008), and invasive monitoring methods can therefore put at risk the very species they aim to protect. Non-invasive survey methods include interviews (Meijaard *et al.* 2011; Remis and Robinson 2012), index counts (Walsh *et al.* 2003; Cronin *et al.* 2013), direct (Teelen 2007; Waltert *et al.* 2008) and indirect (Tutin and Fernandez 1984; van Schaik *et al.* 2005) distance sampling methods, and capture–recapture methods (Arandjelovic *et al.* 2010, 2011; Chancellor *et al.* 2012; Head *et al.* 2013). To date, most primate surveys have relied on line/point and strip/quadrat transects, which count the number of direct or indirect signs of the species of interest in a defined area and then convert these encounters into estimates of abundance. However, these methods typically require site-specific information on parameters such as nest or faecal production and decay rate when indirect signs are counted (Kuehl *et al.* 2007a, 2008; Marshall and Meijaard 2009) and often do not offer enough precision to monitor small-scale changes in abundance (Plumptre 2000; Mathewson *et al.* 2008; Boyko and Marshall 2010). Capture–recapture surveys historically involve a portion of a population being captured, marked, and released, and then later another portion captured and the number of marked individuals within the sample counted, but in the field of primate research, surveys are non-invasive (and 'marked' individuals are captured indirectly through faecal or hair sampling, or photographs). Indirect capture–recapture surveys can provide precise information and enable researchers to follow specific individuals over the long term, but analysis can be expensive and sample collection labour intensive (e.g. genetic sampling; see Arandjelovic *et al.* (2010)). Capture–recapture using remote camera-traps (O'Connell *et al.* 2011) is proving to be an effective tool for monitoring terrestrial mammals, permitting precise density estimates, and enabling comprehensive population assessments including social and demographic structure and movement patterns (Di Bitetti *et al.* 2006; Head *et al.* 2013). Camera-traps have also been used to assess presence–absence of arboreal species (Kierulff *et al.* 2004; Olson *et al.* 2012) and their use as a monitoring tool for these species is in development (Nekaris, pers. comm.).

Monitoring of primates can be done at many different spatial scales—from site-level initiatives that monitor a few groups (Van Krunkelsven *et al.* 2000; Kumara *et al.* 2011), to protected area, regional (Stokes *et al.* 2010) and national surveys (Tutin and Fernandez 1984; Campbell *et al.* 2008;

Tweh *et al.* 2014), and continental trend assessments (Junker *et al.* 2012). Because of the advantages and disadvantages of the various survey techniques, different methods may be appropriate to survey the same species at different scales. For example, camera-traps can be used effectively at small spatial scales (e.g. up to approximately 200 km^2) to enable more precise estimates of density, but this method is not appropriate for covering much larger areas, and in such cases transects or genetic censuses are more cost effective. This chapter discusses the step-by-step process of selecting the appropriate survey method based on the survey objectives, primate species of interest, and survey area. We use three case studies to illustrate different non-invasive monitoring approaches used to address different survey objectives for various primate species.

6.2 Sampling objectives, survey methods, and design

The primary step before undertaking a survey is to gather all available information on the survey species and the area to be covered. This includes information on the behavioural and ecological characteristics of the target species (e.g. ranging patterns and habitat requirements), a detailed map of the survey area (e.g. vegetation, topographical, and hydrological maps), and data from any previous surveys that may have been conducted within the area (Ross and Reeve 2003). This information may help answer the following questions: why, what, how, where, and when to survey (Figure 6.1)? We address each question in more detail in Section 6.3. Two statistical indicators, accuracy and precision, will be referred to throughout this chapter and thus are defined in Section 6.2.1.

6.2.1 Sampling considerations

It is important to understand the difference between accuracy and precision when planning and conducting a primate survey. Accuracy measures the closeness of the computed value to the true value, for example how close the population estimate is to true population size. Accuracy of a population estimate can only be measured if true population size is known. Precision, on the other hand, is the closeness of repeated measures to one another and is especially important when estimating changes in population size over time.

When estimating population abundance, precision is often represented by the coefficient of variation (CV), which is based on the sample variance and the sample mean. This then sets the confidence limits of the estimate, representing a range of values that are expected to contain the true value at a given probability (usually 95% is used). For example, imagine that a line transect survey estimated the population of howler monkeys in a National Park at 129 individuals (confidence limits = 67–191). A total count that was conducted simultaneously yielded a population size of 95 individuals. Therefore, the population estimate obtained from the line transect survey was neither accurate nor precise, because it differed greatly from the true population size and its confidence limits were broad.

A primate survey should aim for high accuracy and precision, which can be obtained by selecting and adjusting several factors (e.g. survey design and sample size) appropriately when planning the survey. These factors will be discussed in more detail in the following sections.

6.3 Why survey?

The reason a survey is conducted is defined by its objectives. Once these are well formulated, the appropriate survey method may be selected and a suitable survey design developed. The following section describes four different survey objectives, namely to estimate: (1) distribution, (2) abundance, (3) population trend, and (4) population structure.

6.3.1 Distribution

Information on the spatial distribution of a species can help answer questions such as whether a species is restricted to a specific area or habitat type, or whether a species' occurrence is influenced by environmental or anthropogenic factors (e.g. Junker *et al.* 2012). Distribution of primate species may also vary temporally. For instance, seasonal variation in food abundance may

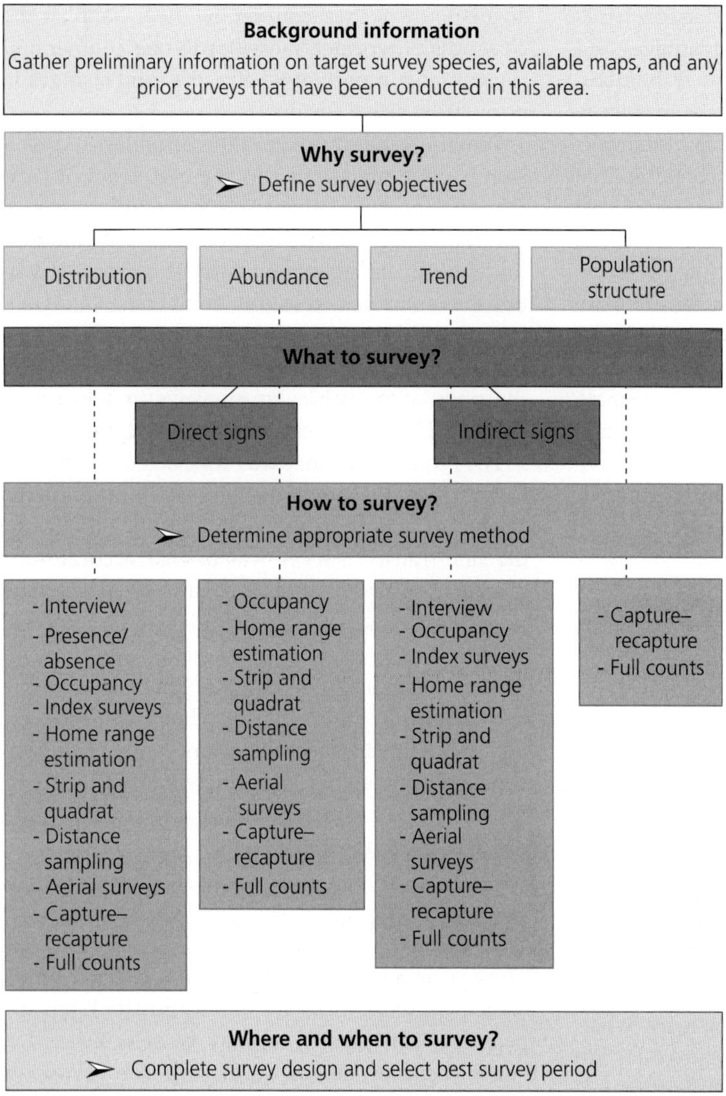

Figure 6.1 Steps to follow in order to plan and design a primate survey.

determine where a species is found at different times of the year (e.g. Grueter *et al.* 2012). Obtaining information on the spatial and temporal distribution of a species is typically less costly than estimating its abundance, because it usually focuses on obtaining presence–absence data over a predetermined area, and thus requires less survey effort, time, human resources, logistical support, and equipment.

6.3.2 Abundance

Different measures of abundance can be estimated: relative abundance, absolute abundance, and density. Relative abundance is an index of abundance, usually presented as an encounter rate of direct or indirect signs recorded per unit of distance (e.g. number of crowned sifaka groups seen per km walked) or per unit of time (e.g. number of black-and-white colobus groups heard per hour walked). Relative abundance is easily obtained and calculated, but caution should be taken when interpreting and comparing these results between sites or habitat types (for an example, see Section 6.2). Absolute abundance refers to the total number of individuals in the survey area, while density equals absolute abundance divided by the size of the survey area (e.g. number of individuals per km^2). Abundance estimates should always be accompanied by a confidence interval, which indicates the level of precision of the estimate. The precision of a once-off estimate is considered reliable at approximately 20% or lower (William *et al.* 2002), although precision may have to be higher when the objective is to detect subtle or relatively short-term population trends (see Section 6.3.3). It is often costly to estimate primate abundance because many primate species occur at low densities, and thus a greater survey effort is required to obtain a representative and large enough sample size to ensure an accurate and precise estimate.

6.3.3 Population trends

Information on population trends (i.e. increasing, stable, or decreasing population size over space or time) is important for informing and assessing the effectiveness of conservation management decisions. For example, if a heavily hunted primate population is decreasing over time, protection measures, such as anti-poaching patrols, may have to be strengthened to prevent further population decline. Monitoring programmes typically aim to estimate population trends over several years. Surveys that form part of such programmes need to be repeated over a sufficiently long period of time in order to distinguish natural fluctuations from human-induced population changes. The freely available software TRENDS (Gerrodette 1993) can help answer such questions as 'How many years are required between two consecutive surveys to detect a trend?' or 'Over how many years in total should surveys be conducted to detect a trend?' given that certain parameters are known (e.g. precision of the baseline estimate). More recently, hierarchical Bayesian models have been developed to estimate population trends, and have proven to be more precise and to allow for a better inclusion of uncertainty in the trend assessment (Kéry *et al.* 2009; Gitzen *et al.* 2012).

6.3.4 Population structure

Acquiring data on the structure of a population is more difficult, because it requires obtaining information on the age and sex classes of the individuals sampled. This may be tractable when surveying directly (e.g. using camera-traps or by direct encounters with primate groups), but is more challenging when surveying for indirect signs, because it is often difficult to assign age and sex class with confidence to these signs. Information on population structure can provide useful insights into the reproductive health of a population, immigration and emigration patterns, population trends, and population viability.

6.4 What to survey? Indirect vs direct signs

The type of information to be recorded during the survey will be influenced by the behaviour of the target species and its environment. Although it would be preferable to collect data on the individuals or primate groups themselves (i.e. direct signs), this is not always possible given that many primate species are shy, fast-moving, and consequently hard to detect, or live at such low densities that a massive sampling effort would be required to permit accurate density assessments. Furthermore, the habitat in which some species are found may be difficult to access, which should also be taken into consideration when planning a survey (Neilson *et al.* 2013). Therefore, a primate species that vocalizes frequently but is otherwise difficult to detect (e.g. gibbons living in rugged terrain) might be better surveyed by recording indirect signs of its presence (i.e. vocalizations). Indirect signs of primate presence may include nests, dung, vocalizations, and tracks and, depending on the type of signs recorded, auxiliary variables may be required to calculate population density, such as the production rate (i.e. the number of signs/cues produced per unit time), the decay rate (i.e. the time the sign/cue stays in the environment after its creation), as well as the proportion of the population that actually leaves the 'detectable' signs (Kuehl *et al.* 2008; Plumptre *et al.* 2013).

To date, most surveys of primates (except for great apes) have used direct counts (e.g. Harding 1984; Galat *et al.* 1994; Peres 1997; Bermejo 1999; Irwin *et al.* 2005; Plumptre and Cox 2006; Medhi *et al.* 2007; Linder and Oates 2011; Aquino *et al.* 2014). These are generally preferred over indirect counts because the latter have the disadvantage that density calculations may be biased by inaccurate measurement of the auxiliary variables (i.e. production and decay rates, and proportion of the population leaving the signs), which are notoriously difficult to obtain. For example, Kouakou *et al.* (2009) showed that chimpanzee densities could be either under- or overestimated by as much as 300% when using imprecise nest decay rates (see also Mathewson *et al.* (2008)). However, direct counts may also be biased by fast-moving primate species that are difficult to detect and count, which can result in the double-counting of some individuals while missing shy animals moving out of the survey area in response to the presence of observers (Buckland *et al.* 2001).

6.5 How to survey?

Once the survey objectives are clearly defined, the next step is to choose the appropriate survey method and develop a survey design to achieve these objectives. Here, again, it is crucial to consider the ecology of the species to be surveyed (e.g. temporal or spatial variation in distribution and density, behaviour) and the precision required from the survey (this depends largely on the study objectives, see Section 6.2) to yield an accurate and precise population estimate.

For example, if you plan to count monkey vocalizations, but females of the species you want to study vocalize considerably less often than the males, you will consistently undercount the number of monkeys in the population. Therefore, the density estimate would have a negative bias, resulting in low accuracy. Accuracy can also be increased by reducing errors in the way information is collected through the careful training of field assistants (e.g. taking accurate measurements, correctly identifying individuals, detecting all individuals present; White and Edwards (2000)).

To ensure sufficient precision of the estimate, a small-scale pilot study can be conducted to estimate the encounter rate of the species of interest and the area/number of samplers to be surveyed (Buckland *et al.* 2001). It is also important to consider the size and physical properties of the study area, and available human resources and financial, logistical, and time constraints. For example, if the study area is very mountainous, rocky, or steep, line transects might not be feasible and point transects or index counts may be more appropriate. The following section provides a brief summary of different survey methods and their objectives, advantages, and challenges.

At the end of each section, we briefly state the context in which each method would be appropriate and provide a summary table (Table 6.1) that can be used to determine which one of the methods to use under which circumstances. We encourage readers to consult the literature for more detailed information on each specific method.

6.5.1 Interview methods

Interviews with hunters, local villagers, and/or officials from in-country governmental organizations have been successfully used to gain information on perceived primate presence and threats, distribution and encounter rates, approximate off-take rates, and population trends (Kuehl *et al.* 2008; Medhi *et al.* 2007; Starr *et al.* 2011). For example, interviews of nearly 7,000 people in 687 villages in Kalimantan, Indonesia, provided estimates of relative presence and encounter rates of orang-utans and allowed for the quantification of the main factors that threaten their survival (Meijaard *et al.* 2011a). Another study in the same area (Meijaard *et al.* 2011b) used interviews to collect information on orang-utan conflicts and killings. The study mapped areas of high conflict potential and showed that killing rates were substantially higher than previously thought—information that is key to informing management decisions on the ground. In Sierra Leone, Brncic and colleagues (2010) conducted interviews across the country to guide survey effort for a subsequent

Table 6.1 Summary of the different survey methods and for which conditions and/or species they are suitable (indicated by 'X'). When a cell is empty, this does not necessarily mean that the method cannot be used under these circumstances, but rather that they are less suitable in these situations.

			Interview method	Presence–absence sampling	Occupancy method	Index surveys	Home range estimation	Strip & quadrat sampling	Line transects (Distance)	Point transects (Distance)	Aerial surveys	Capture-recapture method	Full Counts
Objective	Distribution		X	X	X	X	X	X	X	X	X	X	X
	Abundance				X		X	X	X	X	X	X	X
	Density						X	X	X	X	X	X	X
	Trends		X		X	X	X	X	X	X	X	X	X
Study area	Size	Small			X			X	X			X	X
		Large	X	X		X	X		X	X	X		
	Habitat	Open	X	X	X	X	X	X	X		X	X	X
		Closed	X	X	X	X	X	X	X	X	X	X	
Species characteristics	Appropriate density	Low	X			X	X		X	X		X	X
		High	X	X	X	X	X	X	X		X	X	
	Ranging pattern	Clumped	X	X		X	X		X	X		X	X
		Not clumped	X	X	X	X		X	X	X	X	X	X
	Behaviour	Cryptic	X	X	X	X	X	X	X	X		X	X
		Conspicuous	X	X	X	X	X	X	X		X	X	X
		Slow moving	X	X	X	X	X	X	X	X	X	X	X
		Fast moving	X	X	X		X				X	X	
		Diurnal	X	X	X	X	X	X	X	X	X	X	X
		Nocturnal	X	X	X	X	X		X			X	
		Terrestrial/semi-arboreal	X	X	X	X	X	X	X	X	X	X	X
		Arboreal	X	X	X	X		X	X	X		X	
	Detection method	Direct		X	X	X	X	X	X		X	X	X
		Nests/dung		X	X	X	X	X	X	X	X	X	X
		Indirect vocalizations		X		X	X	X				X	

continued

Table 6.1 *Continued*

			Interview method	Presence–absence sampling	Occupancy method	Index surveys	Home range estimation	Strip & quadrat sampling	Line transects (Distance)	Point transects (Distance)	Aerial surveys	Capture-recapture method	Full Counts
Planning & logistics	Financial resources	Low cost	X	X	X	X		X		X			
		High cost					X		X		X	X	X
	Human resources	Low human investment	X	X	X	X	X				X	X	
		High human investment						X	X	X			X
	Equipment	Technically not challenging	X	X		X		X	X	X			X
		Technically challenging			X		X					X	
	Time	Time efficient	X	X	X	X				X	X		
		Time consum-ing					X	X	X			X	X
	Type of survey	Single-species only					X	X		X	X	X	X
		Multiple spe-cies	X	X	X	X			X				

nationwide chimpanzee population survey using transect methods. The information collected during interviews allowed for the stratification of the sample area into predicted high and low chimpanzee densities. Interviews can be mailed-in questionnaires prepared in advance and sent to interviewees, or face-to-face interviews conducted at randomly or systematically selected sites (Kuehl *et al.* 2008).

Context: Interviews are suitable where rapid information on species distribution over a relatively large area is needed. Interviews are also suitable for collecting pilot data about the study area (e.g. different land-use types, human impact, threats, crude species densities).

6.5.2 Presence–absence sampling

This method is used to record presence or absence of the primate species. Occurrences are recorded in predefined sampling units where the interest is not on the numbers of animals (or their signs), but rather on the numbers of sample units that are occupied by animals (Royle and Nichols 2003). It is not necessary to count the total number of animals observed in sample units, neither is it necessary to identify individuals as in the capture–recapture method. Therefore, presence–absence surveys are relatively easy and efficient to conduct (Royle and Nichols 2003). Various methods exist to obtain presences (and absences), including camera or video devices, automated audio recording units, direct observations, interviews and expert opinion, aerial surveys, and carrion fly-derived DNA surveys (Calvignac-Spencer *et al.* 2013).

The fraction of sampling units in which presences are recorded can later be used to describe species occurrence, range, distribution, and habitat selection (Kuehl *et al.* 2008; Karanth *et al.* 2010). Numerous software packages have been developed to model species occurrence using presence-only data or presence data in addition to some form of absence data, ranging from climatic envelopes and logistic regression, to boosted regression trees and multivariate regression splines (see Elith *et al.* (2006) for a comparison of the various modelling methods). One of the most frequently used methods to model species distributions is called *Maxent* (Phillips *et al.* 2006). *Maxent*, which uses a maximum-entropy machine learning approach, takes as input a set of layers or environmental variables (e.g. elevation, precipitation) and a set of georeferenced occurrence locations to produce a model of the distribution of the given species. *Maxent* has previously been used to successfully predict primate distributions at regional (e.g. Thorn *et al.* 2008; Hickey *et al.* 2013; Chetan *et al.* 2014), national (e.g. Torres *et al.* 2010), and landscape scales (e.g. Lahoz-Monfort *et al.* 2010). Factors that may bias species distribution models using this approach include the non-random distribution of presence points (e.g. mostly within protected areas, or within more accessible areas such as close to roads), a lack of comparable environmental variables, lack of large-scale datasets, or low resolution of spatial covariates (e.g. Junker *et al.* 2012; Kramer-Schadt *et al.* 2013).

Context: Presence–absence sampling is economical, time efficient, and can be used to estimate species distribution over large areas. Numerous methods can be used to collect presence–absence information, which can be either based on direct or indirect signs. The method is especially suitable for multispecies distribution surveys. This method can be particularly suitable for surveying areas that are difficult to access.

6.5.3 Occupancy methods

Occupancy methods, which are an extension of presence–absence sampling, are used to assess distribution (Michalski and Peres 2005; Karanth *et al.* 2010; Coudrat and Nekaris 2013), but may also be used to estimate abundance (Royle and Nichols 2003; Stanley and Royle 2005), density (Rowcliffe *et al.* 2008), and trends (Cassey *et al.* 2007). To estimate abundance from presence–absence data, the occupancy method assumes that there is a correlation between the detection probability of the species of interest (measured by either direct or indirect signs of its presence) and species abundance (MacKenzie and Royle 2005). As abundance may vary considerably among sites and over time (Rowcliffe *et al.* 2008), this heterogeneity in abundance needs to be accounted for during the estimation process (i.e. see Royle and Nichols 2003).

Additionally, occupancy methods that use trapping rates to provide density estimates (e.g. with camera-traps or acoustic monitoring devices) require information on trap-related parameters, such as the radius and angle of detection, as well as animal-related parameters, including speed of movement (i.e. day range) and group size of the focal species (Rowcliffe *et al.* 2008). While modelling trap-related parameters is relatively straightforward, estimating animal-related parameters is more problematic because they are generally difficult to measure without bias. For example, animal surveys frequently suffer from the fact that smaller groups are harder to detect (thus violating the detection assumption), leading to overestimation of group size (Rowcliffe *et al.* 2008). Another important assumption of the detection model is that animals move independently of the detection devices, which is violated if the placement of cameras or acoustic devices either avoids or targets focal species, or if the focal animals themselves avoid the monitoring devices (Rowcliffe *et al.* 2008).

When spatial variation in density is influenced by recognizable zonation on the ground, such as habitat type, stratification by these zones will be desirable to improve the precision of estimates (Rowcliffe *et al.* 2008). The method also assumes that the population surveyed is closed and that no migration occurs.

***Context*:** The occupancy method can be used to obtain accurate abundance, density, and trend estimates by using repeated presence–absence surveys in predefined sampling units. The method should be used only if it is feasible to locate these units randomly across the study area, and if it is possible to accurately measure the speed of animal movement and group size of the focal species. This method can be particularly suitable for surveying areas that are difficult to access.

6.5.4 Index surveys

Index sampling produces encounter rates, where it is assumed that species encounter rates are proportional to actual densities (Plumptre and Cox 2006). However, detection probability, which is influenced by visibility, habitat type, production, and decay rates of presence signs, may also influence encounter rates. Possible sources of variation (other than animal density) should be carefully considered and, if possible, accounted for. For example, if information on habitat differences is available, then encounter rates should be calculated separately for each habitat type. Furthermore, plotting the number of observations per unit distance and sampling until the plotted curve levels out will help to interpret and minimize potential variation in visibility (Kuehl *et al.* 2008). To obtain data on encounter rates, observers usually conduct 'reconnaissance walks' ('recces'), during which they walk in a predetermined direction that they may temporarily deviate from to choose the path of least resistance (White and Edwards 2000; Plumptre and Cox 2006). Here, different approaches may be adopted: (1) observers record all individuals or their signs within a strip of predetermined width without measuring perpendicular distances; (2) observers record all objects detected and measure perpendicular distances to observations; and (3) objects at all distances from the recce line are recorded, but without measuring perpendicular distances to these observations.

In addition to information on the spatial distribution of the focal species and approximate densities, approaches (1) and (2) may also yield abundance estimates and account for habitat differences and associated differences in detection probabilities, while approach (3) cannot be used to estimate abundance. However, it should be kept in mind that the area surveyed during recces is biased towards more accessible areas and that variation in recce encounter rates is very likely to result from a variety of sources other than animal density (Kuehl *et al.* 2008). Recces can easily be combined with other survey methods and are frequently used during pilot studies (e.g. Arandjelovic *et al.* 2010). Repeat index surveys may also provide data on population trends (e.g. Campbell *et al.* 2008).

***Context*:** Index surveys are a good method for obtaining relative abundance (and trends therein) over relatively large areas, in a relatively short amount of time, and with limited financial resources. The method is frequently used in pilot studies to allow surveyors to become acquainted with the study area (vegetation, topography) and to familiarize themselves with detection of different signs. Recces can easily be combined with other methods to provide additional information (e.g. conducted in between transects or listening posts).

6.5.5 Home range estimation

When data are available on home range size of the target species, average number of individuals in a group, and the extent to which home ranges overlap between groups, population size can be estimated (e.g. Fashing and Cords 2000; Head *et al.* 2013). Data on home range size, group/community size, and home range overlap can be collected through direct (e.g. Fashing and Cords 2000; Bermejo *et al.* 2006) or indirect (e.g. camera images, genetic markers; e.g. Head *et al.* (2013)) observations of the primate species of interest. To calculate home range size, several different approaches may be used. Home range estimators commonly used in the literature include minimum convex polygons (MCP) (Mohr 1947), kernel density estimation (KDE) (Worton 1987, 1989), and clusters (Kenward 1987). The MCP method determines the smallest polygon encompassing all external locations of the dataset. While this method is affected neither by animal movements within the home range nor by the statistical distribution of the dataset, it fails to provide information on the variation in spatial and temporal home range use by the animals, and relies solely on outlying points to anchor the corners of the home range, which may considerably inflate home range size. The cluster estimator also uses polygons to delineate the home range (Kenward 1987). Here, polygons are gradually built from clusters that provide the minimal sum of nearest-neighbour distances. Because cluster estimators allow the delineation of multiple core areas, they are more efficient at describing areas used by an animal exhibiting heterogeneous range use (Girard *et al.* 2002). KDE creates contour lines of the intensity of space utilization of individuals by calculating the mean influence of data points at grid intersections, where each contour contains a fixed percentage (e.g. 95%) of the utilization density, suggestive of the amount of time that the animal spends within these areas (Hemson *et al.* 2005).

Differences in accuracy and/or precision of home range estimates may not only arise from the type of estimator used (e.g. Huck *et al.* 2008; Nielsen *et al.* 2008; Grüter *et al.* 2009), but data reliability also depends on sampling intensity, more specifically the number of individuals sampled and the number of location data sampled per individual (Börger *et al.* 2006). Potential bias introduced by spatial and temporal differences in space use (Cushman *et al.* 2005), as well as individual behavioural differences (Morales and Ellner 2002), may be minimized by collecting repeated measures of home range size over time on individual animals (Börger *et al.* 2006).

***Context*:** Estimates of population size can be obtained from home range estimation when home range size of the target species, average group size, and home range overlap can be reliably estimated. Although this method allows for estimating abundance over large areas, collecting accurate data on home range and group size may be costly and time consuming.

6.5.6 Strip and quadrat sampling

The principle of strip and quadrat (plot) sampling is that all animals of interest or their signs (hereafter 'objects') are counted within a series of randomly placed strips (Figure 6.2 c) or quadrats (Figure 6.2d) of predetermined size. For example, a nest survey consisting of 21 plots of 0.2 ha in size, located 200 m apart, aiming to estimate density in the Mawas Reserve in Borneo, provided estimates close to the estimated true density in the area (van Schaik *et al.* 2005). Furthermore, plot counts performed better than line transects, where the latter underestimated true orang-utan densities (van Schaik *et al.* 2005). However, confidence intervals for the density estimate based on plot counts were relatively higher than those for line transects because of the larger number of nests sampled on the line transects. In another study on five primate species in Northern Columbia (Green 1978), fewer animals were detected in quadrats (these were actually circles covering 0.32 km^2) than on strip transects, resulting in an underestimate of true density for all species.

The difference between strips and quadrats is their shape, where narrow strips may be easier and faster to survey than quadrats (Buckland *et al.* 2001). The critical assumption of this method is that all animals or their signs are detected within the sampling units and that none from outside are included. When objects occur in clusters, only those objects that are located within the strip are counted and objects outside the strip, even though belonging to the same cluster, have to be ignored (Buckland *et al.* 2001). The total number of detections within the sample area is then extrapolated to the total study area to obtain abundance estimates. 'Lure strip transects' are a modification of strip transects and can be used when animals can be 'lured in' by playbacks of vocalizations (e.g. Savage *et al.* 2010). In this study, strips were 1.5 km long and 200 m wide (spaced 400 m apart). Two field teams traversed the long edges walking in parallel and undertaking near-continuous playbacks to ensure that all animals within the transect were detected and counted. Upon detection of cotton-top tamarins, observers recorded the location along the transect, the number of animals and their age class, and the direction from which the group approached. This enabled those groups arriving from outside of the transect strip to be excluded from analysis.

When using this method, it is important to carefully consider an appropriate strip width. There may be a trade-off between ensuring that the strips represent all habitats and conditions (e.g. using many narrow strips) and sample size (e.g. if strips are too narrow then many animals may fall outside the strip, and the resulting sample size may not be large enough to assess abundance).

Context: Strip or quadrat sampling is a suitable method to estimate distribution, density, abundance, and trends of primate species that leave behind conspicuous signs or that are relatively slow moving. Strip transects are most efficient when species occur at moderate to high population densities, and where populations are not heavily clumped. Although this method can be used to collect data over large areas, it may be relatively costly, and time and labour intensive.

6.5.7 Distance sampling methods

The most frequently used survey method for diurnal primates is distance sampling. For distance sampling (Buckland *et al.* 2001, 2004), observers detect direct or indirect signs of the primate species on a series of either line (Figure 6.2a, Figure 6.3) or point transects (Figure 6.2b). Line and strip transects have also been used to survey nocturnal primates, where observers use a light source to detect individuals (e.g. Weisenseel *et al.* 1993; Blackham 2004; Kumara and Radhakrishna 2013). This is possible because all nocturnal primates, except owl monkeys and tarsiers, have a reflective layer at the back of their eye, creating 'eyeshine' when flashed with light (Setchell and Curtis 2003).

In contrast to strip transect sampling, where the observer travels down the centreline of a long narrow strip counting all objects within the strip, line transect sampling entails travelling along a line recording all detected objects. One major difference between these two methods is that line transect sampling is not based on the critical assumption that all objects within a specific area are detected, as is the case for strip transect sampling. Rather, the number of objects detected is expected to decrease as visibility decreases, that is, with distance from the line (Buckland *et al.* 2001). Therefore, every time an object is detected, observers measure the shortest (i.e. perpendicular) distance from the location of the detected object to the survey transect line or point.

In the case of primates, line transects are much more frequently used than point transects (Buckland *et al.* 2010). The main problem with point transects is that they are less efficient than line transects for species that occur at low densities or that are cryptic. An extension of this method, which entails luring animals to the point (i.e. by playing calls that the primate will respond to), can be used to overcome this problem (Buckland *et al.* 2006). Here, not the animals themselves, but their call will be recorded. In addition, point transects are less robust to object movement, whether this movement is in response to the observer moving to the point or not (Buckland *et al.* 2001).

Line transects

If strips of width $2w$ and total length L are surveyed, an area of size $a = 2wL$ is censused. In line transect sampling, however, only a proportion (P_a) of objects in the area a is detected. To estimate P_a, one needs to measure perpendicular distance from the transect line to each observed object. A detection function can then be estimated from the frequency distribution of measured distances of observed objects to the line, where P_a is the ratio of the area under the detection function (i.e. area under the curve) to the total area (i.e. area of the rectangle). Once we

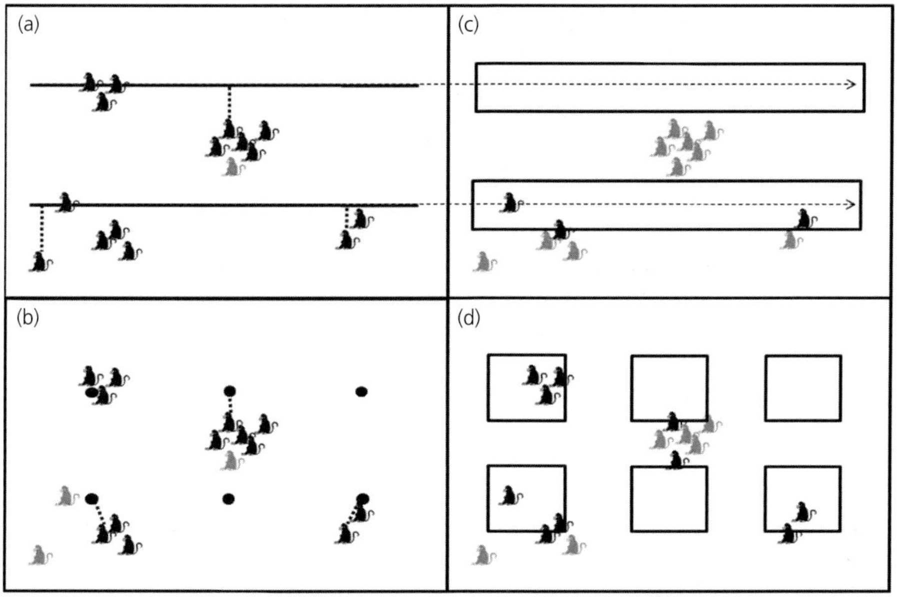

Figure 6.2 Schematic diagram showing four different sampling approaches for surveying the same non-randomly distributed primate population: (a) line transect distance sampling, (b) point transect distance sampling, (c) strip transect sampling, and (d) quadrat sampling. Solid lines or points represent transect locations and stippled lines represent measured distances. Animals in grey were not counted, either because they were not detected or because they were outside the sampling area. Note that when animals occur in clusters, distances can also be measured to the centre of the cluster, instead of each individual in the cluster.

know P_a, we can estimate density (D) by the number of objects detected (n) divided by the surveyed area ($2wL$), in which a certain proportion of objects could be detected (P_a), or $D= n/2wLP_a$ (Buckland *et al.* 2001). Alternatively, we can estimate density if we know the width of the strip within which the number of objects missed is equal to the number seen beyond this distance (μ), also called the 'effective strip half-width'. Then density can be estimated by the number of objects detected divided by the effective strip width (2μ) multiplied by the total length L, or $D= n/2\mu L$ (Buckland *et al.* 2001).

Context: Line transect distance sampling is a suitable method to estimate distribution, density, abundance, and trends of species that leave behind conspicuous signs or that are relatively slow moving. This method is more efficient than strip or quadrat sampling when surveying species that are more sparsely distributed. Although this method can be used to collect data over large areas, it may be relatively costly, and time and labour intensive.

Point transects

Point transects can be thought of as line transects of zero length. As with line transects, a series of points is surveyed during which the radial distances between detected objects and the centrepoint of each transect are measured (Buckland *et al.* 2001). On point transects, the area at the centrepoint should have a detection probability of one (note: for line transects, the area on/near the line should have a detection probability of one), where the probability of detecting objects (P_a) decreases with distance from the point. Thus, if we have a series of point transects (k), each comprising a circular plot of radius w, the sampled area is $k\pi w^2$. As P_a is the expected proportion of objects detected within radius w, then density (D) can be estimated as the number of objects detected (n) divided by the surveyed area ($k\pi w^2$), in which a certain proportion of objects could be detected (P_a), or $D = n/k\pi w^2 P_a$ (Buckland *et al.* 2001). The 'effective radius' p, comparable to the effective

Figure 6.3 Liberian survey team on a line transect during the nationwide chimpanzee and large mammal survey recording chimpanzee nests, perpendicular distances from individual nests to the transect line (which is where the white topofil thread is visible on the photograph), nest decay stage, the first direct or indirect sign of other large mammals including all primates, and all signs of human presence. Note how the individual team members each have their place on the line and a specific task apart from looking for the target species (i.e. from left to right: first position: recording data; second position: marking locations with the GPS and making sure that the compass bearing does not change; third position: ensuring that the compass holder walks on the correct bearing and clearing the path where necessary; fourth position: leading the team on the bearing to ensure straightness of the line). Two additional people who are not depicted on this picture always walked on each side of the team member at the 2nd position at a distance of about 5 m to look for chimpanzee nests and primates right above the line. Copyright: Jessica Junker.

strip width, is the radius for which as many objects beyond p are detected as are missed within p (p can be estimated from the radial distances between detected objects and the centrepoint of the transect). Then, density can be estimated by the number of objects detected in the sampled area, where the known radius is replaced by the effective radius, or $D = n/k\pi p^2$ (Buckland *et al.* 2001).

***Context*:** Point transect distance sampling can be used to estimate distribution, density, abundance, and trends of species that regularly vocalize or those that may respond to playback calls. Point transects are less robust to animal movement than line transects and less efficient for species that occur at low densities and that are cryptic.

Assumptions

To obtain accurate and precise results, it is important that (1) transects are located randomly with respect to the distribution of the study animals, (2) all animals on/near the line/point are detected, (3) distance from the line/point to the animal is measured accurately to the initial location of detection, and (4) sightings of animals and their signs are not influenced by previous detections (Buckland *et al.* 2001).

Sample size

A minimum of between 60 and 80 observations is required to reliably estimate detection probability (Buckland *et al.* 2001). If the population is clustered, as is the case for many primate species that live in social groups, sample size (and thus survey effort) should usually be larger to yield a similar precision for the abundance estimate. Additionally, if the size of individual clusters can be expected to vary greatly, such as in species that exhibit fission–fusion dynamics (e.g. baboons, spider monkeys, uakari monkeys), then sample size should be even larger. If individuals form large aggregations more frequently at certain times of the year (e.g. Guinea baboons), then surveys should avoid those times (for more details, see Buckland *et al.* (2001)).

Individuals occurring in clusters

If the focal species lives in clusters (e.g. social groups), then perpendicular distance to the middle of the group at the point of detection should be measured instead (Buckland *et al.* 2001). In addition, cluster size should be recorded accurately. However, when conducting direct counts of primates that can move readily away from the point where they were first detected, cluster size is usually estimated. Even estimating cluster size and cluster centre can be extremely difficult in practice and usually requires good visibility and tight, habituated groups (Marshall *et al.* 2008). When the group centre is hard to estimate, it is preferable to estimate the centre based on only those individuals whose initial location could be measured during a transect walk (Marshall *et al.* 2008). More than one observer may be required to increase the accuracy of the estimate of cluster size and centre. Furthermore, photographing clusters (e.g. such as during aerial surveys) will allow for accurate estimation of cluster size (Buckland *et al.* 2001).

DISTANCE software

Assuming that the basic assumptions are met in the design and execution of field surveys, the free software package DISTANCE (Thomas *et al.* 2010) is normally used to analyse survey data (i.e. to fit a detection function, to estimate density and sampling variance). Performance/fit of each of four models (uniform, half-normal, hazard-rate, negative exponential) used by DISTANCE to estimate the detection function (to determine detection probability) is evaluated by comparing AIC values. Spatial tools in DISTANCE can also be used to develop an appropriate survey design.

Spatial models

Recently developed methods that combine distance sampling with spatial modelling techniques can be useful to investigate relationships between species distribution and environmental covariates, reliably estimate abundances, and predict population patterns (Fiske and Chandler 2011; Marshall *et al.* 2014). For more details on these methods, see Miller *et al.* (2013).

6.5.8 Aerial survey

Fixed-wing aircrafts, helicopters, microlights, and drones can provide aerial survey platforms to census primate species that live in open habitats, such as savannah baboons (e.g. Wilson 1979; Chase 2011), or those that leave conspicuous signs, such as orang-utans that build nests high up in the canopy (Ancrenaz *et al.* 2004, 2005; Koh and Wich 2012). This method can be used for recording presence-only data to estimate species distribution, or, when collected within strips or near lines, to estimate species abundance.

When used in the context of line transect sampling, perpendicular distances can be measured either by directly flying over the object and recording its geographical position in relation to the line, or by measuring the angle of declination with the aid of a clinometer and computing perpendicular distance from a known altitude (Buckland *et al.* 2001).

Surveying from an aircraft is advantageous in that large areas can be surveyed in a relatively short period of time and that they require lower human investment than typical ground surveys (Buckland *et al.* 2001; Kuehl *et al.* 2008). An aerial view may also provide better visibility through vegetation and a means of surveying remote areas that are not readily accessible from the ground (Kuehl *et al.* 2008). In addition, problems associated with animal movement may be avoided, because the aircraft speed is typically greater than any target animal (Buckland *et al.* 2001). However, aerial surveys present a number of problems with regard to satisfying the assumptions of distance sampling, especially the assumption that all objects on the line are detected. This is because (1) the flight altitude may put too large a distance between the object and the observer making it difficult to detect small objects, (2) the type of aircraft used may not provide downward visibility, (3) the vegetation may obstruct visibility, and (4) flight speed may be too high to detect and identify objects and record data (Buckland *et al.* 2001). Moreover, species that occur at high local densities can be particularly difficult to survey from the air and surveys of only one side of the transect and/or using multiple observers may cope with this problem (Buckland *et al.* 2001).

Helicopters, although more easily manoeuvrable than fixed-wing aircrafts or microlights, may cause a lot of noise and stress to the animals on the ground, and in some areas refuelling may be challenging. Drones can overcome many of these problems, as they fly relatively slowly, can be manoeuvred easily, make little noise, and provide unrestricted forward and downward visibility with the aid of (video-) cameras, which also allow for accurate population estimation even when species locally occur at high densities because photographs can be consulted even after the survey is completed. The use of drones for conservation monitoring purposes is currently still under development (e.g. Koh and Wich 2012; van Andel *et al.* 2015; Wich 2015), but could lead to significant savings in terms of time, manpower, and financial resources. For more details on aerial search and survey protocols and distance measurements, see Buckland *et al.* (2001).

Context: Aerial surveys can be used effectively to survey relatively conspicuous primate species in open habitat or primate nests in the canopy of forested or savannah habitats. Aerial surveys are suitable for covering large areas in a short period of time. Although relatively costly, they require lower human investment than ground surveys.

6.5.9 Capture–recapture methods

The capture–recapture is a sampling method in which the same animals are 'captured' in the same area on multiple sampling occasions (Ross and Reeve 2003) to estimate their abundance. On the first sampling occasion, individuals are marked and released, and then on the following sampling occasion (only one additional occasion is necessary, although more are possible) marked animals (i.e. previous captures) are recorded and released along with unmarked individuals. Based on the proportion of initially marked individuals in the second sampling occasion, a capture probability is calculated, which in turn is used to estimate population size (Kuehl *et al.* 2008). For example, assume that during a first sampling occasion a total of 25 individuals (n) are trapped and identified. In the second round, we catch 30 individuals (n_2), of which 10 individuals (m_2) are individuals that were captured in the first round. In this case, we can expect that the ratio of animals caught during the second round (m_2), to the total number of animals captured on the second round n_2, equals the ratio of the number of animals available for capture (or number caught during the first round) (n) to the total population (N_{total}). Thus, $N_{total} = n^*n_2/m_2$. Applying this to our example would give us a total population estimate of 75 individuals.

In cases where it is impractical and/or unethical to physically capture wild-living primates and mark them, less invasive methods such as remote camera-traps (Figure 6.4 and 6.5) can be used for species that can be individually recognized visually, automatic (audio) recording units for species that are acoustically distinct, or genetic material extracted from hair or faeces (Constable *et al.* 1995; Arandjelovic *et al.* 2010). Here, 'captures' represent pictures or video recordings, acoustic recordings, or organic samples, respectively. Instead of physically marking individuals, their pictures (e.g. Head *et al.* 2013), calls, or genotypes extracted from the samples (e.g. Guschanski *et al.* 2009; Arandjelovic *et al.* 2011) are identified and recorded instead. The proportion of identified (i.e. 'recaptured') individuals can then be calculated from the second sampling occasion.

Context: Although technically challenging in some environments, the capture–recapture method is a cost-efficient method to obtain information on the distribution, density, abundance, trends of species, or their signs in relatively small areas. Additionally, video and genetic data can provide information on the species' population structure. The method is not suitable for arboreal primates for which genetic samples are difficult to obtain.

6.5.10 Full counts

During full counts, all individuals in a population are detected and counted. For sweep surveys, individuals do not need to be identified. Observers simply form a line, move in the same direction, and record all animals (or their signs) in their path (e.g. McNeilage *et al.* 2001, 2006). Alternatively, researchers may count individuals that are habituated to their presence at long-term study sites (e.g. McNeilage *et al.* 2006). Known individuals may also be counted from 'observation platforms' that are under permanent human or camera surveillance and which are located in areas that are frequently visited by the animals (e.g. Gatti *et al.* 2004). To ensure that no individuals in the population have been missed, one can plot the accumulative number of individuals counted over time. At the point in time at which the regression curve flattens out, one can assume that most of the population has been surveyed (Levin *et al.* 2009).

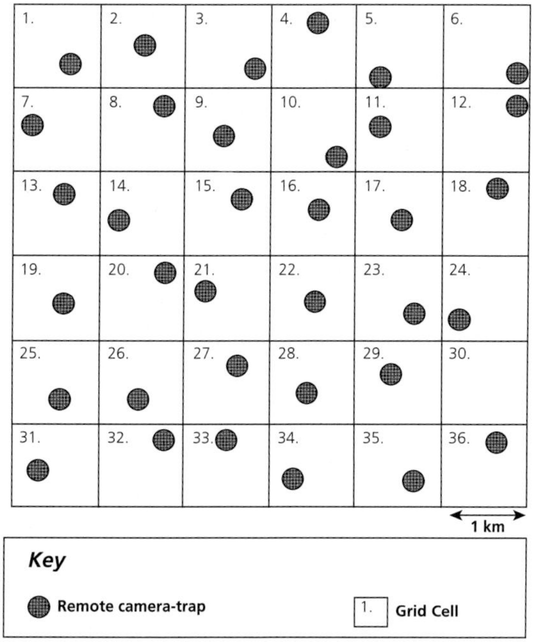

Figure 6.4 Example of a survey design for a remote camera-trapping study. Here cameras have been placed systematically with one camera per 1 km grid cell, but within each grid cell cameras have been placed optimally (e.g. along known animal trails) to increase capture probability.

Figure 6.5 A remote video camera-trap, deployed in Loango National Park, Gabon. The camera had been recently installed and so temporarily camouflaged with leaves in order to reduce the fear response from wildlife encountering it for the first time. Once animals have become accustomed to the cameras, the leaves are removed. Copyright: Josephine Head.

The type of full counts described above can be extremely time consuming, are less practical for large study areas or for counting individuals with large and overlapping home ranges, and are therefore much less frequently used to survey primates than other methods. However, full counts conducted with the aid of fixed-wing aircrafts or helicopters can be used to survey large areas in a relatively short amount of time (Buckland *et al.* 2001). This method could permit rough estimation of population size for species that occur in open habitat, such as baboons (Ross and Reeve 2003).

Context: Full counts typically yield accurate information on species distribution, density, abundance, trends, and population structure. Full counts can be costly and time consuming and are only feasible in small areas.

6.6 Survey design

It is easy to ask: 'how many white-faced capuchins live in Cahuita National Park in Costa Rica?' An exact answer to this question, however, is very difficult to obtain. Instead, numbers are usually estimated using sampling units. The principle behind sampling is to estimate population size by counting the number of animals within a smaller sample area. This number is then extrapolated to the larger study area. Sampling methods assume that (1) all animals in the sample area are detected and accurately counted, and (2) animals are sampled at random. To satisfy the first assumption, all portions of the study area should have an equal probability of being included in the sample. This assumption is frequently ignored where survey data tend to spatially cluster in and around protected areas or in areas that are more easily accessible, such as along roads, paths, rivers, valleys, and so on (e.g. Fowler *et al.* 1989; Sha *et al.* 2008; Bearder and Karlsson 2009).

The second assumption can only be achieved if the study area is sampled randomly, so any form of subjective selection of sampling units should be avoided at all costs. In addition, a sufficiently large sample size will help minimize variance and maximize confidence in the results, and a minimum of 10–20 samplers should be used in any survey (Buckland *et al.* 2001).

To determine survey effort for point or line transects, a pilot study or data collected in the same area during previous surveys can provide valuable information (Maisels and Aba'a 2008). For example, let us assume that 20 objects (n_0) were detected over the course of a line transect of a total length of 5 km (L_0). To obtain a rough estimate of the line length (L) and thus sample size required to achieve a precision in the density estimator of, for example, 10%, one can use the following equation: $L= L_0/n_0=$ 75.0 km (Buckland *et al.* 2001). Considering that at least 10–20 samplers are needed to minimize variance (Buckland *et al.* 2001), a design including a total of 30 transects, each 2.5 km in length, could be developed.

Another aspect to consider when designing a survey is the placement of the sampling units (consisting of transects, interview locations, trapping stations, etc.; see Figure 6.6 for an example of a survey design developed for a camera-trapping project). Ideally, a primate survey would include a sufficiently large number of sampling units to ensure widespread and high coverage of the survey area (Figure 6.6a). However, it is often impossible to survey intensively on a broad scale because of financial, logistical, and time constraints. A sampling design where transects are located randomly across space with the same orientation (e.g. north–south; Figure 6.6b) may be more feasible, but analysis can be hampered by insufficient primate observations on transects where animal densities are low, and the resulting estimate may not be representative of the survey area. Here, a systematic grid of lines (or points) with a random starting position (Figure 6.6 c) or a stratified design (Figure 6.6d) may be more appropriate (Maisels and Aba'a 2008). Stratified random sampling can be used to improve precision of the estimate where densities are expected to vary greatly across the study area (i.e. get density information by consulting the literature, collecting pilot data, acquiring expert knowledge). This may be useful, for instance, when the survey area comprises different habitat types (Ross and Reeve 2003), when parts of the survey area are managed by different land-use strategies (e.g. Stokes *et al.* 2010), and/or are exposed to varying amounts of human impact. For example, Brncic *et al.* (2010) conducted a

Figure 6.6 Sooty mangabey (*Cercocebus atys atys*) photographed by a camera-trap within Taï National Park, Côte d'Ivoire. Copyright: Genevieve Campbell.

nationwide chimpanzee survey in Sierra Leone after confirming their countrywide distribution with a pilot interview survey. Because randomly placed transects alone would have been unlikely to obtain enough observations of chimpanzee nests to produce a reliable abundance estimate, each systematically placed 'survey block' was stratified into low and high chimpanzee density based on information gathered on site. This allowed for transect effort to be targeted to increase sampling efficiency and at the same time reduce the number of transects with zero detections.

Placement of sampling units may also be influenced by topography and vegetation, and will eventually represent a balance between survey effort, time, and financial constraints. When planning a survey design, it is therefore also important to consider available manpower, equipment, and time and money needed to complete the survey. Unfortunately, for most surveys, only limited budgets are available, resulting in a trade-off between survey costs and data quality. However, if sampling procedure does not follow a well-developed sampling design and field protocol to ensure sufficient data quality, statistical analysis of the data and the interpretation of the results will not be possible later on (Kuehl *et al.* 2008).

The final layout of the survey design can be completed and displayed using a Geographic Information System, such as ArcGIS (ESRI 2011), Quantum GIS (QGIS Development Team 2014), Google Earth, or the software DISTANCE (Thomas *et al.* 2010). For more details on survey design, please refer to White and Edwards (2000), Buckland *et al.* (2001, 2004), Strindberg *et al.* (2004), and Maisels and Aba'a (2008).

6.7 Time of survey

Another factor to take into account when planning a one-off survey (or repeat surveys) is the appropriate time of the year when the survey(s) should be conducted, as there may be considerable temporal variability in the detection probability of the study species or the signs it produces (e.g. Mathewson *et al.* 2008). Inter- and intra-observer variability in detection range and probability, environmental or anthropogenic factors, such as climatic conditions, food availability, or hunting pressure, may influence the frequency at which animals or their signs are encountered. A survey should thus be conducted during a time period that maximizes the chances of detecting the study species or its signs. In the case of long-term monitoring programmes that aim at estimating population trends, temporal variability in detection probability can never be completely overcome (e.g. areas that are affected by El Niño), but repeating surveys at the same time of year may minimize temporal variability that could be accounted for.

6.8 Storing data

After a survey is completed, the information should be stored, standardized, and centralized in a central database from which data may be accessed by a third party. These data may then be used to plan a subsequent survey in the same area, model the distribution and trends of primate species on a larger scale (e.g. Junker *et al.* 2012), evaluate the success of conservation activities (e.g. Tranquilli *et al.* 2011), and identify conservation and research gaps (Plumptre *et al.* 2010). Examples of such data platforms are the IUCN/SSC A.P.E.S. (Ape Populations, Environments, and Surveys) database[1] (Kuehl *et al.* 2007b) and the Eastern Africa Primate Diversity and Conservation Programme.[2]

6.9 Case studies

In the following section, three case studies are presented that have different objectives, survey methods, and design.

6.9.1 Spatial distribution of diurnal primate species within Taï National Park, Côte d'Ivoire

Taï National Park (TNP) in Côte d'Ivoire is the largest protected remnant of the Upper Guinean Forest Block, covering 5,400 km^2. Although it has been protected since 1973, poaching remains a serious threat to the viability of the primate populations living within its boundaries (Caspary *et al.* 2001; Refish and Koné 2005). This study aimed to investigate which factors positively or negatively influence the distribution of diurnal primate species within the park. In particular, the impact that long-term research presence had on primate populations was examined. Nine diurnal primate species are present within TNP, including the endangered western red colobus (*Procolobus badius badius*), and the vulnerable Diana monkey (*Cercopithecus diana diana*) and sooty mangabey (*Cercocebus atys atys*) (Oates *et al.* 2008a, b). These species differ in their behaviour, ecological requirements, and both natural and human-influenced spatial distribution (McGraw *et al.* 2007). For example, the western red colobus and Diana monkey vocalize frequently, but it has been suggested that their vocalization frequency can be influenced by hunting pressure (Koné and Refish 2007), which may in turn affect their detection probability and distribution. One species, *Cercopithecus nictitans*, was excluded from the analysis because they occur sporadically throughout their range and are historically absent from large portions of the park (Eckhardt and Zuberbühler 2004), which would bias results on the factors affecting their spatial distribution. Most of the primate species occurring in this area can be surveyed using direct signs of their presence. However one species, the western chimpanzee (*Pan troglodytes verus*), is best surveyed using indirect signs of its presence (i.e. nests) in dense habitat, because individuals are encountered less frequently than their nests. Direct and indirect signs of human presence were also recorded during the survey and included in the analyses.

Survey effort was restricted to areas neighbouring the research site since the influence of the presence of a research area on primate distribution was specifically being investigated. Given the long-term research presence in the area, preliminary data on the predicted primate encounter rate were available and allowed for determining the survey effort needed to obtain accurate and precise survey data. Transect length was set at 1 km, based on the preliminary encounter rate data, to ensure that transects would be long enough to avoid having transects with no observations, which in turn would increase variability. In total 75 transects were systematically placed within a 200 km^2 survey area located in the western part of TNP using the DISTANCE software (Thomas *et al.* 2010; Figure 6.7). Each transect was walked three times over a one-year period to increase sample size and reduce the number of transects with no encounters. The potential bias related to seasonal variation in detection probability was accounted for by randomly selecting the order in which a group of transects would be surveyed (i.e. groups numbered 1–10 on Figure 6.8). In order to understand which factors might influence the spatial distribution of the

[1] http://apes.eva.mpg.de.
[2] http://www.wildsolutions.nl/records.htm.

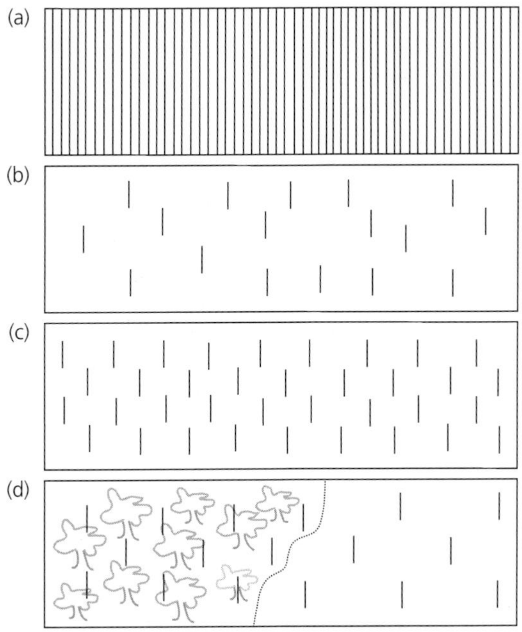

Figure 6.7 Examples of different survey design based on the line transect sampling approach: (a) systematic sampling with high coverage, (b) random line transect sampling with medium coverage, (c) systematic sampling with medium coverage, and (d) stratified sampling where the strata on the left- and right-hand side (divided by the stippled line) include different habitat types that support high and low primate densities, respectively.

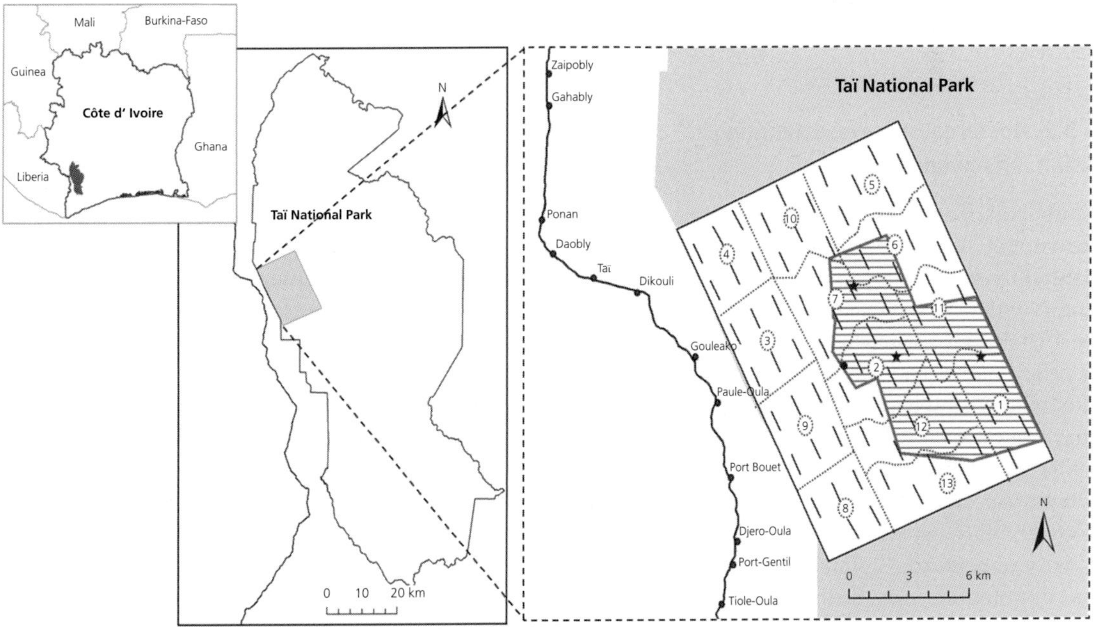

Figure 6.8 Design used to survey eight diurnal primate species within the Taï National Park, Côte d'Ivoire. The research area is represented by dashed lines and the research camps by stars. Groups of transects are delimited by dotted lines and the numbers represent the first sampling order.

eight primate species, Generalized Linear Models (GLMs) were performed and included the main factors (i.e. neighbouring human density, forest type, distance to the border of the park, and distance to the research area) that could affect primate distribution in this area. Proximity to the research area was the best predictor for finding more signs of primate species' presence and a lower encounter rate of poaching signs (Campbell *et al.* 2011). The benefit of the presence of a long-term research area for primate populations living within its limits was demonstrated.

Following the findings of this study and based on the spatial distribution maps, it was suggested that anti-poaching patrol frequency be increased within the entire survey area (i.e. 200 km^2), but particularly in areas of high human activities to protect the resident wildlife population. The design used in this study is now being included in a longer term monitoring programme, with data from the first survey being used as baseline data against which to assess the effectiveness of increased intensity in anti-poaching patrols to evaluate the effectiveness of this conservation measure. Although this monitoring programme is now conducted yearly, several years may be needed before it is possible to detect with certainty a trend in the population size of these species.

6.9.2 Nocturnal surveys: estimating Asian loris abundance

More than one third of all primate species are nocturnal, yet surveys of nocturnal primates still lag behind those of their diurnal counterparts (Nekaris and Nijman 2013). Surveying animals at night can be a challenge, especially when species are rare or cryptic (Marcot and Molina 2007). Due to difficulties with visibility, conventional survey methods may not be ideal (Charles-Dominique and Bearder 1979), and researchers have long suggested that alternative methods should be adopted to ensure comparability across sites (Duckworth 1998). Sensitivity of nocturnal animals to factors like nocturnal illumination may further affect survey design (Nash 2007; Starr *et al.* 2012). Estimating abundance is further exacerbated for species like Asia's lorises (belonging to the genera *Loris* and *Nycticebus*) that are relatively cryptic in both movement and vocalizations.

Asian lorises are a nocturnal species confined to South and Southeast Asia (from Sri Lanka, India, and China south, to Indonesia and the Philippines). Ten species are recognized, most of them living allopatrically. All are listed as threatened on the IUCN Red List, but field data are still scarce despite being vital to assessment of their conservation status. In an attempt to fill this gap, the first systematic studies of loris distribution and abundance were undertaken in the late 1990s and early 2000s, but did not take into account vital variables such as probability of detection or variation in 'density' according to habitat type (e.g. Singh *et al.* 1999; Nekaris and Jayewardene 2004; Rhadakrishna *et al.* 2006). Therefore, these data are not comparable, as they do not control for the environment in which the surveys were conducted.

As many nocturnal primates occur in a wide range of habitat types, it is indeed no longer suitable to provide simple relative abundance estimates, such as the numbers of animals per kilometre walked, as these estimates may not be comparable between different habitat types (e.g. due to differences in detection probability among habitats). Here, two examples are presented where distance sampling was successfully used to estimate Asian loris populations (*Loris* and *Nycticebus*). Both of these studies involved: (1) using red lights at all times for the field work (because red light provides less disturbance to the animals); (2) listening for the calls of lorises to help detect individuals; and (3) scanning all levels of the forest (e.g. no assumption was made that lorises use specific heights or types of trees).

The first study took place in southwestern Sri Lanka, focusing on the endangered red slender loris (*Loris tardigradus tardigradus*) (Nekaris and Stengel 2013). Over 72 days of field effort, the authors traversed 126 km, encountering 44 lorises at six out of eight forest reserves. Density estimates ranged from 3.4 to 28 lorises/km^2 with encounter rates ranging from 0.1 to 1.1 lorises/km depending on the forest reserve surveyed. The authors used the program DISTANCE to determine density over the whole study area. The best-fit model generated a density of 31 lorises/km^2 with a detection probability of 0.45 (confidence intervals 0.48–0.63). Extrapolating across sites, the authors estimated a population of 1,293 red slender lorises for the study area. Forest patch size strongly correlated with loris encounter rate.

In a similar study of the Critically Endangered Javan slow loris (*Nycticebus javanicus*), Nekaris *et al.* (2014) used distance sampling to estimate their population in Mt. Gede-Pangrango National Park, West Java. During 45 nights of transect walks, the authors spent 260 h on 23 transects covering 93 km, with repeats on some transects to compare effect of moonlight and observer speed on detectability, resulting in detection of 37 slow lorises on 25 occasions. Overall density was estimated as 15.6 individuals/km^2 (95% CI = 9.7–25.2 individuals/km^2). Considering only Mt. Gede Pangrango National Park, the authors suggested a population of 70 lorises occurred in the study area. The number of lorises detected per km was strongly influenced by the speed at which transects were walked. The number of slow lorises detected per km surveyed declined progressively as transects were walked faster. Even walking transects at a pace of 500–800 m/h led to a marked decrease in encounter rates per km. Detectability was not affected by lunar luminosity per hour (if the moon was present or not during each hour a transect was walked). Likewise, on darker nights encounter rates did not differ from that of nights with moonlight.

Until recently, all Asian lorises were listed as Least Concern by the IUCN Red List. With extreme habitat fragmentation compounded by the threat of illegal wildlife trade, quantifying the abundance of these species has become essential. After 20 years of trial and error in surveying these animals, it is now possible to gain the sample size needed for most robust analytical techniques. Respect should be shown towards the cautiousness of nocturnal animals and their sensitivity to light, with researchers moving slowly, silently, and never shining white bright lights. In this manner, it may finally be possible to gain an insight into the distribution and conservation status of nocturnal primates.

6.9.3 Estimating density and examining socio-demographic structure of great apes in Loango National Park, Gabon

In 2005, the Max Planck Institute of Evolutionary Anthropology (MPI-EVA) set up a research site in Loango National Park, Gabon, to carry out ecological and behavioural research on sympatric central chimpanzees (*Pan troglodytes troglodytes*) and western lowland gorillas (*Gorilla gorilla gorilla*). One focus of the research was non-invasive surveying to monitor abundance and distribution of the sympatric apes, and to this end the MPI-EVA conducted both line transect and genetic surveys in the area (Arandjelovic *et al.* 2010, 2011). However, behavioural research was also central to the MPI-EVA project, though previous survey approaches were limited in terms of providing socio-demographic information. Therefore, in 2009 a remote video camera-trap study was implemented to estimate the abundance, density, home range size, and social and demographic structure of the resident chimpanzees and gorillas using capture–recapture techniques (see Section 6.4.9; Head *et al.* 2013). The study aimed to link remote video trapping with spatially explicit capture–recapture (SECR) techniques to improve on current monitoring methods and obtain a comprehensive population assessment.

Since one of the goals was to examine the socio-demographic structure of the apes, it was important to design a protocol that included sufficient video camera-traps and a long enough study period to capture the majority of the individuals, while also covering a large enough area to answer questions about home range size and grouping patterns. Consequently, over a 20 month period 45 video camera-traps were used in a systematic grid design covering an area of 60 km^2 (Figure 6.9). The video camera-traps were placed systematically with one per 1 km^2 grid square, but within the square they were placed optimally to increase capture probability (e.g. along known animal trails). This study design resulted in 1,045 chimpanzee and 471 gorilla images being captured on remote video cameras, and of these 439 (42%) and 103 (22%) positive individual identifications were possible for the two species, respectively. Wild apes have long been individually

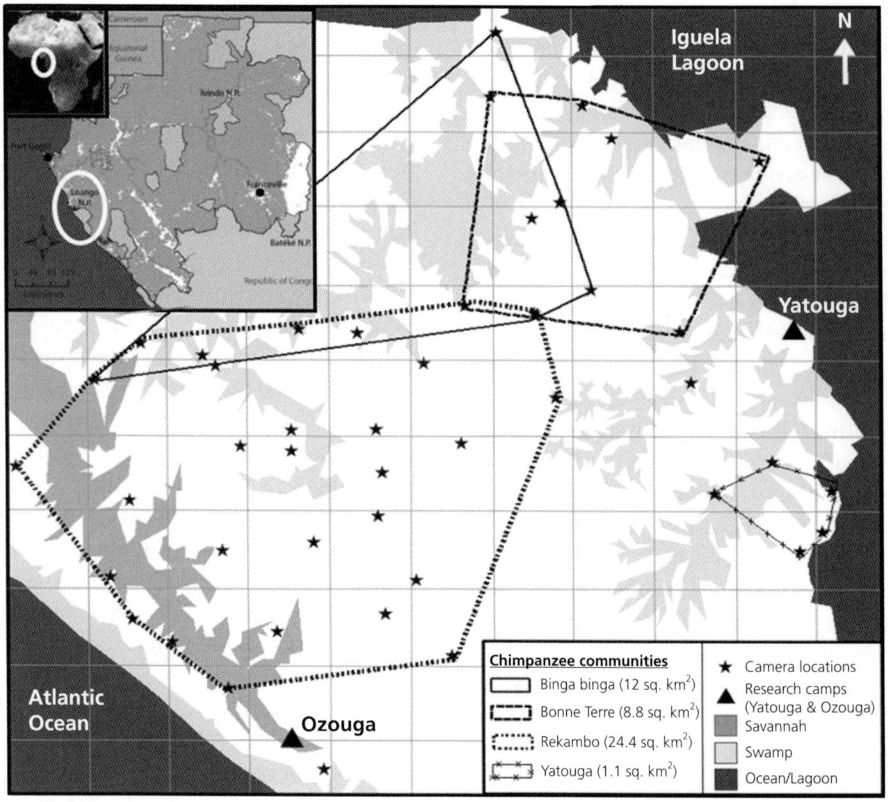

Figure 6.9 Map showing chimpanzee community home ranges, minimum home range size, and camera distribution for the remote camera-trapping study in Loango National Park, Gabon. The inset displays the location of Gabon within Africa, and the location of Loango National Park within Gabon.

identified using a combination of unique facial and body characteristics including shape and coloration of ears, nose, face, and body, in addition to scars or disfigurements (e.g. Fossey 1983; Goodall 1988), and this method is widely used and accepted by primatologists. Therefore, the individual chimpanzees and gorillas filmed were identified using this method by one experienced observer. However, because body scars, size, and coloration can change over time, only captures in which multiple features of an individual were observed were considered. In addition, inter-observer reliability testing was carried out to confirm the reliability of the experienced observer.

Density estimates were calculated in R using the maximum-likelihood-based SECR 2.3.2 package (Efford 2012). This SECR model was chosen over other analytical options because, unlike other likelihood-based capture–recapture models that assume a closed population (defined as one in which immigration or emigration of individuals is not occurring), SECR takes explicit account of temporary emigration of individuals from the sampling area (Borchers and Efford 2008). Given that the video camera-traps covered an area of 60 km^2, and ape home ranges are typically 25–30 km^2, it was thought to be extremely likely that many individual apes would be moving in and out of the sampling area on a regular basis. Therefore, maximum-likelihood-based SECR was considered the most appropriate for the analysis. In addition, the SECR package for R has the added advantage of being relatively user friendly.

In terms of examining group size and ranging patterns, individuals captured together during the same video trigger or within 15 min of other individuals on the same video camera, or else captured independently with a third individual in common, were considered part of the same group or community (Arandjelovic *et al.* 2010). Overall, there were 123 and 52 unique individual chimpanzees and gorillas identified from the video camera-trap footage, and density was estimated at 1.72 and 1.2 individuals per km^2 respectively.

The analysis revealed that the chimpanzee density estimates using video camera-traps and SECR had greater precision (based on the known density of one chimpanzee community within the study area) than the previous studies that used genetic sampling and non-spatial methods, but the gorilla density estimate with SECR had lower precision than the estimate resulting from the genetic survey (Arandjelovic *et al.* 2010). This was due to the low identification rate of gorillas from the video camera-traps, which resulted in a smaller dataset and thus higher uncertainty. However, recent improvements in camera quality (increased megapixels and longer battery life) suggest that this issue could be overcome. Overall this study highlighted the potential of combining video camera-traps with SECR methods to conduct detailed population assessments across multiple terrestrial species, and confirmed the utility of video camera-traps in long-term monitoring for conservation.

6.10 Conclusion

As primate populations continue to dwindle, the need for effective surveys to monitor long-term changes in primate population distribution, abundance, and trends is increasingly urgent. As discussed throughout this book, with the exception of diseases such as Ebola, human activities account for all threats to the long-term survival of primate populations across the globe. Collecting baseline information on primate population parameters not only enables measurement of these threats, but also allows conservationists to monitor and assess the effectiveness of conservation measures over time. Financial constraints result in the majority of surveys being conducted when the primate population in question is already under pressure and likely declining. If surveys were conducted at an earlier point in time and with sufficient frequency to allow proper trend analyses, population declines could be better monitored and their severity potentially reduced. However, detecting subtle trends requires many years of surveys and highly accurate estimates, which are costly and time intensive. Because of the limited resources and the urgency with which data are often required, selecting the correct survey approach is key because even a very well executed survey can still provide data of little or no relevance to the questions being asked if the incorrect method or survey design is used. Which approach is most well suited will depend on the scale of the survey, the resources available, and whether the goal is to obtain data on distribution, abundance, population trend, or socio-demographic structure of the primate population under study.

Continuing to improve the accuracy and precision of monitoring methods is crucial if conservationists are to most effectively use limited resources and offer vulnerable primate species the best chance of long-term survival. Each method outlined in this chapter has its own drawbacks (see Table 6.1), and conservationists are always seeking new ways to improve on current methods and the precision of survey results. For example, recent advances in automated identification software may improve the efficiency and precision of camera trapping analysis for terrestrial primates (Loos and Pfitzer (2012); for a review on recent advances in animal biometrics, see Kühl and Burghardt (2013)), while unmanned aerial vehicles[3] are also rapidly gaining popularity for aerial surveys, even in forested habitats (Ernst and Kühlbeck 2011; Koh and Wich 2012; van Andel *et al.* 2015). Furthermore, other new technologies, such as acoustic recording units currently being used to monitor other vocal species (e.g. elephants, Wrege *et al.* (2011)), have significant potential for use in primate monitoring (Campbell *et al.* unpubl.; Heinicke *et al.* 2015). Primate researchers and conservationists typically aim to strike a balance between cost, effort, accuracy, and precision; and we must continue to develop improved and innovative approaches that are cost effective if we are to effectively protect vulnerable primates from decline and extinction.

[3] http://conservationdrones.org.

References

Ancrenaz, M., Gimenez, O., Ambu, L., Ancrenaz, K., Andau, P., *et al.* (2005). Aerial surveys give new estimates for orangutans in Sabah, Malaysia. *PLoS Biology* **3**: e3. doi:10.1371/journal.pbio.0030003.

Ancrenaz, M., Goossens, B., Gimenez, O., Sawang, A., and Lackman-Ancrenaz, I. (2004). Determination of ape distribution and population size using ground and aerial surveys: a case study with orang-utans in lower Kinabatangan, Sabah, Malaysia. *Animal Conservation* **7**: 375–385.

Aquino, R., López, L., García, G., and Heymann, E. W. (2014). Diversity, abundance and habitats of the primates in the Río Curaray Basin, Peruvian Amazonia. *Primate Conservation* **28**: 1–8.

Arandjelovic, M., Head, J., Kuhl, H., Boesch, C., Robbins, M. M., *et al.* (2010). Effective non-invasive genetic monitoring of multiple wild western gorilla groups. *Biological Conservation* **143**: 1780–1791.

Arandjelovic, M., Head, J., Rabanal, L. I., Schubert, G., Mettke, E., *et al.* (2011). Non-invasive genetic estimation of group number and population size in wild central chimpanzees. *PLoS One* **6**: e14761.

Bearder, S. K. and Karlsson, J. (2009). A Survey of Nocturnal Primates in Malawi—August 2009. Unpublished Report. Available at: <http://www.galagoides.se/media/downloads/Bearder and Karlsson Malawi report Aug 2009.pdf> [Accessed November 2015].

Bermejo, M. (1999). Status and conservation of primates in Odzala National Park, Republic of the Congo. *Oryx* **33**: 323–331.

Bermejo, M., Rodriguez-Teijeiro, J. D., Illera, G., Barroso, A., Vilà, C., *et al.* (2006). Ebola outbreak killed 5000 gorillas. *Science* **314**(5804): 1564.

Blackham, G. (2004). Pilot Survey of Nocturnal primates, Tarsius bancanus borneanus (Western tarsier) and Nycticebus coucang menagensis (Slow loris) in Peat Swamp Forest, Central Kalimantan, Indonesia. MSc thesis. Oxford Brookes University, Oxford, UK.

Borchers, D. L. and Efford, M. G. (2008). Spatially explicit maximum likelihood methods for capture-recapture studies. *Biometrics* **64**: 377–385.

Börger, L., Franconi, N., De Michele, G., Gantz, A., Meschi, F., *et al.* (2006). Effects of sampling regime on the mean and variance of home range size estimates. *Journal of Animal Ecology* **75**: 1393–1405.

Bowen-Jones, E. and Pendry, S. (1999). The threat to primates and other mammals from the bushmeat trade in Africa, and how this threat could be diminished. *Oryx* **33**: 233–246.

Boyko, R. H. and Marshall, A. J. (2010). Using simulation models to evaluate ape nest survey techniques. *PLoS One* **5**: e10754.

Brncic, T. M., Amarasekaran, B., and McKenna, A. (2010). Final Report of the Sierra Leone National Chimpanzee Census Project, Freetown. Unpublished Report. Tacugama Chimpanzee Sanctuary, Freetown, Sierra Leone.

Buckland, S. T., Anderson, D. R., Burnham, K. P., Laake, J. L., Borchers, D. L., *et al.* (2001). *Introduction to Distance Sampling: Estimating Abundance of Biological Populations*. Oxford: Oxford University Press.

Buckland, S. T., Anderson, D. R., Burnham, K. P., Laake, J. L., Borchers, D. L., *et al.* (2004). *Advanced Distance Sampling: Estimating Abundance of Biological Populations*. Oxford: Oxford University Press.

Buckland, S. T., Plumptre, A. J., Thomas, L., and Rexstad, E. (2010). Design and analysis of line transect surveys for primates. *International Journal of Primatology* **31**: 833–847.

Buckland, S. T., Summers, R. W., Borchers, D. L., and Thomas, L. (2006). Point transect sampling with traps or lures. *Journal of Applied Ecology* **43**: 377–384.

Calvignac-Spencer, S., Merkel, K., Kutzner, N., Kühl, H., Boesch, C., *et al.* (2013). Carrion fly-derived DNA as a tool for comprehensive and cost-effective assessment of mammalian biodiversity. *Molecular Ecology* **22**: 915–924.

Campbell, G., Kuehl, H., Diarrassouba, A., N'goran, P. K., and Boesch, C. (2011). Long-term research sites as refugia for threatened and over-harvested species. *Biology Letters* **7**: 723–726.

Campbell, G., Kuehl, H., N'goran, P. K., and Boesch, C. (2008). Alarming decline of West African chimpanzees in Côte d'Ivoire. *Current Biology* **18**: 903–904.

Campbell, G., Mundry, R., Wagner, O., Eckhardt, N., Janmaat, K., *et al.* (unpublished). Validating remote audio-visual wildlife monitoring techniques for obtaining comprehensive population assessments.

Caspary, H. U., Koné, I., Prouot, C., and de Pauw, M. (2001). *La chasse et la filière de la viande de brousse dans l'espace Tai* (Tropenbos Côte d'Ivoire Série 2). Côte d'Ivoire: Tropenbos.

Cassey, P., Lockwood, J. L., and Fenn, K. H. (2007). Using long-term occupancy information to inform the management of Cape Sable seaside sparrows in the Everglades. *Biological Conservation* **139**: 139–149.

Chancellor, R. L., Langergraber, K., Ramirez, S., Rundus, A. S., and Vigilant, L. (2012). Genetic sampling of unhabituated chimpanzees (*Pan troglodytes schweinfurthii*) in Gishwati forest reserve, an isolated forest fragment in western Rwanda. *International Journal of Primatology* **33**: 479–488.

Chapman, C. A., Balcomb, S. R., Gillespie, T. R., Skorupa, J. P., and Struhsaker, T. T. (2000). Long-term effects of logging on African primate communities: a 28-year comparison from Kibale National Park, Uganda. *Conservation Biology* **14**: 207–217.

Charles-Dominique, P. and Bearder, S. K. (1979). Field studies of lorisid behavior: methodological aspects. In: Doyle, G. A. and Martin, R. D. (Eds), *The Study of Prosimian Behavior*. pp. 567–629. New York, NY: Academic Press.

Chase, M. (2011). Fixed-wing aerial survey of wildlife in the Abu Wildlife Management Area (Ng 26) Okavango Delta, Botswana, October 2010. Available at: <http://www.elephantswithoutborders.org/downloadspapers/EWB%20AbuAerialSurveyReport2010.pdf>. [Accessed November 2015].

Chetan, N., Praveen, K. K., and Vasudeva, G. K. (2014). Delineating ecological boundaries of Hanuman Langur Species Complex in peninsular India using MaxEnt modeling approach. *PLoS One* **9**: e87804.

Constable, J. J., Packer, C., Collins, D. A., and Pusey, A. E. (1995). Nuclear DNA from primate dung. *Nature* **373**: 393.

Coudrat, C. N. and Nekaris, K. (2013). Modelling niche differentiation of co-existing, elusive and morphologically similar species: a case study of four macaque species in Nakai-Nam Theun National Protected Area, Laos. *Animals* **3**(1): 45–62.

Cronin, D. T., Riaco, C., and Hearn, G. W. (2013). Survey of threatened monkeys in the Iladyi river valley region, south-eastern Bioko Island. Equatorial Guinea. *African Primates* **8**: 1–8.

Cushman, S. A., Chase, M., and Griffin, C. (2005) Elephants in space and time. *Oikos* **109**: 331–341.

Di Bitetti, M. S., Paviolo, A., and De Angelo, C. (2006). Density, habitat use and activity patterns of ocelots (*Leopardus pardalis*) in the Atlantic forest of Misiones, Argentina. *Journal of Zoology* **270**: 153–163.

Duckworth, J. W. (1998). The difficulty of estimating population densities of nocturnal forest mammals from transect counts of animals. *Journal of Zoology* **246**: 466–468.

ESRI (2011). ArcGIS Desktop: Release 10. Redlands, CA: Environmental Systems Research Institute.

Eckhardt, W. and Zuberbühler, K. (2004). Cooperation and competition in two forest monkeys. *Behavioural Ecology* **15**: 400–411.

Efford, M. G. (2012). SECR-spatially explicit capture-recapture in R. Available at: <http://cran.r-project.org/web/packages/secr/secr.pdf> [Accessed November 2015].

Elith, J., Graham, C. H., Anderson, R. P., Dudík, M, Ferrier, S., *et al.* (2006). Novel methods improve prediction of species distributions from occurrence data. *Ecography* **29**: 129–151.

Ernst, A. and Kühblbeck, C. (2011). Fast face detection and species classification of African great apes. In: Proceedings of the 8th IEEE International Conference on Advanced Video and Signal-Based Surveillance, pp. 279–284.

Estrada, A. and Coates-Estrada, R. (1996). Tropical rainforest fragmentation and wild populations of primates at Los Tuxtlas, Mexico. *International Journal of Primatology* **17**: 759–783.

Fa, J. E., Juste, J., Del Val, J. P., and Castroviejo, J. (1995). Impact of market hunting on mammal species in Equatorial Guinea. *Conservation Biology* **9**: 1107–1115.

Fa, J. E., Peres, C. A., and Meeuwig, J. (2002). Bushmeat exploitation in tropical forests: an intercontinental comparison. *Conservation Biology* **16**: 232–237.

Fashing, P. J. and Cords, M. (2000). Diurnal primate densities and biomass in the Kakamega Forest: an evaluation of census methods and a comparison with other forests. *American Journal of Primatology* **50**: 139–152.

Fiske, I. and Chandler, R. B. (2011). Unmarked: an R package for fitting hierarchical models of wildlife occurrence and abundance. *Journal of Statistical Software* **43**: 1–23.

Fossey, D. (1983). *Gorillas in the Mist*. Boston, MA: Houghton Mifflin.

Fowler, S. V., Chapman, P., Checkley, D., Hurd, S., McHalefl, M., *et al.* (1989). Survey and management proposals for a tropical deciduous forest reserve at Ankarana in Northern Madagascar. *Biological Conservation* **47**: 297–313.

Galat, G., Galat-Luong, A., Benoit, M., Chevilotte, H., Diop, A., *et al.* (1994). Primate density in the Niokolo Koba National Park, Senegal. *Folia Primatologica* **62**: 197.

Gatti, S., Levrero, F., Menard, N., and Gautier-Hion, A. (2004). Population and group structure of western lowland gorillas (*Gorilla gorilla gorilla*) at Lokoue, Republic of Congo. *American Journal of Primatology* **63**: 111–123.

Gerrodette, T. (1993). Trends: software for a power analysis of linear regression. *Wildlife Society Bulletin* **21**: 515–516.

Girard, I., Ouellet, J.-P., Courtois, R., Dussault, C., and Breton, L. (2002). Effects of sampling effort based on GPS telemetry on home-range size estimations. *Journal of Wildlife Management* **66**: 1290–1300.

Gitzen, R. A., Millspaugh, J. J., Cooper, A. B., and Licht, D. S. (Eds) (2012). *Design and Analysis of Long-Term Ecological Monitoring Studies*. Cambridge: Cambridge University Press.

Goodall, J. (1988). *In the Shadow of Man*. London: Weidenfeld and Nicholson.

Green, K. M. (1978). Primate censusing in Northern Colombia: a comparison of two techniques. *Primates* **19**: 537–550.

Grueter, C. C., Li, D., Ren, B., and Wei, F. (2009). Choice of analytical method can have dramatic effects on primate home range estimates. *Primates* **50**: 81–84.

Grueter, C. C., Li, D., Ren, B., Xiang, Z., and Li, M. (2012). Food abundance is the main determinant of high-altitude range use in snub-nosed monkeys. *International Journal of Zoology*: doi:10.1155/2012/739419.

Guschanski, K., Vigilant, L., McNeilage, A., Gray, M., Kagoda, E., *et al.* (2009). Counting elusive animals: comparing field and genetic census of the entire mountain gorilla population of Bwindi Impenetrable National Park, Uganda. *Biological Conservation* **142**: 290–300.

Harding, R. S. O. (1984). Primates of the Kilimi Area, Northwest Sierra Leone. *Folia Primatologica* **42**: 96–114.

Hart, T. and Mwinyihali, R. (2001). *Armed Conflict and Biodiversity in Sub-Saharan Africa: The Case of the Democratic Republic of Congo*. Washington, DC: Biodiversity Support Program.

Head, J., Boesch. C., Robbins, M. M., Rabanal, L. I., Makaga, L., *et al.* (2013). Effective socio-demographic population assessment of elusive species in ecology and conservation management. *Ecology and Evolution* **3**: 2903–2916.

Heinicke, S., Kalan, A., Wagner, O., Mundry, R., Lukashevich, H., *et al.* (2015). Assessing the performance of a semi-automated acoustic monitoring system for primates. *Methods in Ecology and Evolution*: doi: 10.1111/2041–2210X.12384.

Hemson, G., Johnson, P., South, A., Kenward, R., Ripley, R., *et al.* (2005). Are kernels the mustard? Data from global positioning system (GPS) collars suggests problems for kernel home-range analyses with least-squares cross-validation. *Journal of Animal Ecology* **74**: 455–463.

Hickey, J. R., Nackoney, J., Nibbelink, N. P., Blake, S., Bonyenge, A., *et al.* (2013). Human proximity and habitat fragmentation are key drivers of the rangewide bonobo distribution. *Biodiversity and Conservation* **22**: 3085–3104.

Huck, M., Davison, J., and Roper, T. J. (2008). Comparison of two sampling protocols and four home-range estimators using radio-tracking data from urban badgers *Meles meles*. *Wildlife Biology* **14**: 467–477.

Irwin, M. T., Johnson, S. E., and Wright, P. C. (2005). The state of lemur conservation in south-eastern Madagascar: population and habitat assessments for diurnal and cathemeral lemurs using surveys, satellite imagery and GIS. *Oryx* **39**: 204–218.

IUCN Red List of Threatened Species. Available at: <http://www.iucnredlist.org/> [Accessed November 2015].

Jimoh, S. A., Ikyaagba, E. T., Alarape, A. A., Obioha, E. E., and Adeyemi, A. A. (2012). The role of traditional laws and taboos in wildlife conservation in the Oban Hill sector of Cross River National Park (CRNP), Nigeria. *Journal of Human Ecology* **39**: 209–219.

Jones, J. P. G., Andriamarovololona, M. M., and Hockley, N. (2008). The importance of taboos and social norms to conservation in Madagascar. *Conservation Biology* **22**: 976–986.

Junker, J., Blake, S., Boesch, C., Campbell, G., du Toit, L., *et al.* (2012). Recent decline in suitable environmental conditions for African great apes. *Diversity and Distributions* **18**: 1077–1091.

Kalpers, J., Williamson, E. A., Robbins, M. M., McNeilage, A., Nzamurabaho, A., *et al.* (2003). Gorillas in the crossfire: assessment of population dynamics of the Virunga mountain gorillas over the past three decades. *Oryx* **37**: 326–337.

Karanth, K. K., Nichols, J. D., and Hines, J. E. (2010). Occurrence and distribution of Indian primates. *Biological Conservation* **143**: 2891–2899.

Kenward, R. E. (1987). *Wildlife Radio Tagging, Equipment, Field Techniques and Data Analysis*. London: Academic Press.

Kéry, M., Dorazio, R. M., Soldaat, L., Van Strien, A., Zuiderwijk, A., *et al.* (2009). Trend estimation in populations with imperfect detection. *Journal of Applied Ecology* **46**: 1163–1172.

Kierulff, M. C. M., Dos Santos, G. R., Canale, G., Guidorizzi, C.E., and Cassano, C. (2004). The use of camera-traps in a survey of the buff-headed capuchin monkey, Cebus xanthosternos. *Neotropical Primates* **12**: 56–59.

Koh, L. P., Dunn, R. R., Sodhi, N. S., Colwell, R. K., Proctor, H. C., *et al.* (2004). Species coextinctions and the biodiversity crisis. *Science* **305**: 1632–1634.

Koh, L. P. and Wich, S. A. (2012). Dawn of drone ecology: low-cost autonomous aerial vehicles for conservation. *Tropical Conservation Science* **5**: 121–132.

Koné, I. and Refish, J. (2007). Can monkey behaviour be used as an indicator for poaching pressure? A case study of the Western red colobus (Procolobus badius) and the Diana guenon (Cercopithecus Diana) in the Tai National Park. In: McGraw, W. S., Zuberbühler, K., and Noë, R. (Eds), *Monkeys of the Taï Forest: An African Primate Community*. Cambridge: Cambridge University Press.

Köndgen, S., Kühl, H. S., Ngoran, P. K., Walsh, P. D., Schenk, S., *et al.* (2008). Pandemic human viruses cause decline of endangered great apes. *Current Biology* **18**: 260–264.

Kouakou, C. Y., Boesch, C., and Kuehl, H. (2009). Estimating chimpanzee population size with nest counts: validating methods in Taï National Park. *American Journal of Primatology* **71**: 447–457.

Kramer-Schadt, S., Niedballa, J., Pilgrim, J. D., Schröder, B., Lindenborn, J., *et al.* (2013). The importance of correcting for sampling bias in MaxEnt species distribution models. *Diversity and Distribution* **19**: 1366–1379.

Kuehl, H. S., Todd, A., Boesch, C., and Walsh, P. D. (2007a). Manipulating decay time for efficient large-mammal density estimation: gorillas and dung height. *Ecological Applications* **17**: 2403–2414.

Kuehl, H. S., Williamson, L., Sanz, C., Morgan, D., and Boesch, C. (2007b). Launch of A.P.E.S. database. *Gorilla Journal* **34**: 20–21.

Kuehl, H. S, Maisels, F., Williamson, L., and Ancrenaz, M. (2008). Best Practice Guidelines for Surveys and Monitoring of Great Ape Populations. Occasional papers of the IUCN Species Survival Commission.

Kühl, H. S. and Burghardt, T. (2013). Animal biometrics: quantifying and detecting phenotypic appearance. *Trends in Ecology and Evolution* **28**: 432–441.

Kumara, H. N. and Radhakrishna, S. (2013). Evaluation of census techniques to estimate the density of slender Loris (*Loris lydekkerianus*) in Southern India. *Current Science* **104**: 25.

Kumara, H. S., Sasi, R., Suganthasakthivel, R., and Srinivas, G. (2011). Distribution, abundance and conservation of primates in the Highwavy Mountains of Western Ghats, Tamil Nadu, India and conservation prospects for lion-tailed macaques. *Current Science* **100**: 1063–1067.

Lahoz-Monfort, J. J., Guillera-Arroita, G., Milner-Gulland, E. J., Young, R. P., and Nicholson, E. (2010). Satellite imagery as a single source of predictor variables for habitat suitability modelling: how Landsat can inform the conservation of a critically endangered lemur. *Journal of Applied Ecology* **47**: 1094–1102.

Levin, S. A., Carpenter, S. R., Godfray, H. C. J., Kinzig, A. P., Loreau, M., *et al.* (2009). *The Princeton Guide to Ecology*. Princeton, NJ: Princeton University Press.

Linder, J. M. and Oates, J. (2011). Differential impact of bushmeat hunting on monkey species and implications for primate conservation in Korup National Park, Cameroon. *Biological Conservation* **144**: 738–745.

Loos, A. and Pfitzer, M. (2012). Towards automated visual identification of primates using face recognition. In: *Systems, Signals and Image Processing (IWSSIP), 2012 19th International Conference*. pp. 425–428. Vienna: IEEE.

MacKenzie, D. I. and Royle, J. A. (2005). Designing efficient occupancy studies: general advice and tips on allocation of survey effort. *Journal of Applied Ecology* **42**: 1105–1114.

Maisels, F. and Aba'a, R. (2008). Section 3: Survey design. In: Best Practice Guidelines for Surveys and Monitoring of Great Ape Populations. Occasional papers of the IUCN Species Survival Commission.

Marcot, B. G. and Molina, R. (2007). Special considerations for the science, conservation, and management of rare or little-known species. In: Raphael M. and Molina N. (Eds), *Conservation of Rare or Little-Known Species: Biological, Social, and Economic Considerations*. pp. 93–124. Washington DC: Island Press.

Marshall, A. J., Beaudrot, L., and Wittmer, H. U. (2014). Responses of primates and other frugivorous vertebrates to plant resource variability over space and time at Gunung Palung National Park. *International Journal of Primatology* **35**: 1178–1201.

Marshall, A. R., Lovett, J. C., and White, P. C. L. (2008). Selection of line-transect methods for estimating the density of group-living animals: lessons from the primates. *American Journal of Primatology* **70**: 452–462.

Marshall, A. J. and Meijaard, E. (2009). Orangutan nest surveys: the devil is in the details. *Oryx* **43**: 416–418.

Mathewson, P. D., Spehar, S. N., Meijaard, E., Nardiyono, Purnomo, *et al.* (2008). Evaluating orangutan census techniques using nest decay rates: implications for population estimates. *Ecological Applications* **18**: 208–221.

McGraw, W. S., Zuberbühler, K., and Noë, R (Eds) (2007). *Monkeys of the Taï Forest: An African Primate Community*. Cambridge: Cambridge University Press.

McNeilage, A., Plumptre, A. J., Brock-Doyle, A., and Vedder, A. (2001). Bwindi Impenetrable National Park, Uganda: gorilla census 1997. *Oryx* **35**: 39–47.

McNeilage, A., Robbins, M. M., Gray, M., Olupot, W., Babaasa, D., *et al.* (2006). Census of the mountain gorilla *Gorilla beringei beringei* population in Bwindi Impenetrable National Park, Uganda. *Oryx* **40**: 419–427.

Medhi, R., Chetry, D., Basavdatt, C., and Bhattacharjee, P. C. (2007). Status and diversity of temple primates in Northeast India. *Primate Conservation* **22**: 135–138.

Meijaard, E., Nurcahyo, A., Mengersen, K., Buchori, D., Nurcahyo, A., *et al.* (2011a). Why don't we ask? A complementary method for assessing the status of great apes. *PLoS One* **6**: e18008.

Meijaard, E., Buchori, D., Hadiprakarsa, Y., Utami-Atmoko, S. S., Nurcahyo, A., *et al.* (2011b). Quantifying killing of orangutans and human-orangutan conflict in Kalimantan, Indonesia. *PLoS One* **6**: e27491.

Michalski, F. and Peres, C. A. (2005). Anthropogenic determinants of primate and carnivore local extinctions in a fragmented forest landscape of southern Amazonia. *Biological Conservation* **124**: 383–396.

Miller, D. L., Burt, M. L., Rexstad, E. A., and Thomas, L. (2013). Spatial models for distance sampling data: recent developments and future directions. *Methods in Ecology and Evolution* **4**: 1001–1010.

Mohr, C. (1947). Table of equivalent populations of North American small mammals. *The American Midland Naturalist Journal* **37**: 233–249.

Morales, J. M. and Ellner, S. P. (2002). Scaling up animal movements in heterogeneous landscapes: the importance of behavior. *Ecology* **83**: 2240–2247.

Nash, L. T. (2007). Moonlight and behavior in nocturnal and cathemeral primates, especially *Lepilemur leucopus*: illuminating possible anti-predator efforts. In: Gursky, S. L. and Nekaris, K. A. I. (Eds), *Primate Anti-predator Strategies*. pp. 173–205. New York, NY: Springer.

Neilson, E., Nijman, V., and Nekaris, K. A. I. (2013). Conservation assessments of arboreal mammals in difficult terrain: occupancy modeling of pileated gibbons

(*Hylobates pileatus*). *International Journal of Primatology* **34**(4): 823–835.

Nekaris, K. A. I. and Jayewardene, J. (2004). Survey of the slender loris (Primates, Lorisidae Gray, 1821: *Loris tardigradus* Linnaeus, 1758 and *Loris lydekkerianus* cabrera, 1908) in Sri Lanka. *Journal of Zoology* **262**(04): 327–338.

Nekaris, K. A. I. and Nijman, V. (2013). The ethics of conducting field research: do long-term great ape field studies help to conserve primates? In: MacClancy, J. and Fuentes, A. (Eds), *Ethics in the Field*. pp. 108–123. Oxford: Berghahn Books.

Nekaris, K. A. I., Pambudi, J. A. A., Susanto, D., Ahmad, R. D., and Nijman, V. (2014). Densities, distribution and detectability of a small nocturnal primate (Javan slow loris *Nycticebus javanicus*) in a montane rainforest. *Endangered Species Research* **24**: 95–103.

Nekaris, K. A. I. and Stengel, C. J. (2013). Where are they? Quantification, distribution and microhabitat use of fragments by the red slender loris (*Loris tardigradus tardigradus*) in Sri Lanka. In: Marsh, L. and Chapman, C. (Eds), *Primates in Fragments*. pp. 371–384. New York, NY: Springer.

Nichols J. D. and Williams B. K. (2006). Monitoring for conservation. *Trends in Ecology and Evolution* **21**: 668–673.

Nielsen, E. B., Pedersen, S., and Linnell, J. D. C. (2008). Can minimum convex polygon home ranges be used to draw biologically meaningful conclusions? *Ecological Research* **23**: 635–639.

Oates, J. F., Gippoliti, S., and Groves, C. P. (2008a). *Cercopithecus diana*. In: IUCN Red List of Threatened Species. Version **2013**.2.

Oates, J. F., Struhsaker, T., McGraw, S., Galat-Luong, A., and Ting, T. (2008b). *Procolobus badius*. In: IUCN Red List of Threatened Species. Version **2013**.2.

Olson, E. R., Marsh, R. A., Bovard, B. N., Randrianarimanana, H. L. L., Ravaloharimanitra, M., *et al.* (2012). Arboreal camera trapping for the Critically Endangered greater bamboo lemur *Prolemur simus*. *Oryx* **46**: 593–597.

O'Connell, A. F., Nichols, J. D., and Karanth, K. U. (2011). *Camera Traps in Animal Ecology*. New York, NY: Springer.

Peres, C. A. (1997). Primate community structure at twenty western Amazonian flooded and unflooded forests. *Journal of Tropical Ecology* **13**: 381–405.

Phillips, S. J., Anderson, R. P., and Schapire, R. E. (2006). Maximum entropy modeling of species geographic distributions. *Ecological Modelling* **190**: 231–259.

Plumptre, A. J. (2000). Monitoring mammal populations with line transect techniques in African forests. *Journal of Applied Ecology* **37**: 356–368.

Plumptre, A. J. (2003). Lessons learned from on-the-ground conservation in Rwanda and the Democratic Republic of the Congo. *Journal of Sustainable Forestry* **16**: 69–88.

Plumptre, A. J. and Cox, D. (2006). Counting primates for conservation: primate surveys in Uganda. *Primates* **47**: 65–73.

Plumptre, A. J., Rose, R., Nangendo, G., Williamson, E. A., Didier, K., *et al.* (2010). Eastern Chimpanzee (*Pan troglodytes schweinfurthii*): Status Survey and Conservation Action Plan 2010–2020, 52. Gland, Switzerland: IUCN. Available at: <http://data.iucn.org/dbtwwpd/html/African%20primates/cover.html.> [Accessed November 2015].

Plumptre, A. J., Sterling, E. J., and Buckland, S. T. (2013). Primate census and survey techniques. In: Sterling, E. J., Bynum, N., and Blair M. E. (Eds), *Primate Ecology and Conservation: A Handbook of Techniques*. pp. 10–26. Oxford: Oxford University Press.

QGIS Development Team (2014). QGIS Geographic Information System. Open source Geospatial Foundation Project. Available at: <http://qgis.osgeo.org> [Accessed November 2015].

Rabanal, L. I., Kuehl, H. S., Mundry, R., Robbins, M. M., and Boesch, C. (2010). Oil prospecting and its impact on large rainforest mammals in Loango National Park, Gabon. *Biological Conservation* **143**: 1017–1024.

Refish, J. and Koné, I. (2005). Impact of commercial hunting on monkey populations in the Tai region, Côte d'Ivoire. *Biotropica* **37**: 136–144.

Remis, M. J. and Robinson, C. A. (2012). Reductions in primate abundance and diversity in a multiuse protected area: synergistic impacts of hunting and logging in a Congo basin forest. *American Journal of Primatology* **74**: 602–612.

Rhadakrishna, S., Goswami, A. B., and Sinha, A. (2006). Distribution and conservation of *Nycticebus bengalensis* in northeastern India. *International Journal of Primatology* **27**(4): 971–982.

Rosen, G. E. and Smith, K. F. (2010). Summarizing the evidence on the international trade in illegal wildlife. *Ecohealth* **7**: 24–32.

Ross, C. and Reeve, N. (2003). Survey and census methods: population distribution and density. In: Setchell, J. M. and Curtis, D. J. (Eds), *Field and Laboratory Methods in Primatology*. pp. 90–109. Cambridge: Cambridge University Press.

Rowcliffe, J. M., Field, J., Turvey, S. T., and Carbone, C. (2008). Estimating animal density using camera traps without the need for individual recognition. *Journal of Applied Ecology* **45**: 1228–1236.

Royle, J. A. and Nichols, J. D. (2003). Estimating abundance from repeated presence absence data or point counts. *Ecology* **84**: 777–790.

Savage, A., Thomas, L., Leighty, K. A., Soto, L. H., and Medina, F. S. (2010). Novel survey method finds dramatic decline of wild cotton-top tamarin population. *Nature Communications* **1**: 1–7.

Setchell, J. M. and Curtis, D. J. (2003). *Field and Laboratory Methods in Primatology*. Cambridge: Cambridge University Press.

Sha, J. C. M., Bernard H., and Nathan, S. (2008). Status and conservation of proboscis monkeys (*Nasalis larvatus*) in Sabah, East Malaysia. *Primate Conservation* **23**: 107–120.

Singh, M., Udhayan, A., Kumar, M. A., Kumara, H. N., and Lindburg, D. G. (1999). Status survey of slender loris *Loris tardigradus lydekkerianus* in Dindigul, Tamil Nadu, India. *Oryx* **33**: 30–36.

Stanley, T. R. and Royle, J. A. (2005). Estimating site occupancy and abundance using indirect detection indices. *Journal of Wildlife Management* **69**: 874–883.

Starr, C., Nekaris, K. A. I., and Leung, L. (2012). Hiding from the moonlight: luminosity and temperature affect activity of Asian nocturnal primates in a highly seasonal forest. *PLoS One* **7**: e36396.

Starr, C., Nekaris, K. A. I., Streicher, U., and Leung, L. K. P. (2011). Field surveys of the vulnerable pygmy slow loris *Nycticebus pygmaeus* using local knowledge in Mondulkiri Province, Cambodia. *Oryx* **45**: 135–142.

Stokes, E. J., Strindberg, S., Bakabana, P. C., Elkan, P. W., and Iyengue, F.C. (2010). Monitoring great ape and elephant abundance at large spatial scales: measuring effectiveness of a conservation landscape. *PLoS One* **5**: e310294.

Strindberg, S., Buckland, S. T., and Thomas, L. (2004). Design of distance sampling surveys and Geographic Information Systems. In: Buckland, S. T., Anderson, D. R., Burnham, K. P., Laake, J. L., Borchers D. L., *et al.* (Eds), *Advanced Distance Sampling: Estimating Abundance of Biological Populations*. pp. 190–228. Oxford: Oxford University Press.

Teelen, S. (2007). Primate abundance along five transect lines at Ngogo, Kibale National Park, Uganda. *American Journal of Primatology* **69**: 1030–1044.

Thomas, L., Buckland, S. T., Rexstad, E. A., Laake, J. L., Strindberg, S., *et al.* (2010). Distance software: design and analysis of distance sampling surveys for estimating population size. *Journal of Applied Ecology* **47**: 5–14.

Thorn, J. S., Nijman, V., Smith, D., and Nekaris, K. A. I. (2008). Ecological niche modelling as a technique for assessing threats and setting conservation priorities for Asian slow lorises (Primates: *Nycticebus*). *Diversity and Distributions* **15**: 289–298.

Torres, J., Brito, J. C., Vasconcelos, M. J, Catarino, L., Gonçalves, J., *et al.* (2010). Ensemble models of habitat suitability relate chimpanzee (*Pan troglodytes*) conservation to forest and landscape dynamics in Western Africa. *Biological Conservation* **143**: 416–425.

Tranquilli, S., Abedi-Lartey, M., Amsini, F., Arranz, L., Asamoah, A., *et al.* (2011). Lack of Conservation effort rapidly increases African great ape extinction risk. *Conservation Letters* **5**: 48–55.

Tutin, C. E. G. and Fernandez, M. (1984). Nationwide census of gorilla (*Gorilla. g. gorilla*) and chimpanzee (*Pan. t. troglodytes*) populations in Gabon. *American Journal of Primatology* **6**: 313–336.

Tweh, C., Lormie, M., Kouakou, C. Y., Hillers, A., Kühl, H. S., *et al.* (2014). Conservation status of chimpanzees (*Pan troglodytes verus*) and other large mammals in Liberia: a nationwide survey. *Oryx*: doi:10.1017/S0030605313001191.

van Andel, A. C., Wich, S. A., Boesch, C., Koh, L. P., Robbins, M. M., *et al.* (2015). Locating chimpanzee nests and identifying fruiting trees with an unmanned aerial vehicle. *American Journal of Primatology*: doi: 10.1002/ajp.22446.

van Krunkelsven, E., Bila-Isia, I., and Draulans, D. (2000). A survey of bonobos and other large mammals in the Salonga National Park, Democratic Republic of Congo. *Oryx* **34**: 180–187.

van Schaik, C. P., Wich, S., Utami, S., and Odom, K. (2005). A simple alternative to line transects of nests for estimating orangutan densities. *Primates* **46**: 249–254.

Wallis, J. and Lee, D. R. (1999). Primate conservation: the prevention of disease transmission. *International Journal of Primatology* **20**: 803–826.

Walsh, P., Abernethy, K., Bermejo, M., Beyers, R., De Wachter, P., *et al.* (2003). Catastrophic ape decline in western equatorial Africa. *Nature* **422**: 611–614.

Waltert, M., Abegg, C., Ziegler, T., Hadi, S., Priata, D., *et al.* (2008). Abundance and community structure of Mentawai primates in the Peleonan forest, North Siberut, Indonesia. *Oryx* **42**: 1–5.

Weisenseel, K., Chapman, C. A., and Chapman, L. J. (1993). Nocturnal primates of Kibale Forest: effects of selective logging on Prosimian densities. *Primates* **34**: 445–450.

White, L. and Edwards, A. (2000). *Conservation Research in the African Rain Forests. A Technical Handbook*. New York, NY: The Wildlife Conservation Society.

Wich, S. A. (2015). Drones and conservation. In: Kakaes, K. (Ed.), *Drones and Aerial Observation: New Technologies for Property Rights, Human Rights, and Global Development. A Primer*. Available at: <http://drones.newamerica.org/primer/> [Accessed November 2015].

Wich, S. A., Garcia-Ulloa, J., Kühl, H. S., Humle, T., Lee, J. S., *et al.* (2014). Will oil palm's homecoming spell doom for Africa's great apes? *Current Biology* **24**: 1659–1663.

Williams, B. K., Nichols, J. D., and Conroy, M. (2002). *Analysis and Management of Animal Populations*. San Diego, CA: Academic Press.

Williams, J. M., Lonsdorf, E. V., Wilson, M. L., Schumacher-Stankey, J., Goodall, J., *et al.* (2008). Causes of death in the Kasekela chimpanzees of Gombe National Park, Tanzania. *American Journal of Primatology* **70**: 766–777.

Wilson, R. T. (1979). The primates in Darfur, Republic of Sudan. *Folia Primatologica* **31**: 219–226.

Worton, B. J. (1987). A review of models of home range for animal movement. *Ecological Modelling* **38**: 277–298.

Worton, B. J. (1989). Kernel methods for estimating the utilization distribution in home-range studies. *Ecology* **70**: 164–168.

Wrege, P. H., Rowland, E. D., Bout, N., and Doukaga, M. (2011). Opening a larger window onto forest elephant ecology. *African Journal of Ecology* **50**: 176–183.

CHAPTER 7

Habitat change: loss, fragmentation, and degradation

Mitchell Irwin

Recent deforestation of rainforest habitat at Tsinjoarivo, eastern Madagascar. Photo copyright: Mitch Irwin

7.1 Introduction

The very definition of ecology (from the Greek word 'Oikos', meaning house or family) is the study of an organism's interactions with its habitat. It is little surprise that as habitat changes, this relationship also changes, leaving the organism facing potential challenges related to its nutritional state, health, predation risk, and parasitism, as well as altering trophic, mutualistic, and competitive relationships with other species. Such changes may induce behavioural responses such as altered ranging, habitat use, and social behaviour, which may fully or partially mitigate the impacts (but in some cases may exacerbate them).

Habitat change is one of the most important pressures primate populations face today, yet the term itself is hard to define, as it encompasses a broad range of processes resulting from an even broader range of proximate and ultimate causes. Even habitat itself is hard to define, as in reality primates vary in habitat requirements, and do not divide landscapes dichotomously into either 'habitat' or 'non-habitat'. As a necessary simplification, this chapter will tend to treat 'forest' as primate habitat, and 'non-forest' as non-habitat, even though this is untrue for several primate species, and oversimplified for the rest.

Here, habitat change is defined as including three major forces (Figure 7.1): habitat loss (conversion of

Irwin, M.T., *Habitat change: loss, fragmentation, and degradation*. In: *An Introduction to Primate Conservation*. Edited by: Serge A. Wich and Andrew J. Marshall, Oxford University Press (2016). © Oxford University Press.
DOI 10.1093/acprof:oso/9780198703389.003.0007

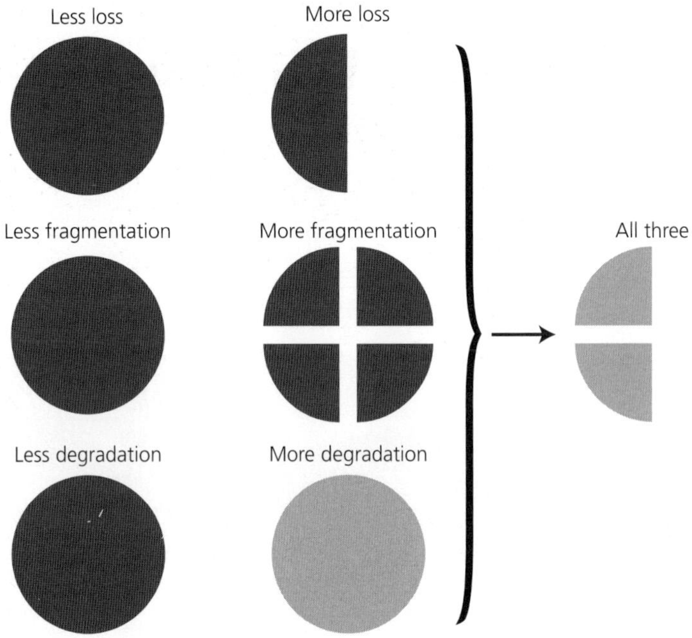

Figure 7.1 Schematic depiction of the three types of habitat change discussed in this chapter; black indicates undisturbed habitat and grey indicates disturbed habitat.

habitat to non-habitat), habitat fragmentation (habitat subdivision and increase in the number and isolation of habitat remnants), and habitat degradation (detrimental changes in the local characteristics of the habitat). The causes and prevalence of each of these processes is discussed, as well as the specific impacts they can have on primate individuals and populations, and the theoretical frameworks we can use to explain their effects and predict future impacts.

7.2 Habitat loss

7.2.1 The problem: patterns and scope of deforestation

Destruction and conversion of habitat is a major cause of biodiversity loss throughout the world. The most recent assessments of global forest cover (FAO 2010, 2012) found that just over 4 billion hectares of forest remain, equivalent to roughly 30% of Earth's land area, and this area is steadily decreasing. Self-reported statistics provided by countries suggest that the annual deforestation rate declined from the 1990s (net loss of 8.3 million hectares per year) to the decade 2000 to 2010 (5.2 million hectares per year). However, FAO's remote sensing survey suggests the opposite, with net annual losses of 2.7 million hectares in the 1990s and 6.3 million hectares between 2000 and 2005—which adds up to a net loss of 66.4 million hectares over that 15-year period (1.7% of all forest).

However, the reality of deforestation's impacts on primates in particular is much more serious than these statistics imply, for two reasons. First, only 36% of the 'forest cover' in these assessments is primary forest—the remainder includes forested areas resulting from natural regeneration and planted forests (which have utility for such things as supplying construction wood, but are less likely to support primate populations). Second, there is a vast discrepancy among climatic domains (FAO 2012). Boreal, temperate, and subtropical forests all saw net gains in forest cover over 1990 to 2005. Primates are largely restricted to the fourth climatic domain, tropical forest, which is the most extensive (1.6 billion hectares in 2005), but suffered net losses of 5.6 million hectares per year between 1990 and 2000, and 9.1 million hectares per year between 2000 and 2005 (the latter time period representing 0.6% lost per year). These losses are much higher than the global average and are the primary drivers of net global losses. The factors that have caused this difference among climatic domains are complex, but largely arise from the economic disparities between temperate countries, which tend to have stronger and more mature economies, and tropical countries, which are more often poverty-stricken and/or 'emerging' economies, a category characterized by growing populations, a high appetite for raw materials, and less central control. If the current rates of tropical forest loss continue, it's easy to extrapolate that only a shockingly short time (167 years) remains until all tropical forest will be gone.

Even within tropical regions, the differences among primate range countries are striking. Some inherited a large share of tropical forest and primate diversity: Brazil, the Democratic Republic of the Congo, and Indonesia are the three top tropical countries in forest area, collectively having 768,000,000 hectares (FAO 2010), or nearly half of all tropical forest (Table 7.1). Primate range countries also vary considerably in the extent of past and ongoing deforestation within their borders (Table 7.2); this varies widely, based on socio-economic and political conditions, population density, economic growth, poverty, political stability, corruption, and global economic demands, some of which differ among countries in complex and hard-to-quantify ways (Estrada 2013; Galinato and Galinato 2013). The situation continues to evolve: although Brazil still leads in absolute annual forest loss, its deforestation rate is declining, while Indonesia's annual loss has been increasing in recent years (Hansen *et al.* 2013; Gaveau *et al.*, Chapter 12, this volume).

7.2.2 How can habitat loss cause primate declines and extinctions?

Both the area affected by habitat destruction and the spatial configuration of remaining forest can be easily and accurately assessed using remote sensing (Harper *et al.* 2007; Hansen et al. 2013; Gaveau *et al.* 2014). The more difficult task is to build theoretical frameworks to translate different degrees of habitat loss and landscape configurations into

Table 7.1 The top ten countries ranked by forest area include six primate range countries (countries with native non-human primate populations) (Source: FAO 2010).

Country	2010 reported forest area (million ha)	% of land area	Average annual loss/gain 2005–10 (million ha and percent change)	Non-human primate species richness[1]
Russian Federation	809	49	+ 0.060 (+ 0.01%)	0
Brazil	520	62	−2.194 (−0.42%)	111
Canada	310	34	0 (0%)	0
United States of America	304	33	+ 0.383 (+ 0.13%)	0
China	207	22	+ 2.763 (+ 1.39%)	20
Democratic Republic of the Congo	154	68	−0.311 (−0.20%)	35
Australia	149	19	−0.924 (−0.61%)	0
Indonesia	94	52	−0.685 (−0.71%)	46
Sudan[2]	70	29	−0.054 (−0.08%)	12
India	68	23	+ 0.145 (+ 0.21%)	22

[1] Primate species richness figures (total number of species) from www.iucnredlist.org.
[2] Figures for Sudan predate the political succession of South Sudan.

Table 7.2 Primate range countries ranking highest in (a) average absolute annual forest loss and (b) average annual percent forest loss 2005–10 (Source: FAO 2010).

(a)

Country	**2010 reported forest area (million ha)**	**% of land area**	**Average annual loss/gain 2005–10 (million ha and percent change)**
Brazil	520	62	−2.194 (−0.42%)
Indonesia	94	52	−0.685 (−0.71%)
Nigeria	9	10	−0.410 (−4.00%)
Tanzania	33	38	−0.403 (−1.16%)
Zimbabwe	16	40	−0.327 (−1.97%)
Democratic Republic of the Congo	154	68	−0.311 (−0.20%)

(b)

Country	**2010 Reported forest area (million ha)**	**% of land area**	**Average annual loss/gain 2005–10 (million ha and percent change)**
Togo	0.3	5	−0.020 (−5.75%)
Nigeria	9	10	−0.410 (−4.00%)
Uganda	3	15	−0.088 (−2.72%)
Pakistan	1.7	2	−0.043 (−2.37%)
Ghana	5	22	−0.115 (−2.19%)
Honduras	5	46	−0.120 (−2.16%)

likely outcomes in terms of primate population decline and extinction risk. The most obvious effect of habitat loss for a primate species (not yet considering the effects of fragmentation and degradation) is a simple reduction in population size; assuming no change in population density, loss of 30% of a species' natural habitat should cause a 30% reduction in the wild population. The impact of such population reductions is highly dependent on the precise numbers: dropping from 100 to 70 is clearly more dire than dropping from 100,000 to 70,000. Similarly, losing the initial 30% is not as serious as a subsequent decline from 31 to 1% of the original population, even though the absolute number lost is the same.

Both these examples illustrate the principle that extinction risk increases in a non-linear fashion as populations decrease—that is, the risk stays low over much of the decline, then increases rapidly at the end. The trickier question is whether there is a 'point of no return'—a population size below which recovery becomes impossible due to the depletion of genetic variation. Conservation biologists in the 1980s introduced the term 'minimum viable population size' (Franklin 1980)—defined probabilistically as the smallest population size that can assure a 95% certainty of survival over a defined time period. Early attempts to quantify minimum viable population size were almost certainly oversimplifications of a complicated issue, but one of the charges of the growing field of conservation genetics is to define how extinction risk and recovery potential change as population size drops (see Salgado Lynn *et al.*, Chapter 5, this volume).

Moving from the level of the individual species to the primate community, the next question is how habitat loss might reduce species richness in a given area. The logic of the preceding paragraphs might be easily applied when the question is asked at a particular scale—for example, if a small private reserve decreased from 1,000 to 800 ha, the resident primate species might each suffer a population reduction of 20%, and persist without problem.

However, at other spatial scales, impacts may vary across species, with some going extinct.

At larger scales, species exist in allopatry (i.e. they have non-overlapping geographic ranges), meaning that the degree of habitat reduction is not the same for all species. For example, Brazil is down to 62% forest cover at a national level, yet most deforestation has happened in eastern and southern areas closer to the coast. Therefore, some interior Amazon species might have suffered zero deforestation within their geographic ranges, while some coastal species have lost most of their habitat—such as the golden lion tamarin, which came perilously close to extinction (Kierulff *et al.* 2012). On much smaller scales, for example within a coastal Brazilian forest fragment, some primate species in a community may be lost when habitat reduction drives their population below viable levels, while others survive. In both cases a species-by-species approach is needed—and the extinctions might take several primate generations to unfold.

Ecologists' understanding of how habitat loss can lead to species extinctions has largely been guided by a phenomenon known as the 'species–area relationship'. Numerous studies dating back to the early twentieth century noted the positive, curvilinear relationship between area sampled and species richness (Connor and McCoy 1979): typical examples include different-sized sampling areas within a broader expanse of habitat, such as the Amazon forest, or isolated habitats of different sizes, such as oceanic islands. Although various mathematical functions have been fit to this relationship, the most commonly used is the double logarithmic function, whereby the relationship between species richness and area is 'linearized' by plotting log (species) against log (area). For a given habitat area (or isolated patch), the 'species–area relationship' predicts that species richness (S) within a community is related to area (A) according to the relationship:

$$\log S = \log c + z \log A,$$

or

$$S = cA^z$$

where c and z are constants (the logarithmic y-intercept and slope). Although the biological underpinnings of this relationship are still debated, it's clear that several forces are at work; the four most frequently cited explanations are: (1) the 'sampling hypothesis', a null model stating that smaller areas have fewer species due to sampling considerations alone; (2) the 'habitat-diversity hypothesis', maintaining that larger areas enclose more discrete habitat types and therefore more species; (3) the 'area-per-se hypothesis', which follows classical Island Biogeography Theory (MacArthur and Wilson 1963, 1967) and explains species richness based on the local balance between extinction and colonization rates, which are impacted by the size and isolation of the habitat patch; and (4) the 'disturbance' hypothesis, which posits that a larger patch is more likely to have interior areas free from disturbance (Lindenmayer and Fischer 2006). Which of these explanations drives any particular species–area relationship (and it may be more than one) is likely to depend on many factors. One such factor is spatial scale (1 and 2 fit the Brazil-wide example above, while 3 better fits the coastal forest fragment example), and another is whether one is considering samples within contiguous habitat or isolated fragments.

Empirical estimates of the constants c and z may be informative in comparisons among systems, regions, or guilds—and judging the likely extinction risk resulting from different habitat loss scenarios (but see Connor and McCoy (1979) for warnings). The logarithmic slope 'z' varies considerably among island studies, usually falling between 0.20 and 0.40 (Connor and McCoy 1979), but non-isolated systems, where 'patches' are samples within continuous habitat, have lower z-values (0.12–0.19; Connor and McCoy (1979)); in these systems, species which could not survive in isolated patches do exist in the contiguous landscape, leading to a shallower slope.

Conservation biologists were quick to realize, beginning in the 1970s, that this simple relationship could be applied to the problem of reserve design and habitat management: it was seen as a way to predict the expected number of species extinctions (Wilcox 1980) resulting from the past or planned reduction of a protected area or forest block. Generally, empirical evidence upholds the existence of this relationship in primates. Marshall *et al.* (2010) examined monkey communities in forest fragments of the Udzungwa Mountains of Tanzania; although additional climate, landscape, and hunting variables had significant impacts, by far the top predictor of species richness was the area of the fragment. Benchimol and Peres (2013) also found area was a strong predictor of species richness for primates in Neotropical forests, while identifying hunting as an additional pressure. Cowlishaw (1999) applied this principle on a continental scale, examining the primate species richness of African countries and their historic and current forest cover. He found a strong positive relationship between a country's forest area and its primate species richness, but the relationship was stronger for historic, rather than current, forest cover. This suggests that these faunas have yet to 'relax' to new equilibria, raising the possibility of an 'extinction debt'—inevitable extinctions that will unfold in the generations to come. The average loss predicted at the country level was 30% of its species (1–8 species each). But is the inference really that simple? One major complication is that this seemingly simple relationship is seen in so many different situations (samples within continuous habitat, isolated habitat remnants, or oceanic islands) and scales (from continental to very small scale), and the drivers of the relationship vary. Although these studies highlight the importance of habitat area, the mechanisms at work are almost certainly different. On smaller scales, such as that in the Marshall *et al.* study, the driving forces are likely pressures causing local extinction in vulnerable small populations and reducing species richness within local communities. On the continental scale, the number of species under consideration is much higher than any local species richness, and the mechanism by which the 'extinction debt' will be paid is less clear.

More generally, it's important to keep in mind that modelling species richness as a single number addresses the bigger picture (community ecology), but in the process it sacrifices precision and detail. Although the species–area relationship is useful as a guiding principle, primatologists should seek to examine and understand the extinction risk faced by individual species whenever the available data allow such an approach.

7.3 Habitat fragmentation

7.3.1 The problem: patterns and scope of habitat fragmentation

The second major type of habitat change that can affect primates is habitat fragmentation, or the subdivision of remaining habitat into discrete patches embedded in a 'matrix' of non-habitat (Fahrig 2003). Like habitat loss, fragmentation can be assessed via remote sensing and GIS, and various methods exist to quantify both fragmentation and its corollary, connectivity (Jorge and Garcia 1997; Moilanen and Nieminen 2002). We can conceptualize habitat loss and fragmentation separately, but in the real world they are intertwined, since it is usually loss that drives fragmentation in the first place. It is therefore extremely difficult to measure the effects of fragmentation without a confounding impact of habitat loss. The repercussions of fragmentation itself are best envisioned by asking: 'from the point of view of the target species, how does this subdivided habitat differ from a landscape with an equal amount of similar habitat that is not subdivided?' Under this scenario, the two habitats can in theory support the same number of primate species or individuals, assuming the amount of food produced is equal (which may not be true—see Section 7.4). The difference is simply that the populations are subdivided, rather than continuous, thereby restricting individuals' movement through the landscape.

This concept can be best illustrated as a continuum between two extremes (Figure 7.2). On the left, individuals range only within fragments, and never cross the matrix, yielding four separate populations—the strongest possible impact. On the right, individuals range freely across the landscape, meaning it is simply a population and the fragmentation itself has not had any negative impacts (though the loss of habitat might). In the intermediate scenario, shown in the middle, fragmentation inhibits movement among patches, but some amount of movement still occurs. For primates, this might represent a situation in which groups occupy single fragments (perhaps sharing with other groups) but individuals are motivated enough to cross to other fragments during dispersal from natal groups.

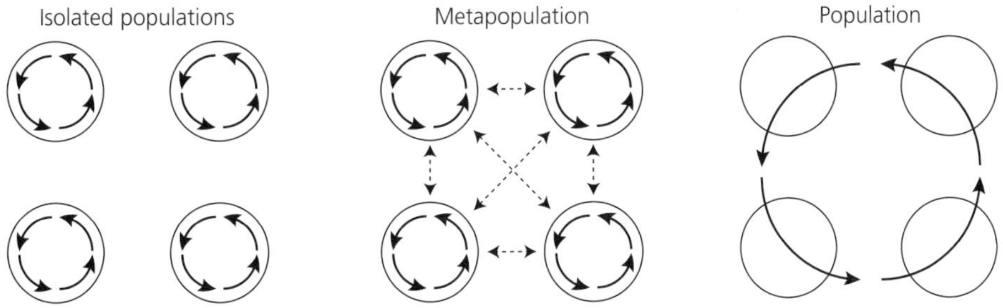

Figure 7.2 Schematic representation of three responses primates might exhibit in a landscape of four habitat fragments. On the left, individuals move only within fragments; on the right individuals move freely among patches (even though we perceive the habitat as patchy, the animals do not). Between these two extremes are metapopulations, in which most ranging takes place within habitat fragments, but animals do sometimes cross the matrix.

7.3.2 How can habitat fragmentation cause primate declines and extinctions?

The most immediate threat of fragmentation is when animals become stranded in a habitat fragment smaller than an ideal home range, which can lead to extirpation or attrition of individuals (if food supply is insufficient), or a more precarious existence over the longer term (i.e. less 'insurance' in terms of the ability to 'habitat shift' in years of a bad food crop). Incorporating multiple habitat patches into a home range is one way to mitigate this threat, though this is possible only if the species is tolerant of traversing the matrix between patches. On the other hand, if the scale of fragmentation is such that groups' or individuals' home ranges are not restricted, existence over the short term may not be affected, but individuals wishing to disperse may face problems. Animals may either show delayed or reduced rates of dispersal, or may suffer disproportionately higher mortality during dispersal due to their vulnerability in matrix. Either scenario may have detrimental demographic and population genetic consequences.

These examples underscore the fact that the impact of fragmentation on primates is highly scale dependent: what is a huge fragment to a mouse lemur might not support even a subgroup of chimpanzees. All else being equal, the negative impacts should be roughly proportional to the ratio between fragment size and home range size for an individual or group. A high ratio means a greater abundance of individuals/groups per fragment, meaning that subpopulation should persist longer even in the absence of dispersal.

For primate populations whose landscape has restricted, but not eliminated, movement among fragments (Figure 7.2, centre), metapopulation theory can be an extremely useful conceptual tool. The idea of a 'metapopulation', or a 'set of local populations which interact via individuals moving between local populations' (Hanski and Gilpin 1991) was first introduced (Levins 1969, 1970) in the context of pest control problems (Hanski 1999; Hanski and Gilpin 1997). The original formulation of the model included several spatially discrete habitat patches, each of which is suitable habitat for the focal species but not continuously occupied (but the model was spatially implicit, meaning differences among patches were not considered). Any individual population has a finite lifetime (i.e. time to extinction), while the metapopulation as a whole can be maintained through a balance between these extinctions and subsequent re-colonization from other patches—in much the same way as a population can persist even though no individual lives forever. In its simplest form, the change in metapopulation occupancy (p, proportion of patches occupied) over time can be expressed as:

$$dp/dt = mp(1 - p) - ep$$

where 'm' is the rate of colonization of empty patches and 'e' is the rate of extinction of occupied patches. The equilibrium occupancy p^* can be solved by setting dp/dt to 0:

$$p^* = 1 - e/m$$

Although originally formulated for single-species models, metapopulation models gradually expanded to include two-species and multispecies/'metacommunity' models (Holyoak *et al.* 2005). In addition, the earliest models suffered from the oversimplification of being spatially implicit (i.e. patches were assumed to have equal probability of extinction and colonization), thereby ignoring important aspects of landscape configuration. Later models, however, became spatially explicit and considered the area and isolation of each patch, with isolation modelled as a function accounting not just for the distance to the nearest occupied patch, but to all occupied patches discounted by their distance from the focal patch ('Incidence Function Models': Hanski (1998, 1999)). In addition, the rescue effect (Brown and Kodric-Brown 1977) was included to account for the effects of dispersal to already occupied patches, which can fortify those populations (Hanski 1991).

These more complex models have generated predictions for landscape occupancy that match empirical observations closely (e.g. Wahlberg *et al.* 1996). When incorporated as a fundamental part of population viability analyses (PVAs), such models can be extremely useful in predicting the effects of future habitat destruction in real landscapes, or the future effects of existing habitat destruction (Lande 1988); such models are already being applied to primates, yielding extremely useful recommendations for future management of fragmented landscapes (Zeigler *et al.* 2013). In addition, metapopulation models can also be used for sensitivity analyses, to ascertain which species characteristics and landscape parameters are most important in determining the likelihood of species' persistence (Swart and Lawes 1996).

Metapopulation models, therefore, go further than species–area relationships in allowing more realistic models of landscape structure, and using observed rates of dispersal and habitat preferences to guide simulations. This can be extremely useful both in estimating the long-term chances of persistence for species whose landscape has become fragmented, and in modelling the effects of future reductions in the size and number of habitat patches. However, this increased sophistication comes at a high cost; it requires fairly intimate knowledge of these species' habitat requirements and proclivity for dispersal. As with any model, no matter how sophisticated, it is only as good as the data used to parameterize it.

One key point about the metapopulation approach is that different species will experience the same landscape in different ways. Across animals more broadly, some of these differences are obvious: in Figure 7.2, the right panel might represent a bird, the left panel a snail, and the middle panel a forest-dependent primate whose subadults cross between patches in search of mating opportunities. Within primates, it is possible to imagine similar diversity: chimpanzees' large size and terrestrial locomotion seems to enable them to cross through matrix relatively easily (Tutin 1999; Tutin *et al.* 1997), and even some larger monkeys such as mandrills and orangutans are similarly mobile (Ancrenaz *et al.* 2015). Long-term studies that have proven a lack of mobility are hard to find in primate species, but this is well imaginable for smaller, strictly arboreal primates in patches that are relatively far apart—and community ecology work has pointed to dispersal as a major limitation shaping primate community assembly (Beaudrot and Marshall 2011).

A second key point is that we must not forget that 'the matrix matters' (Ricketts 2001; Jules and Shahani 2003). Even if one can make the simplifying assumption that the 'habitat' itself is homogeneous and not degraded, the matrix in which fragments are embedded is tremendously variable, and its nature directly affects whether animals will cross (Prugh *et al.* 2008). A 'harder' matrix might be agricultural fields or low grassland, which offer no natural foods and a high predation risk, while a 'softer' matrix might be secondary forest, in which primates can move arboreally, have some protection from predators, and possibly even consume natural foods. In sum, both intrinsic (size, locomotion, and personality in terms of risk avoidance), and extrinsic factors (especially the distance between fragments and the nature of the matrix) combine to influence the degree to which primates will choose to cross matrix to reach another fragment—and the realized rates of patch crossing determine the strength of fragmentation's impacts.

A final consideration that may prove more troublesome is that the sociality of many primates violates one of the main assumptions of metapopulation theory: that re-colonization of an unoccupied patch requires only the arrival of a dispersing individual. In reality, group-living primates seek groups (not only because they seek mates, but also because animals often infer habitat quality from the presence of conspecifics); this 'conspecific attraction' (Smith and Peacock 1990; Zeigler *et al.* 2011; Mihoub *et al.* 2011) means that dispersers arriving at unoccupied, but otherwise suitable, fragments are unlikely to settle. This can be a major obstacle to the persistence of a primate metapopulation, potentially causing landscape extinction even when models predict persistence.

Until now this section has been focused at the level of individual species, yet in reality conservation strategies in fragmented landscapes should address all species together, with the overarching goal of minimizing species extinctions. One example of the complexity involved is the need to balance measures of success on different scales, exemplified by the 'SLOSS' debate (Single Large Or Several Small). Conservation biologists have long recognized the conflict between maximizing species richness in individual habitat fragments or protected areas, and maximizing the species collectively preserved on a regional or continental level—the debate asked: given a fixed amount of land to be protected, should conservationists focus on a single large reserve (to keep population numbers high) or several small ones (to spread protected land across different species' ranges and different habitat types)? The short answer is: it depends—on to what degree reserves would be redundant, or complementary, in the species they contain (Gilpin and Diamond 1980).

The degree of redundancy can be measured via 'nestedness' (Atmar and Patterson 1993), an important parameter of fragmented landscapes that reflects to what extent species-poorer reserves contain communities that are 'nested subsets' of species-richer reserves (Figure 7.3). In a perfectly nested system, any species found in a habitat fragment of species richness '*s*' should also be present in all patches with species richness '*s*' or higher—reduction in size extirpates species deterministically, in order from most to least vulnerable. The higher the nestedness, the stronger the motivation to maximize reserve area rather than the number of reserves. Empirical evidence has shown a high nestedness in primate communities (Dehgan 2003)—in

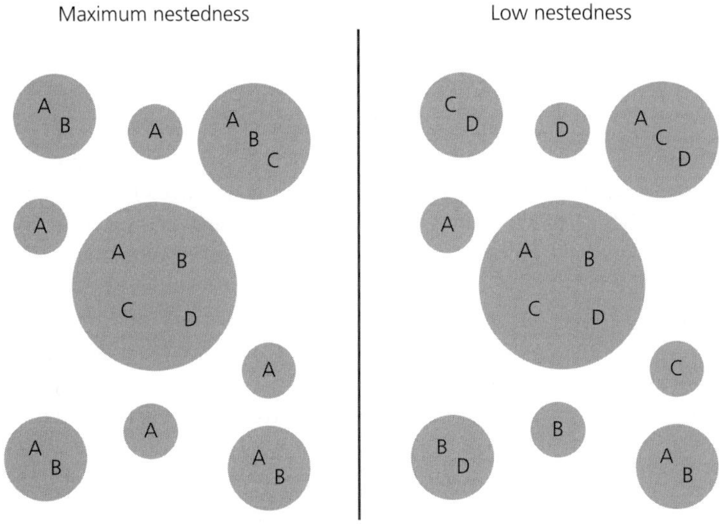

Figure 7.3 Schematic representation of 'nestedness'. In this network of isolated forest fragments (in grey), the central large fragment contains the full primate community (four species: A, B, C, D) but the smaller fragments have suffered local extinctions due to decreased fragment area. The fragment network on the left shows perfect nestedness, with species D, C, and B dropping out in order with decreasing area. The fragment network on the right shows low nestedness, with varying subsets of the four species persisting in smaller fragments.

other words, species vary systematically from more to less tolerant of area reduction (Benchimol and Peres 2014), as expected based on their substantial differences in home range, ecology, and locomotion. Although the SLOSS concept by itself is too simplistic to be of much use, like the species–area relationship, it has helped frame key issues, such as the concept of nestedness—and these simple concepts are gradually yielding to more complex approaches that model each species individually, yet integrate results quantitatively to better guide landscape-level conservation decisions.

7.4. Habitat degradation

7.4.1 The problem: patterns and scope of habitat degradation

The amount of remaining habitat and its spatial configuration are of great importance in predicting species' declines; however, assuming that all remaining habitat is equivalent in quality to the original state may lead to an underestimate of the impacts. The third key question that must be addressed, and the hardest to assess from satellite images, is the 'quality' (i.e. internal characteristics) of the remaining habitat—in other words, its level of degradation. This degradation typically falls in two main categories: 'edge effects', which are the naturally occurring changes that unfold at newly created habitat edges, and anthropogenic extraction.

7.4.2 Edge effects

'Edge effects' (Harper *et al.* 2005) are the physical and biotic changes that occur at a habitat edge (i.e. where habitat meets non-habitat): 'the result of the interaction between two adjacent ecosystems, when the two are separated by an abrupt transition' (Murcia 1995: 58). Forest destruction typically causes edge creation as areas previously surrounded by forest find themselves adjacent to cleared land. However, this definition is relevant to both newly created edges and naturally occurring, more mature edges in mosaic landscapes.

If edges are unsuitable, or less suitable, habitat for primates, then the landscape's effective carrying capacity should likewise be adjusted downward to reflect this. However, the manner in which this adjustment should be made is no simple matter. Empirical observations of both biotic and abiotic conditions near edges and the consequent development of spatial models are the best tools currently available to aid in this process.

Edge effects in forest remnants can be divided into three categories (Murcia 1995): abiotic effects, direct biological effects, and indirect biological effects. Abiotic edge effects result from the striking difference in microclimate between the two adjacent ecosystems. In areas from which forest has been cleared, the reduced or absent plant cover causes increased penetration of solar radiation during the day and re-radiation at night, causing higher daytime temperatures and greater daily fluctuations.

Kapos (1989) found that recently created Amazonian forest fragment edges within agricultural matrix had increased air temperature, decreased air and soil humidity, and increased photosynthetically active radiation relative to fragment interiors, with a penetration distance of 20–60 m. Kapos *et al.* (1997) studied the same 100-ha fragment four years later and found that regrowth had effectively 'sealed' the edge; microclimatic conditions showed reduced deviations from interior conditions, but some variables also showed more complex (non-monotonic) responses to distance from edge.

In much the same way as the 'matrix' matters for animal movement, it also matters for the penetration of edge effects. Didham and Lawton (1999) examined the effect of the regrowth type in surrounding pastures on the magnitude and penetration of microclimatic edge effects. Where matrix had been repeatedly burned to maintain suitable grazing for cattle, edges were dominated by *Vismia* sp., a low, xeric-adapted species, which inhibits the growth of other species (producing 'open edges'). Where matrix regrew naturally, edges were dominated by the taller *Cecropia* spp. (producing 'closed edges'). The open edges had higher air temperatures, higher rates of evaporative water loss, lower leaf litter moisture, and much steeper gradients for all variables. Although regrowth pathways are region specific due to both natural and anthropogenic variation (Laurance *et al.* 2007), these results show the potential impact of human activities on edge regeneration, and of matrix vegetation on edge effects.

The second type of effect, 'direct biological effects', comprises changes in the physical structure of the vegetation, such as increased leafing and subcanopy closure at edges in response to increased light levels, and species turnover (i.e. loss of some species and invasion by others). This turnover can occur via damage and mortality of some plant species, typically caused by windfall, water stress, or fire. Laurance *et al.* (1998) documented increased mortality, damage, and turnover for trees at the edges of Amazonian fragments. Finally, tree recruitment patterns may also be affected, with altered seed production, germination, and seedling survival leading to successional changes in the plant community, aided by seed rain of successional species from nearby matrix (Janzen 1983).

Finally, the changes summarized above can cause indirect biological effects on other species in the community via trophic or competitive relationships—this is the class most likely to impact primates. For example, populations of frugivorous primates dependent on trees that suffer increased mortality may decline, while those of folivores who find the leaves of pioneer species palatable may increase. These effects are likely to be complex, as the diverse and as-yet largely unknown species interactions cause variation among species and regions that confounds generalizations.

The little empirical work available already emphasizes the diversity of reactions primates have to edges. Lehman and coworkers (Lehman 2007; Lehman *et al.* 2006a, b, c) used a census-based approach and found that most lemurs were either omnipresent (detected equally frequently at edges vs interiors), or 'edge-tolerant' (detected at higher frequencies within 100 m of the edge). Only the most frugivorous species, the red-fronted lemur (*Eulemur rufifrons*), was detected less frequently near edges. Irwin (2008) used GIS analysis of focal animal data to quantify edge use in four diademed sifaka groups (*Propithecus diadema*), and found that these mostly folivorous primates avoided edges when habitat was more degraded, but used edges disproportionately often in less-disturbed habitat, where they ate leaves and fruit of two trees (one of which was invasive) found only at edges.

In general, the factors that determine how edge effects will impact primates are varied, and include characteristics of the primate (diet, predation risk, etc.), the forest type and climate, the geometry of the edge (including its aspect, or the direction it faces), and the characteristics of the matrix (Figure 7.4). On a landscape scale, the cumulative impact of edge effects in a landscape will depend strongly on how much edge there is in that landscape—meaning that management efforts should minimize the amount of edge created, or prioritize less 'edgy' habitat for protection.

7.4.3 Anthropogenic disturbance within forests

Humans are impressively industrious and mobile animals, and this is a major reason that habitat degradation and alteration is not restricted to forest edges. The most pervasive impact humans have on forests is through their collection of resources, ranging from small-scale collection of medicinal plants and other non-timber forest products (NTFPs) all the way to intensive selective logging. This removal of habitat elements can reduce the 'quality' of habitat for a given primate, via several mechanisms. First, changes in forest structure can impact the microclimate, generally by reducing the insulating characteristics of the forest, and this could have energetic consequences as primates expend more energy on thermoregulation. Second, a change in the forest structure may force a change in the primate's locomotion (Manduell *et al.* 2011)—an obvious example being a newly discontinuous forest canopy forcing primates to come to the ground to cross gaps. Third, changes in visibility can alter predation risk. Fourth, human presence and/or other ecological changes could affect the risk and consequences of infectious disease, especially zoonotic pathogens (Gillespie *et al.* 2009; Chapman *et al.* 2005). Fifth, if the changes reduce the availability of food for the primate, that habitat's carrying capacity (on a per area basis) will be reduced (Felton *et al.* 2010; Husson *et al.* 2009; Balko and Underwood 2005; Rode *et al.* 2006). Finally, one might also count the indirect effect of hunting as a byproduct of other forms of forest use (Remis and Robinson 2012). It is important to remember that some of these changes need not be negative: for example, increased sun penetration may aid in thermoregulation, and plant

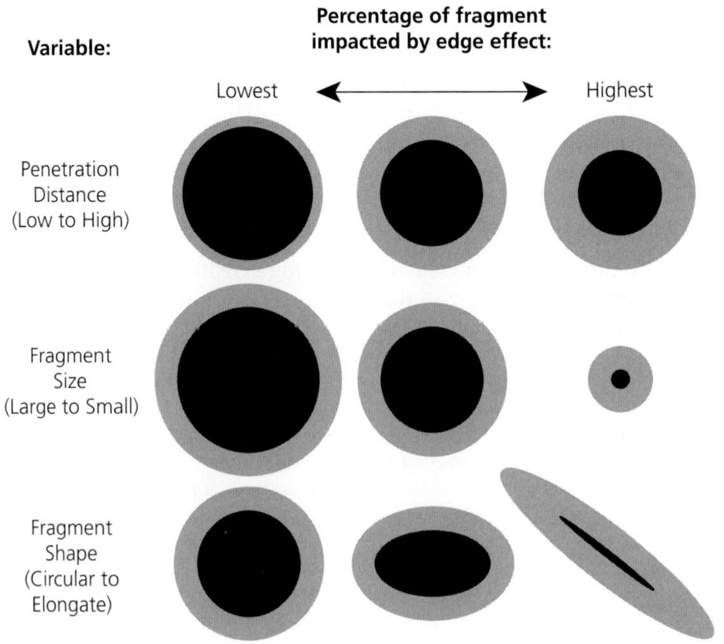

Figure 7.4 Schematic representation of how three spatial variables affect the percentage of a given forest fragment impacted by an edge effect (shown in grey). In reality, the effect would not be dichotomous (i.e. either affected or not), but the edge effect would diminish from the edge towards the interior; however, the same spatial principles should hold.

community changes may actually increase available food.

One of the more extreme forms of habitat disturbance, selective logging, has been shown to affect primates in many ways. Following the immediate impacts of the disturbance, which can be a stressful experience due to noise and human presence (often causing movement out of the area), the primates must adjust to an altered habitat. Mounting evidence since the 1980s is showing impacts of habitat disturbance on population density (Phoonjampa *et al.* 2011; Chapman *et al.* 2010; Felton *et al.* 2003; Herrera *et al.* 2011), behavioural responses in affected primate species (Guo *et al.* 2008; Arnhem *et al.* 2008), and a striking variation among species in the direction and magnitude of the changes.

In one of the only surveys of how habitat disturbance affects primates across the entire order, Johns and Skorupa (1987) compiled and examined population density estimates for 37 rainforest species represented in both intact and degraded forest. A species' 'survival ratio' (density in disturbed forest divided by density in intact forest) was negatively impacted by its degree of frugivory, particularly when the effect of body size was accounted for—in other words more folivorous species seemed to be most resilient to disturbance. Harcourt (1998) performed a similar, but phylogenetically controlled, analysis and found that only the species' home range size was correlated with the survival ratio (negatively); diet and other ecological characteristics, somewhat surprisingly, had no predictive power. More recently, Meijaard and colleagues (Meijaard *et al.* 2008) found that the most important predictor of tolerance in Bornean mammal species (including primates) was phylogenetic age—more ancient species tend to be more specialized ecologically and less tolerant, while more recently radiated species are more generalized and more tolerant. Together these results suggest that primate species' susceptibility to logging is highly species specific, and difficult to predict based on simple species characteristics.

Irwin and Raharison (in press) also point out that population density can actually be a misleading indicator of habitat quality (van Horne 1983)—meaning that 'survival ratios' only capture a snapshot of local abundance in a disturbed area, in some studies only shortly after logging. These snapshots may be useful as a quick look, but more detailed study of ecological changes following disturbance is starting to yield suggestions that animals packed in at higher density might not be healthier (as individuals, or as a population) than counterparts in undisturbed forest (Irwin 2008; Irwin *et al.* 2010)—perhaps reflecting only short-term compression of populations. Only through more detailed (and longer term) ecological study of primates in degraded habitat (which will represent a much more substantial investment than censusing) can we gain a better understanding of these complexities.

7.5 Complications

7.5.1 Complication 1: temporal lags in the repercussions arising from habitat change

The forces impacting primates described in this chapter do not happen instantaneously. Many of these changes take years or decades to unfold, including such changes as reduced recruitment of important food tree species, and demographic and genetic collapse of the primate population itself via reduced dispersal of subadults. This causes, and almost certainly has already caused, 'extinction debts' to be incurred (Cowlishaw 1999); habitat changes that have already occurred are slowly but inexorably going to cause species extinctions in the near future (Marshall *et al.* 2009).

One major stumbling block that is particularly important in primate conservation lies in the inconvenient but inescapable fact that primates have unusually long generation times. This likely prolongs both the intrinsic causes of extinction (such as inbreeding) as well as many of the extrinsic causes: for example reduced food supply in degraded habitat may be enough for adults to survive, but not to reproduce. In browsing the metapopulation literature, for example, it can be puzzling how few researchers have studied primates within this framework until one realizes the considerable logistical challenges such a study would pose, since metapopulation dynamics act on an intergenerational timescale. Quite logically, most metapopulation studies have focused on invertebrates, birds, or micromammals, all of which have fast generational turnover. It is a sobering prospect indeed to think that a primatologist could spend five years (and a great deal of money and effort) studying multiple groups of an endangered primate in a fragmented landscape and conclude that no dispersal among patches occurred—yet this was simply because it happens on average every six years. Such a rate may well be enough to keep those subpopulations connected from a genetic point of view (Marshall *et al.* 2009).

The varying threats resulting from habitat change are best viewed as sequential, rather than simultaneous, challenges, much like a set of hurdles to be jumped over one by one. Although this is likely to differ among regions, a common pattern might be that habitat loss acts most quickly (if a habitat is reduced below a viable size), degradation might act on medium time scales (if a reduced food availability causes compromised health and reproductive output), and fragmentation on the longest timescales (if groups are healthy enough within fragments but demographic/genetic implosion unfolds over generations). This is especially important to consider when studies use presence–absence data to infer habitat suitability or species resilience; counting a population as 'present' may mean only that they have passed the earliest hurdles, yet may still stumble at one of the upcoming hurdles. Failing to take this into account will cause an overestimation of the viability of primate populations, and we may wind up with more extinctions than our theoretical frameworks predict.

7.5.2 Complication 2: community effects

In addition to the more direct effects of habitat change detailed above, primates could be impacted indirectly through changes in the animal community in altered habitats; this could include trophic (predator–prey or herbivore–plant), competitive, or mutualistic interspecific interactions. Primates may be especially susceptible to changes in predation rate, and although this may be manifested as reduced predation, when top predators are extirpated in altered habitats, predation rates can increase as well, for two reasons. First, smaller primates may suffer from 'mesopredator release' (Crooks and Soulé 1999): extirpation of top carnivores releases competitive pressure on medium-sized predators, allowing their populations to increase. Second, some studies have suggested that discontinuous landscapes may reduce the continuity of predator movements throughout the landscape; when a predator gets 'stuck' in a fragment, even temporarily, it could have a more concentrated and decimating impact on residents (Irwin *et al.* 2009).

7.6 Summary and future directions

Ignoring the consequences of habitat change on primates is dangerous, and scarcely a risk we can afford to take. Many extant primate species exist only in small, threatened forest fragments: lion-tailed macaques *Macaca silenus* (Umapathy and Kumar 2000), Northern muriquis *Brachyteles hypoxanthus* (Strier and Ives 2012), Central American squirrel monkeys *Saimiri oerstedi* (Boinski and Sirot 1997), golden-crowned sifaka *Propithecus tattersalli* (Meyers and Ratsirarson 1989), and many others inhabit ranges incorporating significant (and increasing) degrees of fragmentation. Even the vast Amazon Basin is rapidly being subdivided by roads and development projects (Laurance *et al.* 2001, 2014).

If the extinction spasm is to be minimized for primates, the following four important and poorly explored initiatives should be increasingly incorporated into the paradigm of primate conservation.

(1) *Understand diversity in landscape use*: primates treat their landscape in a diversity of ways, ranging from strict patch residents to transients visiting several patches in a landscape, and often suffer extirpation from fragments or disturbed areas; this diversity depends on both landscape characteristics (e.g. patch area, isolation, and matrix type), as well as inherent differences between primate species. Perhaps most importantly, the use and tolerance of matrix in both foraging and dispersal, hitherto unexplored in primate studies, needs to be addressed. Studies of Australian marsupials (Laurance 1991) and Amazonian vertebrates (Gascon *et al.* 1999) suggest that matrix tolerance may be the most important factor determining how animals use their landscape, and their chances of survival. Field studies directly quantifying matrix tolerance in primate species are urgently needed to build these theoretical underpinnings.

(2) *Understand processes leading to extinction*: it is important to distinguish the 'declining population' forces from 'small population processes' that cause already vulnerable populations to go extinct; Caughley (1994) suggests that the forces causing extinction in small subpopulations remain poorly understood. Demographic stochasticity has the potential to disrupt reproduction in small populations, primarily by skewed sex ratios; genetic stochasticity can lead to loss of heterozygosity due to genetic drift or inbreeding, leaving populations with reduced prospects of both survival and future adaptability; and environmental stochasticity can wipe out a small population through a localized natural disaster. As conservation managers are increasingly forced to make trade-offs among populations and species, understanding the vulnerability of increasingly smaller populations will be critically important.

(3) *Understand the effects of behaviour on extinction*: most models of extinction in small populations ignore the effects of behaviour, especially social behaviour. Conspecific attraction, described above, is an example of such an effect that would cause difficulties in primate persistence in fragmented landscapes. As primates exhibit a diversity of complex social systems and interactions, they may prove to be especially vulnerable to these effects. In particular, the mechanics of dispersal between social groups may exhibit significant interactions with the effects of habitat discontinuity.

(4) *Slow and reverse habitat change (and continual monitoring from above)*: perhaps as important as understanding the effects of past habitat change is the prevention or slowing of present and future loss, fragmentation, and degradation; unfortunately this is a complex social and economic issue that requires a careful trade-off between human development needs and the needs for intact, functioning ecosystems. One important component of such efforts is the up-to-date monitoring of the rate and spatial distribution of deforestation through remote sensing.

7.7 The future: primate conservation in reduced, fragmented, and degraded habitat

Extant primates face a diversity of threats to their continued existence, of which habitat change is only one. Understanding the nature of primate populations' responses to these changes, including the interspecific differences in tolerance, may allow for wiser application of resources to conservation problems. It must be re-emphasized here that, while signs of failure are self-evident (extirpation/extinction), signs of success should always be regarded with caution. Merely detecting a species in a given fragment or fragmented landscape, even if that species is reproducing, does not necessarily indicate that the species is capable of long-term persistence. Primates, being especially long-lived mammals, may persist in areas incapable of supporting a population in the long term. Even more insidiously, many trees upon which primates rely for food are even longer lived still; the floristic changes in fragments as they relax to equilibrium may take decades or even centuries. As a result, primates will be detected, and even seen to reproduce, in fragments in which their long-term persistence is impossible. A deeper understanding of these changes in both population and community ecology is necessary if we are to prevent or slow primate extinctions.

Acknowledgements

I wish to thank Drs Wich and Marshall for the invitation to participate in this important volume, the institutions that supported me as I developed these ideas (Stony Brook University, McGill University, University of Queensland, and Northern Illinois University), Karen Samonds and two anonymous reviewers for useful feedback, and NSERC for postdoctoral fellowship funding. I also wish to thank the many field assistants who have contributed to the datasets I have compiled at Tsinjoarivo, Madagascar, the Malagasy scientists and government agencies who make such work possible, and those groups' counterparts in other primatology projects around the world.

References

Ancrenaz, M., Oram, F., Ambu, L., Lackman, I., Ahmad, E., *et al.* (2015). Of Pongo, palms and perceptions: a multidisciplinary assessment of Bornean orangutans *Pongo pygmaeus* in an oil palm context. *Oryx* 49: 465–472.

Arnhem, E., Dupain, J., Drubbel, R. V., Devos, C., and Vercauteren, M. (2008). Selective logging, habitat quality and home range use by sympatric gorillas and chimpanzees: a case study from an active logging concession in Southeast Cameroon. *Folia Primatologica* **79**: 1–14.

Atmar, W. and Patterson, B. D. (1993). The measure of order and disorder in the distribution of species in fragmented habitat. *Oecologica* **96**: 373–382.

Balko, E. A. and Underwood, H. B. (2005). Effects of forest structure and composition on food availability for *Varecia variegata* at Ranomafana National Park, Madagascar. *American Journal of Primatology* **66**: 45–70.

Beaudrot, L. H. and Marshall, A. J. (2011). Primate communities are structured more by dispersal limitation than by niches. *Journal of Animal Ecology* **80**: 332–341.

Benchimol, M. and Peres, C. A. (2013). Anthropogenic modulators of species-area relationships in Neotropical primates: a continental-scale analysis of fragmented forest landscapes. *Diversity and Distributions* **19**: 1339–1352.

Benchimol, M. and Peres, C. A. (2014). Predicting primate local extinctions within 'real-world' forest fragments: a pan-neotropical analysis. *American Journal of Primatology* **76**: 289–302.

Boinski, S. and Sirot, L. (1997). Uncertain conservation status of squirrel monkeys in Costa Rica, *Saimiri oerstedi oerstedi* and *Saimiri oerstedi citrinellus*. *Folia Primatologica* **68**: 181–193.

Brown, J. H. and Kodric-Brown, A. (1977). Turnover rates in insular biogeography: effect of immigration on extinction. *Ecology* **58**: 445–449.

Caughley, G. (1994). Directions in conservation biology. *Journal of Animal Ecology* **63**: 215–244.

Chapman, C. A., Gillespie, T. R., and Goldberg, T. L. (2005). Primates and the ecology of their infectious diseases: how will anthropogenic change affect host-parasite interactions? *Evolutionary Anthropology* **14**: 134–144.

Chapman, C. A., Struhsaker, T. T., Skorupa, J. P., Snaith, T. V., and Rothman, J. M. (2010). Understanding long-term primate community dynamics: implications of forest change. *Ecological Applications* **20**: 179–191.

Connor, E. F. and Mccoy, E. D. (1979). The statistics and biology of the species-area relationship. *American Naturalist* **113**: 791–833.

Cowlishaw, G. (1999). Predicting the pattern of decline of African primate diversity: an extinction debt from historical deforestation. *Conservation Biology* **13**: 1183–1193.

Crooks, K. R. and Soulé, M. E. (1999). Mesopredator release and avifaunal extinctions in a fragmented system. *Nature* **400**: 563–566.

Dehgan, A. (2003). The Behavior of Extinction: Predicting the Incidence and Local Extinction of Lemurs in Fragmented Habitats of Southeastern Madagascar. Ph.D., University of Chicago.

Didham, R. K. and Lawton, J. H. (1999). Edge structure determines the magnitude of changes in microclimate and vegetation structure in tropical forest fragments. *Biotropica* **31**: 17–30.

Estrada, A. (2013). Socioeconomic contexts of primate conservation: population, poverty, global economic demands, and sustainable land use. *American Journal of Primatology* **75**: 30–45.

Fahrig, L. (2003). Effects of habitat fragmentation on biodiversity. *Annual Review of Ecology, Evolution and Systematics* **34**: 487–515.

FAO (2010). Global Forest Resources Assessment 2010. Food and Agriculture Organization of the United Nations.

FAO (2012). Global Forest Land-use Change 1990–2005. Food and Agriculture Organization of the United Nations.

Felton, A. M., Engstrom, L. M., Felton, A., and Knott, C. D. (2003). Orangutan population density, forest structure and fruit availability in hand-logged and unlogged peat swamp forests in West Kalimantan, Indonesia. *Biological Conservation* **114**: 91–101.

Felton, A. M., Felton, A., Foley, W. J., and Lindenmayer, D. B. (2010). The role of timber tree species in the nutritional ecology of spider monkeys in a certified logging concession, Bolivia. *Forest Ecology and Management* **259**: 1642–1649.

Franklin, I. R. (1980). Evolutionary change in small populations. In: Soulé, M. E. and Wilcox, B. A. (Eds), *Conservation Biology: An Evolutionary-Ecological Perspective*. Sunderland, MA: Sinauer.

Galinato, G. I. and Galinato, S. P. (2013). The short-run and long-run effects of corruption control and political stability on forest cover. *Ecological Economics* **89**: 153–161.

Gascon, C., Lovejoy, T. E., Bierregaard Jr., R. O., Malcolm, J. R., Stouffer, P. C., *et al.* (1999). Matrix habitat and species richness in tropical forest remnants. *Biological Conservation* **91**: 223–229.

Gaveau, D. L. A., Sloan, S., Molidena, E., Yaen, H., Sheil, D., *et al.* (2014). Four decades of forest persistence, clearance and logging on Borneo. *PLoS One* **9**: 11.

Gillespie, T. R., Morgan, D., Deutsch, J. C., Kuhlenschmidt, M. S., Salzer, J. S., *et al.* (2009). A legacy of low-impact logging does not elevate prevalence of potentially pathogenic protozoa in free-ranging Gorillas and Chimpanzees in the Republic of Congo: logging and parasitism in African apes. *Ecohealth* **6**: 557–564.

Gilpin, M. E. and Diamond, J. M. (1980). Subdivision of nature reserves and the maintenance of species diversity. *Nature* **285**: 567–568.

Guo, S. T., Ji, W. H., Li, B. G., and Li, M. (2008). Response of a group of Sichuan snub-nosed monkeys to commercial logging in the Qinling Mountains, China. *Conservation Biology* **22**: 1055–1064.

Hansen, M. C., Potapov, P. V., Moore, R., Hancher, M., Turubanova, S. A., *et al.* (2013). High-resolution global maps of 21st-century forest cover change. *Science* **342**: 850–853.

Hanski, I. (1998). Metapopulation dynamics. *Nature* **396**: 41–49.

Hanski, I. (1999). *Metapopulation Ecology*. Oxford: Oxford University Press.

Hanski, I. and Gilpin, M. (1991). Metapopulation dynamics: brief history and conceptual domain. *Biological Journal of the Linnean Society* **42**: 3–16.

Hanski, I. A. and Gilpin, M. E. (Eds) (1997). *Metapopulation Biology: Ecology, Genetics, and Evolution*. San Diego: Academic Press.

Harcourt, A. (1998). Ecological indicators of risk for primates, as judged by species' susceptibility to logging. In: Caro, T. (Ed.), *Behavioral Ecology and Conservation Biology*. Oxford: Oxford University Press.

Harper, G. J., Steininger, M. K., Tucker, C. J., Juhn, D., and Hawkins, F. (2007). Fifty years of deforestation and forest fragmentation in Madagascar. *Environmental Conservation* **34**: 325–333.

Harper, K. A., Macdonald, S. E., Burton, P. J., Chen, J. Q., Brosofske, K. D., *et al.* (2005). Edge influence on forest structure and composition in fragmented landscapes. *Conservation Biology* **19**: 768–782.

Herrera, J., Wright, P., Lauterbur, E., Ratovonjanahary, L., and Taylor, L. (2011). The effects of habitat disturbance on lemurs at Ranomafana National Park, Madagascar. *International Journal of Primatology* **32**: 1091–1108.

Holyoak, M., Leibold, M. A., and Holt, R. D. (Eds) (2005). *Metacommunities: Spatial Dynamics and Ecological Communities*. Chicago: University of Chicago Press.

Husson, S. J., Wich, S. A., Marshall, A. J., Dennis, R. D., Ancrenaz, M., *et al.* (2009). Orangutan distribution, density, abundance and impacts of disturbance. In: Wich, S. A., Utami Atmoka, S. S., Mitra Setia, T., and Van Schaik, C. P. (Eds), *Orangutans: Geographic Variation in Behavioral Ecology and Conservation*. Oxford: Oxford University Press.

Irwin, M. T. (2008). Diademed sifaka (*Propithecus diadema*) ranging and habitat use in continuous and fragmented forest: higher density but lower viability in fragments? *Biotropica* **40**: 231–240.

Irwin, M. T. and Raharison, F. J.-L. (in press). Interpreting small-scale patterns of ranging by primates: what does it mean, and why does it matter? In: Shaffer, C. A.,

Dolins, F. L., Hickey, J. R., Nibbelink, N. P., and Porter, L. M. (Eds), *GPS and GIS for Primatologists: A Practical Guide to Spatial Analysis*. Cambridge: Cambridge University Press.

Irwin, M. T., Raharison, J.-L., and Wright, P. C. (2009). Spatial and temporal variability in predation on rainforest primates: do forest fragmentation and predation act synergistically? *Animal Conservation* **12**: 220–230.

Irwin, M. T., Junge, R. E., Raharison, J. L., and Samonds, K. E. (2010). Variation in physiological health of diademed sifakas across intact and fragmented forest at Tsinjoarivo, eastern Madagascar. *American Journal of Primatology* **72**: 1013–1025.

Janzen, D. H. (1983). No park is an island: increase in interference from outside as park size decreases. *Oikos* **41**: 402–410.

Johns, A. D. and Skorupa, J. P. (1987). Responses of rainforest primates to habitat disturbance: a review. *International Journal of Primatology* **8**: 157–191.

Jorge, L. A. B. and Garcia, G. J. (1997). A study of habitat fragmentation in Southeastern Brazil using remote sensing and geographic information systems (GIS). *Forest Ecology and Management* **98**: 35–47.

Jules, E. S. and Shahani, P. (2003). A broader ecological context to habitat fragmentation: why matrix habitat is more important than we thought. *Journal of Vegetation Science* **14**: 459–464.

Kapos, V. (1989). Effects of isolation on the water status of forest patches in the Brazilian Amazon. *Journal of Tropical Ecology* **5**: 173–185.

Kapos, V., Wandelli, E., Camargo, J. L., and Ganade, G. (1997). Edge-related changes in environment and plant responses due to forest fragmentation in central Amazonia. In: Laurance, W. F. and Bierregaard Jr., R. O. (Eds), *Tropical Forest Remnants: Ecology, Management, and Conservation of Fragmented Communities*. Chicago: University of Chicago Press.

Kierulff, M. C. M., Ruiz-Miranda, C. R., Procopio De Oliveira, P., Beck, B. B., Martins, A., *et al.* (2012). The Golden lion tamarin *Leontopithecus rosalia*: a conservation success story. *International Zoo Yearbook* **46**: 36–45.

Lande, R. (1988). Demographic models of the northern spotted owl (*Strix occidentalis caurina*). *Oecologia* **75**: 601–607.

Laurance, W. F. (1991). Ecological correlates of extinction proneness in Australian tropical rain forest mammals. *Conservation Biology* **5**: 79–89.

Laurance, W. F., Ferreira, L. V., Rankin-De Merona, J., and Laurance, S. G. (1998). Rain forest fragmentation and the dynamics of Amazonian tree communities. *Ecology* **79**: 2032–2040.

Laurance, W. F., Cochrane, M. A., Bergen, S., Fearnside, P. M., Delamônica, P., *et al.* (2001). The future of the Brazilian Amazon. *Science* **291**: 438–439.

Laurance, W. F., Nascimento, H. E. M., Laurance, S. G., Andrade, A., Ewers, R. M., *et al.* (2007). Habitat fragmentation, variable edge effects, and the landscape-divergence hypothesis. *PLoS One* **2**: 8.

Laurance, W. F., Clements, G. R., Sloan, S., O'Connell, C. S., Mueller, N. D., *et al.* (2014). A global strategy for road building. *Nature* **513**: 229.

Lehman, S. M. (2007). Spatial variations in *Eulemur fulvus rufus* and *Lepilemur mustelinus* densities in Madagascar. *Folia Primatologica* **78**: 46–55.

Lehman, S. M., Rajaonson, A., and Day, S. (2006a). Edge effects and their influence on lemur density and distribution in southeast Madagascar. *American Journal of Physical Anthropology* **129**: 232–241.

Lehman, S. M., Rajaonson, A., and Day, S. (2006b). Edge effects on the density of *Cheirogaleus major*. *International Journal of Primatology* **27**: 1569–1588.

Lehman, S. M., Rajaonson, A., and Day, S. (2006c). Lemur responses to edge effects in the Vohibola III Classified Forest, Madagascar. *American Journal of Primatology* **68**: 293–299.

Levins, R. (1969). Some demographic and genetic consequences of environmental heterogeneity for biological control. *Bulletin of the Entomological Society of America* **15**: 237–240.

Levins, R. (1970). Extinction. In: Gerstenhaber, M. (Ed.), *Some Mathematical Problems in Biology*. Providence, RI: American Mathematical Society.

Lindenmayer, D. B. and Fischer, J. (2006). *Habitat Fragmentation and Landscape Change*. Washington, DC: Island Press.

Macarthur, R. H. and Wilson, E. O. (1963). An equilibrium theory of insular zoogeography. *Evolution* **17**: 373–387.

Macarthur, R. H. and Wilson, E. O. (1967). *The Theory of Island Biogeography*. Princeton, NJ: Princeton University Press.

Manduell, K. L., Morrogh-Bernard, H. C., and Thorpe, S. K. S. (2011). Locomotor behavior of wild Orangutans (*Pongo pygmaeus wurmbii*) in disturbed peat swamp forest, Sabangau, Central Kalimantan, Indonesia. *American Journal of Physical Anthropology* **145**: 348–359.

Marshall, A. J., Lacy, R., Ancrenaz, M., Byers, O., Husson, S. J., *et al.* (2009). Orangutan population biology, life history and conservation: Perspectives from population viability analysis models. In: Wich, S. A., Utami Atmoka, S. S., Mitra Setia, T., and Van Schaik, C. P. (Eds), *Orangutans: Geographic Variation in Behavioral Ecology and Conservation*. Oxford: Oxford University Press.

Marshall, A. R., Jorgensbye, H. I. O., Rovero, F., Platts, P. J., White, P. C. L., *et al.* (2010). The species area relationship and confounding variables in a threatened monkey community. *American Journal of Primatology* **72**: 325–336.

Meijaard, E., Sheil, D., Marshall, A. J., and Nasi, R. (2008). Phylogenetic age is positively correlated with sensitivity to timber harvest in Bornean mammals. *Biotropica* **40**: 76–85.

Meyers, D. M. and Ratsirarson, J. (1989). Distribution and conservation of two endangered sifakas in northern Madagascar. *Primate Conservation* **10**: 81–86.

Mihoub, J. B., Robert, A., Le Gouar, P., and Sarrazin, F. (2011). Post-release dispersal in animal translocations: social attraction and the 'vacuum effect'. *PLoS One* **6**: 10.

Moilanen, A. and Nieminen, M. (2002). Simple connectivity measures in spatial ecology. *Ecology* **83**: 1131–1145.

Murcia, C. (1995). Edge effects in fragmented forests: implications for conservation. *Trends in Ecology and Evolution* **10**: 58–62.

Phoonjampa, R., Koenig, A., Brockelman, W. Y., Borries, C., Gale, G. A., *et al.* (2011). Pileated gibbon density in relation to habitat characteristics and post-logging forest recovery. *Biotropica* **43**: 619–627.

Prugh, L. R., Hodges, K. E., Sinclair, A. R. E., and Brashares, J. S. (2008). Effect of habitat area and isolation on fragmented animal populations. *Proceedings of the National Academy of Sciences of the United States of America* **105**: 20770–20775.

Remis, M. J. and Robinson, C. A. J. (2012). Reductions in primate abundance and diversity in a multiuse protected area: synergistic impacts of hunting and logging in a Congo Basin forest. *American Journal of Primatology* **74**: 602–612.

Ricketts, T. H. (2001). The matrix matters: effective isolation in fragmented landscapes. *American Naturalist* **158**: 87–99.

Rode, K. D., Chapman, C. A., Mcdowell, L. R., and Stickler, C. (2006). Nutritional correlates of population density across habitats and logging intensities in redtail monkeys (*Cercopithecus ascanius*). *Biotropica* **38**: 625–634.

Smith, A. T. and Peacock, M. M. (1990). Conspecific attraction and the determination of metapopulation colonization rates. *Conservation Biology* **4**: 320–323.

Strier, K. B. and Ives, A. R. (2012). Unexpected demography in the recovery of an endangered primate population. *PLoS One* **7**: 11.

Swart, J. and Lawes, M. J. (1996). The effect of habitat patch connectivity on samango monkey (*Cercopithecus mitis*) metapopulation persistence. *Ecological Modelling* **93**: 57–74.

Tutin, C. E. G. (1999). Fragmented living: behavioural ecology of primates in a forest fragment in the Lopé Reserve, Gabon. *Primates* **40**: 249–265.

Tutin, C. E. G., White, L. J. T., and Mackanga-Missandzou, A. (1997). The use by rain forest mammals of natural forest fragments in an equatorial African savanna. *Conservation Biology* **11**, 1190–1203.

Umapathy, G. and Kumar, A. (2000). The occurrence of arboreal mammals in the rain forest fragments in the Anamalai Hills, south India. *Biological Conservation* **92**: 311–319.

Van Horne, B. (1983). Density as a misleading indicator of habitat quality. *Journal of Wildlife Management* **47**: 893–901.

Wahlberg, N., Moilanen, A., and Hanski, I. (1996). Predicting the occurrence of endangered species in fragmented landscapes. *Science* **273**: 1536–1538.

Wilcox, B. A. (1980). Insular ecology and conservation. In: Soulé, M. E. and Wilcox, B. A. (Eds), *Conservation Biology: An Evolutionary-Ecological Perspective*. Sunderland, MA: Sinauer.

Zeigler, S. L., De Vleeschouwer, K. M., and Raboy, B. E. (2013). Assessing extinction risk in small metapopulations of golden-headed lion tamarins (*Leontopithecus chrysomelas*) in Bahia State, Brazil. *Biotropica* **45**: 528–535.

Zeigler, S. L., Neel, M. C., Oliveira, L., Raboy, B. E., and Fagan, W. F. (2011). Conspecific and heterospecific attraction in assessments of functional connectivity. *Biodiversity and Conservation* **20**: 2779–2796.

CHAPTER 8

Present-day international primate trade in historical context

Vincent Nijman and Aoife Healy

Infant greater slow loris Nycticebus coucang offered for sale at a wildlife market in Jakarta; despite legal protection slow lorises are traded internationally. Photo copyright: Andrew Walmsley.

8.1 Introduction

Since prehistoric times, humans have had a plethora of uses for primates. Fossil evidence from East Africa suggests that *Homo erectus* hunted the ancient giant gelada *Theropithecus oswaldi* between 400,000 and 700,000 years ago (Shipman *et al.* 1981). Prehistoric hunting has also been implicated in the extirpation of the orang-utan *Pongo* spp. from Java and of several giant lemur species from Madagascar (Von Koeningswald 1982; Perez *et al.* 2005; Godfrey and Jungers 2003). Archaeological sites at the earliest centres of human civilization—Ancient Egypt and even earliest Mesopotamia—show evidence of primate use (Kletter 2002; Dunham 1985). Over time these uses have been many and varied with both our historic and current relationships with primates being paradoxical. Primates have been attributed great medicinal properties, eaten, feared, detested, persecuted, exploited, enjoyed, protected, sacrificed as tributes, and worshipped as gods. The most immediately apparent and oldest use for primates is as food, though their uses go well beyond their basic value as a protein source, and while a great many uses of primates have been domestic, others have stimulated international trade since the very earliest days of modern human civilization. Some of these drivers of international trade have persisted, though on a different scale to historical trade, others have fallen out of fashion, while others have newly emerged. Here we provide an overview of some of the historical drivers of international

Nijman, V. and Healy, A., *Present-day international primate trade in historical context.* In: *An Introduction to Primate Conservation.* Edited by: Serge A. Wich and Andrew J. Marshall, Oxford University Press (2016). © Oxford University Press.
DOI 10.1093/acprof:oso/9780198703389.003.0008

primate trade and the transition from those historical, societal, and fashion trends to the modern international primate trade, which has been largely driven by the demands of the biomedical research industry.

8.2 Primates worshipped

Primates have been worshipped as sacred animals across many cultures over time. For instance, Japan gave us the legend of the three wise monkeys, inspired by the Japanese macaque *Macaca fuscata* (Smith 1993). The story of the introduction of Buddhism to China features a monkey as the hero (Hargett 1988) and throughout India Hanuman was the monkey ally of the God Rama (Wolcott 1978). For the most part these practices did not stimulate international trade.

The origins of the Barbary macaque population on Gibraltar are unclear, but genetic evidence supports stories that they were introduced to the rock from North Africa (Modolo *et al.* 2005, 2008) and at least the maintenance of the population has been due if not to worship, then at least to superstition. During their War of the Spanish Succession in 1704 the British, who then gained control over Gibraltar, came to think of the monkeys as lucky. This started the belief that British reign on Gibraltar would last as long as the monkeys remained. This was the stimulus for the importing of macaques from North Africa to maintain the colony during the Second World War. The monkeys were put under official protection by the British army and then later by the Gibraltar regiment (Stockey and Grocott 2012).

For a much earlier and definitive example we must look to Ancient Egypt where some of the earliest monkey worship was practised (DuQuesne 2007; Von den Driesch 1993; Nerlich *et al.* 1993; Goudsmith and Brandon-Jones 1999, 2000). Some of the earliest Egyptian deities were depicted as monkeys, including: 'God of Creation' Atum, depicted as a vervet monkey *Chlorocebus* spp., and the 'son of Horus', Hapi, the 'God of the Sun' Re (Ra), and the 'God of the Moon', Thoth, which were depicted as baboons *Papio* spp. In the hall of judgement in the underworld four baboons guarded the lake of fire. The hamadryas baboon, *P. hamadryas*, and some African guenons, *Cercopithecus* spp., were sacred in Ancient Egypt. Male hamadryas baboons were worshipped as an embodiment of masculine sexuality with depictions of the baboon god Bebon usually emphasizing the phallus, whereas the occurrence of overt sexual swellings in female hamadryas baboons were linked to the lunar cycle, making them subject of worship as well.

It is possible that these primates did occur in parts of Egypt until the Old Kingdom (~ 2700–2100 BC), but it is a certainty that they were being imported from the land of Punt (Somalia, Djibouti, Eritrea, Northeast Ethiopia, and the coast of Sudan) by the time of the New Kingdom (1500–1000 BC) (Goudsmit and Brandon-Jones 1999, 2000; Masseti and Bruner 2009). By tradition, when a new baboon was brought to the temples where the sacred troops were housed they would be presented with a tablet, reed pen, and ink well. Those that passed this literacy test would be kept in the temples and their upkeep paid for by worshippers. When they died they would be embalmed and mummified. Those that failed were trained to work picking fruit (Deputte and Anderson 2009). Such scenes have been depicted in rock paintings, carvings, and sculptures at burial sites dating from as early as the 12th Dynasty (2500–3450 BC).

8.3 Primates in traditional medicine

There is evidence from as early as Ancient Mesopotamia of the bones of imported primates being used in drugs and potions (Morris and Morris 1966). Later, in second-century Europe, the idea spread that there were health benefits to eating primates. At the same time, a Greek philosopher described a treatment for 'scrofulous tumours' that involved a poultice of monkey faeces (Morris and Morris 1966). According to a survey by Alves *et al.* (2010), primates are still commonly used in traditional folk medicine, with over 70 species used in over 50 range countries. These range from squirrel monkeys, *Saguinus* spp. (Tejada *et al.* 2006), and angwantibos, *Arctocebus* spp. (Svensson and Friant 2014), to common chimpanzees, *Pan troglodytes* (Sa *et al.* 2012), and purportedly cure a wide range of ailments (Alves *et al.* 2010).

In Cambodia, for instance, traditional Khmer medical practice uses pygmy and Bengal slow lorises (*Nycticebus pygmeaus* and *N. bengalensis*) to treat 100 ailments (Starr *et al.* 2010). Slow lorises are among the most commonly observed protected mammals in traditional medicine shops in Cambodia (Box 8.1). However, the international component of this trade seems to be largely restricted to East Asia and Indochina. Here, primates are threatened by collection to supply the insatiable demand

Box 8.1 Domestic trade of slow lorises in Cambodia and Indonesia

It has been asserted that wildlife trade could be a more severe threat in Asia than any other continent due to economic incentives and cultural drivers, as well as changing agricultural practices resulting in increased forest access, large markets, weak law enforcement, and poor policing (Corlett 2007; Nekaris *et al.* 2010). Asian primates have traditionally been traded as pets and for use in traditional medicine. Though the CITES-reported trade in live slow lorises *Nycticebus* spp. is relatively low (~1,750 live individuals between 1975 and 2011) they are frequently seen in the wildlife markets of Southeast Asia (Nekaris *et al.* 2010; Shepherd *et al.* 2005; Nijman *et al.* 2014). Modern hunting practices, which involve the use of high-powered search lights as well as improved transport infrastructure and more efficient weapons, have made modern extraction practices unsustainable and increasingly threaten the survival of these nocturnal forest-dwellers.

Nekaris and colleagues (2010) carried out an ethnoprimatological survey of the cultural perceptions and folklore history of slow lorises in Cambodia and Indonesia. Combined with trade data from selected sites, these findings provided an estimate of the documented domestic trade in these animals, which facilitated inter-specific and inter-country comparisons, and furthermore they informed an understanding of the cultural drivers of the trade in these two range countries.

There is a high volume of trade in Bengal and pygmy slow lorises (*N. bengalensis* and *N. pygmaeus*) in Cambodia where slow lorises are among the most highly sought-after and most commonly observed mammals in traditional medicine shops (Figure 8.1). The demand for traditional medicines containing slow loris is the primary driver of hunting of slow lorises in the country and has been on the rise in

Figure 8.1 Dried pygmy slow lorises *Nycticebus pygmaeus* for sale in a shop in Cambodia (photo Chris R Shepherd, TRAFFIC).

continued

Box 8.1 *Continued*

recent years due to the growing demand of an increasing urban population.

In Indonesia, slow lorises are the most commonly traded legally protected primates (Nekaris and Nijman 2007). Here slow lorises are popular pets, despite them being legally protected. Market surveys in the Medan, North Sumatra, wildlife market showed the greater slow lorises *N. coucang* to be both common and numerous (i.e. they were present at most, if not all, surveys with often multiple individuals observed at a time), recorded in 90% of survey counts, and with a high rate of turnover. Almost 700 animals were recorded between January 1997 and December 2001 (Shepherd *et al.* 2005). Given that slow lorises are venomous and their bite can cause serious harm people, the anterior teeth of many slow lorises are removed with pliers or nail clippers as to make them more suitable as pets (Nekaris 2014).

of the traditional Asian medicine industry, and this includes international cross-border trade (Fitch-Snyder 2001; Nekaris *et al.* 2010).

In the recent past there was a great international trade in bezoar or geliga/geligu stones that are found in the lower digestive tract of folivorous colobines. Bezoar stones are found as a calculus or concretion in the stomachs or intestines of various ruminants (antelopes, deer, goats), porcupines, and colobines. Bezoar stones are principally calcium phosphate, but the active ingredient is the crystalline mineral brushite. The name bezoar comes from the Persian padzahr, meaning 'to expel poison'. Once it was speculated that the bezoar stone originated from the Unicorn and would protect its possessor from evil, and the stone would be especially effective in preventing poisoning. From the Middle Ages down to our own time, the bezoar stone's reputed efficacy has grown to include all manner of diseases and maladies. Bezoar stones continue to be traded to supply the Traditional Asian Medicine market (Nijman 2010). In the last century, bezoar stones were exported from Southeast Asia to present-day northern India in great quantities to be used as antidotes to snake bites and for other complaints, including fever and asthma. In 1949 Tom Harrison, curator of the Sarawak Museum, was dismayed by the international demand for these stones and the pressure it was putting on the survival of the Hose's langur *Presbytis hosei*. There were complaints that the stones were more rare and considerably smaller than they had been. He surmised that unsustainable demand meant that the monkeys seldom reached a sufficient age to grow the stones to an acceptable size. Demand for bezoar stones fluctuates, and even in the present day can affect populations of Hose's and other langurs on Borneo (Nijman 2005).

8.4 Primates in fashion

Primates have played their part in fashion, with pelts being commercially traded internationally since at least medieval times. Pelts of the black-and-white colobus, *Colobus guereza*, were used domestically for ceremonial garb, to cover shields, and to make decorative costumes (Mittermeier 1973; Morris and Morris 1966; Oates 1977). During the Middle Ages these skins were exported by Arab traders to Europe, where there was much curiosity about the extraordinary and never before seen furs. European furriers, when they first saw these furs in Italy made up as shoulder capes, were sure there was some new and as yet unlearned skill involved in inserting the long white hairs into the black skins (Morris and Morris 1966). They were increasingly sought after in women's fashion through the latter half of the eighteenth century, and as the demand grew so too did the numbers hunted and exported from range countries. They were hunted in every part of their range with increasing efficiency as rifles replaced bows and arrows. In 1892, 175,000 colobus skins reached Europe alone (Morris and Morris 1966). Many pelts were damaged with shotgun pellets and these flawed items were not accepted. Perfect specimens only are counted in this given figure. It is estimated that over 2 million colobus were killed while their skins were at the height of fashion.

The pelt of the golden snub-nosed monkey, *Rhinopithecus roxellana*, was another popular fashion accessory (Kirkpatrick 1995). The golden hair of the males would be plucked and woven into fabric for officials' robes. In the mid 1980s Wang and Quan (1986) estimated that some 10,000–20,000 primate skins were traded annually in China. Monkey skins made popular lampshades, rugs, and mats. Primate pelts have fallen out of fashion since the twentieth century, and as a fashion statement are now more niche than in previous centuries. Overall the fashion industry is no longer an international driver of commercial trade in primate products, though they are still traded as antiquities.

8.5 Primates as entertainers, pets, and status symbols

Primates have been popular entertainers and pets, both domestically and internationally, throughout history (Morris and Morris 1966; Burgess 1959; Radhakrishna 2013; Kanagavel *et al.* 2013; Yang-Martinez 2011; Osterberg and Nekaris 2015). The Ancient Egyptians imported pet primates, mostly African guenons. The Romans imported pet primates from 50 BC. Barbary macaques and African guenons were popular pets of affluent Greek households from at least the seventh century BC, with some additional species arriving in Europe in trade caravans from India. By the third century BC, primate pets were widespread across all walks of Greek life. Roman and Greek writings also talk of monkey entertainers that could 'play' musical instruments such as the harp and lyre, shoot a bow and arrow, and walk on stilts. Unlike the majority of exotic animals imported to Ancient Rome, primates generally escaped the fate of having to take part in the gladiator fights in the arenas, though there are some ambiguous writings that suggest it is likely that either gorillas or large-bodied baboons did fight in the games (Morris and Morris 1966).

Monkeys were not generally known in Western Europe until the eleventh or twelfth centuries AD, when Barbary macaques and their trainers were imported from North Africa across the Mediterranean. Medieval European royalty would customarily keep monkeys that would be carried around by court jesters for entertainment. They became expensive status symbols kept by the rich and influential and were even kept on display in the cloisters of Notre Dame in Paris, though this was eventually discouraged by the papal legate. By the Middle Ages monkeys were popular among the citizenry, the trade being so strong in Paris that a tax was levied on every monkey entering the city to be sold. As in Ancient Rome and Greece, monkeys were trained as entertainers by travelling minstrels. Depictions of monkeys were incorporated into family crests. In thirteenth-century Ireland the Earls of Kildare adopted a monkey as their family emblem.

Monkey enthusiasts abounded in Western Europe, particularly among the politically influential, when new trade routes opened up to the New World. The variety of species on offer increased greatly at this time and the smaller New World monkeys, the marmosets and tamarins (*Callithrix, Cebuella, Callibella, Mico, Saguinus*), became the newest fashion craze. By the middle of the seventeenth century monkeys were so popular in the UK, as entertainers and pets, that the Chancellor of the time considered a tax on monkeys as a source of significant revenue. The fondness for monkey pets persisted and Victorian owners even had their dead monkeys stuffed and kept as mementos. Inevitably, when they could be widely acquired they were no longer great status symbols, and they fell out of high fashion.

In the late nineteenth century, minstrels were replaced by travelling musicians who were, like the travelling minstrels of the seventeenth and eighteenth centuries, accompanied by monkeys. These 'musicians' were in truth often disguised beggars who played the street organ for donations. Their monkeys, often capuchins and rhesus macaques, would collect money from onlookers in the street. Performing monkeys developed from street performances to more sophisticated entertainment with primates making their way onto the stage by the 1900s, with the first chimpanzee stage show in 1926, and eventually onto television screens by the latter half of the twentieth century.

The international trade in primates for the pet trade is not as great as it was, though commercial trade in live primates is on the rise (see Section 8.7). However the proportion of this trade composed of pets is not readily quantifiable, and the source of pet primates in non-range countries is generally unknown.

8.6 Primates hunted

People have hunted primates for meat since prehistoric times. While most of this trade in wild primate meat is domestic (Fa *et al.* 2006), there is an international component in areas close to borders, for instance the active bushmeat trade across the Nigeria–Cameroon border (Fa *et al.* 2006; Fa and Tagg, Chapter 9, this volume). While the (international) bushmeat trade in primates has been well documented and researched (Fa and Tagg, Chapter 9, this volume), primates are additionally traded internationally as trophies. This trade happens legally as part of the ever-increasing trophy hunting industry where primates are often shot free of charge as 'targets of opportunity'. Trophy hunting outfitters seem to be almost exclusive to sub-Saharan Africa with a predominantly American and European clientele, with Old World Monkeys being killed and traded in the largest numbers. In the last 30 years almost 30,000 primate trophies have been exported from over 40 countries, with the USA importing the largest volumes by far (Box 8.2).

8.7 Primates in research and twentieth-century trade

Many historical uses for primates that stimulated significant international trade have not persisted to the current day on the same scale. What has undoubtedly grown as a driver of international trade is the use of primates in research. Primates have been vital to medical research since Roman times. For instance, Galen's famous writings on human anatomy were based on macaque dissections. Barbary macaques were the most popular research animals of the time. Since Roman times primates have continued to become increasingly useful models and research tools. Their prominence as a biomedical model was solidified in the 1870s when researchers turned their attention to the primate brain, when Ferrier described the similarities between the monkey and human brains and when the neuroscientists Horsley and Beevor mapped the brain of the bonnet macaque *M. radiata* and later the orang-utan *Pongo* spp. By the late nineteenth century their usefulness in disease research became apparent, and since then they have served in biomedical research in ever increasing numbers (Mack and Mittermeier 1984).

By the twentieth century, monkeys were being exported from range countries, primarily from India, Colombia, and Peru, in their hundreds of thousands to supply the biomedical research industry, the trade reaching its peak in the mid 1900s (Wolfheim 1983; Mack and Mittermeier 1984). In 1938 alone, 250,000 rhesus macaques from India were imported to the USA. The 1950s saw India exporting 100,000 to 200,000 rhesus macaques per year. By the end of the 1950s there were reports of commercial exporters having trouble filling demand. Exports of rhesus macaques declined through the following two decades and by 1975 annual exports had been cut to 20,000. In 1978 India banned the export of its macaques to the USA and later to all other countries. Between 1950 and 1975 at least 4 million primates, almost exclusively wild caught, were traded internationally. Rhesus macaques dominated the trade with 2 million monkeys traded over the 25 years: long-tailed macaques, *M. fascicularis*, from Peninsular Malaysia and common squirrel monkeys and black-capped squirrel monkey, *S. boliviensis*, from Colombia and Peru, accounted for 1 million and 400,000 individuals, respectively.

Though it may seem obvious to us now that there is a need to monitor trade in endangered species of fauna and flora, when discussions began in the first half of the last century about a global international multilateral agreement for the regulation and monitoring of this sort of trade it was a relatively novel idea. Agreements of the earlier portion of the 1900s were regionally limited, had too little national support to be effective, and having being written for a more colonial world they in part had lost their relevance. Global concern for species survival and the conservation impact of overexploitation for international trade was first expressed in 1960 at the seventh general assembly of the IUCN in Poland. In 1973 a global plenipotentiary conference was attended by 80 countries. The Convention was signed by 21 parties, ratified by 10, and in 1975 the Convention on International Trade in Endangered Species of Wild Fauna and Flora (CITES) became a reality.

Box 8.2 Primate trophy hunting

Trophy hunting is the selective hunting of wild game animals, often with parts of the slain animal being kept as hunting trophies or memorials. The practice is typically legal with the hunter (or their representative) having obtained prior permission to take a certain number of animals. Hunts are frequently conducted on private reserves specifically tailored to trophy hunting tourism. Hunters will order target animals in advance—in the form of a 'wish list'—or will select a pre-packaged tour. The sport is generally restricted to sub-Saharan Africa.

The volume of primate trophies traded internationally can be determined through analysis of the CITES trade database.[1] Additional data on trophy prices were obtained from the online profiles of hunting outfitters. In 2014 at least 400 operators in 12 countries included primates on their pricelists, with South Africa having the greatest number of trophy hunting outfitters (~200), while Ethiopia offers the most primate species (nine species). The average trophy fee for a primate is USD195 but ranges from 'free of charge opportunity kills' (*Chlorocebus* spp. and *Papio* spp.) to a maximum of USD3,000 for a gelada *Theropithecus gelada*.

Between 1975 and 2011 almost 28,000 primates from 73 taxa were killed and traded internationally as trophies. CITES reports that the number of primates killed for the trophy trade has increased over time, rising from an annual average of 300 trophies in the 1980s, to 700 in the 1990s, to almost 1,500 in the 2000s. Chacma baboons *P. ursinus* and vervet monkeys *C. pygerythrus* are traded in the largest volumes, accounting for 48% and 18% of the trade respectively. They have been exported from 41 countries (including most range countries and a small number of countries where neither baboons or vervets occur naturally) and imported in to 103 countries: South Africa is the largest exporter of primate trophies exporting 41% of all trophies, while the United States is the leading importer, importing 70% of the total trade (Figure 8.2).

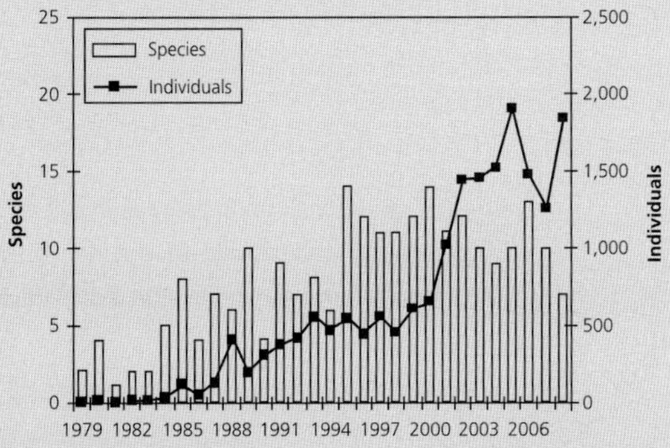

Figure 8.2 Export of primates as trophies from African range countries as reported in the CITES trade database: while the number of species exported has remained relatively stable, the total number of individuals has increased steadily over time.

As of 2016 the Convention has been ratified by 182 parties (parties are mostly sovereign states, but also include the European Union as the first regional economic integration organization to join the Convention). The primary aim of the Convention is to ensure that international trade does not threaten the survival of CITES-listed species. Primates are listed in two of the three Appendices according to the degree of protection they need. The approximately 1,000 species that are listed in Appendix I are threatened with extinction and commercial international trade in them is precluded. For the ~35,000 species listed in Appendix II, international trade is regulated below levels of extraction that would threaten their survival. As of 2013, about 180 species of primates are listed in Appendix I (most *strepsirhines* are listed at the family level and the total number thus depends on the taxonomy one adopts) and all the remaining ones are listed in Appendix II. The CITES Trade Database, managed by the United

[1] http://trade.cites.org/.

Nations Environment Programme (UNEP) World Conservation Monitoring Centre (WCMC) on behalf of the CITES Secretariat, holds approximately 10 million records of trade in wildlife listing 50,000 scientific names of taxa. Approximately 700,000 records of international trade in CITES-listed species are reported annually and entered into the database—a requirement of all 181 parties. Each member party's management authority is responsible for the issuing of the necessary permits and for the compilation of annual reports, though there are acknowledged weaknesses in this reporting system (for detailed discussions see Phelps *et al.* 2011; Smith *et al.* 2011; Bickford *et al.* 2011; Bowman 2013). However, these annual reports are the only available means of consistent monitoring of the global international wildlife trade and the implementation of the Convention.

The CITES Trade Database lists 60,000 primate trade transactions between 1975 and 2011, totalling over 9 million items. The majority of these comprise international trade in parts and derivatives (e.g. skin pieces, bones, bone pieces, derivatives) that cannot reliably be converted to estimates of numbers of individual primates. Other entries can be unambiguously identified as—or, in the case of hands, tails, skulls, and so on, equated to—individual animals that have been removed from a (wild or captive) population.

Almost 1.5 million individual primates were traded internationally between 1975 and 2011, with volumes of trade increasing over time. As shown in Figure 8.3, a brief dip in the mid 1990s

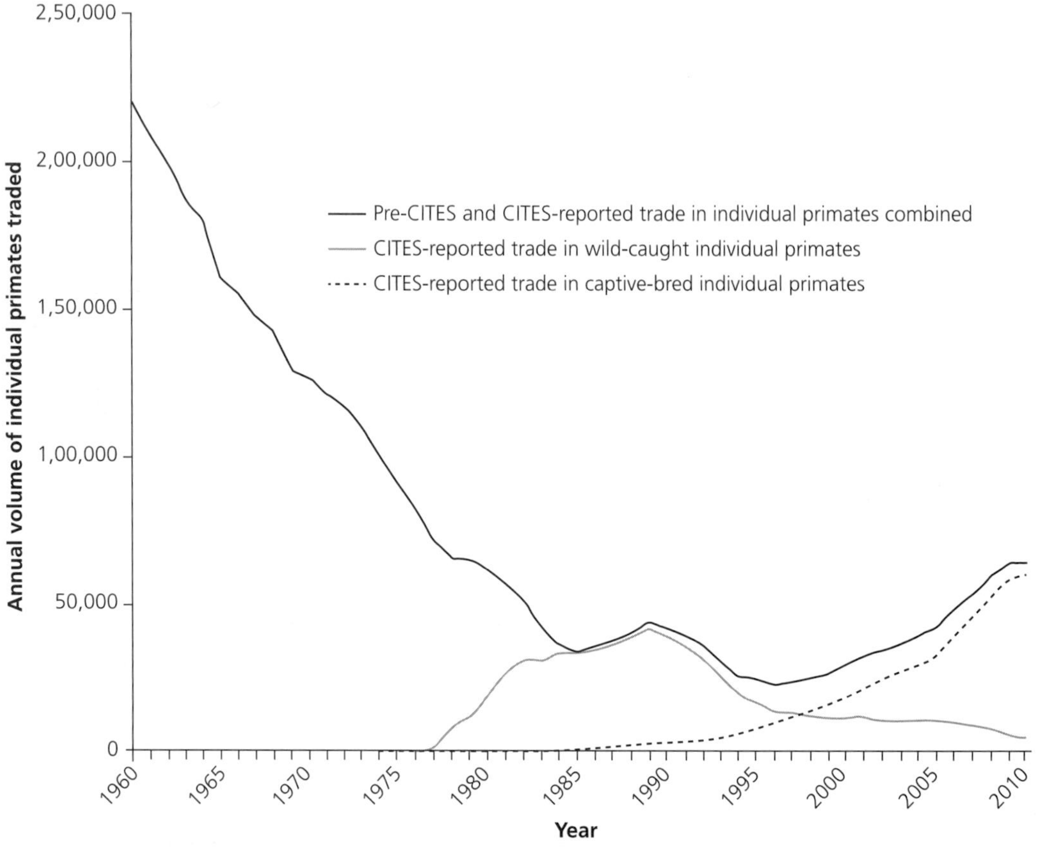

Figure 8.3 Annual volumes of trade in primate-derived products from 1960 to 2011. Pre-CITES trade puts CITES-reported trends in a temporal context showing an overall decline in the number of individual primates traded internationally since pre-CITES years. CITES records are incomplete prior to 1985. For annual volumes of trade for years prior to 1985, data were sourced from Wolfheim (1983) and Mack and Mittermeier (1984).

was followed by an increase by the latter half of the decade and a further rise that has continued into the 2000s. The number of taxa traded—203 in total—has remained relatively stable with an average of 107 taxa traded annually. Asian taxa have been traded in the largest quantities by far, making up 76% of total individual trade. African and Neotropical taxa are traded in smaller quantities, accounting for 15 and 9% of trade respectively. The five taxa traded in the largest volumes, cumulatively accounting for 89% of individual trade, are long-tailed macaques (69%), vervet monkeys (7%), rhesus macaques (6%), common squirrel monkeys (5%), and olive baboons (2%). The five foremost exporters, accounting for 77% of the trade, are China (19%), Indonesia (14%), the Philippines (12%), Mauritius (11%), and Viet Nam (6%). The largest importer by far is the USA (46%) followed by Japan (11%), China (7%), the UK (7%), and France (5%).

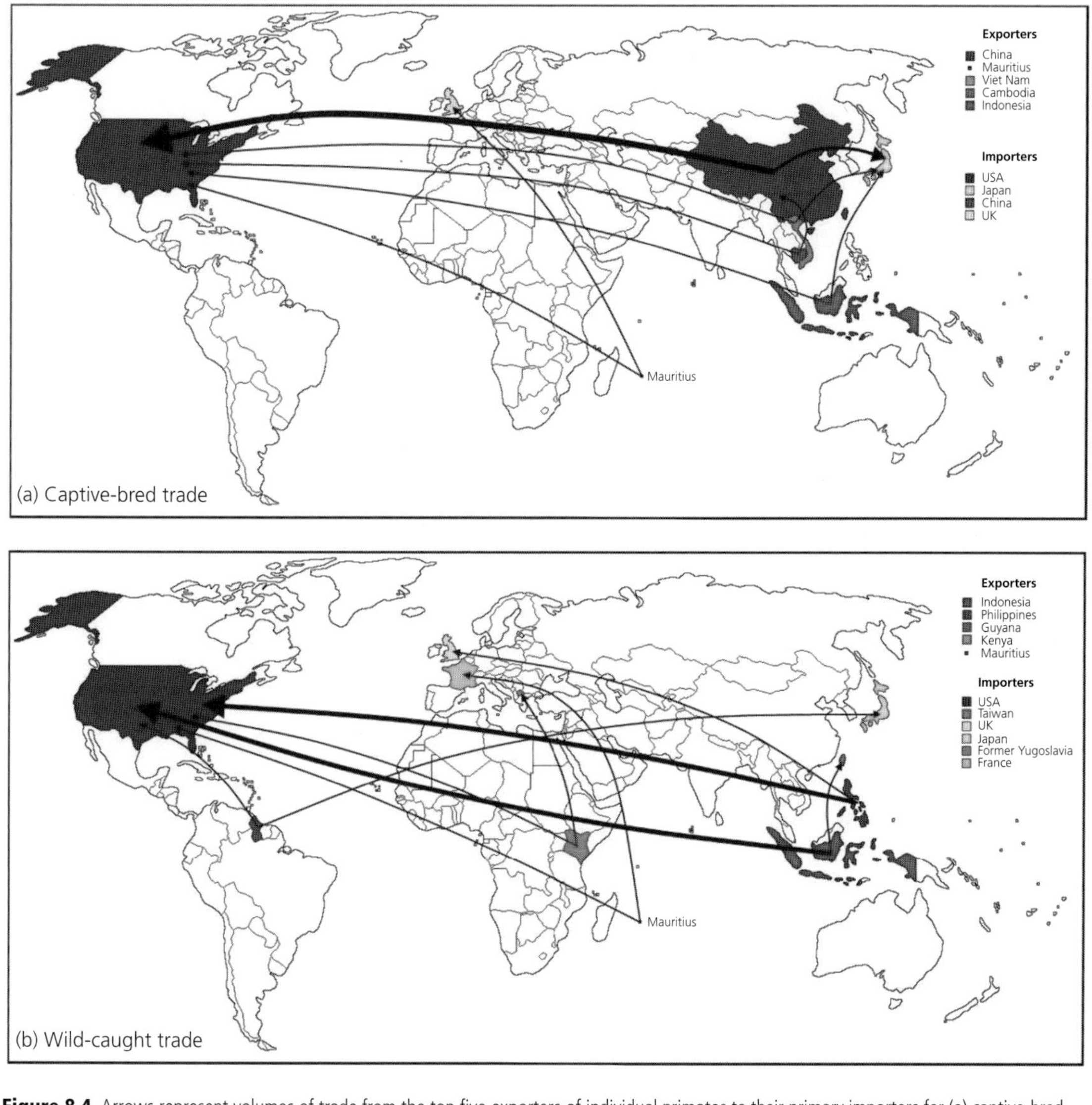

Figure 8.4 Arrows represent volumes of trade from the top five exporters of individual primates to their primary importers for (a) captive-bred trade and (b) wild-caught trade. Composition of these top exporters and importers vary between captive-bred and wild-caught primates, with the USA consistently featuring as a top importer.

Table 8.1 Overview of the composition all individual primate trade, captive-bred trade, and wild-caught trade, including details on the taxa traded, the top exporters and importers of individual primates, and the drivers of the trade, highlighting changes in composition across these three categories. Percentages refer to the proportion of all individuals, captive-bred individuals or wild-caught individuals.

	All primates		Captive-bred primates		Wild-caught primates	
Individuals	1,336,000		646,000		689,500	
Taxa	**273**		**210**		**256**	
Top 5 taxa	*M. fascicularis*	69%	*M. fascicularis*	86%	*M. fascicularis*	53%
	Chlorocebus spp.	7%	*M. mulatta*	7%	*Chlorocebus* spp.	14%
	M. mulatta	6%	*C. jacchus*	2%	*S. sciureus*	9%
	S. sciureus	5%	*M. nemestrina*	1%	*M. mulatta*	4%
	P. anubis	2%	*Chlorocebus* spp.	<1%	*P. anubis*	4%
Origin						
Asian taxa	1,017,000	76%	608,000	94%	409,000	59%
African taxa	196,500	15%	13,500	2%	183,500	27%
Neotropical taxa	122,000	9%	24,500	4%	97,500	14%
Exporters	**174**		**147**		**164**	
Top 5 exporters	China	19%	China	37%	Indonesia	21%
	Indonesia	12%	Mauritius	15%	Philippines	18%
	Philippines	12%	Viet Nam	12%	Guyana	8%
	Mauritius	11%	Cambodia	10%	Kenya	7%
	Viet Nam	6%	Indonesia	7%	Mauritius	7%
Importers	**197**		**151**		**189**	
Top 5 importers	USA	46%	USA	44%	USA	47%
	Japan	11%	Japan	15%	Japan	8%
	China	7%	China	13%	UK	8%
	UK	7%	France	6%	Russia	5%
	France	5%	UK	6%	France	4%
Terms						
Live	1,284,000	96%	638,500	99%	645,500	94%
Dead	52,000	4%	8,000	1%	44,000	6%

There has been a relatively even representation of captive-bred and wild-caught animals across the trade, accounting for 48% (646,000) and 52% (690,000) respectively. However, as shown in Figure 8.4, this distribution has not been even over time. Between 1975 and 1994 approximately 90% of individual trade was in wild-caught primates. The period of the early 1990s saw a shift away from wild-caught trade in favour of captive breeding and from 1995 to 2011 captive-bred animals have supplied over 80% of the trade (note that these figures are based on self-reporting by countries, and an unknown proportion of reported captive-bred individuals in fact comprise wild-caught individuals). Furthermore, as highlighted in Table 8.1, the proportion of captive-bred versus wild-caught trade differ depending on what species are considered, on the exporting country, and depending on the purpose of the trade. Over 645,000 captive-bred individual primates of 210 known taxa were exported from 147 countries/states to at least 151 importing countries/states. The dominance of Asian taxa is most pronounced in the captive-bred trade, accounting for an overwhelming 94% of the volume. The long-tailed macaque is by far the most traded captive-bred primate, accounting for 86% of captive-bred trade.

Almost 700,000 wild-caught individual animals, of 256 taxa, were traded between 164 exporting countries/states and 189 importing countries/states. Asian taxa make up a less pronounced but still a clear majority of wild-caught trade, at 59%. African and Neotropical taxa account for much larger portions than in wild-caught trade at 27 and 14% respectively. Long-tailed macaques account for the majority of wild-caught trade, though by a lesser margin than that of captive-bred trade, at 53% (to the captive-bred trade's 86%). The top five taxa account for 84% of wild-caught trade with vervets contributing a greater portion at 14%. Notably, China is not the primary exporter of wild-caught individuals, nor is it one of the top five exporters. Rather, Indonesia exports the largest volume of wild-caught individual primates and accounts for 21% of wild-caught trade. The Philippines, not featured as a top exporter of captive-bred primates, accounts for 18% of wild-caught trade.

The trade of 1.5 million primates that we have described is but a fraction of the overall trade in primate-derived products traded over the same period, which totals almost 9 million items. Furthermore, though there has been an increase in volume of individual primates traded internationally since CITES came into effect, the last 60 years has in fact seen a substantial decline in the overall volume of trade, especially where it pertains to the trade in wild-caught primates. The documented trade in parts and derivatives has been increasing dramatically. This trade in parts took off in earnest in the late 1990s and now forms the overwhelming majority of the trade and continues to grow: in 2011 alone 1 million primate parts/derivatives were traded.

8.8 Conclusions

The nature of international primate trade has changed over the years, though our relationships to primates remain diverse and contradictory. Commercial trade, while not now on the scale it once was, remains a major impediment to the conservation of selected primate species. Unfortunately, a more detailed account of the components of this commercial trade is not readily available, though primates do still make popular pets. There has been a decline in the trade in individuals as well as a shift away from wild-caught animals in favour of captive-bred ones. But most notable is the recent rapid ascent of trade in primate parts and derivatives—a comprehensive analysis of this dimension would be highly beneficial to contextualize the data presented here.

The trade in primates cannot be viewed in isolation from the trade in other wildlife, which is a global industry. Wildlife trade is considered a leading threat to biodiversity conservation (Mittermeier 2009; Sutherland *et al.* 2009; Phelps *et al.* 2011) and it is imperative that as such it is monitored and regulated. Many factors conspire to stimulate illegal wildlife trade and undermine the legal frameworks that prevent trafficking, including lack of resources, a dearth of alternative economic activities, weak law enforcement, and widespread corruption. Under these conditions wildlife trafficking can and does thrive.

References

Alves, R. N., Souto, W. M. S., and Barboza, R. R. D. (2010). Primates in traditional folk medicine: a world view. *Mammal Review* **40**: 155–180.

Bickford, D., Phelps, J., Webb, E. L., Nijman, V., and Sodhi, N. S. (2011). Boosting CITES through research: a response. *Science* **331**: 857–858.

Bowman, M. (2013). A tale of two CITES: divergent perspectives upon the effectiveness of the wildlife trade convention. *Review of European, Comparative and International Environmental Law* **22**: 228–238.

Burgess, M. (1959). Fairs and entertainers in 18th-century Russia. *The Slavonic and East European Review* **38**: 95–113.

Corlett, R. T. (2007). The impact of hunting on the mammalian fauna of tropical Asian forests. *Biotropica* **39**: 292–303.

Deputte, B. L. and Anderson, J. R. (2009). Baboon palm nut harvesters in Ancient Egypt: new (ancient) evidence, new questions. *Folia Primatologica* **80**: 70–73.

Dunham, S. (1985). The monkey in the middle. *Zeitschrift für Assyriologie und Vorderasiatische Archäologie* **75**: 234–264.

DuQuesne, T. (2007). *Anubis, Upwawet, and other Deities: Personal Worship and Official Religion in Ancient Egypt*. Cairo: American University in Cairo Press.

Fa, J. E., Seymour, S., Dupain, J., Amin, R., Albrechtsen, L., *et al.* (2006). Getting to grips with the magnitude of exploitation: Bushmeat in the Cross-Sanaga Rivers region, Nigeria and Cameroon. *Biological Conservation* **129**: 497–510.

Fitch-Snyder, H. (2001). Vietnam: Joining Forces to Save Vietnam's Lorises. CRES Report.

Godfrey, L. R. and Jungers, W. L. (2003). The extinct sloth lemurs of Madagascar. *Evolutionary Anthropology* **12**: 252–263.

Goudsmit, J. and Brandon-Jones, D. (1999). Mummies of olive baboons and Barbary macaques in the Baboon Catacomb of the Sacred Animal Necropolis at North Saqqara. *Journal of Egyptian Archaeology* **41**: 45–53.

Goudsmit, J. and Brandon-Jones, D. (2000). Evidence from the baboon catacomb in North Saqqara for a west Mediterranean monkey trade route to Ptolemaic Alexandria. *The Journal of Egyptian Archaeology* **43**: 111–119.

Hargett, J. M. (1988). Monkey madness in China. *Merveilles and Contes* **2**: 159–162.

Kanagavel, A., Sinclair, C., Sekar, R., and Raghavan, R. (2013). Moolah, misfortune or spinsterhood? The plight of slender loris *Loris lydekkerianus* in southern India. *Journal Threatened Taxa* **5**: 3585–3588.

Kirkpatrick, R. C. (1995). The natural history and conservation of the snub-nosed monkeys (Genus *Rhinopithecus*). *Biological Conservation* **72**: 363–369.

Kletter, R. (2002). A monkey figurine from Tel Beth Shemesh. *Oxford Journal of Archaeology* **21**: 147–152.

Mack, D. and Mittermeier, R. A. (Eds) (1984). *The International Primate Trade, Volume 1: Legislation, Trade and Captive Breeding*. Washington, DC: TRAFFIC.

Masseti, M. and Bruner, E. (2009). The primates of the western Palaearctic: a biogeographical, historical, and archaeozoological review. *Journal of Anthropological Sciences* **87**: 33–91.

Mittermeier, R. A. (1973). Colobus monkeys and the tourist trade. *Oryx* **12**: 113–117.

Mittermeier, R. A., Wallis, J., Rylands, A. B., Ganzhorn, J. U., Oates, J. F., *et al.* (Eds) (2009). Primates in Peril: The World's 25 Most Endangered Primates 2008–2010. IUCN/Species Survival Commission (SSC) Primate Specialist Group (PSG), International Primatological Society (IPS), and Conservation International (CI), Arlington, VA.

Modolo, L., Martin, R. D., van Schaik, C. P., van Noordwijk, M. A., and Kruetzen, M. (2008). When dispersal fails: unexpected genetic separation in Gibraltar macaques (*Macaca sylvanus*). *Molecular Ecology* **17**: 4027–4038.

Modolo, L., Salzburger, W., and Martin, R. D. (2005). Phylogeography of Barbary macaques (*Macaca sylvanus*) and the origin of the Gibraltar colony. *Proceedings of the National Academy of Sciences of the United States of America* **102**: 7392–7397.

Morris, D. and Morris, R. (1966). *Men and Apes*. London: Hutchinson.

Nekaris, K. A. I. (2014). Extreme primates: ecology and evolution of Asian lorises. *Evolutionary Anthropology* **23**: 177–187.

Nekaris, K. A. I. and Nijman, V. (2007). CITES proposal highlights rarity of Asian nocturnal primates (Lorisidae: *Nycticebus*). *Folia Primatologica* **78**: 211–214.

Nekaris, K. A. I., Shepherd, C. R., Starr, C. R., and Nijman, V. (2010). Exploring cultural drivers for wildlife trade via an ethnoprimatological approach: a case study of slender and slow lorises (*Loris* and *Nycticebus*) in South and Southeast Asia. *American Journal of Primatology* **72**: 877–886.

Nerlich, A. G., Parsche, F., Von Den Driesch, A., and Löhrs, U. (1993). Osteopathological findings in mummified baboons from ancient Egypt. *International Journal of Osteoarchaeology* **3**: 189–198.

Nijman, V. (2005). Decline of the endemic Hose's langur *Presbytis hosei* in Kayan Menterang National Park, East Borneo. *Oryx* **39**: 223–226.

Nijman, V. (2010). Ecology and conservation of the Hose's langur group (Colobinae: *Presbytis hosei, P. canicrus, P. sabana*): a review. In: Gursky, S. and Supriatna, J. (Eds), *Behavior, Ecology and Conservation of Indonesian Primates*. pp. 269–284. New York: Springer.

Nijman, V., Shepherd, C. R., and Nekaris, K. A. I. (2014). Trade in Bengal slow lorises in Mong La, Myanmar, on the China Border. *Primate Conservation* **28**: 139–142.

Oates, J. F. (1977). The guerza and man. In: Prince Rainier of Monaco and Bourne, G. H. (Eds), *Primate Conservation*. pp. 419–467. New York: Academic Press.

Osterberg, P. and Nekaris, K. A. (2015). The use of animals as photo props to attract tourists in Thailand: a case study of the slow loris *Nycticebus* spp. *TRAFFIC Bulletin* **27**: 13–18.

Perez, V. R., Godfrey, L. R., Nowak-Kemp, M., Burney, D. A., Ratsimbazafy, J., *et al.* (2005). Evidence of early butchery of giant lemurs in Madagascar. *Journal of Human Evolution* **49**: 722–742.

Phelps, J., Webb, E. L., Bickford, D., Nijman, V., and Sodhi N. S. (2011). Boosting CITES. *Science* **330**: 1752–1753.

Radhakrishna, S. (2013). The gulf between men and monkeys. In: Radhakrishna, S., Huffman, M. A., and Sinha, A. (Eds), *The Macaque Connection*. pp. 3–15. New York: Springer.

Sá, R. M. M., Ferreira da Silva, M., Sousa, F. M., and Minhós, T. (2012). The trade and ethnobiological use of chimpanzee body parts in Guinea-Bissau: implications for conservation. *TRAFFIC Bulletin* **24**: 31–36.

Shepherd, C., Sukumaran, J., and Wich S. A. (2005). *Open Season: An Analysis of the Pet Trade in Medan, North Sumatra, 1997–2001*. Petaling Jaya: TRAFFIC Southeast Asia.

Shipman, P., Bosler, W., Davis, K. L., Behrensmeyer, A. K., Dunbar, R. I. M., *et al.* (1981). Butchering of giant geladas at an Acheulian site. *Current Anthropology* **22**: 257–268.

Smith, A. W. (1993). On the ambiguity of the three wise monkeys. *Folklore* **104**: 144–150.

Smith, M. J., Benítez-Díaz, H., Clemente-Muñoz, M. Á., Donaldson, J., Hutton, J. M., *et al.* (2011). Assessing the impacts of international trade on CITES-listed species: current practices and opportunities for scientific research. *Biological Conservation* **144**: 82–91.

Starr, C., Nekaris, K. A. I., Streicher, U., and Leung, L. (2010). Traditional use of slow lorises *Nycticebus bengalensis* and *N. pygmaeus* in Cambodia: an impediment to their conservation. *Endangered Species Research* **12**: 17–23.

Stockey, G. and Grocott, C. (2012). *Gibraltar: A Modern History*. Cardiff: University of Wales Press.

Sutherland, W. J., Adams, W. M., Aronson, R. B., Aveling, R., Blackburn, T. M., *et al.* (2009). One hundred questions of importance to the conservation of global biological diversity. *Conservation Biology* **23**: 557–567.

Svensson, M. S. and Friant, S. C. (2014). Threats from trading and hunting of pottos and angwantibos in Africa resemble those faced by slow lorises in Asia. *Endangered Species Research* **23**: 107–114.

Tejada, R, Chao, E., Gómez, H., Painter, R. E. L., and Wallace, R. B. (2006). Wildlife use survey in the Tacana comunitary land of origin, Bolivia. *Ecología en Bolivia* **41**: 138–148.

Von den Driesch, A. (1993). The raising and worship of monkeys in the late period of ancient Egypt. *Tierarztliche Praxis* **21**: 95–101.

Von Koenigswald, G. H. (1982). Distribution and evolution of the orang utan, *Pongo pygmaeus* (Hoppius). In: de Boer, L. E. M. (Ed), *The Orang-utan, Its Biology and Conservation*. pp. 1–15. The Hague: Junk.

Wang, S. and Quan, G. (1986). Primate status and conservation in China. In: Benirschke, K. (Ed.), *Primates: The Road to Self-sustaining Populations*. pp. 213–220. New York: Springer.

Wolcott, L. T. (1978). Hanuman: the power-dispensing monkey in North Indian folk religion. *Journal of Asian Studies* **37**: 653–661.

Wolfheim, J. H. (1983). *Primates of the World: Distribution, Abundance and Conservation*. Seattle and London: University of Washington Press.

Yang-Martinez, S. (2011). An Investigation of Tarsier Tourism in Bohol, Philippines: Assessments of 11 Tarsier Exhibits, A Worry for Tarsier Welfare and Conservation. MSc thesis. Oxford Brookes University, Oxford.

CHAPTER 9

Hunting and primate conservation

John E. Fa and Nikki Tagg

A dead Drill monkey (Mandrillus leucophaeus) on sale in Malabo market, Bioko Island, Equatorial Guinea. Photo copyright: Lise Albretchsen.

Hunting for meat and profit threatens a very large number of primate species. This chapter presents evidence of the extent of hunting of wild animals, and primates in particular, across the tropical areas of the globe. We review studies investigating whether estimated extraction rates of mammal species across the tropics exceed the production levels for these taxa. In particular, data on primate extraction levels in Latin America, Africa, and Asia are presented, along with an assessment of their sustainability. We subsequently review the impact of hunting on game mammals in general, and primates in particular, and the implications of the possible local and global extirpation of primates on ecosystems. Finally, ways to mitigate primate hunting are reviewed.

9.1 Defining hunting

Hunting is the practice of killing or trapping any animal, or pursuing it with the intent of doing so. Hunting wildlife or feral animals is most commonly undertaken by humans for food, recreation, or trade. In present-day use, lawful hunting is distinguished from illegal poaching. Hunting can be broadly classified into categories including subsistence, commercial, and recreational. While there is substantial overlap between these categories, the main distinction concerns the benefits obtained. Recreational hunting refers to activities in which the main objective is the personal enjoyment of the hunter, rather than food or profit (e.g. trophy lion hunting in Tanzania; Whitman *et al.* (2004)). In subsistence hunting, the benefits are predominantly in the form of food and are directly consumed or used by—and play a very significant role in the sustenance of—the harvester and their family (e.g. Peres 2000). Commercial hunting takes place when the majority of the animals or their products are sold for profit.

In most parts of the world, hunting is often undertaken for the acquisition of wild meat, either directly or as a byproduct of sports hunting. People hunt for other reasons as well, including the killing of wild species to obtain live young for pets (e.g. Carpenter *et al.* 2004), the acquisition of hides, ivory, and horns (e.g. Iriarte and Jaksic 1986), or body parts for medicines, or substances traditionally

Fa, J.E. and Tagg, N., *Hunting and primate conservation*. In: *An Introduction to Primate Conservation*.
Edited by: Serge A. Wich and Andrew J. Marshall, Oxford University Press (2016).
DOI 10.1093/acprof:oso/9780198703389.003.0009

considered to have medicinal value (e.g. Cross River gorillas; Etiendem *et al.* (2011)). Animals can also be hunted in situations where they become a source of conflict with humans, as in the case of orang-utans in Indonesia (Meijaard *et al.* 2011), or out of fear or self-defence (Davis *et al.* 2013).

Because most hunting of wild species, including primates, occurs in pursuit of meat for human consumption, often referred to as bushmeat (Fa *et al.* 2003; Nasi *et al.* 2008), the analyses that follow in this chapter refer to hunting in this particular context, with a focus on primates. Bushmeat is broadly defined as any non-domesticated terrestrial mammal, bird, reptile, or amphibian harvested for food in tropical and subtropical forests. Primates are a source of bushmeat wherever they occur naturally.

9.2 Hunting of wildlife for food in the tropics

The importance of bushmeat as food is particularly high throughout the tropics, with some forest-living peoples obtaining up to 90% of their animal protein from wild animals (Fa *et al.* 2003). Studies have mostly focused on the offtake of large vertebrates (>1 kg), which seem to contribute the most to wild meat provision and comprise in particular mammals, but also birds, reptiles, and amphibians. Among mammals, primates generally represent relatively large-bodied species, exhibit slow life-history characteristics, often live in large, noisy groups, and frequently travel on the ground; as a result, primates as a taxon are particularly vulnerable to hunting (Linder and Oates 2011).

Wildlife consumption rates are driven by local, national, and global economics as well as geographic location (Brashares *et al.* 2011). Extractive industries, such as logging and mining, increase wildlife consumption, both directly and indirectly. Such operations result in the immigration into remote forest areas of large numbers of people with little incentive to conserve the locality, as well as the extension of road networks into previously inaccessible forest tracts, and the introduction of channels for the regular transport of a considerable proportion of the bushmeat that arrives in towns and cities (Robinson *et al.* 1999). Indirectly, the arrival of extractive operations in rural localities changes local economies; it improves wealth and facilitates the purchase of firearms and ammunition, thus encouraging forest-living peoples into a wildlife market economy. In turn, this increased income further drives demand for bushmeat, thus influencing patterns of resource consumption (Robinson *et al.* 1999).

9.2.1 Availability of wild meat: standing mammalian biomass

Large-bodied mammals are the main source of wild meat in many regions of tropical Africa, South America, and Asia (Robinson and Bennett 2004). The numbers and biomass of hunted species within a habitat determine its value to hunters. Data on global patterns in density and biomass for the main mammal groups are available for a wide range of non-hunted and lightly hunted forest sites, mostly for Africa and South America, but some data for Asian communities also exist (Robinson and Bennett 2004). In African sites, available figures for total mammalian biomass are around 3,000 kg/km^2 and about 1,000 kg/km^2 in Neotropical sites (Fa and Peres 2001). Distribution of mammalian biomass in non-hunted Asian forests, although less studied, is likely to be more similar to that observed in African forests, given the presence of large-bodied species such as elephants, tapirs, and rhinoceros (Corlett 2007).

The three most important taxonomic groups for human consumption in the tropics—ungulates, primates, and rodents (Figure 9.1)—occur at different relative and absolute densities in different ecosystems; for example, ungulates predominate in open habitats and primates occur most commonly in moist tropical forests. This can be explained by the relationship between aboveground net primary production and the amount and distribution of annual precipitation. Thus, 'wet' and 'moist' forests occur in areas with >2,000 mm of rainfall; 'dry' forests in those with between 1,000–2,000 mm; and savannahs, scrub, and dry woodlands in those localities receiving 100–1,000 mm. Mammalian standing biomass increases with rainfall until the point where grassland is substituted by trees, thus

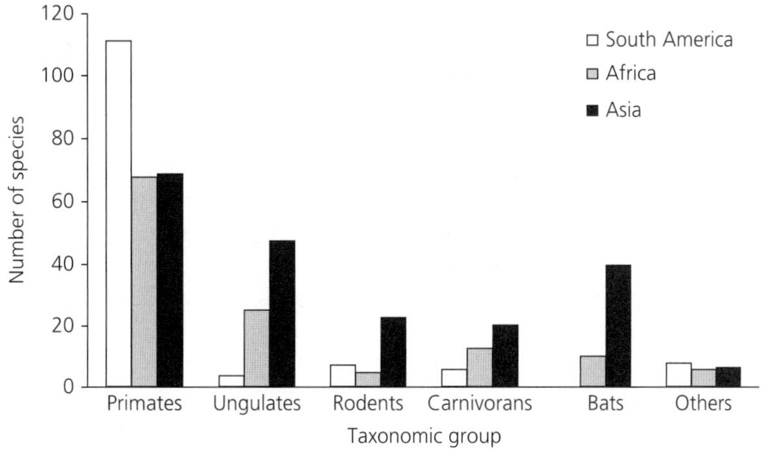

Figure 9.1 Number of hunted mammal species in tropical moist forests in South America, Asia, and Africa. Data from Fa *et al.* (2013).

lowering food availability for herbivores (Robinson and Bennett 2004). Hence, tropical forest ungulates, in particular, have a lower biomass than savannah or grassland ungulates, because in tropical forests the majority of primary production occurs in forms that are inaccessible to terrestrial mammals: inedible tree trunks (Bodmer 1989) or leaves that are located high in the canopy or heavily defended by plant secondary compounds, such as unpalatable lignins and toxins (McKey *et al.* 1981; Waterman *et al.* 1988; Waterman and McKey 1989). Arboreal primary production in tropical rainforests is more accessible to arboreal mammals such as monkeys, sloths, and rodents.

Biomass of taxa, however, will vary between sites within each ecosystem because differences in ecology (such as soil type, elevation, and climate) affect food supply; the picture is further complicated by the influence of human disturbance (logging and hunting), and historical and biogeographical factors (Oates 1996). Thus, in some African sites for example, duikers (*Philantomba* spp. and *Cephalophus* spp.) can attain a biomass greater than that of elephants (Dubost 1978, 1979), while in other sites primates dominate (Oates *et al.* 1990). Furthermore, mammalian species richness is also affected by whether the species are arboreal or terrestrial (Fa *et al.* 2004). For example, differences between Africa and South America, shown by Fa and Peres (2001), suggest that terrestrial species dominate in Afrotropical forests, accounting for almost 80% of the mammalian biomass, whereas in Neotropical forests this trend is reversed towards a predominance of arboreal taxa, accounting for 50–90% (Fa and Peres 2001). This distribution of biomass can explain the tendency for hunters to focus much more on snaring and netting in the Afrotropics as compared to the more common use of projectiles in Neotropical forests (Fa and Peres 2001).

9.2.2 Hunting levels

Although humans have hunted wildlife for millennia, studies that have calculated the sustainability of wildlife extraction harvests in recent decades have almost invariably found that rates of extraction far exceed those of production (Robinson and Bodmer 1999), even in the case of traditional societies still using rudimentary hunting technology (Alvard *et al.* 1997; Noss 1995).

Estimates of wild animal offtake in tropical forests range from global appraisals of what proportion wild animal protein contributes to people's diets (Prescott-Allen and Prescott-Allen 1982), to extrapolations of numbers and biomass consumed within particular regions. Fa and Peres (2001) and Fa *et al.* (2002) recently estimated extraction rates ($kg/km^2/year$) for 57 and 31 mammalian taxa in the

Congo and Amazon Basins, respectively. In terms of actual yields of edible meat (given that muscle mass and edible viscera account on average for 55% of body mass), 62,808 tons was estimated to be consumed in the Amazon, compared to approximately 4 million tons of dressed weight[1] of wild meat in the Congo Basin (Fa *et al.* 2002). Species exploitation rates at specific body masses were significantly greater in the Congo than in the Amazon, with an extraction to production ratio for the Congo of 2.4; 30 times greater than the ratio for the Amazon, calculated at 0.081. Thus, Congo Basin mammals must annually produce approximately 93% of their body mass to balance current extraction rates, whereas Amazonian mammals must produce only 4% of their body mass.

An important reason for harvesting bushmeat in many areas is income generation (Wright and Priston 2010). In terms of marketing, there are also significant differences between Africa and South America, both in emphasis and volume of wild meat sold (Nasi *et al.* 2008). In the Amazon Basin, commercialization occurs in largely hidden markets and bushmeat consumption in urban areas is unevenly studied (Nasi *et al.* 2011). Here, the scale of the bushmeat trade is presumed to be smaller than that recorded for areas such as the Congo Basin (Bodmer and Lozano 2001). Most towns and cities in African rainforest regions operate bushmeat markets that regularly trade large volumes of wild meat per year (Colyn *et al.* 1987; Juste *et al.* 1995; Fa *et al.* 2006). Such produce constitutes an important, and often underestimated, part of the economy of many African countries (Butynski and von Richter 1974; Feer 1993).

9.3 Primate hunting

As a result of the improved road networks carving up remaining forest tracts (Robinson *et al.* 1999) and easier access to firearms and ammunition (Wright 2010), hunters are increasingly able to target species, including primates, that were previously more difficult to capture by snaring and other traditional methods of hunting (Fa *et al.* 2004). Given the greater value of the meat of species such as arboreal monkeys and large-bodied great apes, and the often strong demand of such highly prized meat from people of high standing in African cities, for example, primates are increasingly targeted and their meat traded (Starkey 2004).

As with many large-bodied, slow reproducing mammals, many primate species are recognized as threatened or endangered by the International Union of the Conservation of Nature (IUCN) and are prohibited by national laws from being taken as quarry by hunters. However, widespread hunting of primates both for subsistence and for trade continues (in many cases at an increasing rate) throughout the tropics, with species that are protected by national legislation being sold either on the black market or quite openly in bushmeat markets with no adherence to or enforcement of the law.

9.3.1 Primate abundance in tropical forests

Correlations of primate and ungulate biomass with rainfall across environments show that overall primate abundance increases with rainfall but ungulates decline (Figure 9.2) (Kay *et al.* 1997). In tropical rainforests with high annual rates of precipitation, primates may comprise up to 40% of the non-flying mammalian biomass, as calculated for Africa (Emmons and Gentry 1983; Oates *et al.* 1990) and South America (Peres 1999a). In African rainforests, mean primate densities of 195 individuals/km^2 (858 kg/km^2) are typical (Chapman *et al.* 1999), whereas South American sites exhibit lower densities and biomass, around 124 individuals/km^2 (277 kg/km^2) (Peres 1999b). Most African forest sites are dominated by relatively large-bodied folivorous colobines (for example, colobus monkeys and langurs), thus increasing the number of animals present and their biomass. In Africa and Asia, colobines account for an average of 60% (range 28–91%, $N = 10$) of the primate community biomass (Oates *et al.* 1990; Oates 1996). In Neotropical forests, the equivalent arboreal folivorous primates often represent over half of the biomass of non-flying mammals (Eisenberg and Thorington 1973; Peres 1997).

[1] Dressed weight (also called carcass weight) refers to the weight of an animal after being partially butchered, removing all the internal organs and often the head, as well as inedible (or less desirable) portions of the tail and legs.

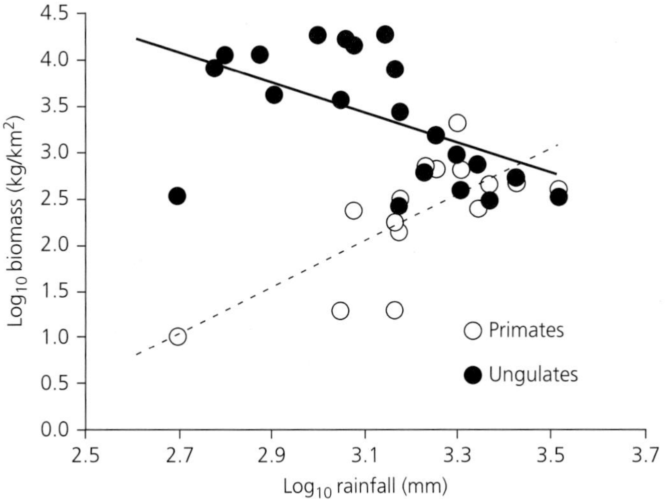

Figure 9.2 Percentage contribution of primates and ungulates to the total biomass of large-bodied mammals (>1 kg) at African and South American sites of varying rainfall. In all sites hunting is negligible, so mammals are assumed to be at or near carrying capacity (K). Data from Robinson and Bennett (2004).

In Africa, forest type also correlates with primate biomass, but forest productivity is a less important determinant (Chapman *et al.* 1999). For example, soil chemistry has been shown to be less important than growth stage, heterogeneity, taxonomic composition, and history of the vegetation in determining the abundance of colobines (Oates *et al.* 1990). In fact, Maisels *et al.* (1994) and Maisels and Gautier-Hion (1994) indicated that primate biomass can still be high in forests on nutrient-poor white-sand soils, where legume seeds and young leaves become prominent in primate diets. A similar pattern is evident in Borneo, where peat swamps offer a low folivore biomass, but are important habitats for frugivores such as gibbons and orang-utans (Posa *et al.* 2011). In the case of African monkeys, their foraging plasticity may also explain why no clear relationship between frugivore primate biomass (guenons and mangabeys) and fruit availability has been found (Tutin *et al.* 1997); frugivorous primates will increase their seed and leaf intake in forests where fleshy fruits are less diverse or absent (Maisels and Gautier-Hion 1994).

In the case of Amazonia, Peres (1999b) showed that forest type, hydrology, and geochemistry were key determinants of primate biomass. Thus, forests on nutrient-rich soils, and perhaps with a higher fruit production, sustain a greater primate biomass, even when differences in hunting pressure are taken into account (Peres 1999b).

9.3.2 Primate extraction

Estimates of numbers of primates hunted throughout the tropics are scant. For the Brazilian Amazonia, Peres (2000) estimated that subsistence hunting by low-income rural populations extracts up to a maximum of 5.3 million primates. In the Congo Basin, however, the number of primates taken by commercial and subsistence hunters is considerably higher than for the Amazon, calculated at over 103 million individual animals hunted annually (Fa and Peres 2001). Data for both basins indicate that a wide range of species of different body sizes is hunted. Medium- and small-bodied taxa, such as capuchins (*Cebus* spp.) in South America and guenons (*Cercopithecus* spp.) in Africa, are more commonly harvested due to their higher relative abundance and ease of capture (Fa and Peres 2001). Large-bodied species, although more economically cost effective for hunters (since the ratio of benefits to costs equals or exceeds the average return for all hunted species), are also more likely to disappear faster because of their associated lower densities and vulnerability to hunting pressure, and therefore often constitute a smaller proportion of hunters' catch or of carcasses on sale. This is supported by data from market sales in 80 bushmeat markets in the Cross-Sanaga region in Nigeria and Cameroon, with which Fa *et al.* (2006) showed that primates

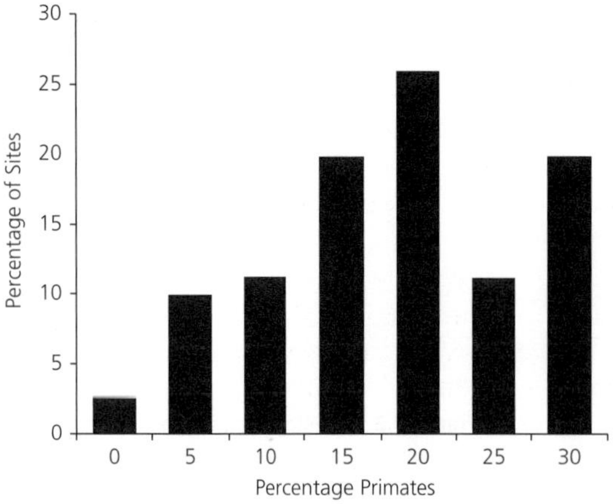

Figure 9.3 Percentage contribution of primates to animals sold in 80 bushmeat markets in the Cross-Sanaga region, Nigeria, and Cameroon. Data from Fa *et al.* (2006).

accounted for on average only 13.5% of all carcasses on sale (Figure 9.3); a disproportionately low percentage given their predominant representation in the overall biomass. Similarly, other data from hunter studies throughout the Congo Basin (Fa *et al.* 2004) indicate the percentage contribution of this taxonomic group to be only about 8%.

9.4 Consequences of hunting

Unsustainable hunting results in the local decline and extirpation of wildlife populations, thus reducing overall species population size and leading to isolation of sub-populations and a loss of genetic diversity; all of which, particularly for species existing at low densities, can reduce the viability of overall survival. Coupled with threats from habitat loss (see Irwin, Chapter 7, this volume)—even from historical deforestation (Cowlishaw 1999)—disease (Nunn and Gillespie, Chapter 10, this volume), and climate change (Korstjens and Hillyer, Chapter 11, this volume), such uncontrolled exploitation through hunting will no doubt threaten to bring about marked population declines and local disappearances, and subsequently the eventual global extinction of the most sensitive taxa, including many primates (Struhsaker 1999).

9.4.1 Impact of hunting on wildlife

Hunting is arguably the greatest threat to a number of animal populations and species in the tropics. For example, while habitat loss as a result of agriculture and extractive industries is a real and growing threat to the future survival of forest-living species (see Irwin, Chapter 7, this volume), in many cases selective and sustainable operational practices are not necessarily incongruent with persistence of wildlife populations. This is supported by the observations that gorillas in Africa thrive on the terrestrial herbaceous vegetation that grows in abundance in logging gaps (Oates 1996), and that orang-utans in eastern Borneo survive at high densities in plantation-dominated landscapes (Meijaard *et al.* 2010).

The impact of hunting on mammal communities in tropical forests is difficult to evaluate given the additional influence of natural factors on primate abundance (Peres and Palacios 2007). If hunting pressure is not too heavy, and large neighbouring tracts of undisturbed forest can buffer and replenish hunted areas, it is possible that fast-reproducing game populations can bounce back after exploitation. However, studies suggest the impact of hunting on some game mammals, such as primates, to be greater than that of moderate habitat disturbance such as logging: as demonstrated in the case of colobus monkeys and gorillas in Africa (Oates 1996). Similarly, orang-utan density in Indonesia is not affected by light levels of logging, but is lower nearer to villages known to hunt (Marshall *et al.* 2006). It is suggested that even when hunting occurs at a low rate and annual harvests remain stable, the exploitation is likely to be unsustainable, and could result in population collapse at some point and a 'bushmeat bust' following the current 'bushmeat boom' (Barnes 2002: 241). For example, it has been demonstrated that even rates of offtake of orang-utans as low as 2% can lead to local extinction (Marshall *et al.* 2009).

Studies demonstrate that vertebrate game biomass declines dramatically as hunting pressure increases. For example, in South American sites, overall game biomass can drop from 1,200 kg/km^2 in non-hunted sites, to around 200 kg/km^2 in heavily hunted areas (Peres 2000). The available data for

Africa, although limited in comparison to the Neotropics, suggest a similar pattern of depletion. From estimates of mammal abundance in non-hunted and hunted sites in Makokou, Gabon, Lahm (1993) showed that body mass and population density were negatively correlated with impact on the species. Data from 20 forest sites in Africa clearly demonstrated decreases in primate numbers as hunting pressure became heavier (Oates 1996). An investigation into the impact of hunting on monkey species in Korup National Park, Cameroon, using data from field surveys and bushmeat markets, demonstrated that primate species richness declined over time in a heavily hunted site, and that hunting was driving the local extinction of certain species (i.e. Preuss's red colobus, *Procolobus pennantii preussi*, and drill, *Mandrillus leucophaeus*) (Linder and Oates 2011).

The vulnerability to hunting and the degree to which hunting affects abundance varies between primate species, due to differing life-history characteristics, such as body size or reproductive potential (Isaac and Cowlishaw 2004; Kumpel *et al.* 2008). Slowly reproducing species may struggle to recover from population reductions. In the Udzungwa Mountains, Tanzania, canopy-dwelling colobines (*Colobus* spp. and *Procolobus* spp.) were dramatically negatively affected by hunting, while guenons were less affected (Revero *et al.* 2012) due to a higher ecological adaptability. Such ecological flexibility, for example in terms of dietary breadth (Cowlishaw and Dunbar 2000), may enable some species to move into the available niche space arising from the removal of another species, and to increase in density due to the resulting competitive release (Linder and Oates 2011; Isaac and Cowlishaw 2004).

In the study by Peres (2000), although primate biomass dropped significantly as a result of more hunting (Figure 9.4), overall primate densities were not affected. This can be explained as follows: population sizes of small- and medium-bodied primates increased as a result of competitive release; therefore, despite the reduction in certain large-bodied species, the positive numerical response of small- and medium-bodied primate species may cancel out the effect of hunting on primate densities overall—a phenomenon known as density compensation. These observed trends are largely a consequence of differences in abundance of large-bodied genera (i.e. howler monkeys *Alouatta* spp., spider monkeys *Ateles* spp., and woolly monkeys *Lagothrix* spp.), accounting for the bulk of the primate biomass in non-hunted sites, but being overharvested or becoming extinct in sites hunted by humans. These responses of the residual assemblage of non-hunted medium-bodied species, where their large-bodied (ateline) counterparts had been severely reduced in numbers, have been demonstrated by Peres and Dolman (2000).

Whether or not density compensation occurs in hunted sites in Africa or Asia—as observed in the Neotropics—remains to be demonstrated. Potential evidence for this phenomenon is provided in studies that demonstrate a shift to smaller prey, which would be the expected scenario in overhunted areas. For example, Peres (1999b) and Jerozolimski and Peres (unpublished data) counted the number of animal carcasses arriving at Malabo market, Bioko Island, Equatorial Guinea, during two eight-month study periods in 1991 and 1996. Between-year comparisons of these harvests showed that the number of species and carcasses in 1996 was 60% higher than in 1991; whereas in terms of biomass, the increase between study years was relatively significantly less, at only 12.5%. A larger number of carcasses of smaller-bodied species, such as rodents and blue duiker (*Philantomba monticola*), were recorded in the later study period; while there was

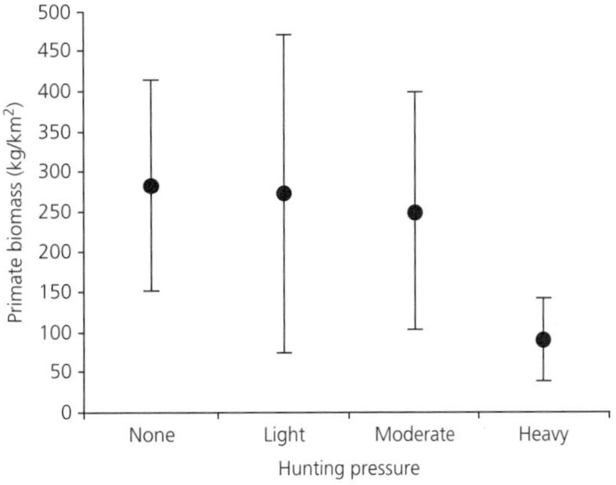

Figure 9.4 Average primate biomass at 56 northern Neotropical sites with varying degrees of hunting pressure. Data from Peres (1999c).

a dramatic reduction in larger-bodied species, including Ogilby's duiker (*Cephalophus ogilbyi*) and the seven studied diurnal primates (Fa *et al.* 2000). This may demonstrate that the mean body mass of all targeted species combined was significantly reduced in response to hunting pressure; alternatively, it can be argued that this simply demonstrates a reduction in the availability of larger prey, forcing hunters to target less-preferred, and often smaller-bodied, game species as a result.

Studies of how hunting might affect the population dynamics of game species are sorely lacking for Africa, Asia, and the Neotropics. Some data are, however, available on the impact of hunting on animals of different age classes in Africa. For example, Dubost (1978, 1980) and Feer (1988) concluded that hunting and trapping by Gabonese villagers most severely affect young adult chevrotain and duikers; the age class with the greatest reproductive potential. However, population age structures and demographics of hunted versus non-hunted sites are rarely available (but see Bodmer *et al.* (1997)). The fact remains that all investigations into the sustainability of game hunting in South America and Africa have shown that most species have been overexploited (Fa *et al.* 1995; Fa 2000; Fitzgibbon *et al.* 1995; Ráez Luna 1995; Noss 1997; Muchaal and Ngandjui 1999; Townsend 1999); an observation that likely applies across the tropics.

9.4.2 Impact of hunting on ecosystems

The majority of tree species in the tropics disperse their seeds by frugivore-mediated seed dispersal (i.e. zoochory), resulting in interdependency between plants and animals (Jordano *et al.* 2007). Reduction or loss of seed dispersers, large granivores and frugivores, and 'habitat landscapers' (Dirzo and Miranda 1991; Chapman and Onderdonk 1998; Wright *et al.* 2000) will impact dependent floral taxa (Sodhi *et al.* 2009). Subsequently, the depletion of wildlife can alter regeneration patterns of plant communities (e.g. Terborgh *et al.* 2008), resulting in long-term changes in tropical forest dynamics. In these forests, overhunting of populations of key vertebrate species alongside forest fragmentation or other forms of habitat deterioration likely underlie the 'half-empty' forest syndrome (Redford and Feinsinger 2001).

Primates are the most common frugivores in tropical forests (Garber and Lambert 1998); they therefore arguably constitute the most important taxon of seed dispersers (Poulsen *et al.* 2001; Lambert 2011; Petre *et al.* 2013). Evidence of the important role of primates in seed dispersal exists from numerous tropical rainforest sites across the world; here, we focus on one example from Africa to demonstrate the phenomenon. Due to their large gut size and high daily fruit consumption, western lowland gorillas (*Gorilla gorilla gorilla*) alone could be responsible for over 40% of all seeds dispersed by the entire primate community (Petre *et al.* 2013). Seeds often pass through the gorilla digestive tract undamaged (Petre *et al.* 2015), and may experience directed deposition at open-canopy forest sites (e.g. treefall gaps) that are often selected by gorillas as nesting sites and offer favourable conditions for seedling recruitment (Petre *et al.* 2013). In addition, western lowland gorillas may be the only disperser of at least one tree species, *Cola lizae*, endemic to central Gabon (Tutin *et al.* 1991), demonstrating interdependency of plants and animals within ecosystems.

Furthermore, there may be a low level of redundancy among primates, meaning that the loss of the dispersal services of one species may not be replaced by the actions of another. In the Dja Biosphere Reserve, Cameroon, for example, primates (including the genera *Gorilla, Cercopithecus, Lophocebus*, and *Colobus*) exhibit only a small overlap in the seed species they disperse (Poulsen *et al.* 2001; Petre *et al.* 2013). Smaller-bodied frugivores cannot swallow and disperse larger seeds (Vanthomme *et al.* 2010). Non-dispersed seeds will fall in high densities beneath the crown of the parent plant, and experience density-dependent mortality and a reduced chance of encountering favourable conditions and surviving to germination.

The loss of a species of disperser, therefore, may influence patterns of tree recruitment. A study in southeastern Peru demonstrated a correlation between the extirpation of large- and medium-bodied primates by hunting and a dramatic lack of seedlings of primate-dispersed trees, with abiotically-dispersed trees subsequently displaying a disproportionately high abundance (Nunez-Iturri *et al.* 2008). The extirpation of primates by hunting, even on a local scale, constitutes a diminution of zoochoric processes, leading to changes in plant demography and dynamics, succession, and spatial distribution (Gillespie *et al.* 2012); subsequent shifts in community structure and composition (Wright 2010); and an ultimate reduced adult tree diversity (Blake *et al.* 2009). These changes in natural forest regeneration and dynamics may constitute a major threat for the future of remaining forests. In turn, these threats to forests also cast into doubt the ecosystem services that forests provide, in terms of the production of timber and non-timber forest products, local economic and cultural uses (Lambert 1998), and the globally relevant issues of carbon and water storage.

9.5 Mitigation of hunting

Results of bushmeat studies overwhelmingly demonstrate that bushmeat hunting affects distribution and density of hunted species and is unsustainable for most large-bodied animals. Coupled with other threats, current bushmeat hunting foretells catastrophic results for African forest wildlife and for people dependent on wild meat (Fa *et al.* 2003). However, control and mitigation of hunting is extremely difficult. In the case of hunting for bushmeat, a number of incentives and approaches have been discussed and tested, and mostly comprise lowering the demand (Wilkie and Carpenter 1999).

Altering demand through social marketing may be effective with those for whom bushmeat consumption is a luxury, for example capital city elites and restaurants; but will not be so in cases where bushmeat consumption meets basic needs, and therefore represents an inelastic commodity (Wilkie *et al.* 2005). In these cases, it is necessary to develop substitutes to bushmeat (Wilkie and Carpenter 1999). The provision of alternative sources of protein, for example through small-animal breeding (cane rats, poultry), and simultaneous technical training and support in areas where wild bushmeat species are already depleted (Wilkie and Carpenter 1999), may effectively help to decrease unsustainable wildlife consumption in some areas (e.g. Foerster *et al.* 2012). If a household becomes engaged in alternative income-generating activities—which may be more resource efficient in the long term compared to bushmeat hunting—this may serve to devalue the benefits to the household of bushmeat hunting, thereby reducing the importance of bushmeat as a source of cash income and consequently reducing its role in alleviating poverty. However, given the observation that bushmeat is often a preferred food (Robinson and Bennett 2000), at least in rural communities (Schenck *et al.* 2006), 'alternatives' often do not replace bushmeat hunting as intended, but either complement it or are neglected in favour of the relatively less labour-intensive hunting. Furthermore, bushmeat cannot be considered a generic food source, as people demonstrate preferences for different species, warning that ill-informed initiatives offering captive-raised, less-preferred bushmeat species will have little impact on hunting (Schenck *et al.* 2006). Alternative sources of income may contribute to reducing the reliance of local people on trading bushmeat, but at best may not lower local demand, and at worst may facilitate hunting through the increased funds available to purchase firearms and ammunition.

Projects that offer alternatives should be coupled with initiatives to raise community awareness: through sensitization to conservation issues and environmental education, local people can become more aware of the implications of unsustainable hunting on wildlife and subsequently on their own livelihoods (Milner-Gulland *et al.* 2003) and the potential benefits of successful management of their surrounding natural resources. Community participation in local wildlife management, whereby local authorities and experts can assist community-appointed groups to assign hunting and no-take areas (Milner-Gulland *et al.* 2003) and set restrictions on when and how much hunting is permitted, can increase motivation to adhere to restrictions. However, zoning is hard to enforce and, considering that time is the main expense of hunters, hunters are unlikely to bypass a no-take area to hunt elsewhere (Wilkie and Carpenter 1999). Similarly, it is difficult to enforce seasonal or species quotas, as hunters are not easily convinced to limit hunting as a function of the number of hunters in the community, or the abundance of hunted species in the forest, as any given individual will strive to take the maximum possible for their own households, regardless of the needs of the community or the impact on future game availability.

Improved law enforcement is particularly important to preserve endangered species on the brink of extirpation (Milner-Gulland *et al.* 2003). In particular, effective management of viable protected areas is crucial, given that protected areas are the source of much wildlife trade (Macdonald *et al.* 2012). Interdiction can exert a strong incentive to avoid the risk of fines or imprisonment and to participate instead in development initiatives. Furthermore, bushmeat price manipulation through fines and taxation, as well as supply reduction through enforcement of bushmeat market regulations, can act as an economic lever to reduce the demand for bushmeat and regulate exploitation (Wilkie and Carpenter 1999; Wilkie *et al.* 2005). Inversely, pricing of domestic meat may also play a role in driving demand for bushmeat. However, governments have little incentive to discourage hunting and further reduce the welfare of their poorest citizens (Wilkie and Carpenter 1999), and resources are commonly lacking across the tropics, where corruption and poor governance are commonplace.

There is an urgent need for research into ways of mitigating the problem of defaunation at a continental scale (Wilkie and Carpenter 1999; Milner-Gulland *et al.* 2003). For example, it is essential to identify and understand drivers of wildlife consumption, and to know population densities and reproductive rates of a species, in order to estimate sustainability and evaluate intervention effectiveness (e.g. Brashares *et al.* 2011), particularly in tropical areas where data are lacking. Research seeking to distinguish the influence of natural and life-history drivers of primate distribution and abundance from the direct effects of hunting is crucial in order to assess the sustainability of hunting and to guide conservation initiatives (Linder and Oates 2011). Projects that offer alternatives should evaluate their effect on demand. Monitoring surveys can investigate the influence of the regulation and policing of protected areas on wildlife trade (Macdonald *et al.* 2012). As an example of a more specific research initiative, bushmeat genetic barcoding technology can be used to identify and monitor non-recognizable or non-distinguishable bushmeat or species among hunters' catch or on market tables (Eaton *et al.* 2010).

Mitigation of hunting must be viewed on a landscape scale, implicating the participation of all stake-holders, including local people, resource extraction companies, conservationists, research bodies, and governments (Milner-Gulland *et al.* 2003). Extractive industries commonly constitute the only significant institutional presence in remote forests, and are therefore best equipped to deal with the issue of wildlife management (Robinson *et al.* 1999). For example, national law can enforce logging companies to ban any trade in wildlife products in their concessions (Wilkie and Carpenter 1999). Economic incentives and public awareness can contribute to putting pressure on range state governments to impose and improve conservation policies for endangered primate species.

9.6 Summary

In this chapter we have considered the extent of wild animal hunting across the tropical areas of the globe, with a particular focus on primates and on subsistence and commercial hunting for food. We have argued that hunting may be one of the greatest threats to many primate species, and that rates of hunting across the tropics are likely to be unsustainable. Bushmeat species—particularly large-bodied mammals in tropical rainforests, including primates—are at risk of local extirpation in many areas, and in some cases the extinction of the species is not unfathomable. Furthermore, there will be considerable knock-on effects for other interdependent wildlife species (e.g. predators and prey species), thus affecting the balance of the wildlife community. Wildlife loss also disrupts ecological balance: for example, a reduction in seed dispersal through the loss of frugivores will alter plant community composition and forest structure. As a result of these changes, the ecosystem services that the forest offers to forest-living people and to our global society can be affected, reduced, or eliminated. Although the short-term benefits of bushmeat hunting to rural people are important, in terms of protein and financial income, ways to mitigate hunting that consider the socio-economics of local communities are available, predominantly revolving around collaborative efforts to reduce the bushmeat demand.

References

Alvard, M. S., Robinson, J. G., Redford, K. H., and Kaplan, H. (1997). The sustainability of subsistence hunting in the Neotropics. *Conservation Biology* **11**(4): 977–982.

Barnes, R. F. (2002). The bushmeat boom and bust in West and Central Africa. *Oryx* **36**(03): 236–242.

Blake, S., Deem, S. L., Mossimbo, E., Maisels, F., and Walsh, P. (2009). Forest elephants: tree planters of the Congo. *Biotropica* **41**(4): 459–468.

Bodmer, R. E. (1989). Ungulate biomass in relation to feeding strategy within Amazonian forests. *Oecologia* **81**: 547–550.

Bodmer, R. E., Eisenberg, J. F., and Redford, K. H. (1997). Hunting and the likelihood of extinction of Amazonian mammals. *Conservation Biology* **11**(2): 460–466.

Bodmer, R. E. and Lozano, E. P. (2001). Rural development and sustainable wildlife use in Peru. *Conservation Biology* **15**: 1163–1170.

Brashares, J. S., Golden, C. D., Weinbaum, K. Z., Barrett, C. B., and Okello, G. V. (2011). Economic and geographic drivers of wildlife consumption in rural Africa. *Proceedings of the National Academy of Sciences* **108**(34): 13931–13936.

Butynski, T. and von Richter, W. (1974). Wildlife utilization in Botswana: a review and evaluation of hunters' returns as a source of administrative and biological data? *South African Wildlife Management Association* **4**: 167–176.

Carpenter, A. I., Rowcliffe, J. M., and Watkinson, A. R. (2004). The dynamics of the global trade in chameleons *Biological Conservation* **120**: 291–301.

Chapman, C. A., Gautier-Hion, A. N. N. I. E., Oates, J. F., and Onderdonk, D. A. (1999). African primate communities: determinants of structure and threats to survival. In: Feagle, J. G., Janson, C., and Reed, K. E. (Eds), *Primate Communities*. pp. 1–37. Cambridge: Cambridge University Press.

Chapman, C. A. and Onderdonk, D. A. (1998). Forests without primates: primate/plant codependency. *American Journal of Primatology* **45**: 127–141.

Colyn, M., Dudu, A., Mankoto, M., and Mbaelele, M. A. (1987). Exploitation du petit et moyen gibier des forets ombrophiles du Zaire. *Nature et Faune* **3**: 22–39.

Corlett, R. T. (2007). The impact of hunting on the mammalian fauna of tropical Asian forests. *Biotropica* **39**: 292–303.

Cowlishaw, G. (1999). Predicting the pattern of decline of African primate diversity: an extinction debt from historical deforestation. *Conservation Biology* **13**: 1183–1193.

Cowlishaw, G. and Dunbar, R. (2000). *Primate Conservation Biology*. Chicago: University of Chicago Press.

Davis, J. T., Mengersen, K., Abram, N. K., Ancrenaz, M., Wells, J. A., *et al.* (2013). It's not just conflict that motivates killing of orangutans. *PLoS One* **8**(10): e75373.

Dirzo, R. and Miranda, A. (1991). Altered patterns of herbivory and diversity in the forest understory: a case study of the possible consequences of contemporary defaunation. In: Price, P. W., Lewinsohn, T. W., Benson, W. M., and Fernandes, G. W. (Eds), *Plant-Animal Interactions: Evolutionary Ecology in Tropical and Temperate Regions*. pp. 273–287. New York: John Wiley & Sons.

Dubost, G. (1978). Account on biology of African chevrotain, *Hyemoschus aquaticus-ogilby*, Artiodactyl traguilid. *Mammalia* **42**: 1.

Dubost, G. (1979). The size of African forest artiodactyls as determined by the vegetation structure. *African Journal of Ecology* **17**: 1–17.

Dubost, G. (1980). L 'écologie et la vie sociale du Céphalophe bleu (*Cephalophus monticola* Thunberg), petit ruminant forestier africain. *Zeitschrift für Tierpsychologie* **54**: 205–266.

Eaton, M. J., Meyers, G. L., Kolokotronis, S.-O., Leslie, M. S., Martin, A. P., *et al.* (2010). Barcoding bushmeat: molecular identification of Central African and South American harvested vertebrates. *Conservation Genetics* **11**: 1389–1404.

Eisenberg, J. F. and Thorington, R. W. (1973). A preliminary analysis of a Neotropica mammal fauna. *Biotropica* **5**: 150–161.

Emmons, L. H. and Gentry, A. H. (1983). Tropical forests structure and the distribution of gliding and prehensile-tailed vertebrates. *The American Naturalist* **121**: 513–524.

Etiendem, D. N., Hens, L., and Pereboom, Z. (2011). Traditional knowledge systems and the conservation of Cross River gorillas: a case study of Bechati, Fossimondi, Besali, Cameroon. *Ecology and Society* **16**: 22: http://dx.doi.org/10.5751/ES-04182–160322.

Fa, J. E. (2000). Hunted animals in Bioko Island, West Africa: sustainability and future. In: Robinson, J. G. and Bennett, E. L. (Eds), *Hunting for Sustainability in Tropical Forests*. pp. 168–198. New York: Columbia University Press.

Fa, J. E., Currie, D., and Meeuwig, J. (2003). Bushmeat and food security in the Congo Basin: linkages between wildlife and people's future. *Environmental Conservation* **30**: 71–78.

Fa, J. E., Farfán, M. A., Márquez, A. L., Duarte, J., and Vargas, J. M. (2013). Reflexiones sobre el impacto y manejo de la caza de mamíferos silvestres en los bosques tropicales. *Ecosistemas* **22**: 76–83.

Fa, J. E., Garcia Yuste, J. E., and Castelo, R. (2000). Bushmeat markets on Bioko Island as a measure of hunter presence. *Conservation Biology* **14**: 1602–1613.

Fa, J. E., Juste, J., Perez del Val, J., and Castroviejo, J. (1995). Impact of market hunting on mammal species in Equatorial Guinea. *Conservation Biology* **9**: 1107–1115.

Fa, J. E. and Peres, C. A. (2001). Game vertebrate extraction in African and Neotropical forests: an intercontinental comparison. In: Reynolds, J. D., Mace, G. M., Robinson, J. G., and Redford, K. H. (Eds), *Conservation of Exploited Species*. pp. 203–241. Cambridge: Cambridge University Press.

Fa, J. E., Peres, C. A., and Meeuwig, J. (2002). Bushmeat exploitation in tropical forests: an intercontinental comparison. *Conservation Biology* **16**: 232–237.

Fa, J. E., Ryan, S. F., and Bell, D. J. (2004). Hunting vulnerability, ecological characteristics and harvest rates of bushmeat species in afrotropical forests. *Biological Conservation* **121**: 167–176.

Fa, J. E., Seymour, S. Dupain, J., Amin, R., Albrechtsen, L., *et al.* (2006). Getting to grips with the magnitude of exploitation: bushmeat in the Cross-Sanaga rivers region, Nigeria and Cameroon. *Biological Conservation* **129**: 497–510.

Feer, F. (1988). Stratégies écologiques de deux espèces de Bovidés sympatriques de la forêt sempervirente africaine (Cephalophus callipygus et C. dorsalis): influence du rythme d/'activité. Doctoral dissertation: Universite Pierre et Marie Curie, Paris.

Feer, F. (1993). The potential for sustainable hunting and rearing of game in tropical forests. *Man and the Biosphere Series* **13**: 691–691.

Fitzgibbon, C. D., Mogaka, H., and Fanshawe, J. H. (1995). Subsistence hunting in Arabuko-Sokoke Forest, Kenya, and its effects on mammal populations. *Conservation Biology* **9**(5): 1116–1126.

Foerster, S., Wilkie, D. S., Morelli, G. A., Demmer, J., Starkey, M., *et al.* (2012). Correlates of bushmeat hunting among remote rural households in Gabon, Central Africa. *Conservation Biology* **26**(2): 335–344.

Garber, P. A. and Lambert, J. E. (1998). Introduction to primate seed dispersal. *American Journal of Primatology* **45**: 3–8.

Gillespie, R. G., Baldwin, B. G., Waters, J. M., Fraser, C. I., Nikula, R., *et al.* (2012). Long-distance dispersal: a framework for hypothesis testing. *Trends in Ecology and Evolution* **27**(1): 47–56.

Iriarte, J. A. and Jaksic, F. M. (1986). The fur trade in Chile: an overview of seventy-five years of export data (1910–1984). *Biological Conservation* **38**: 243–253.

Isaac, N. J. B. and Cowlishaw, G. (2004). How species respond to multiple extinction threats. *Proceedings of the Royal Society Series B* **271**: 1135–1141.

Jordano, P., Garcia, C., Godoy, J. A., and García-Castaño, J. L. (2007). Differential contribution of frugivores to complex seed dispersal patterns. *Proceedings of the National Academy of Sciences* **104**: 3278–3282.

Juste, J., Fa, J. E., P´erez Del Val, J., and Castroviejo, J. (1995). Market dynamics of bushmeat species in Equatorial Guinea. *Journal of Applied Ecology* **32**: 454–467.

Kay, R. F., Madden, R. H., Van Schaik, C., and Higdon, D. (1997). Primate species richness is determined by plant productivity: implications for conservation. *Proceedings of the National Academy of Sciences* **94**(24): 13023–13027.

Kümpel, N. F., Milner-Gulland, E. J., Rowcliffe, M. J., and Cowlishaw, G. (2008). Impact of gun-hunting on diurnal primates in continental Equatorial Guinea. *International Journal of Primatology* **29**: 1065–1082.

Lahm, S. A. (1993). Ecology and Economics of Human/Wildlife Interaction in Northeastern Gabon. Doctoral dissertation: New York University, Graduate School of Arts and Science.

Lambert, J. E. (1998). Primate frugivory in Kibale National Park, Uganda, and its implications for human use of forest resources. *African Journal of Ecology* **36**: 234–240.

Lambert, J. E. (2011). Primate seed dispersers as umbrella species: a case study from Kibale National Park, Uganda, with implications for Afrotropical forest conservation. *American Journal of Primatology* **73**: 9–24.

Linder, J. M. and Oates, J. F. (2011). Differential impact of bushmeat hunting on monkey species and implications for primate conservation in Korup National Park, Cameroon. *Biological Conservation* **144**: 738–745.

Macdonald, D. W., Johnson, P. J., Albrechtsen, L., Seymour, S., Dupain, J., *et al.* (2012). Bushmeat trade in the Cross–Sanaga rivers region: evidence for the importance of protected areas. *Biological Conservation* **147**: 107–114.

Maisels, F. and Gautier-Hion A. (1994). Why are Caesalpinioideae so important for monkeys in hydromorphic rainforests of the Zaire basin? In: Sprent, J. I. and McKey, D. (Ed.), *Advances in Legume Systematics: Part 5 The Nitrogen Factor*. pp. 189–204. Kew: Royal Botanic Gardens.

Maisels, F., Gautier-Hion, A., and Gautier, J. P. (1994). Diets of two sympatric colobines in Zaire: more evidence on seed-eating in forests on poor soils. *International Journal of Primatology* **15**: 681–701.

Marshall, A. J., Engström, L. M., Pamungkas, B., Palapa, J., Meijaard, E., *et al.* (2006). The blowgun is mightier than the chainsaw in determining population density of Bornean orangutans (*Pongo pygmaeus morio*) in the forests of East Kalimantan. *Biological Conservation* **129**: 566–578.

Marshall, A. J., Lacy, R., Ancrenaz, M., Byers, O., Husson, S., *et al.* (2009). Orangutan population biology, life history, and conservation: perspectives from PVA models. In: Wich, S. A., Utami, S., Mitra Setia, T., and van Schaik, C. P. (Eds), *Orangutans: Geographic Variation in Behavioral Ecology and Conservation*. pp. 311–326. Oxford: Oxford University Press.

McKey, D. B., Gartlan, J. S., Waterman, P. G., and Choo, G. M. (1981). Food selection by black colobus monkeys

(*Colobus satanas*) in relation to food chemistry. *Biological Journal of the Linnaen Society* **16**: 115–146.

Meijaard, E., Albar, G., Rayadin, Y., Ancrenaz, M., and Spehar, S. (2010). Unexpected ecological resilience in Bornean Orangutans and implications for pulp and paper plantation management. *PLoS One* **5**(9): e12813.

Meijaard, E., Buchori, D., Hadiprakarsa, Y., Utami-Atmoko, S. S., Nurcahyo, A., *et al.* (2011). Quantifying killing of orangutans and human-orangutan conflict in Kalimantan, Indonesia. *PLoS One* **6**(11): e27491.

Milner-Gulland, E. J., Bennett, E. L., and the SCB Annual Meeting Wild Meat Group (2003). Wild meat: the bigger picture. *Trends in Ecology and Evolution* **18**: 351–357.

Muchaal, P. K. and Nandjui, G. (1999). Impact of village hunting on wildlife populations in the Western Dja Reserve, Cameroon. *Conservation Biology* **13**: 385–396.

Nasi, R., Brown, D., Wilkie, D., Bennett, E., Tutin, C., *et al.* (2008). Conservation and Use of Wildlife-based Resources: The Bushmeat Crisis. Secretariat of the Convention on Biological Diversity and Center for International Forestry Research (CIFOR), Bogor, Indonesia and Montreal, Canada.

Nasi, R., Taber A., and Van Vliet N. (2011). Empty forests, empty stomachs? Bushmeat and livelihoods in the Congo and Amazon Basins. *International Forestry Review* **13**: 355–366.

Noss, A. J. (1995). Duikers, Cables, and Nets: A Cultural Ecology of Hunting in a Central African Forest. Doctoral dissertation: University of Florida.

Noss, A. J. (1997). Challenges to nature conservation with community development in Central African Forests. *Oryx* **31**: 180–188.

Nunez-Iturri, G., Olsson, O., and Howe, H. F. (2008). Hunting reduces recruitment of primate-dispersed trees in Amazonian Peru. *Biological Conservation* **141**: 1536–1546.

Oates, J. F. (1996). Habitat alteration, hunting and the conservation of folivorous primates in African forests. *Australian Journal of Ecology* **21**: 1–9.

Oates, J. F., Whitesides, G. H., Davies, A. G., Waterman, P. G., Green, S. M., *et al.* (1990). Determinants of vairation in tropical forets primate biomess: new evidence from West Africa. *Ecological Society of America* **71**: 328–343.

Peres, C. A. (1997). Effects of habitat quality and hunting pressure on arboreal folivore densities in Neotropical forests: a case study of howler monkeys (*Aouatta* spp.). *Folia Primatologica* **68**: 199–222.

Peres, C. A. (1999a). Evaluating the sustainability of subsistence hunting at multiple Amazonian forest sites. In: Robinson, J. G. and Bennett, E. L. (Eds), *Hunting for Sustainability in Tropical Forests*. pp. 31–56. Columbia: Columbia University Press.

Peres, C. A. (1999b). Non-volant mammal community structure in different Amazonian forests types. In: Eisenberg, J. F. and Redford, K. H. (Eds), *Mammals of the Neotropics, Vol* **5**. pp. 564–581. Chicago: Chicago University Press.

Peres, C. A. (1999c). Effects of subsistence hunting and forest types on the structure of Amazonian primate communities. In: Fleagle, J. G., Janson, C., and Reed, K. (Eds), *Primate Communities*. pp. 268–283. Cambridge: Cambridge University Press.

Peres, C. A. (2000). Effects of subsistence hunting on vertebrate community structure in Amazonian forests. *Conservation Biology* **14**: 240–253.

Peres, C. A. and Dolman, P. M. (2000). Density compensation in Neotropical primate communities: evidence from 56 hunted and nonhunted Amazonia forests of varying productivity. *Oecologia* **122**: 175–189.

Peres, C. A. and Palacios, E. (2007). Basin-wide effects of game harvest on vertebrate population densities in Amazonian forests: implications for animal-mediated seed dispersal. *Biotropica* **39**: 304–315.

Petre, C.-A., Tagg, N., Haurez, B., Beudels-Jamar, R., Huynen, M.-C., *et al.* (2013). Role of the western lowland gorilla (*Gorilla gorilla gorilla*) in seed dispersal in tropical forests and implications of its decline. *Biotechnologie, Agronomie, Société et Environnement* **17**: 517–526.

Petre C-A., Tagg N., Beudels-Jamar R., Haurez B. & Doucet J-L., (2015). Western lowland gorilla seed dispersal: are seeds adapted to long gut retention times? *Acta Oecologica* **67**: 59–65.

Posa, M. R. C., Wijedasa, L. S., and Corlett, R. T. (2011). Biodiversity and conservation of tropical peat swamp forests. *BioScience* **61**(1): 49–57.

Poulsen, J. R., Clark, C. J., and Smith, T. B. (2001). Seed dispersal by a diurnal primate community in the Dja Reserve, Cameroon. *Journal of Tropical Ecology* **17**: 787–808.

Prescott-Allen, R. and Prescott-Allen, C. (1982). *What's Wildlife Worth? Economic Contributions of Wild Plants and Animals to Developing Countries*. London: Earthscan.

Ráez-Luna, E. F. (1995). Hunting large primates and conservation of the Neotropical rain forests. *Oryx* **29**: 43–48.

Redford, K. H. and Feinsinger, P. (2001). The half-empty forest: sustainable use and the ecology of interactions. *Conservation Biology Series Cambridge*: 370–400.

Revero, F., Mtui, A. S., Kitegile, A. S., and Nielsen, M. R. (2012). Hunting or habitat degradation? Decline of primate populations in Udzungwa Mountains, Tanzania: an analysis of threats. *Biological Conservation* **146**: 89–96.

Robinson, J. G. and Bodmer, R. E. (1999). Towards wildlife management in tropical forests. *Journal of Wildlife Management* **63**(1): 1–13.

Robinson, J. G. and Bennett, E. L. (2000). Will alleviating poverty solve the bushmeat crisis? *Oryx* **36**: 332.

Robinson, J. G. and Bennett, E. L. (2004). Having your wildlife and eating it too: an analysis of hunting sustainability across tropical ecosystems. *Animal Conservation* **7**: 397–408.

Robinson, J. G., Redford, K. H., and Bennett, E. L. (1999). Wildlife harvest in logged tropical forests. *Science, New Series* **284**(5414): 595–596.

Schenck, M., Effa, E. N., Starkey, M., Wilkie, D., Abernethy, K., *et al.* (2006). Why people eat bushmeat: results from two-choice, taste tests in Gabon, Central Africa. *Human Ecology* **34**: 433–445.

Sodhi, N. S., Brook, B. W., and Bradshaw, C. J. A. (2009). Causes and consequences of species extinctions. In: Levin, S. (Ed.), *The Princeton Guide to Ecology*. pp. 514–520. Princeton, NJ: Princeton University Press.

Starkey, M. (2004). Commerce and Subsistence: The Hunting, Sale and Consumption of Bushmeat in Gabon. PhD thesis: University of Cambridge.

Struhsaker, T. T. (1999). Primate communities in Africa: the consequence of long term evolution or the artifact of recent hunting? In: Feagle, J. G., Janson, C., and Reed, K. E. (Eds), *Primate Communities*. pp. 289–294. Cambridge: Cambridge University Press.

Terborgh, J., Nuñez-Iturri, G., Pitman, N. C. A., Cornejo Valverde, F. H., Alvarez, P., *et al.* (2008). Tree recruitment in an empty forest. *Ecology* **89**: 1757–1768.

Townsend, R. F. (1999). Agricultural Incentives in Sub-Saharan Africa: Policy Challenges (**Vol. 23**). World Bank Publications.

Tutin, C. E. G., Ham, R. M., White, L. J. T., and Harrison, M. J. S. (1997). The primate community of the Lopé Reserve, Gabon: diets, responses to fruit scarcity, and effects on biomass. *American Journal of Primatology* **42**: 1–24.

Tutin, C. E. G., Williamson, E. A., Rogers, M. E., and Fernandez, M. (1991). A case study of a plant-animal relationship: *Cola lizae* and lowland gorillas in the Lope Reserve, Gabon. *Journal of Tropical Ecology* **7**: 181–199.

Vanthomme, H., Bellé, B., and Forget, P.-M. (2010). Bushmeat hunting alters recruitment of large-seeded plant species in Central Africa. *Biotropica* **42**: 672–679.

Waterman, P. G. and McKey, D. (1989). Herbivory and secondary compounds in rainforest plants. *Ecosystems of the World, AGRIS* **148**: 513–546.

Waterman, P. G., Ross, J. A., Bennett, E. L., and Davies, A. (1988). A comparison of the floristics and leaf chemistry of the tree flora in two Malaysian rain forests and the influence of leaf chemistry on populations of colobine monkeys in the Old World. *Biological Journal of the Linnean Society* **34**: 1–32.

Whitman, K., Starfield, A. M., Quadling, H. S., and Packer, C. (2004). Sustainable trophy hunting of African lions. *Nature* **428**: 175–178.

Wilkie, D. S. and Carpenter, J. F. (1999). Bushmeat hunting in the Congo Basin: an assessment of impacts and options for mitigation. *Biodiversity and Conservation* **8**: 927–955.

Wilkie, D. S., Starkey, M., Abernethy, K., Effa, E. N., Telfer, P., *et al.* (2005). Role of prices and wealth in consumer demand for bushmeat in Gabon, Central Africa. *Conservation Biology* **19**(1): 268–274.

Wright, J. H. and Priston, N. E. (2010). Hunting and trapping in Lebialem Division, Cameroon: bushmeat harvesting practices and human reliance. *Endangered Species Research* **11**: 1–12.

Wright, S. J. (2010). The future of tropical forests. *Annals of the New York Academy of Sciences* **1195**: 1–27.

Wright, S. J., Zeballos, H., Dominguez, I., Gallardo, M. M., Moreno, M. C., *et al.* (2000). Poachers alter mammal abundance, seed dispersal, and seed predation in a Neotropical forest. *Conservation Biology* **14**: 227–239.

CHAPTER 10

Infectious disease and primate conservation

Charles L. Nunn and Thomas R. Gillespie

Monitoring infectious disease and health status in wild primates requires special consideration to ensure the safety and wellbeing of the primates and investigators as illustrated in these images (from left to right) of Dr. DeAnna Bublitz in Madagascar, Abel. Photo copyright: Dr. DeAnna Bublitz, Abel Nzeheke, and Dr. Rodolfo Martinez-Mota in Mexico.

10.1 Introduction

Infectious disease plays a major role in the lives of wild animals, including primates. In wild primates, we find an incredible diversity of parasites, defined as organisms that live in or on another host, at some cost to the host. The organisms include both microparasites, such as viruses and bacteria, and macroparasites, such as helminths and arthropods (Nunn and Altizer 2006). The *Global Mammal Parasite Database* (Nunn and Altizer 2005) identifies more than 603 unique parasite records in wild primates. Importantly, however, this number represents only a small fraction of the remarkable diversity of parasites that infect primates (Cooper and Nunn 2013), and this count may be biased towards organisms that are easy to study and that also cause disease in humans (e.g. Calvignac-Spencer *et al.* 2012). If you were to examine a single primate host in the wild, you would expect to find multiple helminth species. For example, one study documented that baboons at Gombe were infected by a median of four different helminth species (Muller-Graf *et al.* 1996); no individuals were uninfected in that study, and

Nunn, C.L. and Gillespie, T.R., *Infectious Disease and Primate Conservation*. In: *An Introduction to Primate Conservation*. Edited by: Serge A. Wich and Andrew J. Marshall, Oxford University Press (2016). © Oxford University Press.
DOI 10.1093/acprof:oso/9780198703389.003.0010

one individual had seven identifiable helminth species. Similarly, chimpanzees at the same site were infected by a median of five different helminth species (Gillespie *et al.* 2010); no individuals were uninfected, and one individual had eight morphologically distinct helminths. This level of infection seems about on par with other studies of parasitism in wild primates (Nunn and Altizer 2005, 2006).

Parasites vary in their effects on primate hosts. Some infectious agents have devastating impacts on their primate hosts, including the agents that cause Ebola (Leroy *et al.* 2004; Caillaud *et al.* 2006), anthrax (*Bacillus anthracis*, Leendertz *et al.* (2004)), and yellow fever (Holzmann *et al.* 2010). Similarly, respiratory pathogens can 'spill over' from humans to wild primates and cause substantial mortality (Wallis and Lee 1999; Koendgen *et al.* 2008; Palacios *et al.* 2011). While cases of sudden die-offs caused by viruses and bacteria attract much attention, even typical macro-parasites—which are thought to have minor effects on a healthy host—can actually have tremendous impacts on host population dynamics. Population-level effects were shown elegantly in the case of red grouse, where anti-helminthic treatment eliminated cyclical population fluctuations (Hudson *et al.* 1998), and in experiments on a breeding colony of captive mice in which introduction of a nematode generated striking population declines (Scott 1987). These examples foreshadow many important themes that follow in this chapter, including the striking effects that pathogens can have on hosts, and the importance of parasites in healthy ecosystems and as a component of biodiversity.

A critical question at the interface of infectious disease and conservation is, '*how commonly* do parasites drive hosts to extinction?' Many reasons exist to be concerned about infectious agents in primate conservation—as described in Section 10.3—yet epidemiological modelling and empirical evidence suggest that parasite-generated extinction is unlikely, especially for parasites that are found naturally in a population of wild hosts. Most epidemiological models of socially transmitted organisms under 'density dependent transmission' assume that higher host densities result in higher rates of transmission, and that a threshold density exists below which the parasite will go extinct (e.g. Dobson and Meagher 1996). Pathogens causing substantial mortality also result in host population declines, making it more difficult for the parasite to persist as contact among susceptible and infected hosts declines. Thus, harmful parasites will generally go extinct before the host does, at least under the assumptions of density dependent transmission (McCallum and Dobson 1995). The critical question should therefore be, 'under what conditions *could* a parasite cause host extinction?' Included here is whether the host population can decline substantially enough that other factors lead to its demise (such as the effects of reduced genetic diversity or demographic stochasticity).

The overarching goal of our chapter is to synthesize research on conservation and disease ecology from diverse avenues that include field evidence of parasite-driven declines, co-extinction of hosts and parasites, and theoretical and empirical research from other systems. We also consider the potentially important links between biodiversity and infectious organisms, building on suggestions that biodiversity may buffer human populations from disease risk as an ecosystem service (Keesing *et al.* 2010). Throughout, we identify important future directions that are needed to understand the role of infectious disease in primate conservation.

10.2 Parasite-related threats to biodiversity

Infectious disease has been implicated in a number of species declines and even extinctions, with disease accounting for about 4% of extinctions and 8% of classifications of critical endangerment in all species in the IUCN Red List (Smith *et al.* 2006). Disease-related impacts on individual populations of a species are likely even more common. In most cases, disease is one of multiple factors that influence extinction risk, rather than disease driving risk on its own (de Castro and Bolker 2005). Thus, infectious disease risks are entangled in a syndrome of anthropogenic factors that influence the threat of extinction for many groups of animals; similar effects are likely in primate populations (Chapman *et al.* 2005; Nunn and Altizer 2006).

Examples from diverse vertebrate taxa provide a framework for considering disease-related threats to primates. We provide several illustrative examples here from birds, mammals, and amphibians. The examples we provide are some of the best-documented cases of pathogens that have had dramatic effects on populations, in many cases threatening species with extinction. We can therefore learn from these examples to prevent similar declines in primates. Specifically, we use the examples to identify general factors that influence parasite-related declines, and then apply these concepts to the more specific cases of primates and their parasites in the following section.

Many species of Hawaiian honeycreepers—an adaptive radiation of finches—have been heavily impacted by avian malaria, and this disease is thought to restrict the distribution of native Hawaiian birds to areas that are malaria-free (Van Riper *et al.* 1986; LaPointe *et al.* 2012). West Nile virus represents a new threat to these birds (Kilpatrick *et al.* 2004; LaPointe *et al.* 2009), but so far this mosquito-borne virus has been limited to continental North America. While many factors likely influenced the decline of the Hawaiian honeycreepers, the critical factors for disease-related declines in their population likely involve their isolated evolutionary history and anthropogenic factors, specifically the introduction of new disease agents (including the mosquito vector) through human trade, travel, and land-use change. Similar factors were likely at play in the trypanosome-related extinction of the endemic Christmas Island rat (*Rattus macleari*), which went extinct following the introduction of black rats (*Rattus rattus*) and its trypanosome via human activities at the end of the nineteenth century (Wyatt *et al.* 2008).

Another prominent disease-related decline involves the Tasmanian devil (*Sarcophilus harrisii*). Devil populations have declined precipitously via transmission of a cancerous cell line that spreads from one animal to another through aggressive contact (McCallum 2008). The cancerous cells are transmitted and begin to grow and metastasize in the new host, leading to inability to feed and other consequences, and ultimately resulting in lower fecundity and premature death. The disease is known as Devil facial tumour disease, or DFTD, and as a cancerous cell line it is actually a *mammalian* parasite, similar to canine transmissible venereal tumour (Cohen 1985).

In the case of DFTD, we can again point to general principles that may be useful for understanding disease-related threats. First, the spread of the disease is likely related to reduced genetic diversity in Tasmanian devils' small population sizes and an absence of geographic barriers to the spread of disease throughout the population (which is more likely in a smaller, more geographically restricted population). Second, the host's behaviour likely facilitates the transmission of DFTD even as the host population declines: by seeking out other social partners and interacting aggressively with them (e.g. during mating), Tasmanian devils' behaviour maintains high local density and facilitates the spread of disease, even as the host population declines (McCallum 2008). Such effects are likely to occur in primates, too, with their high degree of sociality and seeking of conspecifics for social interactions.

Finally, consider chytrid fungus (*Batrachochytrium dendrobatidis*) of frogs and white nose syndrome of bats, which illustrate another potential pathway of disease-induced extinction: environmental transmission. Chytrid fungus is devastating amphibian populations on all continents of the world, resulting in massive losses of biodiversity (Daszak *et al.* 1999; Lips *et al.* 2006) and increased threat levels for a wide range of frogs on the IUCN Red List (Smith *et al.* 2006). Its origins appear to involve the worldwide trade in African clawed frogs, which were used for pregnancy tests prior to development of modern diagnostic techniques (Weldon *et al.* 2004). These frogs continue to be used as a model system to investigate diverse biological questions. Importantly, the chytrid fungus can be transmitted through water without requiring direct contact of susceptible frogs, it is long lived in the environment, and it can be carried from one aquatic habitat to another, potentially by humans (see St-Hilaire *et al.* (2009)).

White nose syndrome (*Pseudogymnoascus destructans*) is causing massive declines in North American bats, and is rapidly marching across the United States (Blehert *et al.* 2009). It is likely to drive some species to extinction (Frick *et al.* 2010; Hallam and Mccracken 2011), possibly with massive costs in terms of lost ecosystem services, particularly control of agricultural pests that bats feed upon (Boyles *et al.* 2011). Similar to chytrid, *P. destructans* is a fungus and may be spread from one cave to another via human activities, such as cave tourism and scientific research (Puechmaille *et al.* 2011). White nose syndrome and chytrid fungus illustrate that pathogens with potential for human vectoring and substantial environmental transmission can have devastating impacts on host populations, especially when introduced to new populations that lack immune responses to the organism (Lorch *et al.* 2013).

Collectively, these examples reveal several general principles that we can apply to primate disease-related threats. Specifically, we expect that infectious disease is more likely to be a threat to primate hosts when:

- Overall host genetic diversity is low: this makes it more likely that a large majority of the population will succumb to a new infectious agent.
- Few barriers exist that stop the spread of an infectious agent: this makes it less likely that pockets of the population are protected, resulting in larger declines, and is more likely when the species range is small.
- When humans and human activities introduce infectious diseases for which hosts have few genetic adaptations, including infectious diseases of humans and domesticated animals, and especially when coupled with human vectoring of infectious agents to an area where the organism was previously absent.
- When hosts have behaviours that lead them to seek out sources of infection as populations decline, including social interactions with conspecifics for socially transmitted disease, or indirect interactions at food and water sources for environmental sources of infectious agents. More generally, pathogens with non-density dependent transmission provide greater threats than those with density dependent transmission (McCallum 2008).
- When the infectious agent exists in a reservoir host that is relatively less affected by the organism (McCallum and Dobson 1995), or when the infectious organism can survive in the environment (and remain infectious) for long periods of time.

10.3 Infectious disease threats in primates

With these factors in mind, we turn to specific examples of primates, focusing on four examples that demonstrate one or more of these principles: Ebola virus, yellow fever virus, respiratory pathogens, and environmentally transmitted pathogens, particularly anthrax.

10.3.1 Ebola

First, consider Ebola virus in African apes. Although the source of Ebola has yet to be established, evidence suggests that widely distributed African fruit bats play some role in its maintenance (Leroy *et al.* 2005), and potentially insectivorous bats as well (Saéz *et al.* 2015). In the early years of understanding Ebola, two main hypotheses were proposed for its transmission in apes: *the social hypothesis*, which states that Ebola virus is transmitted socially and spreads in a wave-like fashion among hosts; and *the reservoir hypothesis*, which states that the virus is primarily maintained and spreads in a reservoir host, with occasional spill-overs to apes (and some other mammals). These hypotheses are not mutually exclusive, and evidence exists for both transmission mechanisms. For example, a study of an Ebola outbreak in gorillas at Odzala-Kokoua National Park, Congo, failed to find conclusive evidence for social versus non-social (reservoir) transmission, although this study also found that individuals living in groups were more likely to die from Ebola (Caillaud *et al.* 2006).

Ebola has certainly caused substantial mortality in African apes (Walsh *et al.* 2003; Bermejo *et al.* 2006) and is responsible for including 'disease' as a threat to apes in the IUCN Red List. Yet it currently remains unclear whether Ebola continues to spread through populations as originally proposed, or what its long-term impact is likely to be, resulting in some controversy in the wildlife epidemiological community. Some researchers would argue that we do ourselves a disservice by overstating the risk from Ebola. One concern is that it would be more difficult to convince wildlife managers to invest in controlling the very real threats that they can effectively mitigate—such as human respiratory pathogens transmitting to wildlife—if they are concerned that Ebola may wipe out their entire ape population without warning. Another concern is that effective means of controlling Ebola in wildlife are currently non-existent. While some effort has focused on vaccination strategies (Warfield *et al.* 2014), a viable vaccine has yet to be developed. More importantly, large-scale application of a vaccine may be prohibitively costly and logistically challenging (cf. Carne *et al.* 2013; Rushmore *et al.* 2014), potentially resulting in lost opportunities for other conservation priorities, including conserving habitat.

Ebola exhibits several additional patterns that are relevant to our consideration of general factors that lead infectious agents to cause population declines. First, there are few geographic barriers to the spread of the disease across continental Africa, especially considering that the virus is associated with wide-ranging bat species (which face fewer biogeographic obstacles to movement than terrestrial mammals). Geographic barriers have been an important consideration in controlling the spread of Ebola in African apes—for example to create barriers that halt ape-to-ape social transmission (e.g. Walsh *et al.* 2007)—yet it now appears that this is unlikely to be effective unless the reservoir of Ebola is also controlled (again, a tall order if that reservoir is a bat species).

Second, social mechanisms could play an important role in the continued spread of Ebola, even as populations decline in size. For example, declines in group/community size due to disease could lead individuals to seek new groups, potentially carrying disease with them to those groups (Nunn *et al.* 2008). Such an effect would be especially strong if all individuals of one sex were lost, leading members of the remaining sex to transfer. Nunn *et al.* (2008) showed that such 'pathogen mediated dispersal' was possible in a theoretical model that examined the spread of Ebola in chimpanzees and gorillas. The potential spread was greatest in gorillas, due to their more common single-male mating system, with females more often carrying the disease to other groups when the resident male gorilla died (see also Thrall *et al.* (2000)). This example further highlights the importance of social groups in restricting movement of pathogens (Freeland 1979). In the case of Ebola in African apes, social behaviour may also facilitate the spread of disease, for example through overlap at food sources and observations of close physical inspection of dead animals (Walsh *et al.* 2007).

10.3.2 Yellow fever virus

As a second example, we turn to yellow fever virus, an RNA virus transmitted by mosquitoes. Yellow fever was introduced to the New World through the slave trade more than 300 years ago (Bryant *et al.* 2007), as was the protozoan that causes malaria (Yalcindag *et al.* 2012). We would predict that the initial absence of yellow fever virus in the New World would predispose the naïve population to be especially vulnerable to the virus, and that is what was found for at least some primate populations (Nunn and Altizer 2006). In particular, it appears that howler monkeys succumb especially quickly to the virus (Carpenter 1964; Galindo and Srihongse 1967). Yellow fever epidemics appear to have decimated howler populations in Central America in the 1950s (Collias and Southwick 1962) and Argentina, Paraguay, and Brazil in the 1960s (de Bitetti *et al.* 1994). Mean howler troop size on Barro Colorado Island in Panama was depressed for years following a presumed yellow fever outbreak (Milton 1982).

More recently, howler epidemics in Argentina and Brazil were confirmed to be caused by yellow fever virus. For example, Holzmann *et al.* (2010) describe yellow fever epidemics that led to significant losses of two species of howler monkeys in Misiones, Argentina: the brown howler (*Alouatta guariba*) and the black howler (*Alouatta caraya*). Two outbreaks occurred over a one-year period, resulting in the death of more than 35 howlers. As found in previous studies and reports, other primates were less affected in the Misiones outbreaks, including tufted capuchins (*Cebus nigritus*), which suffered no known mortalities.

Similarly, laboratory analyses of 308 dead howler monkeys (*Alouatta caraya* and *A. guariba clamitans*) associated with a an outbreak in Rio Grande do Sul, Brazil, confirmed 180 yellow fever positive animals (Bicca-Marques 2009). Interestingly, these findings also suggest that about 40% of these monkeys did not die of yellow fever and were likely killed by local people fearing howlers as a source of disease. Fortunately, a campaign has been successful in highlighting that howler monkeys actually provide an early warning system for impending human yellow fever exposure and are not responsible for the emergence of human disease (Bicca-Marques 2009; Bicca-Marques and Freitas 2010). It appears that howlers have a genetic susceptibility to yellow fever virus, and by causing significant die-offs and being harboured in other hosts, the virus may contribute to reduced genetic diversity in affected howler populations (James *et al.* 1997).

10.3.3 Respiratory illness

Respiratory illnesses also have been recognized as an important cause of morbidity and mortality in wild apes, even before diagnostics were available to confirm the cause of illness (Goodall 1986; Watts 1998; Boesch and Boesch-Achermann 2000). At some sites, observational data on coughing and other respiratory symptoms have been collected in a systematic way, allowing for retrospective assessment to determine risk and to guide future conservation strategies (Lonsdorf *et al.* 2006, 2011). In addition, necropsies performed soon after death have provided further insights into the aetiology of syndromic respiratory disease (Travis *et al.* 2008; Köndgen *et al.* 2008). In many cases it has been unclear which agents were responsible for death, or their origins: were they acquired from other apes, other wildlife, domesticated animals that come into protected areas, or from humans, including scientists and field assistants involved in research?

Several recent studies have investigated the sources of respiratory outbreaks in wild apes, with an eye towards identifying whether humans were the original source of the pathogens. For example, Köndgen *et al.* (2008) investigated several outbreaks in the Taï chimpanzees. They found evidence for human metapneumovirus and human respiratory syncytial virus in necropsy samples from the chimpanzees. Phylogenetic analyses revealed that these viruses nested evolutionarily within viruses from humans, including evidence for a genetic insertion in some samples that more firmly linked the viruses to previous human outbreaks. Hence, these results suggest that humans were the source of the viruses that killed the chimpanzees.

Human metapneumovirus has also been implicated as the causal agent of illness in chimpanzees at Mahale, Tanzania (Kaur *et al.* 2008), and in mountain gorillas in Rwanda (Palacios *et al.* 2011). Interestingly, these respiratory virus infections are rarely fatal in great apes on their own, and instead require co-infections with bacteria that contribute to pneumonia. For example, chimpanzees infected with human metapneumovirus and human respiratory syncytial virus at Taï were also infected with the common bacterium *Streptococcus pneumonia*, which is a major cause of pneumonia in humans and other animal species. The mountain gorillas that were infected with human metapneumovirus likely died of *Streptococcus* and *Klebsiella* bacterial pneumonia (Palacios *et al.* 2011).

Mycobacterium tuberculosis (the causal agent of tuberculosis) can drive substantial mortality in non-human primates—a fact well known in captive colonies, where tuberculosis testing is typically required for workers and research scientists and staff (Wolf *et al.* 2014). Interestingly, despite high rates of tuberculosis in human populations overlapping with great apes, until recently evidence of *M. tuberculosis* infection has been lacking in wild apes. Curiously, the one confirmation of *Mycobacterium* in wild chimpanzees was a novel *Mycobacterium* tuberculosis complex (MTC)—related to, but distinct from, all known human complexes (Coscolla *et al.* 2013). Fortunately, a non-invasive diagnostic was recently developed that should allow for broad screening of wild chimpanzees in the near future (Wolf *et al.* 2015).

10.3.4 Anthrax and other environmentally transmitted organisms

As we saw in the examples from non-primates, pathogens that survive long in the environment can pose a considerable threat to wildlife, such as chytrid fungus. In primates, a prominent example of such transmission involves anthrax, which has caused mortality at several primate field sites (Leendertz *et al.* 2004, 2006a, 2006c). Anthrax is caused by the bacterium *Bacillus anthracis*, which can survive for long periods of time in the environment, such as soil. It is also harboured in domesticated animals. Via environmental transmission, anthrax easily transmits across species boundaries to atypical hosts, including humans. Thus, we might expect anthrax to appear in wildlife when livestock occur in or near forests, or when humans carry anthrax spores from livestock into primate habitats. Cases of anthrax die-offs on record involve both chimpanzees and gorillas in at least two different locations in West Africa: Taï National Park (Leendertz *et al.* 2004) and Dja Biosphere Reserve in Cameroon (Leendertz *et al.* 2006c). At Taï, at least six chimpanzees died, and the source remains unknown. In Dja, three chimpanzee and one gorilla death were attributed to anthrax. Of course, these documented cases likely represent only a fraction of deaths due to anthrax.

For many environmentally transmitted pathogens, proximity among wild primates, people, and domesticated animals has often provided realized opportunity for cross-species transmission, such as seen with soil-transmitted helminths and protozoa (Nizeyi *et al.* 2002; Hasegawa *et al.* 2014; Bodager *et al.* 2015; Parsons *et al.* 2015); pathogenic bacteria, such as enterotoxogenic *Escherichia coli*, *Salmonella*, and *Shigella* (Nizeyi *et al.* 2001; Bublitz *et al.* 2015); and diarrhoea-associated viruses, such as adenovirus, enterovirus, norovirus, and rotavirus (Harvala *et al.* 2011, 2014; Zohdy *et al.* 2015). In wild primate populations that have not experienced treatment with antibiotics, findings of bacteria in their gastrointestinal tracts that are resistant to multiple antibiotics used by people in the region, as seen in African monkeys (Goldberg *et al.* 2008) and apes (Goldberg *et al.* 2007; Rwego *et al.* 2008; Janatova *et al.* 2014), suggest transmission from humans to wild primates.

We return to the practical side of these issues later, in the context of ecotourism, sanctuaries, field research, and reintroduction of animals into the wild from sanctuaries.

10.4 Connecting biodiversity to patterns of disease risk

While much of the above considered the impact of disease on primate populations, in this book on primate conservation we should also consider impacts of biodiversity on disease transmission in primates. Does greater biodiversity lead to greater disease risk for humans and wildlife (including primates), or to reduced disease risk? Is there a consistent positive or negative linkage between biodiversity loss and disease risk? What is the size of the effect, and what mechanisms underlie these effects? In this case, the question concerns whether conservation of biodiversity affects disease risk itself, rather than whether specific infectious agents impact particular hosts. These are questions of great importance, and are thus gaining increased attention in research on disease ecology (Keesing *et al.* 2010; Vourc'h *et al.* 2012; Wood and Lafferty 2013). Links between biodiversity and disease can arise through different mechanisms, and these mechanisms are not mutually exclusive. Here, we briefly summarize four key mechanisms, which we call the dilution effect, the density effect, the amplification effect, and the reversed dilution effect.

We start with the dilution effect, which has been the focus of much interest lately because it suggests that biodiversity performs disease-related ecosystem services that benefit humans (Keesing *et al.* 2010). The dilution effect hypothesizes that biodiversity provides a protective effect by 'diluting' the risk through hosts that intercept, and thus slow, transmission of infectious agents; conversely, loss of biodiversity is thought to lead to an increase in disease risk, for both humans and for other animals in the ecological community from which hosts are lost. The basic idea is as follows. We assume that hosts of different species vary in their ability to harbour and transmit a generalist pathogen. We further assume that hosts that are better able to transmit the pathogen (more competent) are also more resilient to human disturbance; for example, these hosts might be smaller in body mass, with faster life histories, and thus invest less in immune defences (Previtali *et al.* 2012), leading to higher levels of pathogens circulating in an individual and potentially spreading to other individuals. Thus, with the commencement of human activities, such as logging, the less competent hosts are lost, leading to a larger proportion of competent hosts in the resulting community (and likely higher disease risk as a result).

The dilution effect is thus akin to the loss of immune individuals from a population, resulting in more disease risk for those hosts that remain in the depauperate community. The dilution effect has some support in the context of Lyme disease (Ostfeld 2010) and West Nile virus (Ezenwa *et al.* 2006) in the United States, but has been less well studied in the tropical areas inhabited by most primate species (Wood *et al.* 2014). Studies of the dilution effect are typically focused on prevalence of a single parasite or pathogen.

A second mechanism also connects biodiversity and disease risk, and has been called the amplification effect (Vourc'h *et al.* 2012). In this case, higher biodiversity is thought to increase risks of disease, including both spill-over of disease and the abundance (prevalence) of particular parasites and pathogens in a more biodiverse community of hosts. This mechanism may operate through a higher number of specialist diseases in a community containing more host species, resulting in more opportunities for spill-over and adaptation of a parasite to a new host species. The amplification effect has found support in studies of human parasitic and infectious diseases, the numbers of which across countries are explained by the richness of mammals and birds and other factors (Dunn *et al.* 2010).

A third connection between biodiversity and disease is similar to the dilution effect, although the connection involves the density of the hosts that better transmit the pathogen or vector abundance, rather than composition of the community. Hence, we call this the density effect. If food webs are disrupted in the community through human activities, such as loss of predators, some host species are likely to increase in abundance (Wood *et al.* 2014). When these remaining hosts transmit pathogens, disease risk is likely to increase (i.e. susceptible host regulation has been relaxed). Similarly, when habitat disturbance reduces the abundance of vectors, pathogen prevalence can decline (Wood *et al.* 2014).

Finally, we should keep in mind that the conditions of the dilution effect could be reversed. Thus, rather than losing the hosts that are good at 'diluting' the risk in the community by intercepting transmission, we may lose hosts that are more competent for the infectious agent in question, resulting in lower disease risk for the community as human disturbance increases. Recent indications that hosts with slower life histories may have stronger immune defences (e.g. Previtali *et al.* 2012) suggest this scenario is unlikely: these hosts are also more likely to be diluting hosts, and to be lost as habitat destruction occurs. However, more research is needed to assess the generality of this pattern.

The association between biodiversity and disease has been addressed recently in studies of primates. Young *et al.* (2013) investigated multiple predictions involving the dilution and amplification effects. First, using data on resilience to human disturbance, they investigated whether primate species showing more resilience have fewer parasites or better immune defences. Young *et al.* (2013) found no evidence for such effects, nor for the opposite pattern; most results were non-significant, including after putting the results collectively into a meta-analysis to increase power and control for multiple testing. Next, they conducted a meta-analysis of 14 studies of the links between human disturbance and parasitism in primates. They found a wide diversity of effects, with disturbance sometimes associated with greater parasitism, and sometimes with less parasitism, but no overall consistent effect. In one group of parasites (indirectly transmitted parasites), a positive effect was found, supporting the dilution effect. Third, they investigated whether prevalence of ape malaria increased with human disturbance and decreased with mammal richness. They found the opposite pattern, suggestive of an amplification effect rather than a dilution effect. Finally, Young *et al.* (2013) investigated links between parasite richness, primate richness, and geographic overlap, and again found evidence more consistent with the amplification effect.

Overall, then, Young *et al.* (2013) found mixed evidence, but most tests were non-significant. These results indicate a lack of consistent association between parasitism and human disturbance in primates, suggesting instead that different mechanisms play a role across locations and biological systems. Another meta-analysis of studies of human infectious diseases found a similar lack of consistent patterns, in a paper with a title that states the main take-home message of the authors: 'A meta-analysis suggesting that the relationship between biodiversity and risk of zoonotic pathogen transmission is idiosyncratic' (Salkeld *et al.* 2013). However, a more recent meta-analysis, using a larger set of studies, found stronger evidence for the dilution effect (Civitello *et al.* 2015), suggesting that a consensus is far from reached, even after multiple meta-analyses. The nature of the disturbance (i.e. habitat fragmentation, selective logging, or hunting) and characteristics of the population or ecological community may determine the directionality of such relationships.

10.5 Connecting infectious disease to the generation of biodiversity

Another way that parasites and biodiversity become linked is when infectious agents actually facilitate speciation of host lineages, and thus contribute to the generation of biodiversity over the long term. In such a case, parasites are important to the maintenance of biodiversity, and more biodiverse lineages should have more parasites. Indeed, evidence for such a correlation has been found in primates. Nunn *et al.* (2004) investigated whether rates of primate host diversification covary with measures of parasitism, and found that lineages with more parasites undergo a higher rate of diversification, measured as speciation rate minus extinction rate (net speciation). However, this result is correlational rather than causal; it is also possible that higher rates of diversification make the resulting lineages particularly well suited for parasites—for example, for generalist species that can infect closely related hosts—or that some other variables influence both diversification and parasitism. Indeed, further tests suggested evidence more in line with diversification, rate as the driver of this pattern (Nunn *et al.* 2004).

More generally, interspecific interactions—such as those involving hosts and parasites—are thought to play an important role in the diversification of natural populations, and thus may maintain or increase biodiversity over evolutionary timescales. Several studies have found that host–parasite co-evolution influences genetic diversity (Dybdahl and Lively 1998; Burdon and Thrall 1999; Altizer 2001), and geographic variation in parasite resistance suggests a role for infectious agents in host divergence (Berenbaum and Zangerl 1998; Kaltz and Shykoff 1998; Thompson 1999). In addition, co-evolution among insect herbivores and their host plants increased rates of diversification among some lineages, driven in part by plant specialization and novel defensive mechanisms (e.g. Mitter *et al.* 1991; Farrell 1998; Percy *et al.* 2004). Experimentally, one study of bacteria and virulent phage demonstrated that parasites can drive divergence among host populations in spatially structured environments, thus increasing host diversification by selecting for anti-parasitic defences linked to different traits in different populations (Buckling and Rainey 2002).

Theoretical and empirical studies have focused on the 'Red Queen' process of host–parasite co-evolution in maintaining genetic diversity within and among populations (Anderson and May 1991; Lively and Apanius 1995; Abrams 2000). For example, theoretical work has shown that frequency dependent selection—in which the fitness of a phenotype is a function of its relative frequency of phenotypes in the population—between prey and natural enemies can lead to evolutionary branching in both the host and enemy populations (Doebeli and Dieckmann 2000).

10.6 Parasites as important components of biodiversity and 'healthy' ecosystems

In efforts to facilitate recovery of the black-footed ferret in the United States, a well-meaning veterinarian treated animals brought into a captive breeding programme with insecticides. The goal was to eliminate ectoparasites, particularly those that are responsible for plague (*Yersinia pestis*). In the process, they may have driven host-specific ectoparasites to extinction (Gompper and Williams 1998). Who are we to decide that fleas, lice, and mites are somehow less important than the hosts on which they are found? (See also Durden and Keirans (1996).) This realization has prompted provocative papers that proclaim, for example, 'equal rights for parasites' (Windsor 1995, 1998).

While concern for conservation of parasites might seem a silly digression from an important issue in primate conservation, it actually has great relevance. In particular, it shows that preserving a primate host not only saves that lineage on the tree of life; it also leads to conservation of many other associate lineages, including parasites. Concern for parasite biodiversity also raises important questions about what constitutes a 'healthy' ecosystem: just as we may wish to preserve a community of organisms and their predators, good reasons exist to also conserve the parasites. We discuss relevant factors involving these concerns in what follows.

A number of authors have investigated co-extinction. For example, Altizer *et al.* (2007) found that more threatened primate hosts had fewer specialist parasites, as one might expect given that these parasites have no other options for alternative hosts as the typical host populations decline and become more fragmented. In essence, the parasites are a bio-indicator, the 'canary in the coal mine' that stops singing when threats start having an impact on host population size and structure. Interestingly, even generalist parasites declined in number as threat status increased. Thus, as a simple epidemiological model might suggest, parasites are not typically capable of causing extinction, and are in fact more likely to go extinct before the host—subject of course to the five factors given above that may lead to host extinction or significant population declines.

Additional analyses across a broad range of host taxa confirm that a substantial proportion of additional biodiversity is lost as affiliates—such as parasites—disappear when their hosts disappear (Koh *et al.* 2004; Dunn *et al.* 2009). Host specificity of the parasite is key in these discussions; if a parasite specializes on a single host, it is more likely to be lost when that host declines in abundance or goes extinct. Importantly for the topic of this chapter and the book, among the most threatened parasites in this context are pinworms and fungi infecting primates. Given the high host specificity of these organisms on primates (e.g. pinworms, Hugot (1999)), loss of 10% of primates will result in an approximately equal number of losses of these parasite groups (Dunn *et al.* 2009).

Good arguments are made for the importance of parasites in maintaining a healthy ecosystem, where 'healthy' is defined as an ecosystem that maintains vigour and is resilient to change (Hudson *et al.* 2006). Indeed, parasites are known to alter interspecific competition, enabling a more diverse assemblage of species in an ecological community. This is readily seen in the case of invasive species, and important findings suggest that 'pathogen release' gives invasive species an advantage in competition with native species. Such an effect has been seen in plants and in animals (Mitchell and Power 2003; Torchin *et al.* 2003; Torchin and Mitchell 2004). Similarly, parasites have been used as a bio-indicator of habitat quality and restoration (Lafferty 1997; Huspeni and Lafferty 2004). It would be interesting to apply this framework to assess the quality of primate habitats via the richness and abundance of parasites that are found. More generally, we need to appreciate that parasites can provide healthy functions, such as regulating the abundance of certain hosts in a community, sustaining a more diverse ecological community, and influencing energy flow.

Lastly, pathogens have the capacity to disrupt ecosystem services. Insights over the past three decades have clarified how the health and persistence of tropical forests depend on critical ecosystem services provided by wildlife. We now know that density dependent seedling mortality presents a strong selective pressure for seed dispersal for tropical tree species to escape seedling pathogens (Augspurger 1984). We also have seen the potential of unsustainable hunting to create 'empty forests', where loss of keystone species results in reduced seed dispersal and altered recruitment and relative abundance of trees (Redford 1992). Terborgh *et al.* (2001) elegantly demonstrated that seedlings and saplings of canopy trees are severely reduced and seed predation exponentially increased when top predators are lost from a tropical system. The decline or loss of the seed dispersers, pollinators, seed predators, herbivores, and predators that shape tropical forest ecosystems has long been attributed to habitat degradation or loss and unsustainable hunting. However, it is now becoming clear that infectious diseases affecting wildlife have the capacity to impact tropical forest systems in similar ways. We have lost pollinators, such as honeycreeper declines in Hawaii, due to introduced malaria (Benning *et al.* 2002); seed dispersers, such as lowland gorillas due to Ebola

(Tutin *et al.* 1991; Voysey *et al.* 1999; Leroy *et al.* 2004); and insectivores that regulate crop pests, as in the case of bat declines in the United States due to white nose syndrome (Boyles *et al.* 2011).

10.7 Practicalities of controlling introduced parasites in ecotourism and research

A major point of the previous sections is that naturally occurring parasites are an important component of ecosystems, and introduced pathogens can alter these ecosystems in negative ways. Aside from being part of biodiversity, naturally occurring parasites are unlikely to drive hosts to extinction, and may even help maintain productive, diverse communities of hosts. Yet we know that parasites play a major role in population declines, including many important primate species, as described in earlier sections (Chapman *et al.* 2005; Nunn and Altizer 2006). In addition, we know that human activities—such as logging and resulting forest fragmentation—are likely to lead to increased abundance of some parasites for some hosts (Gillespie *et al.* 2005; Gillespie and Chapman 2006). These parasites may affect patterns of interspecific competition among hosts, further altering host communities, although we presently lack a predictive framework for understanding when this will occur (Young *et al.* 2013).

As we see it, the major disease-related issue for primate hosts involves the introduction of novel parasites through human activities, including through scientific research. These introductions could involve direct spill-over from humans, as seen in cases of respiratory infections into ape populations (Koendgen *et al.* 2008; Palacios *et al.* 2011). These introductions could alternatively arise from reservoir hosts, such as fruit bats as sources of Ebola outbreaks (Leroy *et al.* 2005), or domesticated animals as sources of anthrax in and around tropical forests (Leendertz *et al.* 2004, 2006a). It is important that we document these sources of disease and their impacts on primate populations and consider the options for intervening effectively (Homsy 1999; Gilardi *et al.* 2015). It is also important to implement barriers to the spread of disease when conducting primate research, including through use of masks by observers, or by maintaining safe distances from the animals under observation. In addition to research activities, we must consider the effects of ecotourism on infectious disease outbreaks. One recent study found, for example, that 15% of visitors to an orang-utan sanctuary had symptoms of gastrointestinal or respiratory infection (Muehlenbein *et al.* 2010).

When a wild primate is rescued from poachers or wildlife smugglers, they end up in a sanctuary (if they are lucky). These facilities offer an essential service for animal welfare, providing veterinary care and rehabilitation in an environment where animals can interact with conspecifics. It is intuitive to think that the best outcome for such an animal would be to eventually return to the wild. Unfortunately, despite the best intentions, reintroductions often result in unintended consequences for the welfare of the reintroduced individual (i.e. difficulty competing with resident populations for food and mates and avoiding unfamiliar predators, see Harrington *et al.* (2013)). In addition, reintroductions may threaten the viability of wild populations by inadvertently exposing wildlife to novel pathogens contracted by sanctuary animals from humans or domestic animals during captivity. This concern is highlighted in a recent study demonstrating that chimpanzees at two geographically distinct sanctuaries were infected with human-associated strains of *Staphylococcus aureus* that have not been detected in wild apes and MRSA-like multi-drug resistant strains of *S. aureus* (Schaumburg *et al.* 2012). In addition to the risks of reintroduced primates exposing wild primates to *S. aureus* and similar pathogens, tourists who visit wild great ape populations after visiting an ape sanctuary may inadvertently transport such pathogens from sanctuary apes to wild apes.

We also need more information on the pathogens responsible for declines, which requires systematic monitoring of parasites and pathogens for effective detection of new, disease-causing organisms (Leendertz *et al.* 2006b). Unlike rodents and bats, it is often both politically and logistically difficult to sample and study non-human primate pathogens due to concerns regarding habituation status and risks associated with immobilizing these often endangered species (Travis *et al.* 2008). Thus, most non-human primate studies are focused on non-invasive sampling techniques (e.g. faeces, urine, saliva, and hair) and complete necropsy and post-mortem examinations following documented die-offs (Gillespie 2006; Gillespie *et al.* 2008). A variety of non-invasive approaches are possible for collecting information on parasites of wild primates, and recent advances make this easier than ever (Gillespie *et al.* 2008; Köndgen *et al.* 2010). Such monitoring will likely identify new organisms that do occur naturally, thus helping to eliminate the vast gaps in our understanding of parasites and pathogens of primates (Hopkins and Nunn 2007, 2010; Cooper and Nunn 2013). Monitoring of potential reservoirs of disease, such as bats and rodents—as well as nearby humans and domesticated animals—is an important component of parasite surveillance (Leendertz *et al.* 2006b).

10.8 Conclusions

The conditions under which a parasite can cause extinction are rare, but they do exist. More importantly, however, we have abundant evidence for the negative effects of parasites on host populations, including in a conservation context. In a habituated population, even the loss of a single animal can have a significant negative impact on future research. If we are concerned about maintaining natural patterns of behaviour and demography, it is important to be aware of the risks of pathogen transmission (just as we might want to maintain naturally occurring behaviour by ensuring that animals forage normally by not provisioning them). Similarly, for ecotourism, loss of animals means loss of income for local people and conservation efforts. Given the abundant evidence that infectious disease negatively impacts primate host populations, we need to be vigilant against all possible introductions of disease, including those that are unlikely to spread and cause extinction. We also need better appreciation for the role of parasites in maintaining or disrupting ecosystem services, and for the links between biodiversity and disease risk in humans and wildlife.

Acknowledgements

We thank the editors, two anonymous reviewers, and Chase Nunez for helpful comments. Alexander Vining assisted with formatting the chapter and references.

References

Abrams, P. A. (2000). Character shifts of prey species that share predators. *American Naturalist* **156**: S46–S61.

Altizer, S. M. (2001). Migratory behaviour and host-parasite co-evolution in natural populations of monarch butterflies infected with a protozoan parasite. *Evolutionary Ecology Research* **3**: 611–632.

Altizer, S., Nunn, C. L., and Lindenfors, P. (2007). Do threatened hosts have fewer parasites? A comparative study in primates. *Journal of Animal Ecology* **76**: 304–314.

Anderson, R. M. and May, R. M. (1991). *Infectious Diseases of Humans: Dynamics and Control*. Oxford University Press.

Augspurger, C. K. (1984). Seedling survival of tropical tree species: interactions of dispersal distance, light-gaps, and pathogens. *Ecology* **65**: 1705–1712.

Benning, T. L., LaPointe, D., Atkinson, C. T., and Vitousek, P. M. (2002). Interactions of climate change with biological invasions and land use in the Hawaiian Islands: modeling the fate of endemic birds using a geographic information system. *Proceedings of the National Academy of Sciences* **99**: 14246–14249.

Berenbaum, M. R. and Zangerl, A. R. (1998). Chemical phenotype matching between a plant and its insect herbivore. *Proceedings of the National Academy of Sciences* **95**: 13743–13748.

Bermejo, M., Rodriguez-Teijeiro, J. D., Illera, G., Barroso, A., Vila, C., *et al.* (2006). Ebola outbreak killed 5000 gorillas. *Science* **314**: 1564–1564.

Bicca-Marques, J. C. (2009). Outbreak of yellow fever affects howler monkeys in southern Brazil. *Oryx* **43**: 173.

Bicca-Marques, J. C. and Freitas, D. S. (2010). The role of monkeys, mosquitoes and humans in the occurrence of a yellow fever outbreak in a fragmented landscape in south Brazil: protecting howler monkeys is a matter of public health. *Tropical Conservation Science* **3**: 78–89.

Blehert, D. S., Hicks, A. C., Behr, M., Meteyer, C. U., Berlowski-Zier, B. M., *et al.* (2009). Bat white-nose syndrome: an emerging fungal pathogen. *Science* **80**: 323–327.

Boesch, C. and Boesch-Achermann, H. (2000). *The Chimpanzees of the Tai Forest*. Oxford: Oxford University Press.

Bodager, J. R., Wright, P. C., Rasambainarivo, F. T., Parsons, M. B., Roellig, D., *et al.* (2015). Complex epidemiology and zoonotic potential for *Cryptosporidium suis* in rural Madagascar. *Veterinary Parasitology* **207**: 140–143.

Boyles, J. G., Cryan, P. M., McCracken, G. F., and Kunz, T. H. (2011). Economic importance of bats in agriculture. *Science* **332**: 41–42.

Bryant, J. E., Holmes, E. C., and Barrett, A. D. T. (2007). Out of Africa: a molecular perspective on the introduction of yellow fever virus into the Americas. *PLoS Pathogens* **3**: 668–673.

Bublitz, D. C., Wright, P. C., Bodager, J. R., Rasambainarivo, F. T., and Gillespie, T. R. (2015). Pathogenic Enterobacteria in lemurs associated with anthropogenic disturbance. *American Journal of Primatology* **77**: 330–337.

Buckling, A. and Rainey, P. B. (2002). The role of parasites in sympatric and allopatric host diversification. *Nature* **420**: 496–499.

Burdon, J. J. and Thrall, P. H. (1999). Spatial and temporal patterns in coevolving plant and pathogen associations. *American Naturalist* **153**: S15–S33.

Caillaud, D., Levrero, F., Cristescu, R., Gatti, S., Dewas, M., *et al.* (2006). Gorilla susceptibility to Ebola virus: the cost of sociality. *Current Biology* **16**: R489–R491.

Calvignac-Spencer, S., Leendertz, S. A. J., Gillespie, T. R., and Leendertz, F. H. (2012). Wild great apes as sentinels and sources of infectious disease. *Clinical Microbiology and Infection* **18**: 521–527.

Carne, C., Semple, S., Morrogh-Bernard, H., Zuberbühler, K., and Lehmann, J. (2013). Predicting the vulnerability of great apes to disease: the role of superspreaders and their potential vaccination. *PLoS One* **8**: e84642.

Carpenter, C. R. (1964) *Naturalistic Behavior of Nonhuman Primates*. University Park, PA: Pennsylvania State University Press.

Chapman, C. A., Gillespie, T. R., and Goldberg, T. L. (2005). Primates and the ecology of their infectious diseases: how will anthropogenic change affect host-parasite interactions? *Evolutionary Anthropology* **14**: 134–144.

Civitello, D. J., Cohen, J., Fatima, H., Halstead, N. T., Liriano, J., *et al.* (2015). Biodiversity inhibits parasites: broad evidence for the dilution effect. *Proceedings of the National Academy of Sciences* **112**: 8667–8671.

Cohen, D. (1985). The canine transmissible venereal tumor: a unique result of tumor progression. *Advanced Cancer Research* **43**: 75–112.

Collias, N. and Southwick, C. (1962). A field study of population density and social organization in howling monkeys. *Proceedings of the American Philosophical Society* **96**: 143–156.

Cooper, N. and Nunn, C. L. (2013). Identifying future zoonotic disease threats: where are the gaps in our understanding of primate infectious diseases? *Evolution, Medicine, and Public Health* **1**: 27–36.

Coscolla, M., Lewin, A., Metzger, S., Maetz-Rennsing, K., Calvignac-Spencer, S., *et al.* (2013). Novel *Mycobacterium tuberculosis* complex isolate from a wild chimpanzee. *Emerging Infectious Diseases* **19**: 969–976.

Daszak, P., Berger, L., Cunningham, A. A., Hyatt, A. D., Green, D. E., *et al.* (1999). Emerging infectious diseases and amphibian population declines. *Emerging Infectious Diseases* **5**: 735–748.

de Castro, F. and Bolker, B. (2005). Mechanisms of disease-induced extinction. *Ecology Letters* **8**: 117–126.

di Bitetti, M. S., Placci, G., Brown, A. D., and Rode, D. I. (1994). Conservation and population status of the brown howling monkey (*Alouatta fusca clamitans*) in Argentina. *Neotropical Primates* **2**: 1–4.

Dobson, A. P. and Meagher, M. (1996). The population dynamics of brucellosis in the Yellowstone National Park. *Ecology* **77**: 1026–1036.

Doebeli, M. and Dieckmann, U. (2000). Evolutionary branching and sympatric speciation caused by different types of ecological interactions. *American Naturalist* **156**: S77–S101.

Dunn, R., Harris, N., Colwell, R., Koh, L., and Sodhi, N. (2009). The sixth mass coextinction: are most endangered species parasites and mutualists? *Proceedings of the Royal Society B: Biological Sciences* **276**: 3037.

Dunn, R. R., Davies, T. J., Harris, N. C., and Gavin, M.C. (2010). Global drivers of human pathogen richness and prevalence. *Proceedings of the Royal Society B: Biological Sciences* **277**: 2587–2595.

Durden, L. A. and Keirans, J. E. (1996). Host-parasite coextinction and the plight of tick conservation. *American Entomologist* **42**: 87–91.

Dybdahl, M. F. and Lively, C. M. (1998). Host-parasite coevolution: evidence for rare advantage and time-lagged selection in a natural population. *Evolution* **52**: 1057–1066.

Ezenwa, V. O., Godsey, M. S., King, R. J., and Guptill, S. C. (2006). Avian diversity and West Nile virus: testing associations between biodiversity and infectious disease risk. *Proceedings of the Royal Society B: Biological Sciences* **273**: 109–117.

Farrell, B. (1998). Inordinate fondness explained: why are there so many beetles? *Science* **281**: 555–559.

Freeland, W. J. (1979). Primate social groups as biological islands. *Ecology* **60**: 719–728.

Frick, W. F., Pollock, J. F., Hicks, A. C., Langwig, K. E., Reynolds, D. S., *et al.* (2010). An emerging disease causes regional population collapse of a common North American bat species. *Science* **329**: 679–682.

Galindo, P. and Srihongse, S. (1967). Evidence of recent jungle yellow-fever activity in Eastern Panama. *Bulletin of the World Health Organization* **36**: 151–161.

Gilardi, K., Gillespie, T. R., Leendertz, F., Travis, D., Whittier, C., *et al.* (2015). *Best Practice Guidelines for Great Ape Health*. Gland: IUCN.

Gillespie, T. R. (2006). Non-invasive assessment of gastrointestinal parasite infections in free-ranging primates. *International Journal of Primatology* **27**: 1129–1143.

Gillespie, T. R. and Chapman, C. A. (2006). Prediction of parasite infection dynamics in primate metapopulations based on attributes of forest fragmentation. *Conservation Biology* **20**: 441–448.

Gillespie, T. R., Chapman, C. A., and Greiner, E. C. (2005). Effects of logging on gastrointestinal parasite infections and infection risk in African primates. *Journal of Applied Ecology* **42**: 699–707.

Gillespie, T. R., Nunn, C., and Leendertz, F. (2008). Integrative approaches to the study of primate infectious disease: implications for biodiversity conservation and global health. *Yearbook of Physical Anthropology* **51**: 53–69.

Gillespie, T. R., Lonsdorf, E. V., Canfield, E. P., Meyer, D. J., Nadler, Y., *et al.* (2010). Demographic and ecological effects on patterns of parasitism in eastern chimpanzees (*Pan troglodytesschweinfurthii*) in Gombe National Park, Tanzania. *American Journal of Physical Anthropology* **143**: 534–544.

Goldberg, T. L., Gillespie, T. R., Rwego, I. B., Wheeler, E., Estoff, E. L., *et al.* (2007). Patterns of gastrointestinal bacterial exchange between chimpanzees and humans involved in research and tourism in western Uganda. *Biological Conservation* **135**: 527–533.

Goldberg, T. L., Gillespie, T. R., Rwego, I. B., Estoff, E. L., and Chapman, C. A. (2008). Anthropogenic disturbance promotes bacterial transmission among primates, humans, and livestock across a fragmented forest landscape. *Emerging Infectious Diseases* **14**: 1375–1382.

Gompper, M. E. and Williams, E. S. (1998). Parasite conservation and the black-footed ferret recovery program. *Conservation Biology* **12**: 730–732.

Goodall, J. (1986). *The Chimpanzees of Gombe: Patterns of Behavior*. Cambridge, MA: Harvard University Press.

Hallam, T. and Mccracken, G. (2011). Management of the panzootic white-nose syndrome through culling of bats. *Conservation Biology* **25**: 189–194.

Harrington, L. A., Moehrenschlager, A., Gelling, M., Atkinson, R. P., Hughes, J., *et al.* (2013). Conflicting and complementary ethics of animal welfare considerations in reintroductions. *Conservation Biology* **27**: 486–500.

Harvala, H., Sharp, C. P., Ngole, E. M., Delaporte, E., Peeters, M., *et al.* (2011). Detection and genetic characterization of enteroviruses circulating among wild populations of chimpanzees in Cameroon: relationship with human and simian enteroviruses. *Journal of Virology* **85**: 4480–4486.

Harvala, H., Van Nguyen, D., McIntyre, C., Imai, N., Clasper, L., *et al.* (2014). Co-circulation of enteroviruses between apes and humans. *Journal of General Virology* **95**: 403–407.

Hasegawa, H., Modrý, D., Kitagawa, M., Shutt, K.A., Todd, A., *et al.* (2014). Humans and great apes cohabiting the forest ecosystem in Central African Republic harbour the same hookworms. *PLoS Neglected Tropical Diseases* **8**: e2715.

Holzmann, I., Agostini, I., Areta, J., Ferreyra, H., Beldomenico, P., *et al.* (2010). Impact of yellow fever outbreaks on two howler monkey species (*Alouatta guariba clamitans* and *A. caraya*) in Misiones, Argentina. *American Journal of Primatology* **72**: 475–480.

Homsy, J. (1999). Ape tourism and human diseases: how close should we get? *Consultancy for the International Gorilla Conservation Programme*. <http://wildpro.twycrosszoo.org/000ADOBES/D133Homsy_rev.pdf> [Accessed 23 June 2015].

Hopkins, M. E. and Nunn, C. L. (2007). A global gap analysis of infectious agents in wild primates. *Diversity and Distributions* **13**: 561–572.

Hopkins, M. E. and Nunn, C. L. (2010). Gap analysis and the geographical distribution of parasites. In: Morand, S. and Krasnov, B. (Eds), *The Biogeography of Host-Parasite Interactions*. Cambridge: Cambridge University Press.

Hudson, P. J., Dobson, A. P., and Newborn, D. (1998). Prevention of population cycles by parasite removal. *Science* **282**: 2256–2258.

Hudson, P. J., Dobson, A. P., and Lafferty, K. D. (2006). Is a healthy ecosystem one that is rich in parasites? *Trends in Ecology and Evolution* **21**: 381–385.

Hugot, J. P. (1999). Primates and their pinworm parasites: the Cameron hypothesis revisited. *Systematic Biology* **48**: 523–546.

Huspeni, T. C. and Lafferty, K. D. (2004). Using larval trematodes that parasitize snails to evaluate a saltmarsh restoration project. *Ecological Applications* **14**: 795–804.

James, R. A., Leberg, P. L., Quattro, J. M., and Vrijenhoek, R. C. (1997). Genetic diversity in black howler monkeys (*Alouatta pigra*) from Belize. *American Journal of Physical Anthropology* **102**: 329–336.

Janatova, M., Albrechtova, K., Petrzelkova, K.J., Dolejska, M., Papousek, I., *et al.* (2014). Antimicrobial-resistant Enterobacteriaceae from humans and wildlife in Dzanga-Sangha Protected Area, Central African Republic. *Veterinary Microbiology* **171**: 422–431.

Kaltz, O. and Shykoff, J. A. (1998). Local adaptation in host–parasite systems. *Heredity* **81**: 361–370.

Kaur, T., Singh, J., Tong, S., Humphrey, C., Clevenger, D., *et al.* (2008). Descriptive epidemiology of fatal respiratory outbreaks and detection of a human-related metapneumovirus in wild chimpanzees (*Pan troglodytes*) at Mahale Mountains National Park, Western Tanzania. *American Journal of Primatology* **70**: 755–765.

Keesing, F., Belden, L. K., Daszak, P., Dobson, A., Harvell, D. C., *et al.* (2010). Impacts of biodiversity on the emergence and transmission of infectious diseases. *Nature* **468**: 647–652.

Kilpatrick, A. M., Gluzberg, Y., Burgett, J., and Daszak, P. (2004). Quantitative risk assessment of the pathways by which West Nile virus could reach Hawaii. *Ecohealth* **1**: 205–209.

Köndgen, S., Kuhl, H., N'Goran, P.K., Walsh, P. D., Schenk, S., *et al.* (2008). Pandemic human viruses cause decline of endangered great apes. *Current Biology* **18**: 260–264.

Köndgen, S., Schenk, S., Pauli, G., Hoesch, C., and Leendertz, F. (2010). Noninvasive monitoring of respiratory viruses in wild chimpanzees. *Ecohealth* **7**: 332–341.

Koh, L. P., Dunn, R. R., Sodhi, N. S., Colwell, R. K., *et al.* (2004). Species coextinctions and the biodiversity crisis. *Science* **305**: 1632–1634.

Lafferty, K. (1997). Environmental parasitology: what can parasites tell us about human impacts on the environment? *Parasitology Today* **13**: 251–255.

LaPointe, D. A., Atkinson, C. T., and Samuel, M. D. (2012). Ecology and conservation biology of avian malaria. *Annals of the New York Academy of Sciences* **1249**: 211–226.

LaPointe, D. A., Hofmeister, E.K., Atkinson, C. T., Porter, R. E., and Dusek, R. J. (2009). Experimental infection of Hawaii amakihi (*Hemignathus virens*) with West Nile virus and competence of a co-occurring vector, Culex quinquefasciatus: potential impacts on endemic Hawaiian avifauna. *Journal of Wildlife Diseases* **45**: 257–271.

Leendertz, F.H., Ellerbrok, H., Boesch, C., Couacy-Hymann, E., Matz-Rensing, K., *et al.* (2004). Anthrax kills wild chimpanzees in a tropical rainforest. *Nature* **430**: 451–452.

Leendertz, F., Lankester, F., Guislain, P., Neel, C., Drori, O., *et al.* (2006a). Anthrax in Western and Central African great apes. *American Journal of Primatology* **68**: 928–933.

Leendertz, F.H., Pauli, G., Maetz-Rensing, K., Boardman, W., Nunn, C. L., *et al.* (2006b). Pathogens as drivers of population declines: the importance of systematic monitoring in great apes and other threatened mammals. *Biological Conservation* **131**: 325–337.

Leendertz, F.H., Yumlu, S., Pauli, G., Boesch, C., Couacy-Hymann, E., *et al.* (2006c). A New *Bacillus anthracis* found in wild chimpanzees and a gorilla from West and Central Africa. *PLoS Pathogens* **2**: 1–4.

Leroy, E. M., Rouquet, P., Formenty, P., Souquière, S., Kilbourne, A., *et al.* (2004). Multiple Ebola virus transmission events and rapid decline of central African wildlife. *Science* **303**: 387–390.

Leroy, E. M., Kumulungui, B., Pourrut, X., Rouquet, P., Hassanin, A., *et al.* (2005). Fruit bats as reservoirs of Ebola virus. *Nature* **438**: 575–576.

Lips, K. R., Brem, F., Brenes, R., Reeve, J. D., Alford, R. A., *et al.* (2006). Emerging infectious disease and the loss of biodiversity in a Neotropical amphibian community. *Proceedings of the National Academy of Sciences of the United States of America* **103**: 3165–3170.

Lively, C. M. and Apanius, V. (1995). Genetic diversity in host-parasite interactions. In: Grenfell, B. T. and Dobson, A. P. (Eds), *Ecology of Infectious Diseases in Natural Populations*. pp. 421–449. Cambridge: Cambridge University Press.

Lonsdorf, E., Travis, D., Pusey, A. E., and Goodall, J. (2006). Using retrospective health data from the Gombe chimpanzee study to inform future monitoring efforts. *American Journal of Primatology* **908**: 897–908.

Lonsdorf, E. V., Murray, C. M., Travis, D. A., Gilby, I. C., *et al.* (2011). A retrospective analysis of factors correlated to chimpanzee (*Pan troglodytes schweinfurthii*) respiratory health at Gombe National Park, Tanzania. *Ecohealth* **8**: 6–35.

Lorch, J. M., Muller, L. K., Russell, R. E., O'Connor, M., Lindner, D. L., *et al.* (2013). Distribution and environmental persistence of the causative agent of White-nose syndrome, *Geomyces destructans*, in bat hibernacula of the eastern United States. *Journal of Applied and Environmental Microbiology* **79**: 1293–1301.

McCallum, H. (2008). Tasmanian devil facial tumour disease: lessons for conservation biology. *Trends in Ecology and Evolution* **23**: 631–637.

McCallum, H. and Dobson, A. (1995). Detecting disease and parasite threats to endangered species and ecosystems. *Trends in Ecology and Evolution* **10**: 190–194.

Milton, K. (1982). Dietary quality and demographic regulation in a howler monkey population. In: Leigh, Jr., E. G., Rand, A. S., and Windsor, D. M. (Eds), *The Ecology of a Tropical Forest: Seasonal Rhythms and Long-Term Changes*. pp. 273–289. Washington, DC: Smithsonian.

Mitchell, C. E. and Power, A. G. (2003). Release of invasive plants from fungal and viral pathogens. *Nature* **421**: 625–627.

Mitter, C., Farrell, B., and Futuyma, D. J. (1991). Phylogenetic studies of insect plant interactions—insights into the genesis of diversity. *Trends in Ecology and Evolution* **6**: 290–293.

Muehlenbein, M. P., Martinez, L. A., Lemke, A. A., Andau, P., Ambu, L., *et al.* (2010). Unhealthy travelers present challenges to sustainable primate ecotourism. *Travel Medicine and Infectious Disease* **8**: 169–175.

Muller-Graf, C. D. M., Collins, D. A., and Woolhouse, M. E. J. (1996). Intestinal parasite burden in five troops of olive baboons (*Papio cynocephalus anubis*) in Gombe Stream National Park, Tanzania. *Parasitology* **112**: 489–497.

Nizeyi, J., Cranfield, M., and Graczyk, T. (2002). Cattle near the Bwindi Impenetrable National Park, Uganda, as a reservoir of *Cryptosporidium parvum* and *Giardia duodenalis* for local community and free-ranging gorillas. *Parasitology Research* **88**: 380–385.

Nizeyi, J. B., Rwego, I. B., Erume, J., Kalema, G., Cranfield, M. R., *et al.* (2001). *Campylobacteriosis, salmonellosis*, and *shigellosis* infections in human habituated mountain gorillas of Uganda. *Journal of Wildlife Diseases* **37**: 239–244.

Nunn, C. L. and Altizer, S. (2005). The Global Mammal Parasite Database: An online resource for infectious disease records in wild primates. *Evolutionary Anthropology* **14**: 1–2.

Nunn, C. L., and Altizer, S. M. (2006). *Infectious Diseases in Primates: Behavior, Ecology and Evolution: Oxford Series in Ecology and Evolution*. Oxford: Oxford University Press.

Nunn, C. L., Altizer, S., Sechrest, W., Jones, K. E., Barton, R. A., *et al.* (2004). Parasites and the evolutionary diversification of primate clades. *American Naturalist* **164**: S90–S103.

Nunn, C. L., Thrall, P. H., Stewart, K., and Harcourt, A. H. (2008). Emerging infectious diseases and animal social systems. *Evolutionary Ecology* **22**: 519–543.

Ostfeld, R. (2010). *Lyme Disease: The Ecology of a Complex System*. Oxford: Oxford University Press.

Palacios, G., Lowenstine, L. J., Cranfield, M. R., Gilardi, K. V., Spelman, L., *et al.* (2011). Human metapneumovirus infection in wild mountain gorillas, Rwanda. *Emerging Infectious Diseases* **17**: 711–713.

Parsons, M. B., Travis, D., Lonsdorf, E. V., Lipende, I., Roellig, D. M. A., *et al.* (2015). Epidemiology and molecular characterization of *Cryptosporidium* spp. in humans, wild primates, and domesticated animals in the Greater Gombe Ecosystem, Tanzania. *PLoS Neglected Tropical Diseases* **10**(2): e0003529.

Percy, D., Page, R., and Cronk, Q. (2004). Plant-insect interactions: double-dating associated insect and plant lineages reveals asynchronous radiations. *Systematic Biology* **53**: 120–127.

Previtali, M. A., Ostfeld, R. S., Keesing, F., Jolles, A. E., Hanselmann, R., *et al.* (2012). Relationship between pace of life and immune responses in wild rodents. *Oikos* **121**: 1483–1492.

Puechmaille, S. J., Frick, W. F., Kunz, T. H., Racey, P. A., Voigt, C. C., *et al.* (2011). White-nose syndrome: is this emerging disease a threat to European bats? *Trends in Ecology and Evolution* **26**: 570–576.

Redford, K. H. (1992). The empty forest. *BioScience* **42**: 412–422.

Rushmore, J., Caillaud, D., Hall, R. J., Meyers, L. A., and Altizer, S. (2014). Network-based vaccination improves prospects for disease control in wild chimpanzees. *Journal of the Royal Society Interface* **11**: 20140349.

Rwego I. B., Isabirye-Basuta G., Gillespie T. R., and Goldberg T. L. (2008). Gastrointestinal bacterial transmission among humans, mountain gorillas, and livestock in Bwindi Impenetrable National Park, Uganda. *Conservation Biology* **22**: 1600–1607.

Saéz, A.M., Weiss, S., Nowak, K., Lapeyre, V., Zimmermann, F., *et al.* (2015). Investigating the zoonotic origin of the West African Ebola epidemic. *EMBO Molecular Medicine* **7**: 17–23.

Salkeld, D. J., Padgett, K. A., and Jones, J. H. (2013). A meta-analysis suggesting that the relationship between biodiversity and risk of zoonotic pathogen transmission is idiosyncratic. *Ecology Letters* **16**: 679–686.

Schaumburg, F., Mugisha, L., Peck, B., Becker, K., Gillespie, T. R., *et al.* (2012). Drug-resistant human *Staphylococcus aureus* in sanctuary apes pose a threat to endangered wild ape populations. *American Journal of Primatology* **74**: 1071–1075.

Scott, M. E. (1987). Regulation of mouse colony abundance by *Heligmosomoides polygyrus*. *Parasitology* **95**: 111–124.

Smith, K. F., Sax, D. F., and Lafferty, K. D. (2006). Evidence for the role of infectious disease in species extinction and endangerment. *Conservation Biology* **20**: 1349–1357.

St-Hilaire, S., Thrush, M., Tatarian, T., Prasad, A., and Peeler, E. (2009). Tool for estimating the risk of anthropogenic spread of *Batrachochytrium denrobatidis* between water bodies. *EcoHealth* **6**: 16–19.

Terborgh, J., Lopez, L., Percy Nunez, V., Rao, M., Shahabudin, G., *et al.* (2001). Ecological Meltdown in Predator-Free Forest Fragments. *Science* **294**: 1923–1926.

Thompson, J. N. (1999). Specific hypotheses on the geographic mosaic of coevolution. *American Naturalist* **153**: S1–S14.

Thrall, P. H., Antonovics, J., and Dobson, A. P. (2000). Sexually transmitted diseases in polygynous mating systems: prevalence and impact on reproductive success. *Proceedings of the Royal Society London B* **267**: 1555–1563.

Torchin, M. E. and Mitchell, C. E. (2004). Parasites, pathogens, and invasions by plants and animals. *Frontiers in Ecology and the Environment* **2**: 183–190.

Torchin, M. E., Lafferty, K. D., Dobson, A. P., McKenzie, V. J., and Kuris, A. M. (2003). Introduced species and their missing parasites. *Nature* **421**: 628–630.

Travis, D. A., Lonsdorf, E. V., Mlengeya, T., and Raphael, J. (2008). A science-based approach to managing disease risks for ape conservation. *American Journal of Primatology* **70**: 745–750.

Tutin, C. E. G., Williamson, E. A., Rogers, M. E., and Fernandez, M. (1991). A case study of a plant-animal relationship: *Cola lizae* and lowland gorillas in the Lope Reserve, Gabon. *Journal of Tropical Ecology* **7**: 181–199.

Van Riper, C., Van Riper, S. G., Goff, M. L., and Laird, M. (1986). The epizootiology and ecological significance of malaria in Hawaiian land birds. *Ecological Monographs* **56**: 327–344.

Vourc'h, G., Plantard, O., and Morand, S. (2012). How does biodiversity influence the ecology of infectious disease? In: Morand, S., Beaudeau, F., and Cabaret, J. (Eds), *New*

Frontiers of Molecular Epidemiology of Infectious Diseases. pp. 291–309. Springer.

Voysey B. C., Mcdonald, K. E., Rogers, M. E., Tutin, C. E. G., and Parnell, R. J. (1999). Gorillas and seed dispersal in the Lope Reserve, Gabon. II: survival and growth of seedlings. *Journal of Tropical Ecology* **15**: 39–60.

Wallis, J. and Lee, D. R. (1999). Primate conservation: the prevention of disease transmission. *International Journal of Primatology* **20**: 803–826.

Walsh, P. D., Abernethy, K. A., Bermejo, M., Beyers, R., De Wachter, P., *et al.* (2003). Catastrophic ape decline in western equatorial Africa. *Nature* **422**: 611–614.

Walsh, P. D., Breuer, T., Sanz, C., Morgan, D., and Doran-Sheehy, D. (2007). Potential for Ebola transmission between gorilla and chimpanzee social groups. *American Naturalist* **169**: 684–689.

Warfield, K. L., Goetzmann, J. E., Biggins, J. E., Kasda, M. B., Unfer, R. C., *et al.* (2014). Vaccinating captive chimpanzees to save wild chimpanzees. *Proceedings of the National Academy of Sciences*: 201316902.

Watts, D. P. (1998). Seasonality in the ecology and life histories of mountain gorillas (*Gorilla gorilla beringei*). *International Journal of Primatology* **19**: 929–948.

Weldon, C., Du Preez, L. H., Hyatt, A. D., Muller, R., and Speare, R. (2004). Origin of the amphibian chytrid fungus. *Emerging Infectious Diseases* **10**: 2100.

Windsor, D. A. (1995). Equal rights for parasites. *Conservation Biology* **9**: 1–2.

Windsor, D. A. (1998). Most of the species on Earth are parasites. *International Journal for Parasitology* **28**: 1939–1941.

Wolf, T. M., Sreevatsan, S., Travis, D., Mugisha, L., and Singer, R. S. (2014). The risk of tuberculosis transmission to free-ranging great apes. *American Journal of Primatology* **76**: 2–13.

Wolf, T. M., Mugisha, L., Shoyama, F. M., O'Malley, M. J., Flynn, J. L., *et al.* (2015). Noninvasive test for tuberculosis detection among primates. *Emerging Infectious Diseases* **21**: 468–470.

Wood, C. L. and Lafferty, K. D. (2013). Biodiversity and disease: a synthesis of ecological perspectives on Lyme disease transmission. *Trends in Ecology and Evolution* **28**: 239–247.

Wood, C. L., Lafferty, K. D., DeLeo, G., Young, H. S., Hudson, P. J., and Kuris, A. M. (2014). Does biodiversity protect humans against infectious disease? *Ecology* **95**: 817–832.

Wyatt, K. B., Campos, P. F., Gilbert, M. T. P., Kolokotronis, S., Hynes, W. H., *et al.* (2008). Historical mammal extinction on Christmas Island (Indian Ocean) correlates with introduced infectious disease. *PLoS One* **3**: e3602.

Yalcindag, E., Elguero, E., Arnathau, C., Durand, P., Akiana, J., *et al.* (2012). Multiple independent introductions of *Plasmodium falciparum* in South America. *Proceedings of the National Academy of Sciences* **109**: 511–516.

Young, H., Griffin, R. H., Wood, C. L., and Nunn, C. L. (2013). Does habitat disturbance increase infectious disease risk for primates? *Ecology Letters* **16**: 656–663.

Zohdy, S., Grossman, M. K., Fried, I. R., Rasambainarivo, F. T., Wright, P. C., *et al.* (2015). Diversity and prevalence of diarrhea-associated viruses in the lemur community and associated human population of Ranomafana National Park, Madagascar. *International Journal for Primatology* **36**: 143–153.

CHAPTER 11

Primates and climate change: a review of current knowledge

Amanda H. Korstjens and Alison P. Hillyer

A Temminckii's red colobus in Bijilo, The Gambia. Photo copyright: A. H. Korstjens

11.1 Introduction to climate change

As much as we know about primate socio-ecology, we are still struggling to find ways to predict how primates will respond to climate change. This chapter introduces some of the current research in this field and gives a short overview of the various ways in which we predict how a species is likely to respond to changes in its environment.

Climate change is a global long-term shift in the Earth's weather patterns[1] as a result of natural or human-induced changes to land-surface properties, solar radiation, or atmospheric gas (greenhouse gases) concentrations (IPCC 2013). The natural greenhouse effect of the Earth's atmosphere is important as it ensures that life on Earth is possible by retaining some of the warmth from solar radiation (Serreze 2010). Greenhouse gases absorb and re-radiate thermal planetary radiation and a human-induced increase in these gases causes the average planetary surface temperature to increase. Human activity has also altered land-surface properties, which changes not only the reflection of solar radiation from the Earth's surface but also the patterns of storage and emission of atmospheric gases (IPCC 2013).

Climate change is by no means a uniform process, as some regions will warm more than others. For example, the Arctic has been warming relatively fast (due to loss of its floating ice cover) while the Antarctic has been cooling in recent years (due to reduction in the ozone layer) and this pattern is predicted

[1] http://www.metoffice.gov.uk/climate-guide/climate-change.

Korstjens, A.H. and Hillyer, A., *Primates and climate change: a review of current knowledge.* In: *An Introduction to Primate Conservation.* Edited by: Serge A. Wich and Andrew J. Marshall, Oxford University Press (2016). © Oxford University Press.
DOI 10.1093/acprof:oso/9780198703389.003.0011

to continue for some time (Serreze 2010; Marshall *et al.* 2014). As a result of such variation in the magnitude and nature of climate change, the temperature difference between regions of the world is changing too. The temperature difference between regions drives circulation of air and water as they both flow from warm to cool regions. Therefore, as the difference in temperature between the warm and cool areas changes, this influences air and sea currents around the globe (Vecchi and Soden 2007). These atmospheric and ocean circulations do not only affect local climate conditions but they are also essential drivers behind global extreme weather events, including El Niño and La Niña events, and cyclones/hurricanes (Petoukhov *et al.* 2013). Therefore, the intensity and frequency of such extreme events is predicted to increase under most future climate projections.

Projections of future rainfall patterns are less consistent than those on temperature changes. Most climate models predict an overall increase of rainfall in the tropics but a decrease in most of tropical Africa (IPCC 2013). Since 1901, rainfall has increased in some and decreased in other areas of the Neotropics and Asia, and has decreased generally in sub-Saharan Africa (Hartmann *et al.* 2013).

Despite extensive understanding of meteorological events, it remains difficult to accurately predict how our climate will change globally and regionally (IPCC 2013). The best information on all aspects of climate change and projections of future changes, that is Global Climate models (GCMs), is available on the website of the Intergovernmental Panel on Climate Change (IPCC).[2] GCMs are predictions of future global climate conditions that are based on projections of human growth, economic conditions, land occupation, land use, and changes in greenhouse gas and aerosol concentrations (Cubasch *et al.* 2013). The most commonly used set of climate scenarios and GCMs can be found on the Data Distribution Centre (DDC) of the IPCC.[3]

The outputs of GCMs are used as a source of climate predictors to help model how species will respond to climate change. For example, GCMs were used to predict which areas are likely to remain suitable habitat for orang-utans (*Pongo pygmaeus* and *Pongo abelii*) (Carne *et al.* 2012; Struebig *et al.* 2015; Gregory *et al.* 2012).

To fully understand how climate change will affect a species, we need to understand *how* different aspects of a species' distribution, behaviour, and reproduction are affected by climate change (see Section 11.2) and what determines the *vulnerability* of a species to the predicted changes (see Section 11.3).

11.2 How climate change affects species

The main changes occurring as a result of climate change relate to: (1) phenology, (2) habitat shifts, (3) community changes, and (4) disease.

11.2.1 Phenological changes in response to climate change

Phenology refers to the timing of naturally recurring events (e.g. flowering, fruiting, hibernation, migration). Climate change is expected to lead to greater fluctuations in the timing of phenological events as monthly climate variation is expected to become more irregular and more extreme (IPCC 2013). Phenological shifts in food sources can lead to asynchrony between life cycles of predators and prey, parasitoids and hosts, pollinators and plants, and primates and their food sources (Both *et al.* 2009; Visser 2008; Butt *et al.* 2015). This decoupling will affect primates who have synchronized their reproductive events to fit seasonal patterns of resource availability (see Section 11.3.2) (van Schaik *et al.* 1993; Post 2013).

11.2.2 Habitat shifts and fragmentation

Climate change can result in a spatial shift of suitable climatic conditions and primate habitats, or cause fragmentation, reduction, or expansion of suitable habitat. As temperature is driving a lot of the changes in climate, most habitat shifts are on the altitudinal gradient (Colwell *et al.* 2008) (e.g. geladas, see Section 11.3.3) and/or on the latitudinal gradient (e.g. Sichuan snub-nosed monkeys: Luo *et al.* (2015)). For example, Freeman and Class Freeman (2014) showed a strong upslope range shift for tropical bird species. Struebig *et al.* (2015) predicted

[2] http://www.ipcc.ch.
[3] http://www.ipcc-data.org.

upslope habitat shifts for primates and carnivores in Borneo under future climatic conditions. It has to be noted that when climate conditions shift regionally, this does not always equate to suitable productive habitat shifting at the same speed unrestricted (Marshall and Wich Chapter 18, this volume). The dispersal of plants, especially long-lived fruiting trees (i.e. primate food sources), often lags behind the change in climate (Corlett and Westcott 2013). In addition, dispersal of plants and animals can be severely restricted by natural or human-made migration barriers (e.g. Yunnan snub-nosed monkeys, *Rhinopithecus bieti*: Wong *et al.* (2013)).

Climate change can also lead to habitat fragmentation, which limits migration (important for gene flow and repopulation of suitable areas) and access to food sources. For example, a major reduction in annual rainfall in Senegal has led to habitat degradation and fragmentation that is severely endangering the survival of the already range-restricted Temminckii's red colobus, *Procolobus badius temminckii* (Galat *et al.* 2009). Gregory *et al.* (2012) predicted severe habitat shifts and fragmentation as a result of climate change for orang-utans.

Finally, climate change can lead to range reduction or expansion depending on the habitat preferences of the species. For example, African guenons, Cercopithecini, form an unusually speciose taxon. The high speciosity is the result of a relatively recent radiation most likely due to a refugia effect around five million years ago when their ancestor's range retracted due to climate change (Tosi 2008). When their ranges expanded again, these sister species were able to share their habitats thanks to slight differences in niche preferences, leading to the species-rich guenon communities currently found in some African forests.

It is important to understand how climate change will reshape the distribution of suitable primate habitats because it has implications for the management and placement of conservation areas and wildlife corridors.

11.2.3 Community changes

Climate change will affect the composition of animal communities because individual species within a community will respond differently to the changes (Walther 2010; Root *et al.* 2003; Thomas *et al.* 2004). For example, human disturbance and climate change have affected forest structure and food availability in Kibale National Park, leading to a change in the densities of different primate species (Chapman *et al.* 2010b). The co-existence of sympatric species that (partly) use the same food sources depends on their ability to cope with inter-specific competition during times when preferred foods are scarce by using different fall-back sources (Cowlishaw and Dunbar 2000). If one species disappears from a forest community, other species may be able to flourish due to reduced competition, a process known as density compensation. For example, when larger bodied Neotropical primates are hunted, medium-sized species become more abundant (Peres and Dolman 2000). The reduction in abundance of one species may also have negative effects on the remaining species because predators (or hunters) may shift prey preference to the more abundant species (Peres and Dolman 2000). The absence of one species may also make another species more vulnerable to predation if the two species depended on each other for protection, for example through polyspecific associations (Bshary and Noë 1997; Buchanan-Smith 1999).

It is, therefore, important to consider the effects of climate change on whole species communities if we want to conserve biodiversity.

11.2.4 Increased disease risk due to climate change

Climate change will influence some host–parasite interactions and is likely to increase the prevalence and severity of disease in primates (Chapman *et al.* 2005b) as both animal and plant pathogen abundances change and hosts may become more susceptible to disease (Nunn and Gillespie, Chapter 10, this volume). Many pathogens are more likely to develop, survive, and spread under warmer and wetter conditions (Harvell *et al.* 2002). Especially diseases spread by vectors like mosquitoes are expected to expand latitudinally with increasing temperatures (Nunn *et al.* 2005). Loss of parasite diversity is also predicted in some cases, which could affect ecosystem functioning (e.g. by releasing hosts from parasite population regulation,

Harvell *et al.* (2002)). Barrett *et al.* (2013) found that projected changes in Madagascar's climate would result in some range contractions but mostly range expansions for lemur parasites (mites, ticks, and worms). The most harmful parasites showed the largest range expansions while most (but not all) primate ecto- and endo-parasite indices, such as richness and abundance, increased during warmer, wetter conditions.

Some regions are predicted to get wetter or seasonally receive extreme rainfall, and various studies show an increase of parasite prevalence under wetter conditions. Parasites were more prevalent during the wet compared with the dry season in chimpanzees, *Pan troglodytes*, in Tanzania (Huffman *et al.* 1997), and ursine colobus, *Colobus vellerosus*, in Ghana (Teichroeb *et al.* 2009). Stoner (1996) showed an increase in parasite prevalence with the proximity to water bodies in mantled howler monkeys, *Alouatta palliata*, and increased overall parasite prevalence in wet compared to dry forests in Costa Rica. Similarly, the frequency with which eastern black-and-white colobus, *Colobus guereza*, groups visited wetter areas of the forest correlated positively with indices of parasite infections in Uganda (Chapman *et al.* 2010a).

A study by Behie *et al.* (2014) shows the complexity of the relationship between climate events and primate health. They found that a hurricane increased the exposure of black howler monkeys, *Alouatta pigra*, to the trematode *Controrchis* spp. Interestingly, this increased exposure was the result of increased consumption of a fast-growing pioneer plant (which thrived after destruction of the forest). Consuming this plant led to exposure to ants that are intermediate hosts of *Controchis* and that have a mutualistic relationship with the pioneer plant.

On top of these direct effects of climate change on host–pathogen relationships, there will be indirect effects as environmental change leads to fragmentation, range reductions, and increased proximity between human settlements and wildlife. These indirect effects of human disturbance and climate change affect pathogen prevalence, development of novel pathogens, host vulnerability to diseases, and the risk of human–primate pathogen transmission (Chapman *et al.* 2005b).

11.3 Vulnerability of a species

IPCC's definition of vulnerability is: 'Vulnerability is the propensity or predisposition to be adversely affected. Vulnerability encompasses a variety of concepts including sensitivity or susceptibility to harm and lack of capacity to cope and adapt' (IPCC 2014: 128). Foden *et al.* (2013) elaborated on this and created a framework to identify the species most vulnerable to extinction from a range of climate change-induced stresses (Figure 11.1). They identified three dimensions of vulnerability: 'sensitivity (the lack of potential for a species to persist *in situ*), exposure (the extent to which each species' physical environment will change), and low adaptive capacity (a species' inability to avoid the negative impacts of climate change through dispersal and/or micro-evolutionary change)' (Foden *et al.* 2013: 2).

The species-typical biological traits associated with vulnerability that are most often mentioned in primate research will be discussed in more detail: (1) diet and dietary specialization, (2) life-history traits and phenology, (3) biogeographical range, rarity, and dispersal system, (4) social system and behaviour, and (5) physical traits. We also discuss examples of whether or not species can adapt to changes through alterations in these traits.

11.3.1 Diet and dietary specialization

The diet and dietary specialization of a species will greatly affect its vulnerability to environmental change because it impacts on the species' ability to adapt to changes through dietary shifts. Compared to many other mammals, primates tend to be reasonably flexible in their dietary choice and often eat a range of food sources. The main dietary categories are: vegetative plant matter (leaves), reproductive plant parts (i.e. fruits, seeds, and flowers), gum and sap, and insects/fauna. Species mostly differ in the proportion of their diet devoted to each food source and their main fall-back food sources.

Specialist folivores (leaf eaters) are potentially particularly vulnerable to climate change as an increase in temperature and reduction in rainfall can lead to lower digestibility, lower quality, and higher concentrations of plant secondary compounds

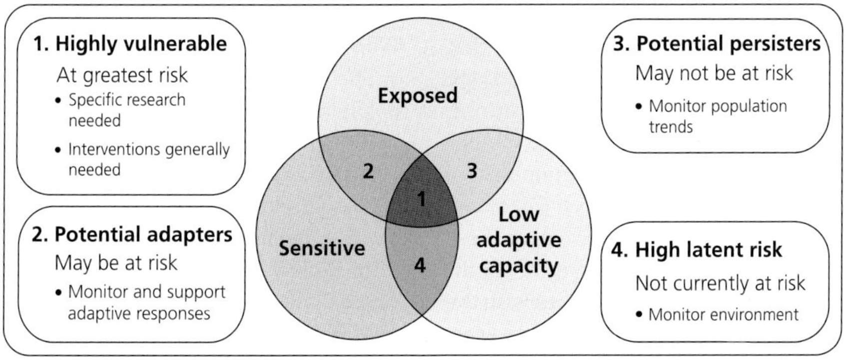

Figure 11.1 A climate change vulnerability framework showing four levels of vulnerability. Foden *et al.* (2013) used biological traits from amphibian, bird, and coral species to illustrate how climate change exposure, species' sensitivity (affected by specialization, rarity, physiological tolerance, and dependence on environmental triggers or inter-specific interactions), and low adaptive capacity (determined by gene flow, genetic diversity, and reproductive rates) influence the vulnerability of a species. Examples for primates for each category could be (see text to understand why species are included in particular categories): 1. Many of the Malagasy lemurs, especially specialist folivores, and *Pongo* spp.; 2. *Colobus* spp.; 3. *Macaca sylvanus*; and 4. few primate habitats are unaffected (low exposure), although Neotropical endemics in the Amazon and *Pan paniscus* may currently fall into this category. Reproduced with permission from Foden and PLoS One.

in leaves (Zvereva and Kozlov 2006; Rothman *et al.* 2014). This means that folivore adaptations to climate change should counteract the increased cost of leaf digestion. For example, a decrease in rainfall has resulted in more fibrous leaves in the diet of the folivorous Milne-Edwards' sifaka, *Propithecus edwardsi*. The increased fibre content wears the animals' teeth down and older females with more worn-down teeth are no longer able to produce sufficient milk for their infants in drier years, leading to infant mortality for older females (King *et al.* 2005). Higher temperatures not only lead to greater toxicity of plant matter, they also directly lead to smaller liver sizes (Moore *et al.* 2015). Since larger livers are needed to deal with more toxic diets, this dual effect of higher temperatures will make folivore specialists particularly vulnerable to global warming. Possible adaptations include: increasing resting/digesting time (Korstjens *et al.* 2010); reducing mobility, exposure to heat, and feeding bout duration to counteract the increased body heat generated by digestive processes (Moore *et al.* 2015); diet shift; or consuming non-food items that counter-balance the effects of the toxins (e.g. charcoal: Struhsaker *et al.* (1997)).

Frugivores (fruit eaters) may benefit from climate change as fruit production is theoretically expected to go up with the projected rising CO_2 levels, but this may be offset by greater climatic stress from drought or intensive rains, (e.g Clark *et al.* 2013). In Kibale National Park, Uganda, rainfall and temperatures have increased in the last 30 years and this has been accompanied by varying responses of the fruiting trees: overall, fruit productivity went up, but some tree species showed a reduction in the number of fruiting individuals (Chapman *et al.* 2005a). In Taï, Côte d'Ivoire, rainfall overall decreased over a 12-year observation period, but overall fruit production went up and fruit production was not related to rainfall variation (Polansky and Boesch 2013).

Dietary specialization and flexibility influence how a species will respond to climate change. For example, Lambert (2002) reports that cercopithecines (Cercopithecinae) showed more regular diet switching than colobines (Colobinae) in a primate community in Kibale, Uganda. She attributed this to greater digestive flexibility of cercopithecines compared to colobines. Cercopithecines have long retention times (allowing for digestion of low-quality foods that requires longer fermentation) but lack the specialized stomachs of colobines. The specialized digestive system of colobines is designed to support leaf eating but limits their exploitation of high-quality, sugar-rich ripe fruits (Lambert 1998). Tutin *et al.* (1997) showed that during the dry season, fruit scarcity in the Lopé Reserve, Gabon, can last for eight months during which two of eight primate species change to non-fruit diets, but chimpanzees and three guenon species (*Cercopithecus nictitans, C. pogonias, C. cephus*) do not. Lack of dietary plasticity would leave these species vulnerable to increased drought conditions. Overall, the most specialized species with the smallest dietary breadth (e.g. *Avahi, Erythrocebus, Hapalemur, Tarsius* spp.) are expected to decrease in abundance under changing environmental conditions (Harcourt *et al.* 2002; Slatyer *et al.* 2013). For example, geladas (*Theropithecus gelada*) are specialist grazers that are restricted to mountains in Ethiopia. As a result of climate change, their suitable habitat has been greatly reduced and increasingly fragmented as it is moving upslope (Dunbar 1998). Their omnivorous–frugivorous relatives, the baboons, are thriving in a wide range of habitats (Dunbar 1998). In Kibale, Uganda, the blue monkey (*Cercopithecus mitis*) is a frugivorous–generalist species that does well in secondary and disturbed forests but is outcompeted by other species in old-growth forests (Chapman *et al.* 2000). This means that blue monkeys are more likely to be colonizers of new habitats that then get replaced by old-growth specialists as the forest matures. Likewise, the black-and-white colobus (*Colobus* spp.) is more able to thrive in forest fragments and secondary forests than its close relative, the red colobus (*Procolobus* spp.), probably thanks to slight dietary differences (Chapman *et al.* 2000; Korstjens and Dunbar 2007).

11.3.2 Life-history traits and phenology

Species with relatively slow life cycles, small litters, and low reproductive rates are particularly vulnerable to extinction. These characteristics are typical of most primates, especially the haplorrhines, Haplorrhini. The fast rate at which the climate is changing means that a slow life history and low reproductive output reduce a species' resilience. Resilience is determined by a species' ability to recover from a population crash and the number of generations available to allow genetic adaptation to suit the new environment (Cowlishaw *et al.* 2009; Williams *et al.* 2008).

The way in which life history events are linked to phenological events of a species' food sources will also put a species at risk when plant phenology patterns change (Brockman and van Schaik 2005). Disruption of synchrony between reproductive events and food abundance may result in mortality of infants and mothers, lower reproductive rates, higher vulnerability to disease or predation, and later age at maturity (e.g. lemurs, see Section 11.4.2). Primates living in low and high latitudes are more likely to have synchronized their breeding events with their relatively seasonal habitats (Janson and Verdolin 2005), for example Japanese macaques (*Macaca fuscata*), South African vervet monkeys (*Chlorocebus pygerythrus*), Barbary macaques (*Macaca sylvanus*), and Malagasy primates (see Section 11.4.2) (van Schaik and Pfannes 2005). The most vulnerable species will be those that are unable to respond to phenological changes by diet switching and range change (van Schaik *et al.* 1993).

11.3.3 Biogeographical range, rarity, and dispersal system

Species with a restricted biogeographical range, those living at low densities, and those with limited dispersal ability are vulnerable to climate change because they will struggle to recruit new individuals after a population crash and have limited ability to colonize new areas (Schloss *et al.* 2012; Chapman *et al.* 2006). Biogeographical ranges and population densities are heavily shaped by random factors (e.g. local extinction events due to a hurricane or disease), historical patterns, dispersal ability, and fragmentation of the habitat through natural or anthropogenic barriers. The latter two factors are particularly prevalent for mountain range restricted species, which will struggle to disperse to new areas as their habitats shift upslope due to global warming and become isolated on mountain tops. These species show severe range contractions (e.g. geladas: Dunbar (1998)) and increased risk of extinction (e.g. amphibians: Alan Pounds *et al.* (2006)) as a result of climate change.

Beaudrot and Marshall (2011) showed that dispersal limitation is a stronger determinant of community composition than ecological niche availability for primates. This means that dispersal limitations will also be important determinants of the ability of species to adapt to climate change by range shifts (Beaudrot and Marshall 2011). Primates with the greatest level of dispersal limitation would be those in which individuals depend on associating with close kin and/or long-term alliance partners (see Section 11.3.4) in home ranges that they are familiar with. Leaving a home range and social group is often costly as it exposes an individual to increased predation risk while travelling alone through unfamiliar terrain, and it reduces the individual's knowledge of the location of food sources (Isbell and van Vuren 1996). In addition, an individual may suffer resistance from established group members when trying to join a new social group. Finally, due to these costs of dispersing between groups, sex-biased dispersal from natal groups is the typical pattern in primates when dispersal mostly functions to avoid inbreeding. Sex-biased dispersal reduces a species' ability to colonize new suitable habitats, as only one sex seeks out new areas to live and multi-sex groups will take a long time to establish themselves in new areas.

Further evidence that primates are particularly vulnerable to climate change due to their distribution patterns, rarity, and dispersal ability comes from a study by Schloss *et al.* (2012). They investigated: (1) the distance between a species' current geographical range and predicted location of suitable habitat in the future (a measure of habitat shift); and (2) how fast a species would be able to disperse to keep up with the required range shift (i.e. dispersal velocity). Primates were most vulnerable to climate change for both these indicators: primate ranges were predicted to shift relatively more than those of other mammals and primates had low dispersal velocity. Dispersal velocity was defined as the distance that a species is likely to travel per year, and it depends on the frequency and distance of dispersal events. Dispersal velocity was calculated using data on body mass, diet, and generation length (based on gestation length, age at sexual maturity, and the period until the next breeding season). Compared to other mammals, primates are relatively dispersal limited, have relatively slow life histories, are of medium size, are often range restricted, and often occur in fragmented habitats. Therefore, nearly all primates would be unable to disperse fast enough to keep pace with the habitat shifts and changes that will occur as a result of climate change. The average range size reduction in primates was predicted to be 75% (Schloss *et al.* 2012).

11.3.4 Social system

The social system of a species influences its vulnerability to climate change because an individual's dependence on others for survival can reduce its flexibility to adapt to change. Most primates depend heavily on living in bonded social groups for benefits like access to food, thermoregulation, information transfer, and reduced predation risk. Climate change can result in increased ecological stress (as food quality, abundance, or distribution change) leading to greater competition among individuals within groups. Increased competition and a reduction in the time available for maintaining social bonds with other group members can result in smaller or more dispersed groups (Lehmann *et al.* 2007b). Such smaller groups may be more vulnerable to predation or thermal stress. For example, species living in areas with low night-time or winter temperatures may depend on maintaining large enough group sizes and strong social bonds for the thermoregulatory benefits of social groups, for example Japanese macaques (e.g. Hanya *et al.* 2007), vervet monkeys (McFarland *et al.* 2015), and Barbary macaques (McFarland and Majolo 2013). Climate change is likely to increase temperature fluctuations and make thermoregulation an even more important factor in species survival, especially for primates living in temperate regions or higher altitudes.

11.3.5 Primate physiological and anatomical features

Physiological and anatomical characteristics can determine how flexibly a species can respond to a fast-changing environment. For example, a species' predominant mode of locomotion affects how well it will cope with reduced connectivity in forest canopy. Highly arboreal primates may struggle to cross open areas in between trees or forest patches to reach emerging habitats or important resources (e.g. Bornean agile gibbon, *Hylobates albibaris*: Cheyne *et al.* (2013)).

Large brain size is another primate characteristic that can affect species' vulnerability to climate change. Whether larger brains make a species more vulnerable or less vulnerable than smaller brained species is, however, still debated and probably highly dependent on the type of change that a species has to deal with. On the one hand, large brains make a species vulnerable to climate change because they are associated with slow life histories (Isaac 2009) (see Section 11.3.2). On the other hand, a larger relative brain size may represent greater ability to develop innovative behavioural adaptations to explore and utilize changing environments (the cognitive buffer hypothesis) (Sol 2009; Reader and Laland 2002). van Woerden *et al.* (2012) suggest that the brain buffers against environmental changes if there are unexplored resources to be found and starvation is not unavoidable (e.g. because the lean period is not too long); but that in other cases large brains will make a species more vulnerable to changes.

Body size is another important determinant of species vulnerability to climate change as it influences life history (see Section 11.3.2), ability to cope with food shortages, dietary requirements (see Section 11.3.1), thermoregulation, and metabolic rate. For example, small body size in mammals is correlated with a fast metabolic rate that increases with rising temperatures (the Arrhenius effect), leading to greater energy expenditure in warmer climates and making small primates more vulnerably to global warming (Lovegrove *et al.* 2014). Hibernation and torpor (a state of decreased physiological activity) have been found to counteract this effect and reduce metabolic rates in small primates, such as the Philippine tarsier, *Tarsius syrichta*, and golden-brown mouse lemur, *Microcebus ravelobensis* (Lovegrove *et al.* 2014). Indeed, extinction risk as a result of climate change may be lower for mammals that can adjust their body temperature (heterothermes) than for homeothermes (Geiser and Turbill 2009). For example, flexible thermoregulatory adaptations like torpor are allowing the reddish-grey mouse lemurs, *Microcebus griseorufus*, to deal with their exceptionally unpredictable environment (Kobbe *et al.* 2011).

11.3.6 Predicting species' responses to climate change

We showed that vulnerability of a species depends on species traits and the type of threat that the primates are facing as a result of climate change (see also Isaac and Cowlishaw (2004)) and discussed a few adaptation strategies. Most often, primate adaptations may include a variety of strategies. For example, long-tailed macaques, *Macaca fascicularis*, successfully adapted to habitat damage caused by severe drought and fire through: diet-shifting from fresh fruit to insects as well as charred or desiccated fruit; increasing the size of their home ranges; and increasing terrestrial travel (Berenstain 1986). Geoffroy's spider monkeys, *Ateles geoffroyi*, responded to habitat change after a hurricane through time budget changes (moving less and resting more), increased fission–fusion dynamics (i.e. reduced subgroup size and fusing less often), and a dietary change from fruits to leaves (as fruit became less abundant) (Schaffner *et al.* 2012).

Various studies try to predict species' responses to environmental change by analysing the relationship of species with current climatic conditions and then modelling their distribution under future conditions based on those observed relationships. The complexity of these models can vary greatly depending on whether or not the mechanisms behind the observed relationships are being investigated as well. More complex mechanistic models incorporate species-typical traits that determine why a species is associated with a particular ecological niche. Understanding the mechanisms behind species–habitat relationships allows us to investigate whether the required adaptations are within the species' behavioural repertoire or physical ability. We already discussed the models by Schloss *et al.* (2012). Another example of mechanistic models are the 'time budget models' that use time budget decisions as indicators of the stress that an environment is putting on a population (Dunbar *et al.* 2009). For example, Korstjens *et al.* (2010) showed that enforced resting time (i.e. minimum time needed for thermoregulation and digestion) increases with average annual temperature, temperature variation, and the amount of leaves in the diet. Therefore, primates are expected to increase resting time with global warming, especially folivores. Similar analyses showed that barbary macaques in high-altitude regions of the Atlas Mountains in Morocco are unable to keep up with the demands on feeding time during extreme winters, leading to high mortality (McFarland and Majolo 2013). A different model suggested that apes, especially gorillas, would need to adopt a greater level of fission–fusion and higher reliance on fruit to be able to adapt to the predicted changes in their habitat (Lehmann *et al.* 2007a, 2010). Harrison and Chivers (2007) suggested that climate change during the Holocene may have caused orang-utans to adopt their extremely dispersed social systems.

How species responded to environmental change in the past can also identify the features that enable a species to adapt and survive in the short or long term. For example, as a result of flooding and agricultural activity, the Tana River red colobus, *Procolobus rufomitratus*, population declined by 80% in the 1970s. Following changes in feeding and ranging behaviour the population increased again in the 1980s (Decker 1994). Chatterjee *et al.* (2012) used niche modelling to examine how modern-day gibbons are likely to respond to climate change based on biogeographical responses of gibbons to historic changes. Bettridge *et al.* (2010) explored the capacity of extinct baboon species to cope with climate changes by looking at fossil baboon distributions and applying knowledge about how climate affects time budgets of extant taxa. Meijaard *et al.* (2008) used phylogenetic age as a proxy for a variety of vulnerability features and showed that phylogenetic age is correlated with resilience to timber extraction in Bornean mammals, suggesting that more recently evolved species are more flexible and less specialized (Meijaard *et al.* 2008).

Overall, the ability to modify behaviour, physiological processes, and timing of life-history events allows species to cope with short-term change, but genetic change is generally required for long-term adaptation at the species level (Hoffmann *et al.* 2011). The level of genetic change that a species can attain within the time available to respond to changes is constrained by the species' resilience (see Section 11.3.2) (Williams *et al.* 2008).

11.4 Differences between regions

Climate change will affect each continent and its forests differently (IPCC 2013: Table 14.3 and TS.2) and tropical forests themselves differ between regions due to divergent historical origins (Zelazowski *et al.* 2010). Therefore, regional variation in the response of forests and their fauna will be considerable (Malhi *et al.* 2014). Malhi *et al.* (2014) highlight the interaction between various past, present, and future disturbances and how tropical forests and their fauna will respond to future environmental change. In this section we explore how primates in different regions are likely to be affected by climate change.

11.4.1 Neotropics

For South and Central America, the IPCC (2013) report predicts increased heat, drier conditions, reduced mean precipitation, and an increase in extreme precipitation in tropical cyclones along the eastern and western coasts. Neotropical primates are likely to lose a lot of their habitat as some forests are predicted to change to savannah woodlands (IPCC 2013). The fully arboreal Neotropical primates may struggle to adapt to a savannah woodland environment that requires regular terrestrial travel.

The predicted increase in frequency and severity of El Niño and La Niña events will also strongly affect primate species. Population size of atelines decreased either immediately or one year after an El Niño event as a result of a reduction in food availability (Wiederholt and Post 2010). The frugivorous atelines (*Ateles, Brachyteles, Lagothrix*) showed a stronger response than the folivorous howler monkeys (*Alouatta*) (Wiederholt and Post 2010). Birth seasonality in Neotropical primates is often linked to seasonality in food sources, especially in the more seasonal higher and lower latitudes (Di Bitetti and Janson 2000). The folivorous howler monkeys are less seasonal than the frugivorous primates and medium-sized primates are more seasonal than larger or smaller species (Di Bitetti and Janson 2000). Thus, changes in seasonality and extreme weather events are likely to affect infant mortality of medium-sized frugivorous Neotropical primates most. Recent dry conditions, El Niño events, and increased mean annual temperature led to lower birth rates and affected the sex ratio at birth in frugivorous woolly monkeys and northern muriquis (Wiederholt and Post 2011). Owl monkeys, *Aotus azarai*, in the Argentinean Chaco responded to drought by increasing the percentage of leaves in their normally very frugivorous diet. Their social system was also affected as inter-birth intervals increased and more individuals dispersed from their natal groups (Fernández-Duque and Heide 2013).

11.4.2 Africa

For Africa, the IPCC (IPCC 2013; Hewitson *et al.* 2014) predicts that temperatures will increase in most areas, and drought/dryness is likely to increase in the West and South but not the East. Summer monsoon precipitation is predicted to increase in West Africa; while short rain will increase in East Africa due to the pattern of Indian Ocean warming. Increased extreme rainfall from landfall cyclones is predicted for the east coast (including Madagascar) (Barros *et al.* 2014: Table 21.7; IPCC 2013: Table 14.3 and TS.2).

Effects of climate changes have already been recorded for various African field sites (see examples from Kibale in Uganda and Taï in Côte d'Ivoire, Section 11.3.1, and Senegal, Section 11.2.2). In Amboseli, Kenya, minimum and maximum daily temperatures increased by 0.071 °C and 0.275 °C per annum respectively between 1976 and 2000; while rainfall varied greatly among years but showed no consistent trend (Altmann *et al.* 2002). At Amboseli, baboons coped with this change in climate through diet shifts, range shifts, and changing time budgets (e.g. less socializing during harsh times) (Alberts *et al.* 2005). In contrast, the local vervet monkeys coped less well and declined in density due to their more specialized diet and tighter link between births and weather seasonality (Alberts *et al.* 2005). Drought caused food shortages for a population of Namibian chacma baboons, *Papio ursinus*, which caused high female, juvenile, and infant mortality and a long-term shift in adult sex ratio (Hamilton 1985).

Madagascar has severe, unpredictable weather and highly seasonal food production (Wright 1999) and is characterized by a high proportion of folivorous primates (Reed and Bidner 2004). Lemur birth, lactation, and weaning patterns are synchronized with the fruiting of their food sources (Wright 2007). Erhart and Overdorff (2008) showed that fruit production was significantly lower between 1988 and 1993 than between 1994 and 2003. This negatively affected reproductive patterns, sex ratio, group sizes, and population density in red-fronted lemurs, *Eulemur fulvus rufus*, during the earlier period. Cyclone frequency increased during this time and fruit availability was markedly less in the month following a cyclone. *E. fulvus rufus* did not change diet but females ranged further outside their home ranges to find food during times of fruit scarcity. There were greater fluctuations in group membership, changes in sex ratio in groups, and higher adult mortality (partly due to increased dispersal as a result of increased within-group competition). In contrast, the densities of competitors—the red-bellied lemur, *Eulemur rubriventer*, and black-and-white ruffed lemur, *Varecia variegata*—increased during this time, although the reason for their increase was not studied (Erhart and Overdorff 2008).

Malagasy primates also regularly experience cyclones. Cyclones and subsequent drought limit first-year offspring survival and reduce the reproductive lifespan of female Milne Edward's sifaka's, *Propithecus edwardsii* (Dunham *et al.* 2011; King *et al.* 2005). During El Niño years this species' fecundity dropped to 65% that of normal years (Dunham *et al.* 2008). Mortality in a ring-tailed lemur, *Lemur catta*, population was elevated after droughts, especially when they were followed by a second year of drought (mortality rates of 80% among infants, 57% among juveniles, and 20–30% among adult females in the birth season) (Gould *et al.* 1999).

Therefore, the predicted overall increase in temperatures, drought, and extreme weather events will negatively effect many African primates. Frugivorous and omnivorous cercopithecine species that can survive in open woodlands (especially the more terrestrial species) and secondary forests are most likely to survive.

11.4.3 Asia

For tropical regions of Asia, temperatures are expected to increase overall, with increased duration, intensity, and frequency of heat waves in East Asia. Precipitation is expected to increase in most Asian areas, except central Asia where it is likely to decrease. The frequencies of intense precipitation events and cyclones are likely to increase in South, Southeast, and East Asia (IPCC 2013).

The dipterocarp forests of Asia have relatively lower fruit production than rainforests in other parts of the world and sustain lower primate biomass (Reed and Bidner 2004). Many Asian primates depend on mast fruiting species, that is species that fruit irregularly but when they fruit there tends to be a large number of individuals fruiting all at once, creating a sudden mass production of fruits and flowers (Butt *et al.* 2015). The El Niño Southern Oscillation (ENSO) system is associated with fruit masting of many important primate fruit trees (Butt *et al.* 2015). It is still unclear whether changes in ENSO and the frequency and intensity of fruit masting will affect frugivorous primates positively or negatively. Drought and cyclones were the cause of a 15% population decline in toque macaques, *Macaca sinica*, in Sri Lanka (Dittus 1988) and increased effects of ENSO will only lead to more such events. Chinese gibbon species are predicted to lose a lot of their current habitat and may not be able to exploit the emerging areas of suitable habitat due to distribution constraints and anthropogenic pressures (Chatterjee *et al.* 2012). Wong *et al.* (2013) showed that black snub-nosed monkeys, *Rhinopithecus bieti*, are vulnerable to climate change despite expansion of the habitat type most suitable for them (mixed forest) because regional forest cover is becoming increasingly fragmented and habitat quality is deteriorating rapidly.

11.5 Climate change mitigation strategies

A multidisciplinary global approach is needed to reduce the emission of greenhouse gases and reduce the effects of climate change on biodiversity. Although there are no mitigation guidelines that have been developed specifically for primates, general strategies for biodiversity conservation in the face of climate change include: (1) forest preservation, restoration, reforestation, and afforestation (Brodie *et al.* 2012; Guariguata *et al.* 2008; Pimm *et al.* 2014; Benchimol and Peres 2014); (2) habitat connectivity (wildlife corridors) among fragmented forests (Hannah *et al.* 2008; Lehman 2006; Shanee *et al.* 2013; Imong *et al.* 2014; Jantz *et al.* 2014); and (3) reintroduction and translocation of animals (IUCN 2012; Loss *et al.* 2011).

1. An estimated two billion hectares of land have the potential to be restored or reforested to increase resilience of forests against climate change. Moreover, planting trees in economically poor but forest-rich countries provides a global environmental service by increasing the number of trees available to act as CO_2 stores (Kaeslin *et al.* 2012). The global initiative by the United Nations' collaborative programme to reduce emissions from deforestation and forest degradation (UN-REDD; launched in 2008) is an effort to create financial value for the carbon stored in forests by offering economical incentives for developing countries to reduce emissions by safeguarding forested lands and investing in low-carbon paths to sustainable development. Biodiversity conservation, sustainable forest management, and enhancement of forest carbon stocks are all components of REDD + (Garcia-Ulloa and Koh, Chapter 16, this volume). REDD + initiatives include: the Forest Carbon Partnership Facility (FCPF) and Forest Investment Program (FIP) hosted by The World Bank.[4] In principle, REDD + was set up to conserve forest cover and carbon stock but it is becoming increasingly important that this should be associated with biodiversity conservation of the wildlife inhabiting the forests (Hinsley *et al.* 2015). Loss of key biodiversity can alter forest composition and the amount of carbon the forest contains. Primates in particular can have strong effects on plant communities and can be considered important ecosystem engineers because they can form a high proportion of the biomass in a forest. For example, the abundance of atelines was positively correlated to plant diversity across 16 Neotropical forest sites (Stevenson 2011). Primates function as seed

[4] www.unep-redd.org.

dispersers (Lambert 2011; Chapman 2006) or by consuming vegetative parts and bark leading to reduced tree production or mortality (Chapman *et al.* 2013).

2. Habitat connectivity is essential to allow for gene flow and avoid inbreeding in small isolated populations in increasingly fragmented landscapes (Brodie *et al.* 2012; Marsh 2003). Small-scale agroforestry can be one mitigation strategy to achieve both forest conservation and habitat connectivity as it aims to preserve forests and maintain connectivity while safeguarding the ecosystem services that forests provide for local communities (Minang *et al.* 2011). Agroforests (or agroecosystems) are important for primates, with at least 57 taxa across primate range countries using or inhabiting agro-forests (Estrada *et al.* 2012). For example, agroforests in Brazil are intensely used by Neotropical primates and function as effective corridors between mature forest areas (Raboy and Dietz 2004).
3. Finally, there is a long history of translocations and reintroductions of rescued primates from the pet trade or from areas of heavy habitat destruction. However, these are not without problems and strict guidelines should be adhered to (IUCN 2012; Campbell *et al.* 2015; Loss *et al.* 2011; Germano *et al.* 2015) and it should often be only considered as a last resort (Kavanagh and Caldecott 2013).

In addition, mitigation plans should consider the effects of anthropogenic factors such as habitat destruction, fragmentation, degradation, pollution, hunting, species introductions (accidental and planned), disease risk, human migration, and habitat disturbance on the ability of animals to adapt to climate change (McCarty 2001; Şekercioğlu *et al.* 2012; Brodie *et al.* 2012). For example, barriers created by humans can impede migration of fauna and flora with changing climate, thus blocking their ability to adapt through habitat shifts (Brodie *et al.* 2012). Similarly, human adaptations to climate change can increasingly lead to conflicts between humans and non-human primates for agricultural land and forest. For example, the Kenyan population of De Brazza's monkey, *Cercopithecus neglectus*, is under immense pressure due to cattle farmers invading their lower elevation habitats during times of drought to feed cattle (Mwenja 2007).

11.6 Conclusion

Most primates are expected to struggle to adapt to rapid climate change because of their limited dispersal ability, long generation times, and restricted ranges. It is, however, very difficult to predict how individual populations or species will respond to climate change because of the interaction between climate change and human habitat disturbance. Species' responses will depend on their species-specific traits, socio-ecology, and phenotypic and genetic plasticity, and ultimately on how humans mitigate climate change through policy, population management, and forest management.

The past and present offer clues to how primates may respond to climate change, which can help the development of next-generation mechanistic biogeography models. More complex models can incorporate climate change and anthropogenic disturbance as well as biotic factors (e.g. species' traits, interactions between species) and historical distribution patterns. Research priorities should include strategies to increase our understanding and knowledge of the relationship between primate foods and climatic conditions, the mechanisms that mediate the effect of climate on species' distributions, the link between life-history traits and population viability, how specific phenological and community shifts affect primates, minimum (genetically *or* demographically) viable population sizes for species, the impacts of climate change on population size, and the patterns that drive primate distributions. Current understanding is especially limited in certain groups such as nocturnal Strepsirhines, Callitrichinae, and *Aotus*.

Many primate species are already threatened by anthropogenic factors, and climate change will only add to the pressure by further reducing habitat availability and shifting habitats. These effects will push primates to their physiological and socio-ecological limits. Central American and Amazonian primates, the great apes, *Colobinae, Theropithecus*, and *Lemuriformes* have all been identified as particularly vulnerable to climate change, whereas the more adaptable *Papio, Chlorocebus*, and *Macaca* may tolerate change better. As primary consumers, ecosystem engineers, and often a significant percentage of a forest's biomass, primate extinctions as a result of climate change may have dire consequences for the forests they inhabit.

References

Alan Pounds, J., Bustamante, M. R. M. R., Coloma, L. A., Consuegra, J. A., Fogden, M. P. L., *et al.* (2006). Widespread amphibian extinctions from epidemic disease driven by global warming. *Nature* **439**: 161–167.

Alberts, S. C., Hollister-Smith, J. A., Mututua, R. S., Sayialel, S. N., Muruthi, P. M., *et al.*. (2005). Seasonality and long-term change in a savanna environment. In: Brockman, D. K. and Vanschaik, C. P. (Eds), *Seasonality in Primates: Studies of Living and Extinct Human and Non-Human Primates*. Cambridge University Press.

Altmann, J., Alberts, S. C., Altmann, S. A., and Roy, S. B. (2002). Dramatic change in local climate patterns in the Amboseli basin, Kenya. *African Journal of Ecology* **40**: 248–251.

Barrett, M. A., Brown, J. L., Junge, R. E., and Yoder, A. D. (2013). Climate change, predictive modeling and lemur health: assessing impacts of changing climate on health and conservation in Madagascar. *Biological Conservation* **157**: 409–422.

Barros, V. R., Field, C. B., Dokken, D. J., Mastrandrea, M. D., Mach, K. J., *et al.* (Eds) (2014). *Climate Change 2014: Impacts, Adaptation, and Vulnerability. Part B: Regional Aspects. Contribution of Working Group II to the Fifth Assessment Report of the Intergovernmental Panel on Climate Change*. Cambridge, UK and New York, NY, USA: Cambridge University Press.

Beaudrot, L. H. and Marshall, A. J. (2011). Primate communities are structured more by dispersal limitation than by niches. *Journal of Animal Ecology* **80**: 332–341.

Behie, A. M., Kutz, S., and Pavelka, M. S. (2014). Cascading effects of climate change: do hurricane-damaged forests increase risk of exposure to parasites? *Biotropica* **46**: 25–31.

Benchimol, M. and Peres, C. A. (2014). Predicting primate local extinctions within 'real-world' forest fragments: a pan-neotropical analysis. *American Journal of Primatology* **76**: 289–302.

Berenstain, L. (1986). Responses of long-tailed macaques to drought and fire in Eastern Borneo: a preliminary report. *Biotropica* **18**: 257–262.

Bettridge, C. M., Lehmann, J., and Dunbar, R. I. M. (2010). Trade-offs between time, predation risk and life history, and their implications for biogeography: a systems modelling approach with a primate case study. *Ecological Modelling* **221**: 777–790.

Both, C., Van Asch, M., Bijlsma, R. G., Van Den Burg, A. B., and Visser, M. E. (2009). Climate change and unequal phenological changes across four trophic levels: constraints or adaptations? *Journal of Animal Ecology* **78**: 73–83.

Brockman, D. K. and Van Schaik, C. P. (2005). *Seasonality in Primates: Studies of Living and Extinct Human and Non-human Primates*. New York: Cambridge University Press.

Brodie, J., Post, E., and Laurance, W. F. (2012). Climate change and tropical biodiversity: a new focus. *Trends in Ecology and Evolution* **27**: 145–150.

Bshary, R. and Noë, R. (1997). Red colobus and diana monkeys provide mutual protection against predators. *Animal Behaviour* **54**: 1461–1474.

Buchanan-Smith, H. M. (1999). Tamarin polyspecific associations: forest utilization and stability of mixed-species groups. *Primates* **40**: 233–247.

Butt, N., Seabrook, L., Maron, M., Law, B. S., Dawson, T. P., *et al.* (2015). Cascading effects of climate extremes on vertebrate fauna through changes to low-latitude tree flowering and fruiting phenology. *Global Change Biology* **21**: 3267–3277.

Campbell, C. O., Cheyne, S. M., and Rawson, B. M. (2015). *Best Practice Guidelines for the Rehabilitation and Translocation of Gibbons*. Gland: IUCN.

Carne, C., Semple, S., and Lehmann, J. (2012). The effects of climate change on orangutans: a time budget model. In: Druyan, L. M. (Ed.), *Climate Models*. Rijeka, Croatia: InTech.

Chapman, C. A. and Russo, S. E. (2006). Primate seed dispersal: linking behavioral ecology with forest community structure. In: Campbell, C. J. F., Mackinnon, K. C., Panger, M., and Bearder, S. (Eds), *Primates In Perspective*. Oxford University Press.

Chapman, C. A., Balcomb, S. R., Gillespie, T. R., Skorupa, J. P., and Struhsaker, T. T. (2000). Long-term effects of logging on African primate communities: a 28-year comparison from Kibale National Park, Uganda. *Conservation Biology* **14**: 207–217.

Chapman, C. A., Chapman, L. J., Struhsaker, T. T., Zanne, A. E., Clark, C. J., *et al.* (2005a). A long-term evaluation of fruiting phenology: importance of climate change. *Journal of Tropical Ecology* **21**: 31–45 M3–10.1017/S0266467404001993.

Chapman, C. A., Gillespie, T. R., and Goldberg, T. L. (2005b). Primates and the ecology of their infectious diseases: how will anthropogenic change affect host-parasite interactions? *Evolutionary Anthropology* **14**: 134–144.

Chapman, C. A., Lawes, M. J., and Eeley, H. A. C. (2006). What hope for African primate diversity? *African Journal of Ecology* **44**: 116–133.

Chapman, C. A., Speirs, M. L., Hodder, S. A. M., and Rothman, J. M. (2010a). Colobus monkey parasite infections in wet and dry habitats: implications for climate change. *African Journal of Ecology* **48**: 555–558.

Chapman, C. A., Struhsaker, T. T., Skorupa, J. P., Snaith, T. V., and Rothman, J. M. (2010b). Understanding long-term primate community dynamics: implications of forest change. *Ecological Applications* **20**: 179–191.

Chapman, C. A., Bonnell, T. R., Gogarten, J. F., Lambert, J. E., Omeja, P. A., *et al.* (2013). Are primates ecosystem engineers? *International Journal of Primatology* **34**: 1–14.

Chatterjee, H. J., Tse, J. S. Y., and Turvey, S. T. (2012). Using ecological niche modelling to predict spatial and temporal distribution patterns in Chinese gibbons: lessons from the present and the past. *Folia Primatologica* **83**: 85–99.

Cheyne, S. M., Thompson, C. J., and Chivers, D. J. (2013). Travel adaptations of Bornean agile gibbons *Hylobates albibarbis* (Primates: Hylobatidae) in a degraded secondary forest, Indonesia. *Journal of Threatened Taxa* **5**: 3963–3968.

Clark, D. A., Clark, D. B., and Oberbauer, S. F. (2013). Field-quantified responses of tropical rainforest aboveground productivity to increasing CO_2 and climatic stress, 1997–2009. *Journal of Geophysical Research: Biogeosciences* **118**: 783–794.

Colwell, R. K., Brehm, G., Cardelús, C. L., Gilman, A. C., and Longino, J. T. (2008). Global warming, elevatonal range shifts, and lowland biotic attrition in the Wet Tropics. *Science* **322**: 258–261.

Corlett, R. T. and Westcott, D. A. (2013). Will plant movements keep up with climate change? *Trends in Ecology and Evolution* **28**: 482–488.

Cowlishaw, G. and Dunbar, R. I. M. (2000). *Primate Conservation Biology*. Chicago: University of Chicago Press.

Cowlishaw, G., Pettifor, R. A., and Isaac, N. J. B. (2009). High variability in patterns of population decline: the importance of local processes in species extinctions. *Proceedings of the Royal Society B-Biological Sciences* **276**: 63–69.

Cubasch, U., Wuebbles, D., Chen, D., Facchini, M. C., Frame, D., *et al.* (2013). Introduction. In: Stocker, T. F., Qin, D., Plattner, G.-K., Tignor, M., Allen, S. K., *et al.* (Eds), *Climate Change 2013: The Physical Science Basis. Contribution of Working Group I to the Fifth Assessment Report of the Intergovernmental Panel on Climate Change (IPCC)*. Cambridge, UK and New York, NY, USA: Cambridge University Press.

Decker, B. S. (1994). Effects of habitat disturbance on the behavioral ecology and demographics of the Tana river red colobus (*Colobus badius ruformitratus*). *International Journal of Primatology* **15**: 703–737.

Di Bitetti, M. S. and Janson, C. H. (2000). When will the stork arrive? Patterns of birth seasonality in neotropical primates. *American Journal of Primatology* **50**: 109–130.

Dittus, W. P. J. (1988). Group fission among wild toque macaques as a consequence of female resource competition and environmental stress. *Animal Behaviour* **36**: 1626–1645.

Dunbar, R. I. M. (1998). Impact of global warming on the distribution and survival of the gelada baboon: a modelling approach. *Global Change Biology* **4**: 293–304.

Dunbar, R. I. M., Korstjens, A., and Lehmann, J. (2009). Time as an ecological constraint. *Biological Reviews* **84**: 413–429.

Dunham, A. E., Erhart, E. M., Overdorff, D. J., and Wright, P. C. (2008). Evaluating effects of deforestation, hunting, and El Niño events on a threatened lemur. *Biological Conservation* **141**: 287–297.

Dunham, A. E., Erhart, E. M., and Wright, P. C. (2011). Global climate cycles and cyclones: consequences for rainfall patterns and lemur reproduction in southeastern Madagascar. *Global Change Biology* **17**: 219–227.

Erhart, E. M. and Overdorff, D. J. (2008). Population demography and social structure changes in Eulemur fulvus rufus from 1988 to 2003. *American Journal of Physical Anthropology* **136**: 183–193.

Estrada, A., Raboy, B. E., and Oliveira, L. C. (2012). Agroecosystems and primate conservation in the tropics: a review. *American Journal of Primatology* **74**: 696–711.

Fernandez-Duque, E. and Heide, G. (2013). Dry season resources and their relationship with owl monkey (aotus azarae) feeding behavior, demography, and life history. *International Journal of Primatology* **34**: 752–769.

Foden, W. B., Butchart, S. H. M., Stuart, S. N., Vié, J.-C., Akçakaya, H. R., *et al.* (2013). Identifying the world's most climate change vulnerable species: a systematic trait-based assessment of all birds, amphibians and corals. *PLoS One* **8**: e65427.

Freeman, B. G. and Class Freeman, A. M. (2014). Rapid upslope shifts in New Guinean birds illustrate strong distributional responses of tropical montane species to global warming. *Proceedings of the National Academy of Sciences* **111**: 4490–4494.

Galat, G., Galat-Luong, A., and Nizinski, G. (2009). Increasing dryness and regression of the geographical range of Temminck's red colobus Procolobus badius temminckii: implications for its conservation. *Mammalia: International Journal of the Systematics, Biology and Ecology of Mammals* **73**: 365–368.

Geiser, F. and Turbill, C. (2009). Hibernation and daily torpor minimize mammalian extinctions. *Naturwissenschaften* **96**: 1235–1240.

Germano, J. M., Field, K. J., Griffiths, R. A., Clulow, S., Foster, J., *et al.* (2015). Mitigation-driven translocations: are we moving wildlife in the right direction? *Frontiers in Ecology and the Environment* **13**: 100–105.

Gould, L., Sussman, R. W., and Sauther, M. L. (1999). Natural disasters and primate populations: the effects of a 2-year drought on a naturally occurring population of ring-tailed lemurs in Southwestern Madagascar. *International Journal of Primatology* **20**: 69–84.

Gregory, S. D., Brook, B. W., Goossens, B., Ancrenaz, M., Alfred, R., *et al.* (2012). Long-term field data and climate-habitat models show that orangutan persistence depends on effective forest management and greenhouse gas mitigation. *PLoS One* **7**: e43846.

Guariguata, M. R., Cornelius, J. P., Locatelli, B., Forner, C., and Sánchez-Azofeifa, G. A. (2008). Mitigation needs adaptation: tropical forestry and climate change. *Mitigation and Adaptation Strategies for Global Change* **13**: 793–808.

Hamilton, W. J. (1985). Demographic consequences of a food and water shortage to desert Chacma Baboons, *Papio ursinus*. *International Journal of Primatology* **6**: 451–462.

Hannah, L., Dave, R., Lowry, P. P., Andelman, S., Andrianarisata, M., *et al.* (2008). Climate change adaptation for conservation in Madagascar. *Biology Letters* **4**: 590–594.

Hanya, G., Kiyono, M., and Hayaishi, S. (2007). Behavioral thermoregulation of wild Japanese macaques: comparisons between two subpopulations. *American Journal of Primatology* **69**: 802–815.

Harcourt, A. H., Coppeto, S. A., and Parks, S. A. (2002). Rarity, specialization and extinction in primates. *Journal of Biogeography* **29**: 445–456.

Harrison, M. E. and Chivers, D. J. (2007). The orang-utan mating system and the unflanged male: a product of increased food stress during the late Miocene and Pliocene? *Journal of Human Evolution* **52**: 275–293.

Hartmann, D. L., Klein Tank, A. M. G., Rusticucci, M., Alexander, L. V., BröNnimann, S., *et al.* (2013). Observations: atmosphere and surface. In: Stocker, T. F., Qin, D., Plattner, G. K., Tignor, M., Allen, S. K., *et al.* (Eds), *Climate Change 2013: The Physical Science Basis. Contribution of Working Group I to the Fifth Assessment Report of the Intergovernmental Panel on Climate Change (IPCC)*. Cambridge, UK and New York, NY, USA: Cambridge University Press.

Harvell, C. D., Mitchell, C. E., Ward, J. R., Altizer, S., Dobson, A. P., *et al.* (2002). Climate warming and disease risks for terrestrial and marine biota. *Science* **296**: 2158–2162.

Hewitson, B., Janetos, A. C., Carter, T. R., Giorgi, F., Jones, R. G., *et al.* (2014). Regional context. In: Barros, V. R., Field, C. B., Dokken, D. J., Mastrandrea, M. D., Mach, K. J., *et al.* (Eds), *Climate Change 2014: Impacts, Adaptation, and Vulnerability. Part B: Regional Aspects. Contribution of Working Group II to the Fifth Assessment Report of the Intergovernmental Panel on Climate Change (IPCC)*. Cambridge, UK: Cambridge University Press.

Hinsley, A. H., Entwistle, A., and Pio, D. V. (2015). Does the long-term success of REDD plus also depend on biodiversity? *Oryx* **49**: 216–221.

Hoffmann, A. A., Sgro, C. M., and Sgrò, C. M. (2011). Climate change and evolutionary adaptation. *Nature* **470**: 479–485.

Huffman, M. A., Gotoh, S., Turner, L. A., Hamai, M., and Yoshida, K. (1997). Seasonal trends in intestinal nematode infection and medicinal plant use among chimpanzees in the Mahale mountains, Tanzania. *Primates* **38**: 111–125.

Imong, I., Robbins, M. M., Mundry, R., Bergl, R., and Kühl, H. S. (2014). Informing conservation management about structural versus functional connectivity: a case-study of Cross River gorillas. *American Journal of Primatology* **76**: 978–988.

IPCC (2013). Climate Change 2013: The Physical Science Basis. Contribution of Working Group I to the Fifth Assessment Report of the Intergovernmental Panel on Climate Change. 1535.

IPCC (2014). Climate Change 2014: Impacts, Adaptation, and Vulnerability. Summaries, Frequently Asked Questions, and Cross-Chapter Boxes. A Contribution of Working Group II to the Fifth Assessment Report of the Intergovernmental Panel on Climate Change. 190.

Isaac, J. (2009). Effects of climate change on life history: implications for extinction risk in mammals. *Endangered Species Research* **7**: 115–123.

Isaac, N. J. B. and Cowlishaw, G. (2004). How species respond to multiple extinction threats. *Proceedings of the Royal Society B-Biological Sciences* **271**: 1135–1141.

Isbell, L. A. and Van Vuren, D. (1996). Differential costs of locational and social dispersal and their consequences for female group living primates. *Behaviour* **133**: 1–36.

IUCN (2012). *IUCN Guidelines for Reintroductions and Other Conservation Translocations*. Gland: IUCN.

Janson, C. H. and Verdolin, J. (2005). Seasonality of primate births in relation to climate. In: Brockman, D. K. and Van Schaik, C. P. (Eds), *Seasonality in Primates: Studies of Living and Extinct Human and Non-human Primates*. Cambridge: Cambridge University Press.

Jantz, P., Goetz, S., and Laporte, N. (2014). Carbon stock corridors to mitigate climate change and promote biodiversity in the tropics. *Nature Climate Change* **4**: 138–142.

Kaeslin, E., Redmond, I., and Dudley, N. (2012). *Wildlife in a Changing Climate*. Rome: Food and Agriculture Organization of the United Nations (FAO).

Kavanagh, M. and Caldecott, J. O. (2013). Strategic guidelines for the translocation of primates and other animals. *The Raffles Bulletin of Zoology*, Supplement: 203–209.

King, S. J., Godfrey, L. R., Arrigo-Nelson, S. J., Pochron, S. T., Wright, P. C., *et al.* (2005). Dental senescence in a long-lived primate links infant survival to rainfall.

Proceedings of the National Academy of Sciences of the United States of America **102**: 16579–16583.

Kobbe, S., Ganzhorn, J. U., and Dausmann, K. H. (2011). Extreme individual flexibility of heterothermy in free-ranging Malagasy mouse lemurs (*Microcebus griseorufus*). *Journal of Comparative Physiology B-Biochemical Systemic and Environmental Physiology* **181**: 165–173.

Korstjens, A. H. and Dunbar, R. I. M. (2007). Time constraints limit group sizes and distribution in red and black-and-white colobus monkeys. *International Journal of Primatology* **28**: 551–575.

Korstjens, A. H., Lehmann, J., and Dunbar, R. I. M. (2010). Resting time as an ecological constraint on primate biogeography. *Animal Behaviour* **79**: 361–374.

Lambert, J. E. (1998). Primate digestion: interactions among anatomy, physiology, and feeding ecology. *Evolutionary Anthropology: Issues, News, and Reviews* **7**: 8–20.

Lambert, J. E. (2002). Resource switching and species coexistence in guenons: a community analysis of dietary flexibility. In: Glenn, M. E. and Cords, M. (Eds), *The Guenons: Diversity and Adaptation in African Monkeys*. New York: Springer US.

Lambert, J. E. (2011). Primate seed dispersers as umbrella species: a case study from Kibale National Park, Uganda, with implications for Afrotropical forest conservation. *American Journal of Primatology* **73**: 9–24.

Lehman, S. M. (2006). Conservation biology of Malagasy strepsirhines: a phylogenetic approach. *American Journal of Physical Anthropology* **130**: 238–253.

Lehmann, J., Korstjens, A., and Dunbar, R. I. M. (2007a). Fission–fusion social systems as a strategy for coping with ecological constraints: a primate case. *Evolutionary Ecology* **21**: 613–634.

Lehmann, J., Korstjens, A., and Dunbar, R. I. M. (2007b). Group size, grooming and social cohesion in primates. *Animal Behaviour* **74**: 1617–1629.

Lehmann, J., Korstjens, A. H., and Dunbar, R. I. M. (2010). Apes in a changing world—the effects of global warming on the behaviour and distribution of African apes. *Journal of Biogeography* **37**: 2217–2231.

Loss, S. R., Terwilliger, L. A., and Peterson, A. C. (2011). Assisted colonization: integrating conservation strategies in the face of climate change. *Biological Conservation* **144**: 92–100.

Lovegrove, B. G., Canale, C., Levesque, D., Fluch, G., Reháková-Petr⊠, M., *et al.* (2014). Are tropical small mammals physiologically vulnerable to Arrhenius effects and climate change? *Physiological and Biochemical Zoology: PBZ* **87**: 30–45.

Luo, Z., Zhou, S., Yu, W., Yu, H., Yang, J., *et al.* (2015). Impacts of climate change on the distribution of Sichuan snub-nosed monkeys (*Rhinopithecus roxellana*) in Shennongjia area, China. *American Journal of Primatology* **77**: 135–151.

Malhi, Y., Gardner, T. A., Goldsmith, G. R., Silman, M. R., and Zelazowski, P. (2014). Tropical forests in the Anthropocene. *Annual Review of Environment and Resources* **39**: 125–159.

Marsh, L. K. (2003). *Primates in Fragments: Ecology and Conservation*. New York: Kluwer Academic/Plenum Publishers.

Marshall, J., Armour, K. C., Scott, J. R., Kostov, Y., Hausmann, U., *et al.* (2014). The ocean's role in polar climate change: asymmetric Arctic and Antarctic responses to greenhouse gas and ozone forcing. *Philosophical Transactions of the Royal Society of London A: Mathematical, Physical and Engineering Sciences* **372**: 20130040.

Mccarty, J. P. (2001). Ecological consequences of recent climate change. *Conservation Biology* **15**: 320–331.

Mcfarland, R., Fuller, A., Hetem, R. S., Mitchell, D., Maloney, S. K., *et al.* (2015). Social integration confers thermal benefits in a gregarious primate. *Journal of Animal Ecology* **84**: 871–878.

Mcfarland, R. and Majolo, B. (2013). Coping with the cold: predictors of survival in wild Barbary macaques, Macaca sylvanus. *Biology Letters* **9**: 20130428.

Meijaard, E., Sheil, D., Marshall, A. J., and Nasi, R. (2008). Phylogenetic age is positively correlated with sensitivity to timber harvest in Bornean mammals. *Biotropica* **40**: 76–85.

Minang, P. A., Bernard, F., Van Noordwijk, M., and Kahurani, E. (2011). Agroforestry in REDD +: Opportunities and Challenges. Nairobi, Kenya.

Moore, B. D., Wiggins, N. L., Marsh, K. J., Dearing, M. D., and Foley, W. J. (2015). Translating physiological signals to changes in feeding behaviour in mammals and the future effects of global climate change. *Animal Production Science* **55:** 272–283.

Mwenja, I. (2007). A new population of De Brazza's monkey in Kenya. *Primate Conservation* **22**: 117–122.

Nunn, C. L., Altizer, S. M., Sechrest, W., and Cunningham, A. A. (2005). Latitudinal gradients of parasite species richness in primates. *Diversity and Distributions* **11**: 249–256.

Peres, C. A. and Dolman, P. M. (2000). Density compensation in neotropical primate communities: evidence from 56 hunted and nonhunted Amazonian forests of varying productivity. *Oecologia* **122**: 175–189.

Petoukhov, V., Rahmstorf, S., Petri, S., and Schellnhuber, H. J. (2013). Quasiresonant amplification of planetary waves and recent Northern Hemisphere weather extremes. *Proceedings of the National Academy of Sciences* **110**: 5336–5341.

Pimm, S. L., Jenkins, C. N., Abell, R., Brooks, T. M., Gittleman, J. L., *et al.* (2014). The biodiversity of species and their rates of extinction, distribution, and protection. *Science* **344**: 1246752: 1–10.

Polansky, L. and Boesch, C. (2013). Long-term changes in fruit phenology in a west African lowland tropical rain forest are not explained by rainfall. *Biotropica* **45**: 434–440.

Post, E. (2013). *Ecology of Climate Change: The Importance of Biotic Interactions*. Monographs in Population Biology. Princeton: Princeton University Press.

Raboy, B. E. and Dietz, J. M. (2004). Diet, foraging, and use of space in wild golden-headed lion tamarins. *American Journal of Primatology* **63**: 1–15.

Reader, S. M. and Laland, K. N. (2002). Social intelligence, innovation, and enhanced brain size in primates. *Proceedings of the National Academy of Sciences* **99**: 4436–4441.

Reed, K. E. and Bidner, L. R. (2004). Primate communities: past, present, and possible future. *American Journal of Physical Anthropology* **47**: 2–39.

Root, T. L., Price, J. T., Hall, K. R., Schneider, S. H., Rosenzweig, C., *et al.* (2003). Fingerprints of global warming on wild animals and plants. *Nature* **421**: 57–60.

Rothman, J. M., Chapman, C. A., Struhsaker, T. T., Raubenheimer, D., Twinomugisha, D., *et al.* (2014). Long-term declines in nutritional quality of tropical leaves. *Ecology* **96**: 873–878.

Schaffner, C. M., Rebecchini, L., Ramos-Fernandez, G., Vick, L. G., and Aureli, F. (2012). Spider monkeys (*Ateles geoffroyi yucatenensis*) cope with the negative consequences of hurricanes through changes in diet, activity budget, and fission–fusion dynamics. *International Journal of Primatology* **33**: 922–936.

Schloss, C. A., Nuñez, T. A., and Lawler, J. J. (2012). Dispersal will limit ability of mammals to track climate change in the Western Hemisphere. *Proceedings of the National Academy of Sciences* **109**: 8606–8611.

Şekercioğlu, Ç. H., Primack, R. B., and Wormworth, J. (2012). The effects of climate change on tropical birds. *Biological Conservation* **148**: 1–18.

Serreze, M. C. (2010). Understanding recent climate change. *Conservation Biology* **24**: 10–17.

Shanee, S., Allgas, N., and Shanee, N. (2013). Preliminary observations on the behavior and ecology of the Peruvian night monkey (*Aotus miconax*: Primates) in a remnant cloud forest patch, *north eastern Peru. Tropical Conservation Science* **6**: 138–148.

Slatyer, R. A., Hirst, M., and Sexton, J. P. (2013). Niche breadth predicts geographical range size: a general ecological pattern. *Ecology Letters* **16**: 1104–1114.

Sol, D. (2009). Revisiting the cognitive buffer hypothesis for the evolution of large brains. *Biology Letters* **5**: 130–133.

Stevenson, P. R. (2011). The abundance of large ateline monkeys is positively associated with the diversity of plants regenerating in neotropical forests. *Biotropica* **43**: 512–519.

Stoner, K. E. (1996). Prevalence and intensity of intestinal parasites in mantled howling monkeys (*Alouatta palliata*) in northeastern Costa Rica: implications for conservation biology. *Conservation Biology* **10**: 539–546.

Struebig, M. J., Fischer, M., Gaveau, D. L. A., Meijaard, E., Wich, S. A., *et al.* (2015). Anticipated climate and land-cover changes reveal refuge areas for Borneo's orangutans. *Global Change Biology* **21**: 2891–2904.

Struhsaker, T. T., Cooney, D. O., and Siex, K. S. (1997). Charcoal consumption by Zanzibar red colobus monkeys: its function and its ecological and demographic consequences. *International Journal of Primatology* **18**: 61–72.

Teichroeb, J. A., Kutz, S. J., Parkar, U., Thompson, R. C. A., and Sicotte, P. (2009). Ecology of the gastrointestinal parasites of *Colobus vellerosus* at Boabeng-Fiema, Ghana: possible anthropozoonotic transmission. *American Journal of Physical Anthropology* **140**: 498–507.

Thomas, C. D., Cameron, A., Green, R. E., Bakkenes, M., Beaumont, L. J., *et al.* (2004). Extinction risk from climate change. *Nature* **427**: 145–148.

Tosi, A. J. (2008). Forest monkeys and Pleistocene refugia: a phylogeographic window onto the disjunct distribution of the *Chlorocebus lhoesti* species group. *Zoological Journal of the Linnean Society* **154**: 408–418.

Tutin, C. E. G., Ham, R. M., White, L. J. T., and Harrison, M. J. S. (1997). The primate community of the Lopé reserve, Gabon: diets, responses to fruit scarcity, and effects on biomass. *American Journal of Primatology* **42**: 1–24.

Van Schaik, C. P. and Pfannes, K. R. (2005). Tropical climates and phenology: a primate perspective. In: Brockman, D. K. and Van Schaik, C. P. (Eds), *Seasonality in Primates: Studies of Living and Extinct Human and Non-human Primates*. New York: Cambridge University Press.

Van Schaik, C. P., Terborgh, J. W., and Wright, S. J. (1993). The phenology of tropical forests: adaptive significance and consequences for primary consumers. *Annual Review of Ecology and Systematics* **24**: 353–377.

Van Woerden, J. T., Willems, E. P., Van Schaik, C. P., and Isler, K. (2012). Large brains buffer energetic effects of seasonal habitats in catarrhine primates. *Evolution* **66**: 191–199.

Vecchi, G. A. and Soden, B. J. (2007). Global warming and the weakening of the tropical circulation. *Journal of Climate* **20**: 4316–4340.

Visser, M. E. (2008). Keeping up with a warming world; assessing the rate of adaptation to climate change. *Proceedings of the Royal Society B-Biological Sciences* **275**: 649–659.

Walther, G.-R. (2010). Community and ecosystem responses to recent climate change. *Philosophical Transactions of the Royal Society B: Biological Sciences* **365**: 2019–2024.

Wiederholt, R. and Post, E. (2010). Tropical warming and the dynamics of endangered primates. *Biology Letters* **6**: 257–260.

Wiederholt, R. and Post, E. (2011). Birth seasonality and offspring production in threatened neotropical primates related to climate. *Global Change Biology* **17**: 3035–3045.

Williams, S. E., Shoo, L. P., Isaac, J. L., Hoffmann, A. A., and Langham, G. (2008). Towards an integrated framework for assessing the vulnerability of species to climate change. *PLoS Biology* **6**: e325.

Wong, H., Li, R., Xu, M., and Long, Y. (2013). An integrative approach to assessing the potential impacts of climate change on the Yunnan snub-nosed monkey. *Biological Conservation* **158**: 401–409.

Wright, P. (2007). Considering climate change effects in lemur ecology and conservation. In: Gould, L. and Sauther, M. (Eds), *Lemurs*. Springer US.

Wright, P. C. (1999). Lemur traits and Madagascar ecology: coping with an island environment. *American Journal of Physical Anthropology*, Supplement **29**: 31–72.

Zelazowski, P., Malhi, Y., Huntingford, C., Sitch, S., and Fisher, J. B. (2010). Changes in the potential distribution of humid tropical forests on a warmer planet. *Philosophical Transactions of the Royal Society of London A: Mathematical, Physical and Engineering Sciences* **369**: 137–160.

Zvereva, E. L. and Kozlov, M. V. (2006). Consequences of simultaneous elevation of carbon dioxide and temperature for plant-herbivore interactions: a metaanalysis. *Global Change Biology* **12**: 27–41.

CHAPTER 12

Are protected areas conserving primate habitat in Indonesia?

D.L.A. Gaveau, Serge A. Wich, and Andrew J. Marshall

Deforestation in Bukit Barisan Selatan National Park, Sumatra. Photo copyright: D.L.A. Gaveau.

12.1 Introduction

The International Union for Conservation of Nature defines terrestrial protected areas as 'parcels of land dedicated to the preservation and maintenance of biological diversity and natural and associated cultural resources that are managed through legal or other means' (IUCN 2008). These modern-day protected areas are classified in different categories (I–VI) depending on their level of protection and objectives (IUCN 2008). Nature reserves (Ia) and wilderness areas (Ib) are areas set aside to conserve biological diversity and wild habitats; human presence is strictly controlled and limited in them. National parks (II) protect large natural ecosystems. Human presence is allowed for educational or recreational purposes in national parks. Categories III and IV protect specific natural monuments and specific endangered species' habitats, respectively. Category V represents areas where the interaction of people and nature has created an area of distinct character, for example hunting parks. Finally, Category VI protected areas allow some levels of natural resource extraction by humans.

While setting aside natural areas to preserve their utilitarian and intrinsic value has been part of the human endeavour for millennia (Borgerhoff-Mulder and Coppolillo 2005), national parks and nature reserves began in the modern era with the

Gaveau, D., Wich, S.A. and Marshall, A.J., *Are protected areas conserving primate habitat in Indonesia?* In: *An Introduction to Primate Conservation*. Edited by: Serge A. Wich and Andrew J. Marshall, Oxford University Press (2016). © Oxford University Press.
DOI 10.1093/acprof:oso/9780198703389.003.0012

establishment of Yellowstone National Park in the United States in 1872.

By the turn of the twenty-first century, national governments around the world had designated more than 122,000 protected areas, covering an estimated 16.94 million km^2 of land or 12.9% of the Earth's land surface (Jenkins and Joppa 2009). Myriad provincial and municipal protected areas, community conserved areas, private reserves, and other formal and informal protected areas complement these nationally designated ones (Berkes 2009). Many conservationists see an effective global network of protected areas as the best hope to preserve representative areas of natural ecosystems and the habitats and species they contain (Le Saout *et al.* 2013; Kramer *et al.* 1997). Primate conservationists generally share this view, arguing that terrestrial protected areas in the tropics are a crucial component of efforts to preserve viable primate populations in the wild (Cowlishaw and Dunbar 2000; Wright 1992).

In this chapter, we examine whether protected areas conserve old-growth forests, arboreal primates' favoured habitats. We examine this question for Indonesia. This country is a high priority centre for primate conservation (Myers *et al.* 2000), ranking third in primate species diversity with 56 species, after Brazil (116 species) and Madagascar (98 species), and second in endemism, with 68% of its species and 77% of its taxa being endemic (cf. Madagascar, with 100% species endemism and 100% taxa endemism, and Brazil with 53% species endemism and 60% taxa endemism) (Mittermeier *et al.* 2013). Indonesia's rainforests are also home to Asia's only great ape, the orang-utan (*Pongo* spp.). Therefore, an effective protected area management is crucial to preserve the country's high primate diversity.

Indonesia's network of protected areas is composed of nature reserves (Category Ia), national parks (Category II), wildlife reserves (Category IV), nature recreation parks, forest parks, and hunting parks (Category V) as well as protection forests (*hutan lindung*), areas with no formal IUCN category that are set aside to sustain ecosystem services, such as the hydrological functions of watersheds. In this chapter we review the major current threats to Indonesian terrestrial protected areas. We describe the challenges that must be overcome if these threats are to be effectively countered, and propose combining protected areas with natural forest timber concessions to sustain larger forest landscapes than would be possible via protected areas alone.

12.2 History of Protected Area establishment in Indonesia

In 1916, the colonial government of the Dutch East Indies introduced the Natural Monuments Ordinance, which led to the creation of Indonesia's first 43 nature reserves in the following decade. By 1940, a network of 101 nature monuments and 35 wildlife sanctuaries had been created. After the Proclamation of Indonesian Independence in 1945, the Indonesian government began logging its vast forest resources on an industrial scale, following the post-colonial development agenda typical of Western industrialized nations (Pretzsch 2005). During this time, logging concessions were sometimes granted within reserves, and reserves were often downsized. However, in the early 1980s the Indonesian government renewed its interest in the protection of its rich biological heritage by promulgating Indonesia's first Environmental Management Act (EMA) (Law of 1982, State Gazette 1982, no 3215). The UNDP/FAO National Parks Development Project (De Wulf *et al.* 1981), and the third World Parks Congress (WPC) held in Bali in 1982 (Sakumoto 1999), were also implemented. These three events strengthened conservation policies in Indonesia and laid the foundation for the country's current reserve system. The Indonesian government expanded the existing network of protected areas. Indonesia further signed the Convention on Biological Diversity in 1993 and was among the first countries to prepare and implement a National Biodiversity Strategy Action Plan, paving the way for financing protected areas and further expanding the protected area network. In a major expansion in 2004, nine new national parks were gazetted, comprising over 1.3 million ha. Today, over 15% of Indonesia's landmass is formally protected, with 50 national parks and 527 reserves (nature reserves, wildlife reserves, nature recreation parks, forest parks, and hunting parks). This is somewhat higher than the global national average of 12.9%.

Despite these laudable conservation commitments, losses of Indonesia's lowland old-growth forests, the preferred habitat of most primates, remain acute today. In 2012, Indonesia surpassed Brazil in deforestation rates, with an estimated 850,000 ha cleared in that year (Margono *et al.* 2014), mainly in the lowlands. Indonesian protected areas have achieved little in preserving lowland habitats because they have disproportionately been established in the highlands (Gaveau *et al.* 2009a, b). This trend in protected area establishment, where primate-rich lowland forests are not targeted for conservation, is not unique to Indonesia. Globally, protected areas tend to be placed 'high and far' (Joppa and Pfaff 2009) because conserving lowland forests lacks support among political elites when faced with the absence of payments for conservation, and when lowland forests offer lucrative near-term opportunities to engage in high-revenue extractive industries, such as oil-palm and monoculture timber plantations (Fitzherbert *et al.* 2008; Venter *et al.* 2009). We note that, although protected areas have arguably been established in less suitable primate habitat (the highlands), climate change may over time render upland protected areas more suitable for primates as species will move upslope to escape global warming (Struebig *et al.* 2015a, b). Indonesian protected areas also lack the effective law enforcement necessary to prevent agricultural encroachment, stop the expansion of road networks, and prevent illegal logging, poaching, and forest fires. There is ample evidence that deforestation, illegal logging, poaching, and fire continue to degrade protected areas in Indonesia (Brandon *et al.* 1998; Curran *et al.* 2004; Gaveau *et al.* 2009b; Leverington *et al.* 2010; Naughton-Treves *et al.* 2006; van Schaik *et al.* 1997; Verissimo *et al.* 2011; Soares-Filho *et al.* 2010; Brun *et al.* 2015). Strict nature reserves (Category Ia) appear to have become particularly exposed to encroachments during the period 2000–2010 (Brun *et al.* 2015). This damage is not unique to Indonesia. Despite their widely acknowledged importance for biodiversity conservation, tropical terrestrial protected areas have become increasingly degraded and fragmented by human farming, hunting, timber extraction, and settlements (Colwell *et al.* 1997; Curran *et al.* 2004; DeFries *et al.* 2005; Southworth *et al.* 2006; Struhsaker *et al.* 2005; Laurance *et al.* 2006; Stoner *et al.* 2007; Leverington *et al.* 2010). Indigenous reserves in Brazil appear to perform better than state-controlled protected area in reducing encroachment and fire (Adeney *et al.* 2009; Nepstad *et al.* 2006). Indonesia does not recognize indigenous reserves, although the Indonesian government has recently begun a process that might change this (Meijaard 2015).

12.3 Major threats to forest cover in Indonesian protected areas

Here we briefly discuss major threats to protected areas in Indonesia. We focus on threats that reduce forest cover or degrade its structure and quality. Although we do not discuss threats that affect the ecological health and functioning of forest ecosystems (e.g. poaching, invasive species), we recognize that they can also have negative effects on primate populations residing within parks.

12.3.1 Agricultural encroachment and illegal logging

Encroachment by independent farmers is the main cause of illegal forest clearing in Indonesian protected areas (Gaveau *et al.* 2012). Agro-industrial companies (e.g. oil-palm and pulp and paper conglomerates) also clear vast areas of forest, but mainly outside protected areas, in unprotected lowlands. Indonesia's human population is growing rapidly and remains largely rural. Independent farmers are seeking land deeper into forest frontiers, where protected areas are located, because prime agricultural land in less remote unprotected areas has become ever scarcer. Much of the lowland is now occupied by large plantations of oil-palm and monoculture timber for the pulp and paper industry. Independent farmers, often migrants, are encouraged to utilize land inside protected areas by local officials and land speculators, many of whom have little interest in biodiversity conservation because protected areas provide few economic returns locally (Levang *et al.* 2007). For example, Bukit Barisan Selatan National Park in southern Sumatra is suffering from massive forest conversion to small (1–2 ha) coffee plantations (Gaveau *et al.* 2009b, 2007) (Figure 12.1). Levang *et al.* (2012) estimated

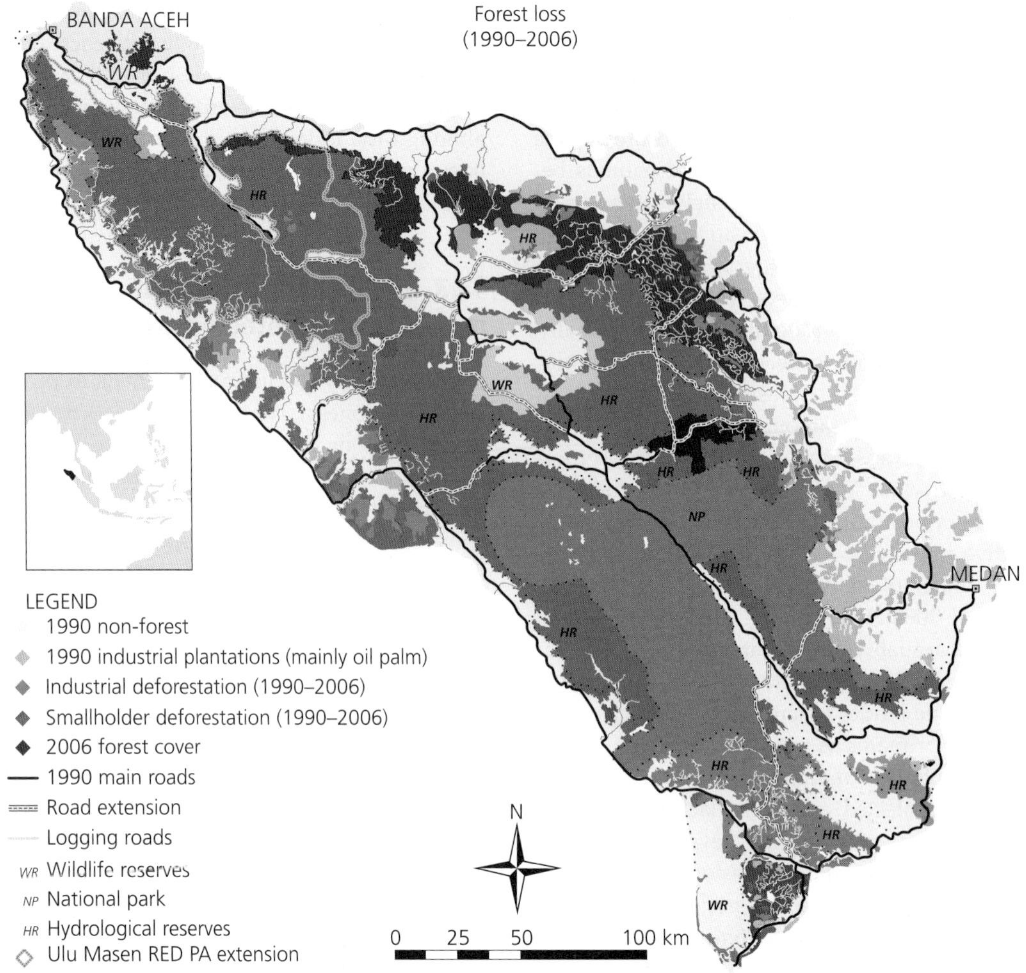

Figure 12.1 Patterns of forest loss and regrowth since 1972 in and around the Bukit Barisan Selatan National Park (BBSNP), including the Gunung Raya Wildlife Sanctuary (GRWS) and Hydrological Reserves (HR). The data are overlaid on a Digital Elevation Model and LANDSAT ETM + satellite imagery (year 2002) with bands 3 (red), 4 (NIR), and 5 (SWIR) combined to generate a false colour background. This figure was originally published in (Gaveau *et al.* 2007). Reproduced with permission from Elsevier.

that more than 100,000 individuals farmed coffee illegally inside this UNESCO-listed park. This park has lost more than 63,700 ha of species-rich *Dipterocarp* forest since 1972, representing a 21% loss in overall forested area (Gaveau *et al.* 2009b). An estimated 20,000 metric tons of *robusta* coffee are produced annually inside this park, representing circa 4% of Indonesia's annual production (WWF 2007a). Most farmers are aware that clearing land within the park is prohibited (Levang *et al.* 2012), but the lack of law enforcement and the knowledge that the practice is encouraged by local elites encourages farmers to believe that they can act with impunity. Heads of villages and wealthy businessmen that are well connected to high ranking officials at the district and provincial level provide key organization skills and financial backing, making encroachment both lucrative and largely risk free (Levang *et al.* 2012). Similarly, Gunung Palung National Park in West Kalimantan was the site of widespread illegal logging during the early 2000s, following an era of breakdown in law and order (Curran *et al.* 2004), although here logging did not cause deforestation, as much of the forest recovered quickly afterwards. In 2013, Tesso Nilo National Park in Riau Province, Sumatra, suffered massive encroachments associated with forest fires. There, migrant communities cultivate oil-palm (Ekadinata *et al.* 2013), with backing from businesses and local authorities. Similar encroachment has occurred and continues to occur in many protected areas across Indonesia, reducing the forest cover inside protected areas and decreasing the available habitat for forest dwelling primates (Wich *et al.* 2008).

Encroachment can also increase the frequency and intensity of conflicts between humans and primates. As primate habitat is decreased, primates are increasingly driven to raid agricultural crops and come into direct contact with humans, which can lead to the persecution and death of primates in these areas (Strum 2010; Campbell-Smith *et al.* 2010, 2011).

12.3.2 Expansion of road networks into protected areas

The threat of agricultural encroachment is exacerbated by the expansion of road networks. The Indonesian government has developed its transportation infrastructure to modernize its economy and to move mass agricultural and timber products from remote areas to their market destinations, a pattern paralleled in Brazil (Ewers *et al.* 2008) and other parts of the tropics (Caro *et al.* 2014; Laurance *et al.* 2009). The expansion of Indonesia's road network has included building inside some its best-known protected areas, which permits the movement of land-seeking farmers into previously inaccessible areas. For example, Northern Sumatra (65,000 km^2), which includes the UNESCO-listed Gunung Leuser National Park and holds around 92% of the world's remaining 6,600 Sumatran orang-utans (Meijaard and Wich 2007; Wich *et al.* 2008), is an area threatened by extensive road developments, collectively dubbed the 'Ladia Galaska' Project (Figure 12.2). As the planned roads are currently being built and paved, conservative estimates suggest that at least 1,384 orang-utans, 25% of the remaining world population, would disappear because of habitat loss, while those remaining would face increased threats from encroachment, conflicts with humans, and extensive habitat fragmentation (Gaveau *et al.* 2009c). These predicted orang-utan population declines are largely due to losses of forest cover in carbon and species-rich Dipterocarp lowland forests (<500 m a.s.l.), which support high densities of orang-utans (van Schaik *et al.* 1995; Wich *et al.* 2004a) and are also targeted by extractive industries. This situation is not unique to Indonesia. Southeast Asia as a whole experiences rapid road development through protected areas (Clements *et al.* 2014). For example, a provincial road was built in 2001 through the Snoul Wildlife Sanctuary, a protected area managed by the Cambodian Ministry of Environment, which triggered unprecedented habitat destruction (Clements *et al.* 2014).

12.3.3 Forest fires

Forest fires are another major source of primate habitat destruction in Indonesian protected areas. Fires are typically lit every year on fallow land during the regular dry season to make way for agricultural plantations. However, their impacts are heightened during years of anomalously low rainfall (Field *et al.* 2009). Drought years in Indonesia occur when cold sea surface temperatures surround Indonesia and warm waters develop in the eastern Pacific Ocean (El Niño Southern Oscillation, ENSO) and in the western Indian Ocean (Positive Indian Ocean Dipole, IOD) (Hendon 2003). During drought years, fires can burn uncontrollably and destroy vast expanses of forest. The fires of 1982–83 and 1997–98, the years that saw the strongest recorded ENSO and IOD, destroyed approximately 10 million ha of lowland forest on the islands of Borneo and Sumatra (Goldammer and Seibert 1990; Yeager *et al.* 2003; Murdiyarso and Adiningsih 2007; Tacconi *et al.* 2007). These were the largest forest fires in the last two centuries of recorded history. They almost completely destroyed the forests of Kutai National Park (100,000 ha), Bukit Suharto Conservation Park (61,850 ha), and Sungai Wain Protection Forest (9,000 ha) in Eastern Kalimantan (Leighton and Wirawan 1986; MacKinnon *et al.* 1994), all areas containing orang-utans (Leighton and Wirawan

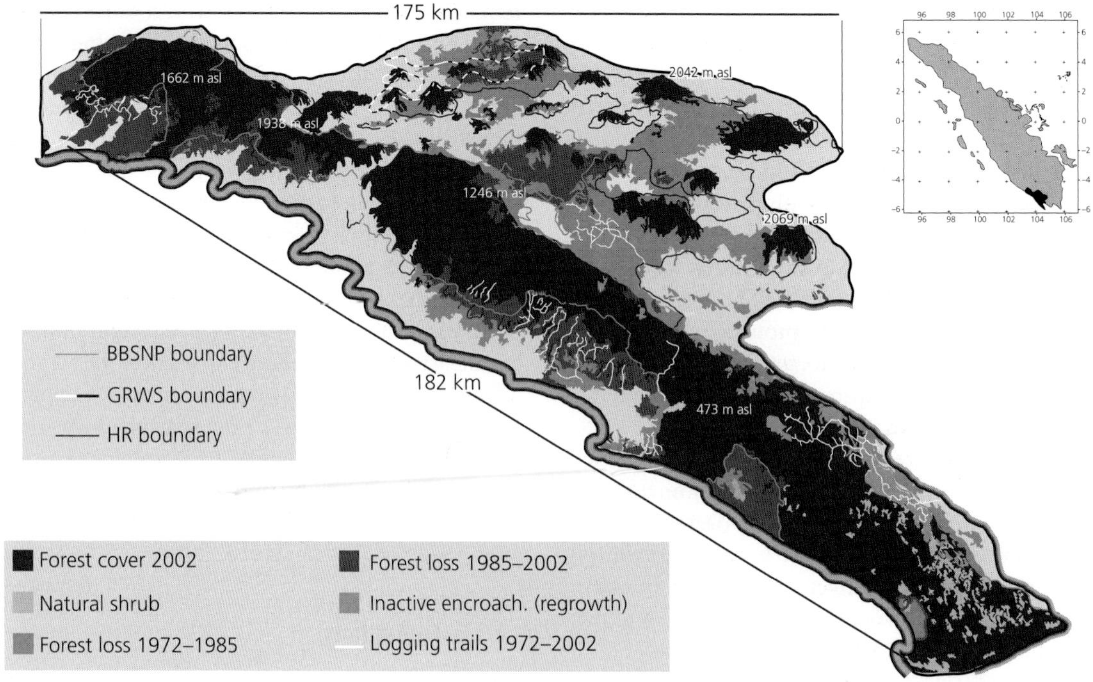

Figure 12.2 Map of deforestation (1990–2006), logging roads, and remaining 2006 forest cover in northern Sumatra derived using LANDSAT satellite imagery with hill shade overlay to reveal the terrain topography. Road extension refers to roads under construction or planned in and out Gunung Leuser National Park. The insert shows the location of the study area in Southeast Asia. This figure was originally published in Gaveau *et al.* (2009c). Reproduced with permission from IOP Publishing.

1986; Goldammer and Siebert 1990; Goldammer 1999; Yeager *et al.* 2003), and the forests of Berbak National Park in Jambi Province, in Southern Sumatra.

12.4 Why are Indonesian protected areas so vulnerable to encroachment and fire?

One reason Indonesian national parks have suffered heavy losses due to agricultural encroachment and fire in recent years is insufficient funding. In 2006, Indonesia's terrestrial protected areas received an average USD1.56/ha in government funding and an estimated USD0.67/ha in funding from non-governmental organizations and international donor agencies (McQuistan *et al.* 2006). This is considerably lower than the average USD13/ha spent on protected area management in countries in the Asia-Pacific Region (WWF 2007b; Emerton 2001). The shortfall in Indonesia's protected area funding—that is, the funds needed to fulfil their legal mandate to protect natural forest—was estimated at USD81.94 million for 2006. This, in turn, makes it extremely difficult to implement management interventions that create disincentives for inappropriate use of resources, for example through law enforcement (Smith and Walpole 2005). The importance of law enforcement in maintaining the integrity of protected areas is well recognized. For example, the strong emphasis placed on law enforcement in the early 1980s following the creation of Indonesia's first national parks greatly reduced deforestation and promoted extensive forest re-growth in Bukit Barisan Selatan National Park (Gaveau *et al.* 2009b) (Figure 12.1). Since then, however, the efficacy of law enforcement has diminished greatly. Indonesian park guards are currently underpaid and ill-equipped and are therefore unable to carry out protracted law enforcement operations. The low pay also results in them being more susceptible to corruption and more willing to turn a blind eye to illegal activities (Terborgh *et al.* 2002).

Equally, attempts to reduce the effects of forest fires using positive incentives, including those implemented through integrated conservation and development projects (Linkie *et al.* 2008), have failed to offer appropriately structured long-term economic incentives nationally and locally to conserve forests (Hutton and Leader-Williams 2003). For example, following a shift in conservation strategies in the late 1980s from a focus on law enforcement to the widespread application of conservation-with-development projects, a USD19 million Integrated Conservation and Development Project was implemented in Kerinci Seblat National Park in the 1990s to reduce threats to the park by helping the economic development of villages bordering it. This project had a focus on alleviating poverty in villages around the national park, and aimed to involve local people through joint decision-making and benefit sharing. This approach failed in its aim of preventing forest loss, perhaps partly due to factors such as the size and complexity of the project, the nature of the funding, and bureaucratic constraints (Linkie *et al.* 2008). In addition, this project took place during a period of decentralization of the authority for managing natural resources, which led to an Indonesia-wide breakdown in law and order and increased illegal logging (Curran *et al.* 2004). Although decentralization was at least partly instituted with the goal of strengthening local civil society and thereby decreasing corruption, it often resulted merely in the decentralization of corruption and an increase in the power of local elites who were often driven largely by self-interest (Levang *et al.* 2012). Local elections have thus become much more important in determining the fate of natural resources and protected areas. While not all local elites are corrupt, this system of decentralization has established perverse incentive systems in which the short-term financial interests of local managers and politicians (and the wealthy families that back them) are pitted against the long-term preservation of biodiversity. Under such systems, it is hardly surprising that politicians seeking power (and the potential for personal enrichment) gain the support of huge numbers of squatters through promises of open access to national parks (Levang *et al.* 2012).

12.5 The way forward

Given economic demands, social pressure on land, and the high cost of forest protection (Carwardine *et al.* 2008; Wilson *et al.* 2010), protected areas in Indonesia are unlikely to ever constitute more than a minor part of the tropical landscape, particularly in lowland areas (Joppa *et al.* 2009; Rodrigues *et al.* 2004; Sloan *et al.* 2012). Some conservation scientists propose combining protected areas with natural forest timber concessions to sustain larger forest landscapes than otherwise possible via protected areas alone (Billand and Nasi 2008; Clark *et al.* 2009; Dickinson *et al.* 1996; Edwards and Laurance 2013; Fisher *et al.* 2011; Meijaard and Sheil 2007; Wilson *et al.* 2010; Gaveau *et al.* 2013). Timber concessions typically generate higher per hectare revenues than neighbouring protected areas. Timber harvesting in natural forests provides one way in which forest lands can provide income and employment while retaining forest: in simple terms, the forest can pay for its own protection—arguably making it easier to gain political and public support for conservation.

Natural forest timber concessions are parcels of natural forest leased out to companies or to communities to harvest timber on a long-term basis. When natural forest timber concessions are adjacent to protected areas, they bring an opportunity to maintain larger and better connected forest landscapes with a greater capacity to maintain low-density, large-range, and high-mobility species (Meijaard and Sheil 2008; Marshall *et al.* 2006). Indeed, natural forest timber concessions are *de facto* a kind of protected area in Indonesia. Conversion of natural forests to plantations in timber concessions is prohibited and concession managers are legally obliged to maintain permanent natural forest cover (Dickinson *et al.* 1996). In addition, laws mandate selective timber harvesting in concessions (Putz *et al.* 2012), whereby only commercially valuable trees above a certain diameter are cut and the remaining trees are left standing to facilitate long-term regeneration. In Indonesia, between two and twenty stems per hectare are typically removed once every few decades (Sist *et al.* 1998a, b). Generally, this leaves more than 90% of the trees standing and remaining vegetation recognizably constitutes a forest. Not only does selective logging maintain a forest structure, logged forests can still be extremely valuable habitats for orang-utans and other species (Ancrenaz *et al.* 2010; Berry *et al.* 2010; Meijaard and Sheil 2008; Wich *et al.* 2012).Thus, a logged tropical forest remains a biologically rich forest (Meijaard and Sheil 2007). We caution, however, that encroachment by independent farmers is as much of a problem in natural forest timber concessions as it is in protected areas (Gaveau *et al.* 2013).

The creation of the 568,700 ha Sabangau National Park in 2004, an area logged throughout the 1990s, but containing the largest contiguous orang-utan population on Borneo, indicates that government of Indonesia is beginning to recognize the value of logged forests for primate conservation. Studies also indicate that on Borneo carefully managed timber concessions can maintain stable orang-utan populations (Ancrenaz *et al.* 2010). Of course this depends on the intensity of logging, and heavily logged areas usually have lower orang-utan densities (Husson *et al.* 2009). Primates' response to logging varies from species to species, with some showing no decline in density or even an increase in logged areas and some showing large declines (Johns and Skorupa 1987; Cowlishaw and Dunbar 2000; Meijaard *et al.* 2005).

We suggest that the government of Indonesia designate its natural forest timber concessions as protected areas under the IUCN Protected Area Category VI (IUCN 2008), because they perform as effectively as protected areas in maintaining forest cover and such a designation would prevent them from being reclassified to other land uses that would not contribute as meaningfully to conservation. The World Database of Protected Areas contains many examples of permanent forest reserves where hardwood extraction is one of the permitted activities. For example, adding Kalimantan's natural forest timber concessions to the protected area network would increase the permanently protected forest in Kalimantan, that is, the area of production forest that legally should remain forested, by 248,305 km^2, which would increase the proportion of the current Bornean orang-utan habitat under protection from currently 22% to 51% (Wich *et al.* 2012). Such a change in policy would be radical, and would require a shift in mindset among concession holders, government policy-makers, and conservation groups—particularly because current government policy does not guarantee timber concessions permanent status as natural forest. Nevertheless, making such a political decision and effectively implementing the policy would have long-term benefits for wildlife and the maintenance of forest ecosystem services, while continuing the generation of income from forests. Such changes are required to achieve sustainable forestry practices, which has long been the stated goal of the Ministry of Forestry. A permanent and inviolate forest estate would certainly also have value under a future of Reducing Emissions from Deforestation and Degradation (REDD) programmes in which Indonesia receives payments for reduced forest loss and damage (Garcia-Ulloa and Koh, Chapter 16, this volume). Areas where primate species such as the Sumatran orang-utan occur often contain high carbon stocks and also provide other ecosystem services, so a system in which there are payments for such services could prove very valuable for primate conservation (Wich *et al.* 2012).

Acknowledgements

We are grateful to Stuart Pimm and Karen Strier for helpful comments that improved our chapter.

References

Adeney, J. M., Christensen Jr., N. L., and Pimm, S. L. (2009). Reserves protect against deforestation fires in the Amazon. *PLoS One* **4**: e5014.

Ancrenaz, M., Ambu, L., Sunjoto, I., Ahmad, E., Manokaran, K., *et al.* (2010). Recent surveys in the forests of Ulu Segama Malua, Sabah, Malaysia, show that orangutans (P. p. morio) can be maintained in slightly logged forests. *PLoS One* **5**: e11510.

Berkes, F. (2009). Community conserved areas: policy issues in historic and contemporary context. *Conservation Letters* **2**: 20–25.

Berry, N. J., Phillips, O. L., Lewis, S. L., Hill, J. K., Edwards, D. P., *et al.* (2010). The high value of logged tropical forests: lessons from northern Borneo. *Biodiversity and Conservation* **19**: 985–997.

Billand, A. and Nasi, R. (2008). Production dans les forêts de conservation, conservation dans les forêts de production: vers des forêts tropicales durables, à partir du cas de l'Afrique centrale.

Borgerhoff-Mulder, M. and Coppolillo, P. (2005). *Conservation: Linking Ecology, Economics, and Culture*. Princeton, NJ: Princeton University Press.

Brandon, K., Redford, K. H., and Sanderson, S. E. (1998). *Parks in Peril: People, Politics, and Protected Areas*. Washington, DC: Island Press.

Brun, C., Cook, A. R., Lee, J. S. H., Wich, S. A., Koh, L. P., *et al.* (2015). Analysis of deforestation and protected area effectiveness in Indonesia: a comparison of Bayesian spatial models. *Global Environmental Change* **31**: 285–295.

Campbell-Smith, G., Simanjorang, H. V., Leader-Williams, N., and Linkie, M. (2010). Local attitudes and perceptions toward crop-raiding by orangutans (Pongo abelii) and other nonhuman primates in northern Sumatra, Indonesia. *American Journal of Primatology* **72**(10): 866–876.

Campbell-Smith, G., Campbell-Smith, M., Singleton, I., and Linkie, M. (2011). Apes in space: saving an imperilled orangutan population in Sumatra. *PLoS One* **6**: e17210.

Caro, T., Dobson, A., Marshall, A. J., and Peres, C. A. (2014). Compromise solutions between conservation and road building in the tropics. *Current Biology* **24**: R722–R725.

Carwardine, J., Wilson, K. A., Ceballos, G., Ehrlich, P. R., Naidoo, R., *et al.* (2008). Cost-effective priorities for global mammal conservation. *Proceedings of the National Academy of Sciences* **105**: 11446–11450.

Clark, C., Poulsen, J., Malonga, R., and Elkan Jr., P. (2009). Logging concessions can extend the conservation estate for Central African tropical forests. *Conservation Biology* **23**: 1281–1293.

Clements, G. R., Lynam, A. J., Gaveau, D., Yap, W. L., Lhota, S., *et al.* (2014). Where and how are roads endangering mammals in Southeast Asia's forests? *PloS One* **9**: e115376.

Colwell, M. A., Dubynin, A. V., Koroliuk, A. Y., and Sobolev, N. A. (1997). Russian nature reserves and conservation of biological diversity. *Natural Areas Journal* **17**: 56–68.

Cowlishaw, G. and Dunbar, R. I. (2000). *Primate Conservation Biology*. Chicago, IL: University of Chicago Press.

Curran, L. M., Trigg, S. N., Mcdonald, A. K., Astiani, D., Hardiono, Y. M., *et al.* (2004). Lowland forest loss in protected areas of Indonesian Borneo. *Science* **303**: 1000–1003.

Defries, R., Hansen, A., Newton, A. C., and Hansen, M. C. (2005). Increasing isolation of protected areas in tropical forests over the past twenty years. *Ecological Applications* **15**: 19–26.

De Wulf, R., Supomo, D., and Rauf, K. (1981). *Barisan Selatan Game Reserve: Management Plan 1982–1987*. Bogor: Food and Agriculture Organisation.

Dickinson, M., Dickinson, J., and Putz, F. (1996). Natural forest management as a conservation tool in the tropics: divergent views on possibilities and alternatives. *Commonwealth Forestry Review* **74**(4): 309–315.

Edwards, D. P. and Laurance, W. F. (2013). Biodiversity despite selective logging. *Science* **339**: 646.

Ekadinata, S., Van Noordwijk, M., Budidarsono, S., and Dewi, S. (2013). Hotspots in Riau, Haze in Singapore: The June 2013 Event Analyzed. ASB Policy Brief No 33.

Emerton, L. (2001). What are Africa's forests worth? *Ecoforum* **24**: 7.

Ewers, R. M., Laurance, W. F., and Souza, C. M. (2008). Temporal fluctuations in Amazionian deforestation rates. *Environmental Conservation* **35**: 303–310.

Field, R. D., Van Der Werf, G. R., and Shen, S. S. (2009). Human amplification of drought-induced biomass burning in Indonesia since 1960. *Nature Geoscience* **2**: 185–188.

Fisher, B., Edwards, D. P., Larsen, T. H., Ansell, F. A., Hsu, W. W., *et al.* (2011). Cost-effective conservation: calculating biodiversity and logging trade-offs in Southeast Asia. *Conservation Letters* **4**: 443–450.

Fitzherbert, E. B., Struebig, M. J., Morel, A., Danielsen, F., Bruhl, C. A., *et al.* (2008). How will oil palm expansion affect biodiversity? *Trends in Ecology and Evolution* **23**: 538–545.

Gaveau, D., Curran, L., Paoli, G., Carlson, K., Wells, P., *et al.* (2012). Examining protected area effectiveness in Sumatra: importance of regulations governing unprotected lands. *Conservation Letters* **5**: 142–148.

Gaveau, D. L., Kshatriya, M., Sheil, D., Sloan, S., Molidena, E., *et al.* (2013). Reconciling forest conservation and logging in Indonesian Borneo. *PloS One* **8**: e69887.

Gaveau, D. L. A., Epting, J., Lyne, O., Linkie, M., Kumara, I., *et al.* (2009a). Evaluating whether protected areas reduce tropical deforestation in Sumatra. *Journal of Biogeography* **36**: 2165–2175.

Gaveau, D. L. A., Linkie, M., Suyadi, S., Levang, P., and Leader-Williams, N. (2009b). Three decades of deforestation in southwest Sumatra: effects of coffee prices, law enforcement and rural poverty. *Biological Conservation* **142**: 597–605.

Gaveau, D. L. A., Wandono, H., and Setiabudi, F. (2007). Three decades of deforestation in southwest Sumatra: have protected areas halted forest loss and logging, and promoted re-growth? *Biological Conservation* **134**: 495–504.

Gaveau, D. L. A., Wich, S., Epting, J., Juhn, D., Kanninen, M., *et al.* (2009c). The future of forests and orangutans

(Pongo abelii) in Sumatra: predicting impacts of oil palm plantations, road construction, and mechanisms for reducing carbon emissions from deforestation. *Environmental Research Letters* **4**: 11.

Goldammer, J. G. (1999). Environmental problems arising from land use, climate variability, fire and smog in Indonesia: development of policies and strategies for land use and fire management. In: Carmichael, G. R. (Ed.), *WMO Workshop on Regional Transboundary Smoke and Haze in Southeast Asia*, Singapore, 2–5 June 1998, Vol. 2. pp. 13–88. World Meteorological Organization, Global Atmosphere Watch Report Series No. 131, WMO TD No. 948, Geneva.

Goldammer, J. G. and Seibert, B. (1990). The impact of droughts and forest fires on tropical lowland rain forest of East Kalimantan. In: J. G. Goldammer (Ed.), *Fire in the Tropical Biota: Ecosystem Processes and Global Challenges*. Berlin: Springer.

Hendon, H. H. (2003). Indonesian rainfall variability: impacts of ENSO and local air–sea interaction. *Journal of Climate* 16: 1775–1790.

Husson, S. J., Wich, S. A., Marshall, A. J., Dennis, R. D., Ancrenaz, M., *et al.* (2009). Orangutan distribution, density, abundance and impacts of disturbance. In: Wich, S. A, Utami Atmoko, S. S., Setia, T. M., and van Schaik, C. P. (Eds), *Orangutans: Geographic Variation in Behavioral Ecology and Conservation*. pp. 77–96. Oxford: Oxford University Press.

Hutton, J. M. and Leader-Williams, N. (2003). Sustainable use and incentive-driven conservation: realigning human and conservation interests. *Oryx* **37**: 215–226.

IUCN (2008). Definition of protected areas [Online]. International Union for Conservation of Nature. Available at: <https://www.iucn.org/about/work/programmes/gpap_home/pas_gpap/>. [Accessed November 2015].

Jenkins, C. N. and Joppa, L. (2009). Expansion of the global terrestrial protected area system. *Biological Conservation* **142**: 2166–2174.

Johns, A. D. and Skorupa, J. P. (1987). Responses of rainforest primates to habitat disturbance: a review. *International Journal of Primatology* **8**(2): 157–191.

Joppa, L. N., Loarie, S. R., and Pimm, S. L. (2009). On population growth near protected areas. *PloS One* **4**: e4279. doi:10.1371/journal.pone.0004279.

Joppa, L. N. and Pfaff, A. (2009). High and far: biases in the location of protected areas. *PLoS One* **4**: 1–6.

Kramer, R. A., Schaik, C. V., and Johnson, J. (1997). *Last Stand: Protected Areas and the Defense of Tropical Biodiversity*. New York and Oxford: Oxford University Press.

Laurance, W. F., Croes, B., Tchignoumba, L., Lahm, S. A., Alonso, A., *et al.* (2006). Impacts of roads and hunting on Central African rainforest mammals. *Conservation Biology* **20**: 1251–1261.

Laurance, W. F., Goosem, M., and Laurance, S. G. (2009). Impacts of roads and linear clearings on tropical forests. *Trends in Ecology and Evolution* **24**: 659–669.

Leighton, M. and Wirawan, N. (1986). Catastrophic drought and fire in Borneo tropical rain forest associated with the 1982–1983 El Nino Southern Oscillation event. In: Prance, G. T. (Ed.), *Tropical Rain Forests and the World Atmosphere*. Colarado: Westview Press.

Le Saout, S., Hoffmann, M., Shi, Y., Hughes, A., Bernard, C., *et al.* (2013). Protected areas and effective biodiversity conservation. *Science* **342**: 803–805.

Levang, P., Sitorus, S., Gaveau, D., and Sunderland, T. (2012). Landless farmers, sly opportunists, and manipulated voters: the squatters of the Bukit Barisan Selatan National Park (Indonesia). *Conservation and Society* **10**: 243.

Levang, P., Sitorus, S., Gaveau, D. L. A., and Abidin, Z. (2007). *Elite's Perceptions about the Bukit Barisan Selatan National Park*. Bogor: Centre for International Forestry Research.

Leverington, F., Costa, K. L., Pavese, H., Lisle, A., and Hockings, M. (2010). A global analysis of protected area management effectiveness. *Environmental Management* **46**: 685–698.

Linkie, M., Smith, R. J., Zhu, Y., Martyr, D. J., Suedmeyer, E., *et al.* (2008). Evaluating biodiversity conservation around a large Sumatran protected area. *Conservation Biology* **22**: 683–690.

Mackinnon, K., Irving, A., and Bachruddin, M. A. (1994). A last chance for Kutai National Park–local industry support for conservation. *Oryx* **28**: 191–198.

Margono, B. A., Potapov, P. V., Turubanova, S., Stolle, F., and Hansen, M. C. (2014). Primary forest cover loss in Indonesia over 2000–2012. *Nature Climate Change* **4**: 730–735.

Marshall, A. J., Engström, L. M., Pamungkas, B., Palapa, J., Meijaard, E., *et al.* (2006). The blowgun is mightier than the chainsaw in determining population density of Bornean orangutans (Pongo pygmaeus morio) in the forests of East Kalimantan. *Biological Conservation* **129**: 566–578.

Mcquistan, C. I., Fahmi, Z., Leisher, C., Halim, A., and Adi, S. W. (2006). Protected Area Funding in Indonesia: A Study Implemented under the Programmes of Work on Protected Areas of the Seventh Meeting of the Conference of Parties on the Convention on Biological Diversity. Jakarta, Indonesia: Ministry of Environment republic of Indonesia, The Nature Conservancy, PHPA, Departemen Kelautan dan Perikanan.

Meijaard, E. (2015). The Final Blow for Indonesia's Forests? *Jakarta Globe*. Available at: <http://thejakartaglobe.beritasatu.com/opinion/commentary-final-blow-indonesias-forests/> [Accessed November 2015].

Meijaard, E. and Sheil, D. (2007). A logged forest in Borneo is better than none at all. *Nature* **446**: 974.

Meijaard, E. and Sheil, D. (2008). The persistence and conservation of Borneo's mammals in lowland rain forests managed for timber: observations, overviews and opportunities. *Ecological Research* **23**: 21–34.

Meijaard, E. and Wich, S. A. (2007). Putting orang-utan population trends into perspective. *Current Biology* **17**: R540.

Meijaard, E., Sheil, D., Nasi, R., Augeri, D., Rosenbaum, B., et al. (2005). Life After Logging: Reconciling Wildlife Conservation and Production Forestry in Indonesian Borneo. Bogor, Indonesia: CIFOR.

Mittermeier, R. A., Wilson, D. E., and Rylands, A. B. (2013). *Handbook of the Mammals of the World: Primates*. Barcelona: Lynx Edicions.

Murdiyarso, D. and Adiningsih, E. S. (2007). Climate anomalies, Indonesian vegetation fires and terrestrial carbon emissions. *Mitigation and Adaptation Strategies for Global Change* **12**: 101–112.

Myers, N., Mittermeier, R. A., Mittermeier, C. G., Da Fonseca, G. A. B., and Kent, J. (2000). Biodiversity hotspots for conservation priorities. *Nature* **403**: 853–858.

Naughton-Treves, L., Alvarez-Berrios, N., Brandon, K., Bruner, A., Buck Holland, M., *et al.* (2006). Expanding protected areas and incorporating human resource use: a study of 15 forest parks in Ecuador and Peru. *Sustainability: Science, Practice and Policy* **2**: 1–13.

Nepstad, D., Schwartzman, S., Bamberger, B., Santilli, M., Ray, D., *et al.* (2006). Inhibition of Amazon deforestation and fire by parks and indegenous lands. *Conservation Biology* **20**: 65–73.

Pretzsch, J. (2005). Forest related rural livelihood strategies in national and global development. *Forests, Trees and Livelihoods* **15**: 115–127.

Putz, F. E., Zuidema, P. A., Synnott, T., Peña-Claros, M., Pinard, M. A., *et al.* (2012). Sustaining conservation values in selectively logged tropical forests: the attained and the attainable. *Conservation Letters* **5**(4): 245–326.

Rodrigues, A. S., Akcakaya, H. R., Andelman, S. J., Bakarr, M. I., Boitani, L., *et al.* (2004). Global gap analysis: priority regions for expanding the global protected-area network. *BioScience* **54**: 1092–1100.

Sakumoto, N. (1999). Development of environmental law and land reform in Indonesia. In: Hardjsoemantri, K. and Sakumoto, N. (Eds), *Reforming Laws and Institutions in Indonesia: An Assessment*. ASEDP n°51, IDE/JETRO.

Sist, P., Dykstra, D., and Fimbel, R. (1998a). Reduced-impact Logging Guidelines for Lowland and Hill Dipterocarp Forests in Indonesia. Occasional Paper No. 15. Bogor, Indonesia: CIFOR.

Sist, P., Nolan, T., Bertault, J.-G., and Dykstra, D. (1998b). Harvesting intensity versus sustainability in Indonesia. *Forest Ecology and Management* **108**: 251–260.

Sloan, S., Edwards, D. P., and Laurance, W. F. (2012). Does Indonesia's REDD + moratorium on new concessions spare imminently threatened forests? *Conservation Letters* **5**: 222–231.

Smith, R. J. and Walpole, M. J. (2005). Should conservationists pay more attention to corruption? *Oryx* **39**: 251–256.

Soares-Filho, B., Moutinhi, P., Nepstad, D., Anderson, A., Rodrigues, H., *et al.* (2010). Role of Brazilian Amazon protected areas in climate change mitigation. *Proceedings of the National Academy of Sciences* **107**: 10821–10826.

Southworth, J., Nagendra, H., and Munroe, D. K. (2006). Introduction to the special issue: are parks working? Exploring human-environment tradeoffs in protected area conservation. *Applied Geography* **26**: 87–95.

Stoner, C., Caro, T., Mlingwa, C., Sabuni, G., and Borner, M. (2007). Assessment of effectiveness of protection strategies in Tanzania based on a decade of survey data for large herbivores. *Conservation Biology* **21**: 635–646.

Struebig, M. J., Fischer, M., Gaveau, D. L., Meijaard, E., Wich, S. A., *et al.* (2015a). Anticipated climate and land-cover changes reveal refuge areas for Borneo's orangutans. *Global Change Biology* **21**(8): 2891–2904.

Struebig, M. J., Wilting, A., Gaveau, D. L., Meijaard, E., Smith, R. J., *et al.* (2015b). Targeted conservation to safeguard a biodiversity hotspot from climate and land-cover change. *Current Biology* **25**(3): 372–378.

Struhsaker, T. T., Struhsaker, P. J., and Siex, K. S. (2005). Conserving Africa's rain forests: problems in protected areas and possible solutions. *Biological Conservation* **123**: 45–54.

Strum, S. C. (2010). The development of primate raiding: implications for management and conservation. *International Journal of Primatology* **31**(1): 133–156.

Tacconi, L., Moore, P., and Kaimowitz, D. (2007). Fires in tropical forests–what is really the problem? Lessons from Indonesia. *Mitigation and Adaptation Strategies for Global Change* **12**: 55–66.

Terborgh, J., Van Schaik, C., Davenport, L., and Rao, M. (2002). *Making Parks Work*. Washington, DC: Island Press.

Van Schaik, C. P., Priatna, A., and Priatna, D. (1995). Population estimates and habitat preferences of orang-utans based on line transects of nests. In: Nadler, R. D., Galdikas, B. M. F., Sheeran, L. K., and Rosen, N. (Eds), *The Neglected Ape*. New York, NY: Plenum Press.

Van Schaik, C. P., Terborgh, J., and Dugelby, B. (1997). The silent crisis: the state of rain forest nature preserves. In: Van Schaik, C. and Johnson, J. (Eds), *Last Stand: Protected Areas and the Defense of Tropical Biodiversity*. New York: Oxford University Press.

Venter, O., Meijaard, E., Possingham, H., Dennis, R., Sheil, D., *et al.* (2009). Carbon payments as safeguard for threatened tropical mammals. *Conservation Letters* **2**: 123–129.

Verissimo, A., Rolla, A., Vedoveto, M., and De Furtada, S. M. (2011). Áreas protegidas na Amazonia Brasileira: avancos e desafios. Belém, Pará, Brazil: Imazon/ISA.

Wich, S. A., Buij, R., and Van Schaik, C. P. (2004a). Determinants of orang-utan density in the dryland forests of the Leuser ecosystem. *Primates* **45**: 177–182.

Wich, S. A., Gaveau, D., Abram, N., Ancrenaz, M., Baccini, A., *et al.* (2012). Understanding the impacts of land-use policies on a threatened species: is there a future for the Bornean orang-utan? *PLoS One* **7**: e49142.

Wich, S. A., Meijaard, E., Marshall, A. J., Husson, S., Ancrenaz, M., *et al.* (2008). Distribution and conservation status of the orang-utan (Pongo spp.) on Borneo and Sumatra: how many remain? *Oryx* **42**: 329–339.

Wilson, K. A., Meijaard, E., Drummond, S., Grantham, H. S., Boitani, L., *et al.* (2010). Conserving biodiversity in production landscapes. *Ecological Applications* **20**: 1721–1732.

Wright, P. C. (1992). Primate ecology, rainforest conservation, and economic development: building a national park in Madagascar. *Evolutionary Anthropology: Issues, News, and Reviews* **1**: 25–33.

WWF. (2007a). *Gone in an instant* [Online]. World Wide Fund For Nature. Available at: < http://wwf.panda.org/about_our_earth/all_publications/?92080/Gone-in-an-instant How-the-trade-in-illegally-grown-coffee-is-driving-the-destruction-of-rhino-tiger-and-elepant-habitat>. [Accessed December 2015].

WWF (2007b). Tracking progress in managing protected areas around the world. An analysis of two applications of the Management Effectiveness Tracking Tool developed by WWF and the World Bank. Gland, Switzerland: WWF International.

Yeager, C. P., Marshall, A. J., Stickler, C. M., and Chapman, C. A. (2003). Effects of fires on peat swamp and lowland dipterocarp forests in Kalimantan, Indonesia. *Tropical Biodiversity* **8**: 121–138.

CHAPTER 13

The role of multifunctional landscapes in primate conservation

Erik Meijaard

Forest and agricultural landscape in North Sumatra, Indonesia. Photo copyright: Conservationdrones.org

13.1 Our future environment

What will our world look like in 50 years from now, or 1,000? It is nearly impossible to even vaguely picture a space view of our globe in, let's say, the year 3000. Glum conservationists or science-fiction writers sometimes depict a dried-out, desert-like, possibly highly urbanized environment, with little green and nearly no wildlife (Figure 13.1). The post-apocalyptic novel, *The Road* (McCarthy 2006), is a good example, with the planet having been reduced to a dusty environment with dead trees and no wildlife apart from groups of cannibalizing humans. Others foresee global ecosystems undergoing major transitions after passing critical tipping points (Barnosky *et al.* 2012). The Easter Island saga is an example of environmental overexploitation apparently contributing to societal collapse (Diamond 2005). While the possibility of such grim outcomes cannot be excluded, environmental changes in the future could also be a more benign reflection of what happened in the past. Much will depend on the ability and willingness of human society to find an effective way to overcome 'the tragedy of the commons' (Hardin 1968), maintain functioning natural environments, and the ecological processes and ultimately the human survival they support.

Like any other species on Earth, humans have influenced and changed their environment and the species with which they have co-existed since they first evolved in southern Africa some 200,000 years ago (Henn *et al.* 2011). Early people hunted and

Meijaard, E., *The role of multifunctional landscapes in primate conservation.* In: *An Introduction to Primate Conservation.*
Edited by: Serge A. Wich and Andrew J. Marshall, Oxford University Press (2016). © Oxford University Press.
DOI 10.1093/acprof:oso/9780198703389.003.0013

Figure 13.1 An orang-utan out of its forest comfort zone. The survival of species like this will depend at least to some extent on how they can adapt to living in multifunctional landscapes. Photo by Serge A. Wich . Reproduced with permission from John Wiley & Sons.

scavenged and are thought to have had a major impact on the megafaunas of most parts of the world (Johnson 2002). Even though natural fires shaped ecosystems for many millions of years, people's use of fire to make the Earth more suited to their lifestyle initiated a new stage, ecologically very different from what was there before (Barbetti 1986; Pausas and Keeley 2009), and driving extinction of species that could not adapt to these new landscapes (Miller *et al.* 2005). Since the start of early agricultural experimentation some 10,000 years ago, people have increased their impact on natural landscapes. Clearance for cropland or permanent pasture has reduced the extent of natural habitats on agriculturally usable land by more than 50% (Green *et al.* 2005), and agriculture is a major cause of land-cover change. Many countries appear to have gone through similar stages of initially rapid deforestation, then a slowing down of these deforestation rates, and ultimately an increase in forest cover. This process is called the forest transition curve. It coincides with the concentration over time of agricultural production in smaller areas of better land, and the agricultural abandonment of larger areas of poorer land, which are then available for reforestation through natural regeneration or planting (Aastrup 2000). So, at least in theory, it is possible that our planet will not end up looking like a wet version of the moon, but rather like a vast, multifunctional landscape with areas of forest and other natural habitats intermingling with areas of more intensive use, and ideally with islands of fully protected 'nature', where the ecologically sensitive species find refuge from people.

Whether or not deforestation will stabilize will depend on a range of factors. First, what is clear is that the world's human population is predicted to exceed 9.6 billion by 2050 (Tilman *et al.* 2011; Baudron and Giller 2014). Feeding and providing livelihoods for this growing population presents a global challenge, particularly given the risks associated with climate change (Bryan *et al.* 2013; Laurance *et al.* 2013). The next question is how much land is needed to produce the goods required by those people, and how these are best produced. There is a longstanding debate between proponents of the land sharing and sparing strategies (Clough *et al.* 2011; Phalan *et al.* 2011b). The first strategy, sharing—or wildlife-friendly farming—envisages that land can deliver a range of goods and services simultaneously, and that mixed land uses such as agroforestry are optimal. The second strategy, sparing, suggests that agriculture should be concentrated on lands that are most suitable for producing agricultural products, so that the rest of the available lands can be kept in natural conditions (Fischer *et al.* 2008). This would result in parts of the world being intensively used for agriculture, with few biodiversity benefits, while the remainder would provide environmental services to the world, such as clean water, temperature regulation, and flood buffering, and also harbour most biodiversity.

Which of the two models, sparing or sharing, or a mixture of them, provides an optimal solution is probably very much dependent on social and environmental contexts (Law *et al.* 2015). Considering solutions as a dichotomous choice might even be counterproductive, especially because it is even less clear how the choice of either model will affect biodiversity (Law *et al.* 2015; Law and Wilson 2015). It is obvious, though, that at a global scale, biodiversity is continuing to decline, with projected losses of up to half of all species by 2050 (Thomas *et al.* 2004). The extinction of species is occurring in relatively undisturbed areas, but also in areas that have a long history of agriculture or forestry use. Such production landscapes therefore face increasing pressure to deliver on multiple outcomes, but land scarcity is forcing trade-offs between provision of food, fibre, energy, and water, conserving remaining habitat of flora and fauna, and mitigating the impacts of climate change through storing and sequestering carbon dioxide (Rockstrom *et al.* 2009; Bullock *et al.* 2011; Phalan *et al.* 2011a; Millennium Ecosystem Assessment 2005; Venter *et al.* 2013; Baudron and Giller 2014; Grau *et al.* 2013).

Although agricultural expansion is an important driver of land-use change almost everywhere, several kinds of data suggest that the effect on wildlife is now greatest in tropical developing countries (Laurance *et al.* 2013). Coarse-scale evidence of changes in forest cover shows that recent net gains in temperate and boreal forest cover are more than offset by continued losses in tropical regions, largely by conversion to agriculture (FAO 2010; Hansen *et al.* 2013). Patchy data on changes in populations of temperate and tropical forest vertebrates confirm this pattern (Jenkins *et al.* 2003).

Major environmental changes seem unavoidable, many of which we will witness within our lifetimes. The effects of forest loss and degradation of forests in tropical zones will likely be exacerbated by climate change, and vice versa: forest loss will speed up climate change by affecting the global carbon balance and, at a more local scale, by influencing temperature, rainfall, humidity, and wind patterns (Danielsen *et al.* 2009; Sheil and Murdiyarso 2009). With regard to the impact on primates, the focus of this book, many species will inhabit heterogeneous landscapes consisting of patches of relatively intact natural habitats surrounded by more intensively used lands, focused on agricultural or silvicultural production. How will primate species cope with these changes?

13.2 What is primate habitat?

Based on molecular studies, the primate lineage is thought to have originated in the Cretaceous, at least 81.5 million years ago (Tavare *et al.* 2002; Janečka *et al.* 2007), or perhaps even earlier (Arnason *et al.* 2001). These dates are much older than the earliest primate fossil finds and it is estimated that only about 7% of the primate species that ever existed are present in the fossil record (Tavare *et al.* 2002). The Cretaceous appears to have been a time of global tropical conditions, an extreme greenhouse world apparently warmer and more humid than our current global environment (Spicer and Corfield 1992). Some areas, like present-day Southeast Asia, appear to have experienced some level of water stress, especially towards the end of the Cretaceous (Morley 2000), but overall it can be said that primates evolved in humid and warm forested environments, generally similar to the lush tropical rainforests of today.

These forest environments are still the main habitat for primates. Out of 419 non-human primate species that exist today, as listed by the International Union for Conservation of Nature (IUCN), 407 use forest habitats, 54 species use savannahs, 30 species shrub and grasslands, 8 species use rocky habitats such as cliffs and mountain tops, and 1 species, the chacma baboon (*Papio ursinus*) also uses desert habitats (IUCN 2015). By and large, primates are forest animals. They mostly occur in the tropical forests of Asia, Africa, Madagascar, South America, and Mesoamerica, but are absent from the Australian tropics. Outside the tropics, primates are rare. The exceptions are the mountain-dwelling Japanese macaques (*Macaca fuscata*) that live in the north of Honshū Island, Japan, where there is snow cover for several months of the year, and the Barbary macaques (*Macaca sylvanus*) that live in the dry Atlas Mountains of Algeria and Morocco, as well as the Rock of Gibraltar (where they were introduced by people, see Modolo *et al.* (2005)), although even Barbary macaques prefer forest habitats. Many primate species, such as the strictly arboreal gibbons, are very unlikely to survive in non-forest habitats.

Maximum primate diversity is found in wet tropical forests, where up to a dozen species co-occur in the richest sites (Corlett and Primack 2011). Most primate communities include specialist leaf eaters, seed predators, frugivores, and insectivores, and many include specialists on plant exudates (Fleagle *et al.* 1999). Where present, primates make up a large proportion of the canopy vertebrate biomass, although there is significant geographical variation in their abundance relative to other taxa (Oates *et al.* 1990; Haugaasen and Peres 2005). Frugivorous primate species are probably third in importance to birds and bats as seed dispersal agents (Fleagle *et al.* 1999), although this can vary extensively based on whether this refers to primary or secondary seed dispersal, or plants that are dispersed by multiple agents or just one.

Despite their preference for forest habitats, some primate species show considerable ecological flexibility, and use other habitat types if forced to do so. Some primates, such as the rhesus macaque (*Macaca mulatta*), can exploit human-modified environments and even live in cities (Southwick and Siddiqi 1966); many species adapt to human presence provided they are not hunted. Bornean orang-utans (*Pongo pygmaeus*) occur in homogenous plantations of oil-palm and acacia, although at much reduced densities compared to natural forests, and they are unlikely to be able to survive in these plantations alone without using surrounding natural forest remnants (Meijaard *et al.* 2010a; Ancrenaz *et al.* 2015). Similarly, chimpanzees (*Pan troglodytes*) use anthropogenic habitat in Mali during at least part of the year (Duvall 2008). The degree to which this happens partly depends on the species' ability to come to the ground and move through non-forest habitat (Ancrenaz *et al.* 2014), as well as on their dietary flexibility. Quite a few primate species are at least partially terrestrial, such as baboons (*Papio* spp.) and Patas monkeys (*Erythrocebus patas*), and a few species are fully terrestrial, such as the geladas (*Theropithecus gelada*) of the Ethiopian highlands (Mittermeier *et al.* 2013). Other species do quite well in human-modified habitats because few people actually know they are there. For example, Dian's tarsiers (*Tarsius dentatus*) prefer agroforestry areas to natural forests, probably because of the higher density of insects, their main prey, and the abundance of trees suitable as sleeping sites (Merker and Mühlenberg 2000). Many villages on the Indonesian island of Sulawesi have tarsiers living in the bamboo stands and fig trees that surround many rural communities, but people are often unaware of their presence (M. Shekkele, pers. comm.).

A recent review (Estrada *et al.* 2012) showed that 57 primate taxa from four regions: Mesoamerica, South America, sub-Saharan Africa, and Southeast Asia used 38 types of agro-ecosystems as temporary or permanent habitats. Fifty-one percent of the taxa recorded in agro-ecosystems were classified as of Least Concern in the IUCN Red List, but the rest were classified as Endangered (20%), Vulnerable (18%), Near Threatened (9%), or Critically Endangered (2%). A survey of all primate species comparing their threat status in both natural forest and savannah environments, and manmade agro-ecosystems and much degraded forest areas (E. Meijaard, unpubl. data.), showed that threat levels in natural and manmade land-cover types were broadly similar, with most species in either ecosystem type falling in the Least Concern category (Figure 13.2). Critically Endangered and Endangered categories have a relatively high representation of species restricted to natural ecosystems, indicating that the forest specialists are seriously threatened by forest loss and fragmentation. As expected from primates' preference for forest, agro-ecosystems are used by a minority of primate species (<14% of the total number of species), probably because these offer suboptimal ecological conditions for their long-term survival. Conflict with people and their agricultural crops, and especially the trapping and hunting of primates associated with this, adds another threat to primates that live outside the forest. It is worth noting that this topic remains much understudied because most primate research is

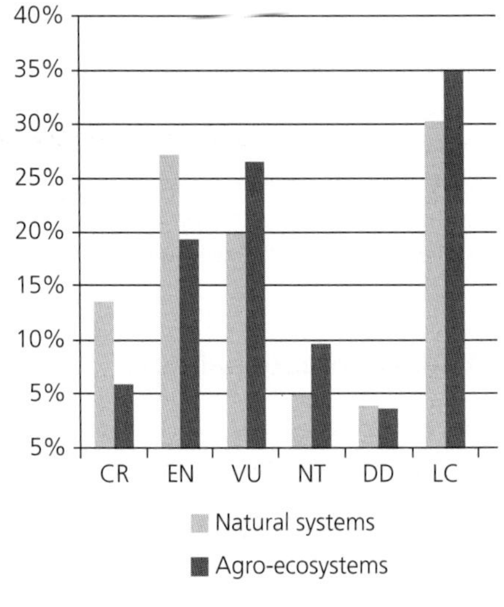

Figure 13.2 A comparison of the threat status of primates that primarily use natural ($n = 423$) or manmade land-cover types ($n = 83$) expressed as the percentage in each threat category compared to their total in either system. CR = Critically Endangered; EN = Endangered; VU = Vulnerable; NT = Near Threatened; DD = Data Deficient; and LC = Least Concern. (E. Meijaard, unpubl. data.) Reproduced with permission from John Wiley & Sons.

conducted on populations inhabiting natural ecosystems. Not much is known about what mix of habitats, and the percentages of each, are needed under different threat levels (e.g. hunting or disease) to maintain viable populations.

13.3 Spatial dynamics and the threat of hunting

Distribution and density variation of primates is determined by a range of factors. Prior to the occurrence of human impacts, the absence or presence of primate species would have been determined by their evolutionary history, the availability of suitable habitats and related environmental factors such as temperature, rainfall, altitude, and so on, and the presence of other species with broadly similar ecologies (i.e. ecological competitors) or potentially negative impacts (i.e. predators, parasites, etc.). Some primate species have always lived alongside humans and their ancestors. In Africa, humans evolved alongside primate species and it has been speculated that non-human species had more time there than in other parts of the world to adapt to the specific ecology and diseases of humans, including their hunting strategies and habitat alterations (Brook and Bowman 2002)—and that Africa may therefore have avoided major megafaunal extinctions. Elsewhere, with humans arriving in primate habitats relatively recently, the impact of humans may have been much more severe. African primates are relatively more threatened than American ones, with 47% of African species listed as Critically Endangered or Endangered as opposed to only 23% of the American species (and 51% of the Asian species) (E. Meijaard, unpubl. data), and it seems obvious that time of contact with the genus *Homo* does not well explain present-day threat patterns in primates.

Still, co-existence with humans is a factor worth exploring. As an example, the arrival of *Homo sapiens* in Southeast Asia some 70,000 years ago coincides with the rapid contraction of the orang-utan range, although climate change (Jablonksi and Whitfort 1999) and even major volcanic events have been implicated as other potential causes (Nater *et al.* 2011). Orang-utan species (note that there were up to seven species, see Harrison *et al.* (2014)) became extinct in South China, Vietnam, Thailand, Peninsular Malaysia, and the Indonesian island of Java; and on the two islands where they remain, Borneo and Sumatra, their range was already much restricted in historic times (Rijksen and Meijaard 1999). Genetic studies indicate a 1,000-fold population decline in northern Borneo over the past several hundred to thousand years (i.e. prior to deforestation) (Goossens *et al.* 2006), while another study estimated a 35–111-fold population decline in several orang-utan populations in different parts of Borneo over the same time period (Sharma *et al.* 2012). It is not possible to confidently attribute these declines to the arrival of humans and the development of increased hunting efficiency, for example through the invention of the blow pipe (Jett 1970). Other factors (e.g. climate, other predators, ecological changes, disease) could have played a role too. Still, the evidence suggests that human hunting likely had a large impact on orang-utan populations (Meijaard *et al.* 2010b). For example, in Late Pleistocene and Holocene deposits, orang-utan teeth are more common than would be expected based on their present-day densities. In six sites in Java, Sumatra, Borneo, Laos, and Vietnam the following numbers of teeth were found: orang-utan (ca. 800), gibbon (ca. 65), macaques (ca. 400), colobines (ca. 325), muntjaks (ca. 750), and pigs (ca. 4,000) (de Vos 1983; Bacon and Long 2001; Long *et al.* 1996; Piper and Rabett 2009; Harriso 1998). With densities of 0.5–5 animals/km^2 orang-utans are, however, much rarer than any of these species now, with, for example, macaques (*Macaca* sp.) often reaching densities of 25–75/km^2, and colobines (*Presbytis* sp.) 20–40/km^2 (E. Meijaard, unpubl. data). Orang-utans could have been much more common in the Late Pleistocene, or easier to catch, or specifically targeted by hunters.

Orang-utans are not the only primate species targeted by hunters and trappers. Primates are hunted for food and killed because of agricultural conflicts and for the pet trade in every geographic location where they co-exist with humans. The trade in primate meat has been identified as a problem for conservation efforts in Africa (Bowen-Jones and Pendry 1999; Fa *et al.* 2008; Hicks *et al.* 2010; Wilkie and Carpenter 1999), Asia (Corlett 2007; Mitchell and Tilson

1968), and Mesoamerica and South America (Bodmer *et al.* 1994; Fa *et al.* 2002; Gonzalez-Kirchner and Sainz de la Maza 1998; Peres 1990). In the 2012–14 *Primates in Peril* report, hunting was identified as a threat to survival for 19 of the top 25 most endangered primates, with 10 primate species explicitly being hunted for meat, including the eastern lowland gorilla (*Gorilla berengei graueri*) (Mittermeier *et al.* 2012). Hunting for meat, not habitat loss, has been identified as the greatest immediate threat to primate conservation in African forests (Gates 2006; Wilkie *et al.* 1998). More than 25 years ago, a worldwide review of primate hunting concluded that the primary reason people in all countries kill primates was to eat them (Mittermeier 1987), and a review 14 years later still concluded that, in most cases, hunting was responsible for primate population declines well before deforestation occurred (Chapman and Peres 2001).

In a landscape context, primate hunting and trapping will likely be focused in areas that are closest to human settlements and agricultural areas (Novaro *et al.* 2000; Marshall *et al.* 2006). This indicates that source/sink dynamics are at play (Novaro *et al.* 2000). Source areas have positive growth rates ($r > 0$), and sink areas have negative growth rates ($r < 0$). For primates this would probably mean that areas with the lowest human population densities maintain the highest primate densities, while primates are scarcer in areas with many people. The sink populations would have high levels of turnover through local extinction and colonization from source areas. It has been suggested that the gaps in the distribution of the Dusky Titi (*Callicebus moloch*) in central and southern Bolivia are an example of this (Cowlishaw and Dunbar 2000). Problematically, however, in primate conservation it may not always be clear which parts of a species' habitat is a source and which is a sink. It would be logical to assume that the areas that are ecologically most suitable for a particular species would be sources with ecologically marginal areas acting as sinks. Hunting may turn things upside down, however. For example, the Bornean orang-utan has been historically extinct in parts of the range that consist of primary forest with very low human population densities (Rijksen and Meijaard 1999). What makes these areas stand out, however, is that the people that did live (and sometimes still live) there were nomadic hunter-gatherers. Nomadic hunters in remote areas would have concentrated around the same ecological resources as orang-utans (e.g. fruiting trees that attract pigs and deer) and could have had a much larger impact on orang-utans than people living in the much more densely settled lowlands (Meijaard *et al.* 2010b). In the latter area, people were primarily farmers of irrigated rice fields and slash-and-burn agriculture areas. Tending fields consumes significant time and limits how much people can hunt and how far they can go. Heavy hunting pressure around fields could indeed turn these areas into local population sinks, but large areas away from villages and fields may have still been population sources. So, in this case, the areas that would ecologically be quite suitable could actually be significant population sinks.

The above example indicates that source–sink dynamics in primate populations can be counterintuitive and difficult to identify. Understanding them may, however, be key to managing primates in multifunctional landscapes (Levi *et al.* 2011). If populations decline in sink populations, then this is less of a conservation problem than if populations decline in source populations, since the local extinction of those source populations could lead to multiple extinctions in dependent sink populations in surrounding areas (Cowlishaw and Dunbar 2000). Unfortunately, sink–source dynamics remain poorly studied in primates, partly because of the difficulty of reliably estimating their densities. Some insights exist from studies on species such as the Gelada baboon (*Theropithecus gelada*) (Ohsawa and Dunbar 1984) that occur in more open areas and are therefore easier to count, or species such as Bornean white-bearded gibbons (*Hylobates albibarbis*) (Marshall 2009). For most other primate species, these dynamics remain poorly understood. One reason for this seems to be that researchers do not systematically sample a wide enough array of habitats of different quality or that experience different levels of human influence required to get the kind of data needed for understanding sink–source dynamics in multifunctional landscapes.

Understanding primate meta-population dynamics, inter-population dispersal and gene flow, and the impacts of local habitat ecology and threats such as hunting are all key to developing effective management strategies. Concentrating limited conservation management resources on the areas that are most important to protect and give the biggest conservation bang for the conservation buck is a wise strategy (Wilson *et al.* 2014). So, as a first priority, we need to be able to identify viable source populations of threatened primate species and focus conservation efforts on reducing whatever threatens them. An example of such an area is the Lower Kinabatangan Wildlife Reserve, in Malaysian Borneo. This area of fragmented forests lies in a much larger landscape dominated by monocultural stands of oil-palm (*Elaeis guinensis*). The remaining forests contain important source populations of a host of primate species, including the enigmatic proboscis monkey (*Nasalis larvatus*), orang-utan, and others. These species use the oil-palm matrix for dispersal, some daring to penetrate deeper than others (Ancrenaz *et al.* 2015). Even within species there are significant differences, with young adult male orang-utans more likely to disperse away from their natal forests than older and young males, and females (Ancrenaz *et al.* 2015). Key questions for such conservation areas include how big they should be to maintain viable primate populations and to what extent the survival of these populations ultimately depends on the ability of animals to cross through the oil-palm matrix and disperse from and to other forest areas.

13.4 'Novel ecosystems' and 'new conservation'

The recognition that humans will modify this planet and that there is therefore an unavoidable need to address the conservation of primates and other wildlife in multifunctional landscapes seems logical. Still, this issue has been the source of heated debate among conservation scientists. 'New conservation' (Kareiva and Marvier 2012; Marvier and Kareiva 2014) suggests we focus on conserving modified landscapes, gaining people's support, working with business, respecting human rights, and maximizing conservation and economic objectives. In other words, it seeks to address conservation issues by fully embedding them in human society, rather than trying to keep the stark distinction between protected nature and the unprotected rest. 'New conservation' has been criticized for compromising traditional conservation approaches of protecting species in the very best wilderness areas (Soulé 2013), or 'uncritically adopting a new human-centred conservation approach based on opinions, untested assumptions, and unwarranted conclusions' (Doak *et al.* 2013). As with many of these black-and-white juxtapositions of different approaches, identifying the best strategy likely depends on local socio-ecological and environmental conditions. Seeking people's support for conservation in areas of warfare and extreme poverty will likely lead to different solutions than in areas of welfare and general societal support for conservation.

Related to the 'new conservation' discussion is the issue of 'novel ecosystems'. A 'novel ecosystem' is 'a system of abiotic, biotic and social components (and their interactions) that, by virtue of human influence, differ from those that prevailed historically, having a tendency to self-organize and manifest novel qualities without intensive human management' (Hobbs *et al.* 2013: 58). It remains a matter of debate whether or not such novel ecosystems actually exist, and how it could be detected whether an irreversible ecological threshold has caused an ecosystem to shift to a new stable state (see Murcia *et al.* (2014), for a summary). At the heart of debates about new conservation and novel ecosystems is the belief by some that no place on Earth can be considered wild anymore and that all ecosystems are fundamentally altered by the 'human footprint', especially in the light of climate change (Vitousek 1996; Doney 2010).

The outcomes of the 'new conservation' and 'novel ecosystems' debates are of importance to primate conservation in landscapes, because they will inform where the conservation movement will decide to spend its resources. Do we recognize that manmade land covers are an integral part of a primate's range and therefore allocate resources to primate management inside and outside natural habitats, or do we concentrate most conservation efforts on the 'wildest', richest, and least touched parts of their range? The example above of the Bornean orang-utan apparently being most threatened or eradicated by hunting in the most pristine part of its range shows the difficulty in finding clear answers to this question, and these answers may further shift with factors like climate change (Struebig *et al.* 2015).

In the light of the discussions above, neither 'traditional conservation' with a focus on protected area establishment and local minimization of the human threat, nor 'new conservation' with its global landscape conservation agenda, will provide a silver conservation bullet. There are strong arguments in support of either approach, but it might be more useful if we see them as extremes along a continuum of conservation strategies (Sheil and Meijaard 2010). Just as football managers, or army generals, adapt their team's composition and tactics to the strengths and weaknesses of the opponent, conservation practitioners should identify which particular strategy is likely to work best given their available resources, the urgency of a particular situation, the extent to which local society is involved or their support needed, and a range of other factors (Game *et al.* 2014).

13.5 The practical reality of managing primates in landscapes with people

Once we start addressing primate conservation in heterogeneous landscapes comprising primate habitats of varying quality and areas of more or less intensive human use, engagement with the people that also use these landscapes will become unavoidable. With regard to often poor communities, a popular approach since the 1980s has been to integrate natural resource and species management with economic development to improve the quality of life of rural people (Kaeslin and Williamson 2010). Effective integration of conservation and development in areas of concern to primate conservation has so far largely failed (van Schaik and Kramer 1997; Wells *et al.* 1999; Oates 1995; Sandker *et al.* 2009; Weber *et al.* 2011), although there are tentative exceptions such as programmes generating major income from ecotourism (Blomley *et al.* 2010). This points to a common constraint on facilitating primate survival in multifunctional landscapes. Unless people are unusually tolerant of primates living where they grow their crops and raise their children (e.g. for langurs and macaques in India, see Pirta (1982)), there is likely to be significant conflict (Estrada *et al.* 2012; Hartter *et al.* 2010; Hockings and Humle 2009; Webber *et al.* 2007). Especially where natural primate habitat is shrinking and being fragmented into increasingly small patches, animals might be forced to feed in agricultural or agroforestry areas. This is well demonstrated for species such as orang-utans and chimpanzees (Davis *et al.* 2013; Campbell-Smith *et al.* 2010; Tweheyo *et al.* 2005) where, despite a certain level of tolerance, people would much rather not have them in their neighbourhoods and often end up killing them. Other primate species are habitually trapped, especially those that come to the ground and feed directly on crops (e.g. species of macaque, see Linkie *et al.* (2007)). Unless such conflicts are resolved, or primates are considered relatively harmless (like the Sulawesi tarsiers), multipurpose landscapes may always act as population sinks, and therefore contribute relatively little to overall conservation.

It is not impossible to reduce human–wildlife conflicts, but much depends on people's attitude towards wildlife. In many European countries, species that were very much in decline in the mid to late twentieth century (e.g. wolves, wild boar, osprey) are making a comeback (Bauer *et al.* 2009). Reasons for this are complex but generally stem from increased societal acceptance or even appreciation for wildlife in close proximity to places where people live and recreate. Commercial or recreational hunting is also in decline, and large areas of farmland are being abandoned (Navarro and Pereira 2012; Bauer *et al.* 2009). The extent to which similar processes can or may take place in the primate areas of Asia, Africa, and the Neotropics remains unclear. Some 32 million primates have been estimated to lose their habitat every year, mostly in the tropics (Chapman and Peres 2001). These losses primarily occur in areas with generally high levels of poverty and the concomitant incentive to develop economically through further deforestation. Increased welfare is associated with greater societal support for environmental conservation (Vincent *et al.* 2014). The question is: how much more destruction of primate habitat needs to happen before increased welfare generates enough societal support to push for conserving primate habitats and reducing other threats?

Large plantations of oil-palm, timber, soy, or other crops or cattle ranges are a different feature of primate survival in multifunctional landscapes. Compared to small-scale agriculture by people, which can often result in quite heterogeneous landscapes, industrial-scale plantations often introduce large areas of monocultural crops into primate habitats. Depending on the type of crop, primates may find some sustenance. For example, orang-utans feed on the cambium of *Acacia mangium* and survive in extensive plantations at low densities along with other primates such as long-tailed and pig-tailed macaques (*Macaca* spp.) (Meijaard *et al.* 2010a). As pointed out earlier, orang-utans also use mature oil-palm areas or feed on young palm shoots (Ancrenaz *et al.* 2015). Because of the damage caused to commercial plantations, large numbers of primates are presumably killed in these areas. Still, there is increasing market pressure on major producers of agricultural and silvicultural commodities to develop practices that minimize impacts of threatened species. The Round Tables for Soy and Palm Oil and the many certification schemes for timber (e.g. Forest Stewardship Council) have started to pressure producers into developing less destructive production systems. Even those companies with the best intentions to be wildlife-friendly will most likely only do so if the benefits (e.g. avoiding negative media or public opinion, avoiding loss of market access) outweigh the costs (e.g. crop losses, reduced areas developed through bigger conservation set asides, hunting control). Still, there is much that can be achieved in these industrial plantations that would increase the likelihood of primate survival in the context of large landscapes. For example, the presence of large natural areas within the plantations, as is often required by law, potentially provides stepping stones and temporal refuges for primates. Similar landscape features, such as forest corridors, or rope bridges and other systems that facilitate river and road crossing, could help animals disperse through these plantations and maintain geneflow between forest areas. Little is known about the effectiveness of landscape features, such as forest corridors that connect various small patches, in maintaining primate populations (Chazdon *et al.* 2009). Obviously much of the success of such interventions will depend on the ecological adaptability of particular primate species, and the extent to which these landscape features can fulfil their ecological needs. Even more importantly for getting primate conservation to work in multifunctional landscapes is that the regulatory and educational actions are put in place that ensure that primates are not killed or hurt in plantations. Ultimately, primates will only be able to survive in multifunctional landscapes if the people, businesses, and governments that also use these landscapes welcome the presence of primates as their neighbours.

13.6 Conclusion

Considering that most primates are forest species and that the tropics and sub-tropics where they occur are undergoing net forest loss (FAO 2010), an increasing number of primates are going to find themselves in rapidly changing new landscapes or even 'novel ecosystems' containing forest fragments, small and large areas of agriculture and silviculture, roads, and a human society generally not very welcoming to their fellow primates. Humans have quite different objectives for the use of these landscapes than other primates. The key to finding better solutions that reconcile the objectives of human development and primate conservation lies in a much better understanding of trade-offs and synergies between different objectives. For example, research has shown that significant areas of planted oil-palm actually cost more to maintain than they generate in income because they were developed in areas affected by frequent flooding, where the palms produce almost no yields (Abram *et al.* 2014). Leaving such flood-prone areas as riverine swamp forests would maintain important primate habitat and deliver services such as flood buffering to human societies, while saving the oil-palm industry money. Such examples indicate that win–win solutions in conservation do exist. Important progress can thus be made in conservation research that helps identify optimal solutions for land and resource uses that simultaneously benefit human development and environmental conservation. Conservation practice has an urgent need for an improved understanding of these trade-off solutions, but unfortunately not enough attention is being paid by conservation scientists to addressing this knowledge gap. In general, conservation students and scientists still prefer to study the behaviour of an endangered primate or the ecological functioning of a primate habitat rather than battle with the complex conservation reality of socio-ecological systems. This is changing, but more people are needed that have the skills for analysing the interactions between manmade and natural ecosystems, and the skills for translating their findings into language that the public and political decision-makers can understand, find compelling, and can turn into meaningful actions. With this chapter I call out to students and scientists to become more deeply engaged in the thinking and practice of how we can help our endangered wildlife survive in a world in which they are increasing rubbing shoulders with people.

Acknowledgements

I thank the Arcus Foundation for supporting our research and policy work under the Borneo Initiative, which has allowed our network to engage in studies on primates in multifunctional landscapes.

References

Aastrup, P. (2000). Responses of West Greenland caribou to the approach of humans on foot. *Polar Research* **19**: 83–90.

Abram, N. K., Xofis, P., Tzanopoulos, J., Macmillan, D. C., Ancrenaz, M., *et al.* (2014). Synergies for improving oil palm production and forest conservation in floodplain landscapes. *PLoS One* **9**: e95388.

Ancrenaz, M., Oram, F., Ambu, L., Lackman, I., Ahmad, E., *et al.* (2015). Of pongo, palms, and perceptions—A multidisciplinary assessment of orangutans in an oil palm context. *Oryx* **49**: 465–472.

Ancrenaz, M., Sollmann, R., Meijaard, E., Hearn, A. J., Ross, J., *et al.* (2014). Coming down the trees: is terrestrial activity in orangutans natural or disturbance-driven? *Scientific Reports* **4**: doi:10.1038/srep04024.

Arnason, U., Gullberg, A., Burguete, A. S., and Janice, A. (2001). Molecular estimates of primate divergences and new hypotheses for primate dispersal and the origin of modern humans. *Hereditas* **133**: 217–228.

Bacon, A.-M. and Long, V. T. (2001). The first discovery of a complete skeleton of a fossil orang-utan in a cave of the Hoa Binh Province, Vietnam. *Journal of Human Evolution* **41**: 227–241.

Barbetti, M. (1986). Traces of fire in the archaeological record, before one million years ago? *Journal of Human Evolution* **15**: 771–781.

Barnosky, A. D., Hadly, E. A., Bascompte, J., Berlow, E. L., Brown, J. H., *et al.* (2012). Approaching a state shift in Earth's biosphere. *Nature* **486**: 52–58.

Baudron, F. and Giller, K. E. (2014). Agriculture and nature: trouble and strife? *Biological Conservation* **170**: 232–235.

Bauer, N., Wallner, A., and Hunziker, M. (2009). The change of European landscapes: human-nature relationships, public attitudes towards rewilding, and the implications for landscape management in Switzerland. *Journal of Environmental Management* **90**: 2910–2920.

Blomley, T., Namara, A., Mcneilage, A., Franks, O., Rainer, H., *et al.* (2010). *Development and Gorillas? Assessing Fifteen Years of Integrated Conservation and Development in South-western Uganda*. Natural Resource Issues No. 23. London: IIED.

Bodmer, R. E., Fang, T. G., Moya I, L., and Gill, R. (1994). Managing wildlife to conserve Amazonian forests: population biology and economic considerations of game hunting. *Biological Conservation Restoration and Sustainability* (**67**): 29–35.

Bowen-Jones, E. and Pendry, S. (1999). The threat to primates and other mammals from the bushmeat trade in Africa, and how this threat could be diminished. *Oryx* **33**: 233–246.

Brook, B. W. and Bowman, D. (2002). Explaining the Pleistocene megafaunal extinctions: models, chronologies, and assumptions. *Proceedings of the National Academy of Sciences* **99**: 14624–14627.

Bryan, B. A., Meyer, W. S., Campbell, C. A., Harris, G. P., Lefroy, T., *et al.* (2013). The second industrial transformation of Australian landscapes. *Current Opinion in Environmental Sustainability* **5**: 278–287.

Bullock, J. M., Aronson, J., Newton, A. C., Pywell, R. F., and Rey-Benayas, J. M. (2011). Restoration of ecosystem services and biodiversity: conflicts and opportunities. *Trends in Ecology and Evolution* **26**: 541–549.

Campbell-Smith, G., Simanjorang, H. V. P., Leader-Williams, N., and Linkie, M. (2010). Local attitudes and perceptions toward crop-raiding by Orangutans (*Pongo abelii*) and other nonhuman primates in Northern Sumatra, Indonesia. *American Journal of Primatology* **72**: 866–876.

Chapman, C. A. and Peres, C. A. (2001). Primate conservation in the new millennium: the role of scientists. *Evolutionary Anthropology* **10**: 16–33.

Chazdon, R. L., Harvey, C. A., Komar, O., Griffith, D. M., Ferguson, B. G., *et al.* (2009). Beyond reserves: a research agenda for conserving biodiversity in human-modified tropical landscapes. *Biotropica* **41**: 142–153.

Clough, Y., Barkmann, J., Juhrbandt, J., Kessler, M., Wanger, T. C., *et al.* (2011). Combining high biodiversity with high yields in tropical agroforests. *Proceedings of the National Academy of Sciences* **108**: 8311–8316.

Corlett, R. T. (2007). The impact of hunting on the mammalian fauna of tropical Asian forests. *Biotropica* **39**: 292–303.

Corlett, R. T. and Primack, R. B. (2011). *Tropical Rain Forests: An Ecological and Biogeographical Comparison*, 2nd edn. Oxford: Wiley-Blackwell.

Cowlishaw, G. and Dunbar, R. (2000). *Primate Conservation Biology.* Chicago and London: University of Chicago Press.

Danielsen, F., Beukema, H., Burgess, N. D., Parish, F., Bruhl, C. A., *et al.* (2009). Biofuel plantations on forested lands: double jeopardy for biodiversity and climate. *Conservation Biology* **23**: 348–358.

Davis, J. T., Mengersen, K., Abram, N., Ancrenaz, M., Wells, J., *et al.* (2013). It's not just conflict that motivates killing of orangutans. *PLoS One* **8**: e75373.

De Vos, J. (1983). The *Pongo* faunas from Java and Sumatra and their significance for biostratigraphical and paleo-ecological interpretations. *Proceedings of the Koninklijke Akademie van Wetenschappen. Series B* **86**: 417–425.

Diamond, J. M. (2005). *Collapse: How Societies Choose to Fail or Succeed.* New York, NY: Viking Penguin.

Doak, D. F., Bakker, V. J., Goldstein, B. E., and Hale, B. (2013). What is the future of conservation? *Trends in Ecology and Evolution* **29**: 77–81.

Doney, S. C. (2010). The Growing Human Footprint on Coastal and Open-Ocean Biogeochemistry. *Science* **328**: 1512–1516.

Duvall, C. S. (2008). Human settlement ecology and chimpanzee habitat selection in Mali. *Landscape Ecology* **23**: 699–716.

Estrada, A., Raboy, B. E., and Oliveira, L. C. (2012). Agroecosystems and primate conservation in the tropics: a review. *American Journal of Primatology* **74**: 696–711.

Fa, J. E., Garcia Yuste, J. E., and Castelo, R. (2008). Bushmeat markets on Bioko Island as a measure of hunting pressure. *Conservation Biology* **14**: 1602–1613.

Fa, J. E., Peres, C. A., and Meeuwig, J. (2002). Bushmeat exploitation in tropical forests: an intercontinental comparison. *Conservation Biology* **16** 232–237.

FAO (2010). Global Forest Resources Assessment 2010. Progress Towards Sustainable Forest Management. FAO Forest Paper 163. Rome, Italy: Food and Agricultural Organization of the United Nations.

Fischer, J., Brosi, B., Daily, G. C., Ehrlich, P. R., Goldman, R., *et al.* (2008). Should agricultural policies encourage land sparing or wildlife-friendly farming? *Frontiers in Ecology and the Environment* **6**: 380–385.

Fleagle, J. G., Janson, C., and Reed, K. (1999). *Primate Communities*. Cambridge: Cambridge University Press.

Game, E., Meijaard, E., Sheil, D., and Mcdonald-Madden, E. (2014). Conservation in a wicked complex world; challenges and solutions. *Conservation Letters* **7**: 271–277.

Gates, J. F. (2006). Habitat alteration, hunting and the conservation of folivorous primates in African forests. *Australian Journal of Ecology* **21**: 1–9.

Gonzalez-Kirchner, J. P. and Sainz De La Maza, M. (1998). Primates hunting by Guaymi Amerindians in Costa Rica. *Human Evolution* **13**: 15–19.

Goossens, B., Chikhi, L., Ancrenaz, M., Lackman-Ancrenaz, I., Andau, P., *et al.* 2006. Genetic signature of anthropogenic population collapse in orang-utans—art. no. e25. *Plos Biology* **4**: 285–291.

Grau, R., Kuemmerle, T., and Macchi, L. (2013). Beyond 'land sparing versus land sharing': environmental heterogeneity, globalization and the balance between agricultural production and nature conservation. *Current Opinion in Environmental Sustainability* **5**: 483.

Green, R. E., Cornell, S. J., Scharlemann, J. P. W., and Balmford, A. (2005). Farming and the fate of wild nature. *Science* **307**: 550–555.

Hansen, M. C., Potapov, P. V., Moore, R., Hancher, M., Turubanova, S. A., *et al.* (2013). High-resolution global maps of 21st-century forest cover change. *Science* **342**: 850–853.

Hardin, G. (1968). The tragedy of the commons. *Science* **162**: 1243–1248.

Harrison, T. (1998). Vertebrate faunal remains from Madai Caves (MAD 1/28), Sabah, East Malaysia. *Indo-Pacific Prehistory Association Bulletin* **17**: 85–92.

Harrison, T., Jin, C., Zhang, Y., Wang, Y., and Zhu, M. (2014). Fossil Pongo from the Early Pleistocene Gigantopithecus fauna of Chongzuo, Guangxi, southern China. *Quaternary International* **354**: 59–67.

Hartter, J., Goldman, A., and Southworth, J. (2010). Responses by households to resource scarcity and human-wildlife conflict: issues of fortress conservation and the surrounding agricultural landscape. *Journal for Nature Conservation* **19**: 79–86.

Haugaasen, T. and Peres, C. A. (2005). Primate assemblage structure in Amazonian flooded and unflooded forests. *American Journal of Primatology* **67**: 243–258.

Henn, B. M., Gignoux, C. R., Jobin, M., Granka, J. M., Macpherson, J. M., *et al.* (2011). Hunter-gatherer genomic diversity suggests a southern African origin for modern humans. *Proceedings of the National Academy of Sciences* **108**: 5154–5162.

Hicks, T. C., Darby, L., Hart, J., Swinkels, J., January, N., *et al.* (2010). Trade in orphans and bushmeat threatens one of the Democratic Republic of the Congo's most important populations of Eastern Chimpanzees (*Pan troglodytes schweinfurthii*). *African Primates* **7**: 1–18.

Hobbs, R. J., Higgs, E. S., and Hall, C. M. (2013). Defining novel ecosystems. In: Hobbs, R. J., Higgs, E. S., and Hall, C. M. (Eds), *Novel Ecosystems: Intervening in the New Ecological World Order*. Chichester: John Wiley.

Hockings, K. and Humle, T. (2009). Best Practice Guidelines for the Prevention and Mitigation of Conflict between Humans and Great Apes. Gland, Switzerland: IUCN SSC Primate Specialist Group.

IUCN (2015). *The IUCN Red List of Threatened Species. Version 2015–4*. <http://www.iucnredlist.org>. [Accessed 19 November 2015].

Jablonksi, N. G. and Whitfort, M. J. (1999). Environmental change during the Quaternary in East Asia and its consequences for mammals. *Records West Australian Museum Supplement* **57**: 307–315.

Janečka, J. E., Miller, W., Pringle, T. H., Wiens, F., Zitzmann, A., *et al.* (2007). Molecular and genomic data identify the closest living relative of primates. *Science* **318**: 792–794.

Jenkins, M., Green, R. E., and Madden, J. (2003). The challenge of measuring global change in wild nature: are things getting better or worse? *Conservation Biology* **17**: 20–23.

Jett, S. C. (1970). The development and distribution of the blowgun. *Annals of the Association of American Geographers* **60**: 662–688.

Johnson, C. N. (2002). Determinants of loss of mammal species during the Late Quaternary 'megafauna' extinctions: life history and ecology, but not body size. *Proceedings of the Royal Society of London. Series B: Biological Sciences* **269** 2221–2227.

Kaeslin, E. and Williamson, D. (2010). Forests, people and wildlife: challenges for a common future. *Unasylva* **16**: 3–10.

Kareiva, P. and Marvier, M. (2012). What is conservation science? *BioScience* **62**: 962–969.

Laurance, W. F., Sayer, J., and Cassman, K. G. (2013). Agricultural expansion and its impacts on tropical nature. *Trends in Ecology and Evolution* **29**(2): 107–116.

Law, E. A., Meijaard, E., Bryan, A. B., Mallawaarachchi, T., Koh, L. P., *et al.* (2015). Better land-use allocation outperforms land sparing and land sharing approaches to conservation in Central Kalimantan, Indonesia. *Biological Conservation* **186**: 276–286.

Law, E. A. and Wilson, K. A. (2015). *Conservation Letters*: doi: 10.1111/conl.12168.

Levi, T., Shepard, G. H., Ohl-Schacherer, J., Wilmers, C. C., Peres, C. A., *et al.* (2011). Spatial tools for modeling the sustainability of subsistence hunting in tropical forests. *Ecological Applications* **21**: 1802–1818.

Linkie, M., Dinata, Y., Nofrianto, A., and Leader-Williams, N. (2007). Patterns and perceptions of wildlife crop raiding in and around Kerinci Seblat National Park, Sumatra. *Animal Conservation* **10**: 127–135.

Long, V. T., De Vos, J., and Ciochon, R. L. (1996). The fossil mammal fauna of the Lang Trang caves, Vietnam, compared with Southeast Asian fossil and recent mammal faunas: the geographical implications. *Bulletin of the Indo-Pacific Prehistory Association (Chaing Mai Papers, Volume 1)* **14**: 101–109.

Marshall, A. J. (2009). Are montane forests demographic sinks for Bornean white-bearded gibbons *Hylobates albibarbis*? *Biotropica* **41**: 257–267.

Marshall, A. J., Nardiyono, Engstrom, L. M., Pamungkas, B., Palapa, J., *et al.* (2006). The blowgun is mightier than the chainsaw in determining population density of Bornean orangutans (*Pongo pygmaeus morio*) in the forests of East Kalimantan. *Biological Conservation* **129**: 566–578.

Marvier, M. and Kareiva, P. (2014). The evidence and values underlying 'new conservation'. *Trends in Ecology and Evolution* **29**: 131–132.

McCarthy, C. (2006). *The Road*. New York, NY: Alfred A. Knopf.

Meijaard, E., Albar, G., Rayadin, Y., Nardiyono, Ancrenaz, M., and Spehar, S. (2010a). Unexpected ecological resilience in Bornean Orangutans and implications for pulp and paper plantation management. *PloS One* **5**: e12813.

Meijaard, E., Welsh, A., Ancrenaz, M., Wich, S., Nijman, V., *et al.* (2010b). Declining orangutan encounter rates from Wallace to the present suggest the species was once more abundant. *PloS One* **5**: e12042.

Merker, S. and Mühlenberg, M. (2000). Traditional land use and tarsiers—human influences on population densities of *Tarsius dianae*. *Folia Primatologica* **71**: 426–428.

Millennium Ecosystem Assessment (2005). *Ecosystems and Human Well-being*. Washington, DC: Island Press.

Miller, G. H., Fogel, M. L., Magee, J. W., Gagan, M. K., Clarke, S. J., *et al.* (2005). Ecosystem collapse in Pleistocene Australia and a human role in megafaunal extinction. *Science* **309**: 287–290.

Mitchell, A. H. and Tilson, R. L. (1968). Restoring the balance: Traditional hunting and primate conservation in the Mentawai Islands, Indonesia. In: Else, J. G. and Lee, P. C. (Eds), *Selected Proceedings of the Tenth Congress of the International Primatological Society Vol 2. Primate Ecology and Conservation*. Cambridge: Cambridge University Press.

Mittermeier, R. A. (1987). Effects of hunting on rain forest primates. In: Marsh, C. W. and Mittermeier, R. A. (Eds), *Primate Conservation in the Tropical Rain Forest*. New York. NY: Alan R. Liss.

Mittermeier, R. A., Rylands, A. B., and Wilson, D. E. (Eds.) (2013). *Handbook of the Mammals of the World—Vol 3. Primates*. Spain: Lynx Edicions.

Mittermeier, R. A., Schwitzer, C., Rylands, A. B., Taylor, L. A., Chiozza, F., *et al.* (Eds) (2012). *Primates in Peril: The World's 25 Most Endangered Primates 2012–2014* Bristol, UK: IUCN/SSC Primate Specialist Group (PSG), International Primatological Society (IPS), Conservation International (CI), and Bristol Conservation and Science Foundation.

Modolo, L., Salzburger, W., and Martin, R. D. (2005). Phylogeography of Barbary macaques (*Macaca sylvanus*) and the origin of the Gibraltar colony. *Proceedings of the National Academy of Sciences* **102**: 7392–7397.

Morley, R. J. (2000). *Origin and Evolution of Tropical Rain Forests.* Chichester: John Wiley & Sons, Ltd.

Murcia, C., Aronson, J., Kattan, G. H., Moreno-Mateos, D., Dixon, K., *et al.* (2014). A critique of the 'novel ecosystem' concept. *Trends in Ecology and Evolution* **29**: 548–553.

Nater, A., Nietlisbach, P., Arora, N., Van Schaik, C. P., Van Noordwijk, M. A., *et al.* (2011). Sex-biased dispersal and volcanic activities shaped phylogeographic patterns of extant orangutans (genus: *Pongo*). *Molecular Biology & Evolution* **28**: 2275–2288.

Navarro, L. M. and Pereira, H. M. (2012). Rewilding abandoned landscapes in Europe. *Ecosystems* **15**: 900–912.

Novaro, A. J., Redford, K. H., and Bodmer, R. E. (2000). Effect of hunting in source-sink systems in the neotropics. *Conservation Biology* **14**: 713–721.

Oates, J. F. (1995). The dangers of conservation by rural development—a case-study from the forests of Nigeria. *Oryx* **29**: 115–122.

Oates, J. F., Whitesides, G. H., Davies, A. G., Waterman, P. G., Green, S. M., *et al.* (1990). Determinants of variation in tropical forest primate biomass: new evidence from West Africa. *Ecology* **71**: 328–343.

Ohsawa, H. and Dunbar, R. I. M. (1984). Variations in the demographic structure and dynamics of gelada baboon populations. *Ecological Sociobiology* **15**: 231–240.

Pausas, J. G. and Keeley, J. E. (2009). A burning story: the role of fire in the history of life. *BioScience* **59**: 593–601.

Peres, C. A. (1990). Effects of hunting on western Amazonian primate communities. *Biological Conservation Restoration and Sustainability* **54**: 47–59.

Phalan, B., Balmford, A., Green, R. E., and Scharlemann, J. P. W. (2011a). Minimising the harm to biodiversity of producing more food globally. *Food Policy* **36**: S62–S71.

Phalan, B., Onial, M., Balmford, A., and Green, R. E. (2011b). Reconciling food production and biodiversity conservation: land sharing and land sparing compared. *Science* **333**: 1289–1291.

Piper, P. J. and Rabett, R. J. (2009). Hunting in a tropical rainforest: evidence from the Terminal Pleistocene at Lobang Hangus, Niah Caves, Sarawak. *International Journal of Osteoarchaeology* **19**: 551–565.

Pirta, R. S. (1982). Conservation note: socioecology and conservation of macaques and langurs in Varanasi, India. *American Journal of Primatology* **2**: 401–403.

Rijksen, H. D. and Meijaard, E. (1999). *Our Vanishing Relative. The Status of Wild Orang-Utans at the Close of the Twentieth Century*. Dordrecht, The Netherlands: Kluwer Academic Publishers.

Rockstrom, J., Steffen, W., Noone, K., Persson, A., Chapin, F. S., *et al.* (2009). Planetary boundaries: exploring the safe operating space for humanity. *Ecology and Society* **14**(2): 32.

Sandker, M., Campbell, B. M., Nzooh, Z., Sunderland, T., Amougou, V., *et al.* (2009). Exploring the effectiveness of integrated conservation and development interventions in a Central African forest landscape. *Biodiversity and Conservation* **18**: 2875–2892.

Schaik van, C. P. and Kramer, R. A. (1997). Toward a new protection paradigm. In: Kramer, R., Van Schaik, C., and Johnson, J. (Eds), *Last Stand. Protected Areas and the Defense of Tropical Biodiversity*. New York, N Y: Oxford University Press.

Sharma, R., Arora, N., Goossens, B., Nater, A., Morf, N., *et al.* (2012). Effective population size dynamics and the demographic collapse of Bornean Orang-Utans. *PLoS One* **7**: e49429.

Sheil, D. and Murdiyarso, D. (2009). How forests attract rain: an examination of a new hypothesis. *BioScience* **59**: 341–347.

Sheil, D. and Meijaard, E. (2010). Purity and prejudice: deluding ourselves about biodiversity conservation. *Biotropica* **42**: 566–568.

Soulé, M. (2013). The 'New Conservation'. *Conservation Biology* **27**: 895–897.

Southwick, C. and Siddiqi, M. R. (1966). Population changes of rhesus monkeys (Macaca mulatta) in India, 1959 to 1965. *Primates* **7**: 303–314.

Spicer, R. A. and Corfield, R. M. (1992). A review of terrestrial and marine climates in the *Cretaceous* with implications for modelling the 'Greenhouse Earth'. *Geological Magazine* **129**: 169–180.

Struebig, M. J., Fischer, M., Gaveau, D. L. A., Meijaard, E., *et al.* (2015). Anticipated climate and land-cover changes reveal refuge areas for Borneo's orang-utans. *Global Change Biology* **21**: 2891–2904.

Tavare, S., Marshall, C. R., Will, O., Soligo, C., and Martin, R. D. (2002). Using the fossil record to estimate the age of the last common ancestor of extant primates. *Nature* **416**: 726–729.

Thomas, C. D., Cameron, A., Green, R. E., Bakkenes, M., Beaumont, L. J., *et al.* (2004). Extinction risk from climate change. *Nature* **427**: 145–148.

Tilman, D., Balzer, C., Hill, J., and Befort, B. L. (2011). Global food demand and the sustainable intensification of agriculture. *Proceedings of the National Academy of Sciences* **108**: 20260–20264.

Tweheyo, M., Hill, C. M., and Obua, J. 2005. Patterns of crop raiding by primates around the Budongo Forest Reserve, Uganda. *Wildlife Biology* **11**: 237–247.

Venter, O., Possingham, H. P., Hovani, L., Dewi, S., Griscom, B., *et al.* (2013). Using systematic conservation planning to minimize REDD plus conflict with agriculture and logging in the tropics. *Conservation Letters* **6**: 116–124.

Vincent, J. R., Carson, R. T., Deshazo, J. R., Schwabe, K. A., Ahmad, I., *et al.* (2014). Tropical countries may be willing to pay more to protect their forests. *Proceedings of the National Academy of Sciences* **111**(28): 10113–10118.

Vitousek, P. M. (1996). Beyond global warming: ecology and global change. *Ecology* **75**: 1861–1876.

Webber, A. D., Hill, C. M., and Reynolds, V. (2007). Assessing the failure of a community-based human-wildlife conflict mitigation project in Budongo Forest Reserve, Uganda. *Oryx* **41**: 177–184.

Weber, J. G., Sills, E. O., Bauch, S., and Pattanayak, S. K. (2011). Do ICDPs work? An empirical evaluation of forest-based microenterprises in the Brazilian Amazon. *Land Economics* **87**: 661–681.

Wells, M., Guggenheim, S., Khan, A., Wardojo, W., and Jepson, P. 1999. *Investing in Biodiversity. A Review of Indonesia's Integrated Conservation and Development Projects*. Washington DC: The International Bank for Reconstruction and Development/The World Bank.

Wilkie, D. S. and Carpenter, J. F. (1999). Bushmeat hunting in the Congo Basin: an assessment of impacts and options for mitigation. *Biodiversity and Conservation* **8**: 927–955.

Wilkie, D. S., Curran, B., Tshombe, R., and Morelli, G. A. (1998). Managing bushmeat hunting in Okapi Wildlife Reserve, Democratic Republic of Congo. *Oryx* **32**: 131–144.

Wilson, H., Meijaard, E., Venter, O., Ancrenaz, M., and Possingham, H. P. (2014). Conservation strategies for orangutans: reintroduction versus habitat preservation and the benefits of sustainably logged forest. *PLoS One* **9**: e102174.

CHAPTER 14

People–primate interactions: implications for primate conservation

Tatyana Humle and Catherine Hill

Sharing landscape features: Bossou chimpanzees in Guinea, West Africa, travelling along a road also utilized daily by local villages and school children. Photo copyright: T. Humle.

14.1 From primate conflict to co-existence

14.1.1 Disciplines and shifts in terminology

Homo sapiens have, since they evolved, shared landscapes and resources with wildlife (Paterson 2005) and, like other animals, compete with those species they share space with. Interactions between people and wildlife may range from being mutually beneficial or benign to harmful for one party or the other. It is these harmful interactions that are frequently referred to as 'human–wildlife conflicts'.

Recent discussion within the wider conservation literature makes a strong case for shifting away from labels such as 'human–wildlife conflict' and 'crop-raiding' (or even 'raider' or 'thief') to terms such as 'co-existence' and 'crop-foraging'. Although these former terms are still commonly used, they are perceived as detrimental to promoting co-existence between people and wildlife, partly because they position wildlife as conscious antagonists of people

Humle, T. and Hill, C.M., *People-primate interactions: implications for primate conservation.* In: *An Introduction to Primate Conservation.* Edited by: Serge A. Wich and Andrew J. Marshall, Oxford University Press (2016). © Oxford University Press.
DOI 10.1093/acprof:oso/9780198703389.003.0014

and they mask the fact that conflicts most often arise as a consequence of stake-holders ascribing different values to different animal species (Madden and McQuinn 2014; Peterson *et al.* 2010; Remis and Hardin 2009; Redpath *et al.* 2013). This shift in terminology has coincided with a notable surge in research on interactions between people and non-human primates (hereafter referred to as primates) (Fuentes 2012; Fuentes and Hockings 2010). This trend parallels recent intensification of human encroachment into natural habitats linked to agricultural, ranching, and extractive industrial activities, as well as increased urbanization across primate range states and aggregations of some primate species around tourist or temple sites (Fuentes 2012). The spatial overlap between people and primates is ever increasing; primates experiencing habitat encroachment and habitat loss face declining availability of their wild food supplies and restrictions in their access to space, water, nesting sites, and familiar and safe habitat (Ancrenaz *et al.* in press). In this context, the people–primate interface poses an increasing challenge for conservation, and studying it requires a cross-disciplinary approach, combining methodological and theoretical approaches from both the social and natural sciences.

In this chapter we review people–primate relationships, patterns, and lessons learned, as well as studies that have sought to mitigate and/or prevent negative interactions between people and primates. We demonstrate that engaging with the challenges of people–primate co-existence requires an ethically sensitive approach, encompassing the active participation and involvement of all stake-holders concerned, and especially those sharing landscapes with their primate relatives (MacKinnon and Riley 2010; Malone *et al.* 2010).

14.2 Characterizing interactions

14.2.1 Types of interactions

Interactions between people and primates can take many different forms, entailing a range of combinations of positive, neutral, or negative outcomes for both parties. For instance, the Zanzibar red colobus (*Procolobus kirkii*) forage on damaged, immature coconuts; this 'pruning' effect promotes palm productivity (Siex and Struhsaker 1999). In some areas the colobus attract tourists, contributing indirectly to stimulating tourism locally (Siex and Struhsaker 1999). Small, faunivorous nocturnal primates may find refuge from predators and/or increase their foraging success around human habitation where lighting attracts preferred prey, including insects and small reptiles (Bearder *et al.* 2002). Frugivorous primates may act as key seed dispersers, thus playing a role in the maintenance of forest habitats that provide key ecosystem services to people (Chapman 1995; Chapman and Onderdonk 1998).

Severe negative outcomes directly impacting human lives, such as predation, whereby people are perceived as prey, are extremely rare. Such events primarily concern adult male chimpanzees (e.g. at Gombe, Tanzania, Frodo reportedly killed and ate a human baby: Kamenya (2002); around Kibale National Park, Uganda, in the mid 1990s, a male chimpanzee, known as Saddam, reportedly attacked seven children and killed two: Goldenberg (2013)). Although predation on people is atypical, reports of primate depredation on livestock are not uncommon. Chimpanzees are reported to eat young goats (*Capra hircus*), sheep (*Ovis aries*), and chickens (*Gallus gallus domesticus*) in parts of Senegal (Carter *et al.* 2003). Baboons are known to predate small livestock around the Pendjari Biosphere Reserve in northwestern Benin (Sogbohossou *et al.* 2011), and young goats and sheep during periods of wild food shortage in Zimbabwe (Butler 2000) and Kibale and Bulindi in Uganda (Naughton-Treves *et al.* 1998; McLennan and Huffman 2012).

Conversely, people hunting primates is a major cause of population decline and can extirpate populations locally, even in areas where suitable habitat is abundant (Fa and Brown 2009; Fa *et al.* 2002; Oates 1996). People hunt primates for a variety of purposes, including for meat, sale as live pets or for use in traditional medicines, rituals or as ornaments (Mittermeier 1987) (see also Fa and Tagg, Chapter 9, this volume). Primates have also long been hunted and persecuted because they are considered to be 'pests' (Davis *et al.* 2013; Hockings and Humle 2009), consuming or destroying crops or property, potentially affecting people's food supplies and economic and psychological wellbeing. During the period 1947–62, the government of Sierra Leone sponsored large-scale monkey drives to reduce the impact of primates on agriculture within the country.

Hunters were paid a bounty for each head or tail; according to government records approximately 245,000 individual primates were killed under this scheme (Tappen 1964). As further discussed in Section 14.2.3, culling campaigns or retributive killings still affect a range of primate species globally.

Indeed, the most challenging types of interactions between people and primates include disease transmission and competition for space and resources. The risk of zoonotic pathogen transmission is elevated in situations where people and primates are in close proximity (see Nunn and Gillespie, Chapter 10, this volume), for example especially in the contexts of tourism or research activities (Fuentes *et al.* 2007; Hockings and Humle 2009; Jones-Engel *et al.* 2008; Macfie and Williamson 2010; Muehlenbein *et al.* 2010). When it comes to competition between people and primates, people often outcompete primates for limited resources, even sometimes causing local extinction. For example, the extinction of an isolated population of proboscis monkeys (*Nasalis larvatus*), numbering around 300 individuals, in the Pulau Kaget Nature Reserve, Indonesia, was attributed to the loss of their habitat caused by illegal agriculture in the reserve (Meijaard and Nijman 2000). This population was reportedly driven to the edge of the reserve and was 'starving to death' (66). On frequent exposure to people, some animals may succumb to disease either through stress or zoonotic transmission (Chapman *et al.* 2005; Gillespie and Chapman 2008; Kondgen *et al.* 2008). Others may be compelled to leave the area, or are extirpated intentionally through hunting, trapping, and/or poisoning (e.g. primates in logged areas in Central Africa: Morgan and Sanz (2007)), or simply captured and translocated elsewhere (e.g. orang-utans in Indonesia: Russon (2009)). In the case of the proboscis monkeys of Pulau Kaget Nature Reserve, nearly a third were eventually captured and moved nearby to unprotected sites. This process resulted in the death of 15% of the translocated individuals, while another 20% of the monkeys were transferred to a zoo where 60% died within four months of their capture (Meijaard and Nijman 2000).

In other cases, however, competition can result in what is known as niche differentiation (Schoener 1974), whereby competing species adapt to occupy different niches or utilize the same resources in different ways. However, such processes typically take place over evolutionary timescales that do not match those facing most competitive interactions between people and wildlife. Nevertheless, management strategies, such as fencing resources, zoning areas of resource use, and switching to agricultural crops not favoured by local primates and other wildlife, could be viewed as measures aimed at eliciting niche differentiation to help foster coexistence between people and wildlife.

14.2.2 Associated costs

People–primate interactions can impose direct, indirect, or hidden costs that can variably affect people's ability to tolerate the presence of primates locally (Hockings and Humle 2009). Direct costs for people include crop losses, property damage or theft, livestock depredation, or attacks on people (Hockings and Humle 2009; Radhakrishna *et al.* 2013). They typically imply some economic loss or social impact, such as human injury or loss of life. The latter, however, as previously discussed, is relatively rare. In contrast, indirect costs include instances of zoonotic disease transmission, fear for safety (especially pertinent where primates occur in large groups, or are large-bodied species such as baboons, orang-utans, chimpanzees, or gorillas), restrictions on people's movement and travel, and time and money spent on protecting crops, livestock, and property against loss or damage.

Indirect costs may also incorporate cultural 'dilemmas' whereby the balance between cultural or religious tolerance, or taboos associated with primates or particular species of primates, is challenged by the cost imposed by these animals on people's financial, physical, and psychological wellbeing. For example, the association between monkeys and the Hindu god, Hanuman, has in the past ensured the conservation of primates across parts of India, especially those populations ranging around temples (Pirta *et al.* 1997). However, Pirta *et al.* (1997) described how local people's perceptions and attitudes towards rhesus macaques (*M. mulatta*) and Hanuman langurs (*Presbytis entellus*) are shifting. The 'conservation ethic' or the 'cultural value' that these species have benefitted from for so long is being eroded, and people's tolerance capacity for wild primates is being severely tested. Cultural values are indeed by no means static, shifting over time with changing experiences and conditions (Hill 2002; Lee and Priston

2005). For example, the way in which a macaque is considered or treated by an individual Balinese can be context specific; the sacredness value of macaques is multifaceted, depending, in part, on their presence and behaviour in temples during rituals. By contrast, macaques found crop-feeding are regarded as a nuisance and an economic liability (Fuentes 2010; Lane *et al.* 2010; Peterson *et al.* 2015; Schillaci *et al.* 2010). A macaque is therefore not provided protection in every context, with individual macaques foraging on crops at risk of being shot, either for crop protection or for sport (Schillaci *et al.* 2010).

Hidden costs associated with investing time and energy in guarding crops include foregoing alternative sources of income, or missing school in the case of children. The extra time and energy spent on preventing animals foraging on crops can be exhausting and, in some cases, may render people more vulnerable to disease either through lack of sleep or exposure to disease vectors such as mosquitoes, tsetse, and *Similium* sp. flies responsible for transmitting malaria, trypanosomiasis (sleeping sickness), and onchocerciasis (river blindness) respectively (e.g. elephant guarding, Kenya: Woodroffe *et al.* (2005)). Finally, more and more research is focusing on such hidden costs, whether indirect or lost opportunity costs, and is improving our understanding of the negative repercussions of interactions between people and wildlife (e.g. Barua *et al.* 2013).

14.2.3 Interactions and balancing values

Mismatches between measured and perceived costs or benefits significantly influence interactions between people and primates. A species or genus of primate may be valued differently by different groups of people. For example, macaques in India, Thailand, Indonesia, and Malaysia are trained to harvest tree crops, such as coconuts and stinking beans (*Parkia* sp.) (the seeds are used for culinary purposes across most of Southeast Asia), and are therefore valued for their utilitarian and/or economic benefits (Richard *et al.* 1989; Sponsel *et al.* 2002). In other regions, macaques are primarily valued for socio-cultural and aesthetic reasons and their ability to generate financial benefits via tourism, as exemplified by macaques on the islands of Bali or Sulawesi, Indonesia (Fuentes *et al.* 2005; Riley and Fuentes 2011; Riley 2010). Other examples where primates are protected because of cultural traditions among different groups of people or localities include Sclater's guenon (*Cercopithecus sclateri*) in Nigeria (Baker *et al.* 2009); chimpanzees (*P. t. verus*) at Bossou, Guinea (Kortlandt 1986; Humle and Kormos 2011); bonobos (*Pan paniscus*) at Wamba, Democratic Republic of Congo (Lingomo and Kimura 2009); gorillas (*Gorilla gorilla*) in the Central African Republic (Remis and Hardin 2009); Geoffroy's black-and-white colobus (*Colobus vellerosus*) at the Boabeng-Fiema Monkey Sanctuary, Ghana (Saj *et al.* 2006); ring-tailed lemurs (*Lemur catta*) and Verreaux's sifaka (*Propithecus verreauxi*) at the Beza Mahafaly Special Reserve, Madagascar (Loudon *et al.* 2006); and a range of Neotropical primates including taxa such as capuchin (*Cebus* sp. and *Sapajus* sp.), howler (*Alouatta* sp.), owl (*Aotus* sp.), tamarin (*Saguinus* sp.), uakari (*Cacajoa* sp.), woolly (*Lagotrix* sp.), and spider (*Ateles* sp.) monkey species across different regions of the Amazon Basin (Cormier 2006).

At first glance, 'conflicts' that arise in association with negative interactions between people and wildlife (referred to as 'human–wildlife impacts' by Redpath *et al.* (2013)) often appear to be about the effects of wildlife behaviour on people's crops, livestock, property, or even human safety. However, the majority of these scenarios can be attributed in some shape or form to conflicts between different groups of people and divergences in how they value a species, power differentials and feelings of exclusion from conservation planning and the political process, lack of autonomy over natural resources, and even aspects of cultural and social identity (Dickman 2010; Knight 2000; Madden and McQuinn 2014; Redpath *et al.* 2013). Unfortunately, a common scenario concerns tension between local people, whose wellbeing and livelihoods are directly affected by interactions with wildlife, and conservationists, who seek to protect biodiversity and ecosystem function. Many such scenarios arise around protected areas. Since protected areas imply legal prohibitions on resource use and on killing wildlife (whether for bushmeat or in the context of protecting one's crops), the presence of protected areas often exacerbates (1) negative perceptions that people hold towards wildlife and local authorities and (2) the impact of wildlife on people's crops and property. For example, post gazetting of the Bwindi Impenetrable National Park, Uganda, East Africa, people were no longer allowed to chase gorillas (*G. beringei beringei*) back into the forest (Hockings and Humle 2009) or pursue commercial activities such as mining or pit sawing within park boundaries (Baker *et al.* 2012). This situation, as well as gorilla damage to banana palms, coffee bushes, and eucalyptus trees at the forest edge (J. Byamukama pers. comm. 2008), exacerbated conflict between local communities and park rangers (Baker *et al.* 2012). More recently, the gorillas have also incorporated maize and beans into their diet and, unfortunately, these two crops are important subsistence crops for people locally (J. Byamukama pers. comm. 2008). However, although farmers do now consider gorillas a problem species, the loss of subsistence crops at the forest edge remains a relatively minor trigger of conflict between park rangers and local people. Indeed, restrictions on extractive commercial endeavours within park boundaries, which are mostly affecting resource-poor households seeking additional sources of income, underlie most social conflict events in the area (Baker *et al.* 2012). In addition, the lack of an effective benefit-sharing scheme for tourism revenues generated from mountain gorilla tracking and viewing is further exacerbating conflict among stake-holders locally, affecting people's value systems and attitudes towards gorillas around Bwindi (Sandbrook and Adams 2012).

14.3 Which primate species and where?

Two thirds of peer-reviewed research articles published between 1990 and 2013 ($N = 75$) returned via a search in the Web of Science, using the following key words (1) 'ethnoprimatology';[1] (2) 'primate' and (a) 'crop damage', (b) 'crop raiding', (c) 'human–wildlife conflict', and (d) 'interactions with humans (or) people', and published between 1990 and 2013 ($N = 75$),[2] focus on cercopithecoids, that is cercopithecine and colobine species, while a quarter focus on great apes. The majority of these studies are from Africa (59%), followed by Asia (29%) and South and Central America (12%), with Uganda (23%) and Indonesia (13%) currently acting as the central foci for research in this area. Although this restricted search failed to capture all the published literature, these results do provide us with a sense of the taxonomic and geographical biases of current research efforts addressing those topics.

Although not all primates are able to survive in close proximity to people, some Old World monkeys, that is cercopithecoids, have adapted particularly well to living alongside humans. Characteristics that influence species' capacity to co-exist successfully with people are outlined in Table 14.1. The most successful extant genera of primates include three omnivorous taxa of cercopithecines, that is macaques (*Macaca* sp.), baboons (*Papio* sp.), and members of the *Chlorocebus* genus (includes vervet, grivet, green, and tantalus monkeys). Members of these genera are predominantly terrestrial, mainly generalist feeders, and occupy a wide range of habitat types and anthropogenic areas, including roadsides, temples, tourist resorts, and urban to semi-urban centres (Lee and Priston 2005; Priston and McLennan 2013). Cercopithecines also have the added advantage of cheek pouches, allowing them to retreat to a safe place to eat, thus potentially alleviating feeding competition and reducing the risk of predation (or retaliation by people) (Lambert 2005). These genera typically exhibit female philopatry (i.e. females remain within their natal group) and male dispersal, with group size varying from approximately 10 to over 200 individuals. Macaque troops occurring in areas of human sympatry, particularly urban or semi-urban areas, tend to be larger and have a higher infant to adult female ratio than those in more remote areas as a result of exploiting human resources and/or being provisioned (e.g. long-tailed macaques (*M. fascicularis*), Sha *et al.* (2009)).

Apes, especially chimpanzees and orang-utans, and to some degree gorillas, have the adaptive and the behavioural flexibility to survive in human-modified landscapes under particular circumstances (Hockings and Humle 2009; Meijaard *et al.* 2010) (see also Meijaard, Chapter 13, this volume). However, their large body size, proclivity

[1] 'theoretically and methodologically interdisciplinary study of the multifarious interactions and interfaces between humans and other primates' (Fuentes 2012: 102)

[2] Each article was then reviewed for relevance and categorized based on focal primate taxon and geographical region.

Table 14.1 Characteristics and factors that influence species-specific vulnerability or resilience to living at the interface with humans and the likelihood that people locally will sustain positive perceptions of the species. (NB: listed characteristics or factors are not mutually exclusive, as some species may exhibit combinations of these.)

More vulnerable to impacts of sharing landscapes	More resilient to impacts of sharing landscapes
Greater age at maturity and longer inter-birth interval	Younger age at maturity and shorter inter-birth interval
Typically single offspring per birth	Typically more than one infant per birth
High susceptibility to stress and/or human diseases	Low susceptibility to stress and/or human diseases
(a) Reduced/limited ability to transport food away from source	(b) High ability to transport food (e.g. hands, cheek pouch)
(c) Vulnerable to social disruption (e.g. demographics, inflexible social organization)	(d) Resilient to social disruption
(e) Limited dispersal capabilities or opportunities	(f) High dispersal capabilities or opportunities
(g) Low degree of behavioural plasticity/flexibility	(h) High degree of behavioural plasticity/flexibility
(i) Narrow diet/Specialist	(j) Broad diet/Generalist
(k) Small home range or territory	(l) Larger home range or territory and flexible ranging patterns
(m) Mainly arboreal	(n) Terrestrial or semi-terrestrial
(o) **Lower likelihood of being viewed positively**	(p) **Higher likelihood of being viewed positively**
(q) Large body size, and/or highly visible	(r) Smaller body size and/or cryptic or secretive
(s) Species that compete with humans for resources (e.g. food, water)	(t) Species that minimally compete with humans for resources
(u) Diurnal activity	(v) Night activity/nocturnal behaviour
(w) Exhibit aggressive/threatening behaviour towards people	(x) Discreet/non-threatening behaviour towards people
(y) Of low economic, aesthetic, or cultural value when alive	(z) Of high economic, aesthetic, or cultural value when alive
(aa) Of high utilitarian and economic value when dead	(bb) Of low utilitarian and economic value when dead
(cc) Unprotected and/or of low conservation status	(dd) Protected by taboos or legislation

to forage on cultivars, and ability to range in human-modified landscapes can also fuel people's negative perceptions and attitudes, putting them at risk of retaliation and of human intolerance (e.g. Campbell-Smith *et al.* 2010; McLennan and Hill 2012) (Table 14.1). Nevertheless, great apes' ability to solve problems, learn socially, cooperate, incorporate diverse food types into their diet (although most are mainly frugivorous), and access embedded or hard to process foods, whether by hand or by using tools, facilitates their ability to survive in anthropogenic landscapes (see Figure 14.1). Other species such as long-tail macaques (e.g. Thailand: Gumert *et al.* (2009)) and capuchin monkeys (e.g. de Freitas *et al.* 2008; McKinney 2011) also share similar abilities and propensities. However, smaller-sized and less behaviourally flexible or cognitively advanced primates, such as guenon species, including red-tailed monkeys (*Cercopithecus ascanius*), also employ effective strategies at the forest edge by adopting solitary, cryptic behaviours when foraging on crops (Baranga *et al.* 2012; Wallace and Hill 2012).

14.4 The changing landscapes of primate interactions and adaptation

14.4.1 Primates' behavioural and social plasticity

Habitat encroachment, destruction, and fragmentation are some of the main drivers forcing primates into competition for resources with humans. Such events may be influenced by overwhelming events,

Figure 14.1 Young Bossou chimpanzee cracking oil-palm nuts using a stone anvil and hammer in a recently cultivated field at the forest edge, with an adult male sitting in the background, in Guinea, West Africa (photo © T. Humle).

such as the dry conditions caused by El Niño and the consequent forest fires that destroyed primate habitat in Indonesia, and impacts of climate change on the environment (Irvin, Chapter 7, this volume; Korstjens and Hillyer, Chapter 11, this volume). However, other human-induced changes to the landscape are responsible for fuelling most negative interactions between primates and people. Increased demand for arable land and rapid encroachment into primate habitat increases the likelihood of encounters between primates and people or property, including crops. Industrial or commercial development projects also exacerbate primate habitat loss, often resulting in large influxes of people, increasing primates' exposure to human activity and the risk of inter-species disease transmission. Additionally, extractive or agricultural industries can intensify the level of hunting in an area, as employees hunt wildlife (including primates) for food, or road infrastructure development facilitates hunter access and commercial trade of bushmeat (e.g. Poulsen *et al.* 2009; Wilkie and Carpenter 1999; Wilkie *et al.* 2000).

The development of commercial plantations and expansion of monocultures, such as oil-palm, rubber, acacia or eucalyptus, across Indonesia and parts of Malaysia, have marginalized and isolated orang-utans, forcing individuals to become obligate or semi-obligate crop feeders to survive (Campbell-Smith *et al.* 2011a, b; Meijaard *et al.* 2010). Some orang-utans occupying these rapidly changing landscapes are unable to meet their nutritional needs (Ancrenaz *et al.* 2008). These individuals experience elevated stress, necessitating rescue and placement in rehabilitation centres that are sometimes unable to cope with this influx of animals while also sustaining release efforts across a limited number of remaining suitable areas (Ancrenaz *et al.* 2008; Robins *et al.* 2013). This is of particular concern because, like chimpanzees in many parts of Africa, the majority of wild orang-utans (approx. 75%) occur outside protected areas, where land is generally managed to meet human needs and its use underpins economic development (Wich *et al.* 2012). However, Campbell-Smith *et al.* (2011b) confirmed that Sumatran orang-utans (*P. abelii*) can adapt to living in some agroforestry landscapes, but only where tolerated by local people. Meijaard *et al.* (2010) also demonstrated that, in the short term at least, Eastern Bornean orang-utans (*P. p. morio*) can survive in relatively high densities in plantation landscapes dominated by *Acacia* spp., though it remains unknown how long they can persist under these conditions, or at what cost to apes, plantation owners, or labourers.

Primates demonstrate an array of ecological and behavioural responses to living at interfaces with humans. Shifts in ranging behaviour as a result of human presence or activity can vary significantly among species. In some cases, chimpanzees may shift their range away from human activity, such as in logging concessions (Morgan *et al.* 2013); by contrast, macaques may restrict their ranging area and inflate their density in response to favourable conditions associated with provisioning, as found around temples and tourist areas (e.g. Fuentes *et al.* 2005; Riley 2008). However, Berman *et al.* (2007) showed that such range restriction among macaque populations around tourist areas with ongoing provisioning also comes at a cost. Aggression rates are generally elevated among individual macaques in restricted provisioned areas and, over time, the rates are positively correlated with infant mortality. Provisioning is well known to influence aspects of population life history (e.g. birth rate, life span, and reproductive parameters) and demography, and to be associated generally with increased intra-group aggression (Hill 1999), reduced activity rates, and obesity, especially among more dominant individuals who have priority access to food (Fuentes *et al.* 2007; Zhao 2005).

Similarly, crops are easily digested, calorie-dense foods, and feeding on crops may benefit the reproductive success of particular populations, providing the species is not locally hunted and there is no retaliation from farmers. A crop-feeding baboon troop (*Papio anubis*) around the Gashaka Gumti National Park in Nigeria, and a crop-feeding community of chimpanzees (*Pan troglodytes verus*) in Bossou, Guinea, West Africa, show significantly shorter inter-birth intervals (IBI) and higher infant survival rates compared with conspecifics more dependent on wild foods (baboons: Higham *et al.* (2009); chimpanzees: Sugiyama and Fujita (2011)).

Primates may also shift their social organization and modify their social associations depending on their environment. For example, forest-dwelling populations of bonnet macaques (*M. radiata*) typically form multi-male and multi-female groups. However, provisioned troops across peninsular India tend to adopt a single-male social structure, with multiple females and a high tendency for female dispersal, atypical of other cercopithecine primates (Sinha *et al.* 2005). Sapolsky and Share (2004) reported an outbreak of tuberculosis among a troop of semi-urban chacma baboons causing a social shift in the troop to a more 'relaxed' dominance hierarchy that persisted 10 years on. The tuberculosis outbreak affected the dominant, more aggressive adult males who had priority access to scavenged garbage; their death selectively resulted in a surviving cohort of atypically unaggressive males. Such social plasticity is not unique to cercopithecines. Chimpanzees at Bossou, Guinea, are more cohesive during crop-feeding and road-crossing events (Hockings *et al.* 2012, 2006). Solitary red-tail monkeys (*C. ascanius*), which are harder to detect than when in social groups (Wallace and Hill 2012), venture further from the forest edge and consequently cause proportionally greater damage than animals foraging in groups (Baranga *et al.* 2012). These results illustrate the high level of behavioural and social plasticity of some primates living in human-modified landscapes and the potential profound impact of zoonotic diseases on their demographics and social behaviour.

14.4.2 Vulnerability, risk perception, and habituation

Farmers sometimes describe primates and their behaviour in anthropomorphic terms, that is, as if they were human. For example, farmers living around the edge of the Budongo Forest Reserve in Uganda refer to baboons as 'vindictive, damaging crops for the sake of it rather than for food alone' (Hill 2000), and as 'enemies' or 'rebels' (Hill and Webber 2010). Where people hold negative attitudes towards primates and other wildlife species, this ultimately intensifies people's perceptions of risk associated with these animals (Naughton-Treves 1997) (see Table 14.1).

Risk perception implies an intuitive assessment of the risks to one's safety, property, wellbeing, and welfare (Smith *et al.* 2000). The concept of risk perception can be applied to both primates and people, as risk perception affects the way people will behave in the presence or hypothetical presence of primates, but also the way primates will behave in response to the presence of people and anthropogenic changes to the landscape. An increasing number of studies have highlighted the importance of evaluating the mismatch between the quantified (real/measured) and subjective (perceived) frequency of incidences (occurrence) of crop-foraging events, and the measured and perceived severity of damage attributed to different species in a specific locality. For example, around the Budongo Forest Reserve, Uganda, baboons are responsible for a large proportion of crop damage in comparison to other species, but the farmers' perception of risk is often disproportionately high when compared with measured damage (Hill and Webber 2010). Such misperceptions are quite common especially when a species may be feared by people (e.g. orang-utans, Sumatra, Indonesia: Campbell-Smith *et al.* (2010); baboons, Kibale National Park, Uganda: Naughton-Treves (1997); red colobus, Zanzibar: Siex and Struhsaker (1999); various primate species, Hoima District, Uganda: Hiser (2012)).

Encouraging people to speak for themselves and voice their worries can eventually help identify interconnected issues that underlie people's insecurities and concerns. This process can also serve to highlight misperceptions and help to inform awareness-raising campaigns and prioritize the management of negative interactions between people and wildlife, including primates. Finally, such an approach can also help us understand how direct negative experiences and socio-economic, cultural, religious, and gender factors influence people's perceptions and behaviour.

Distance to the forest edge or a safe place to retreat and people's response to wildlife foraging on crops may also influence primates' perception of risk. For example, Naughton-Treves (1997, 1998) demonstrated that forest-dwelling wildlife, including chimpanzees, baboons, and red-tailed monkeys living in the Kibale National Park in Uganda, were more likely to damage crops in fields within 500 m of the forest. Similar patterns have been demonstrated across a range of sites and species inhabiting protected areas (e.g. Hill 1997; Sitati *et al.* 2003; Linkie *et al.* 2007; Priston *et al.* 2012). Foraging on crops in fields or orchards where such incursions are not tolerated is potentially a very risky behaviour for primates; they are likely to be chased away, injured, or even killed. This pattern explains why in some cases adult males tend to forage on crops more frequently than adult females or subadults (e.g. Hanuman langurs (*Semnopithecus entellus*), Chhangani and Mohnot (2004); vervets (*Chlorocebus aethiops pygerthrus*), Saj *et al.* (1999); Anubis baboons (*Papio anubis*), Forthman-Quick (1986); chimpanzees (*P. troglodytes*), Hockings (2007) and Wilson *et al.* (2007)). However, if risk associated with foraging on crops is small (e.g. low levels of retaliation and intolerance from farmers), and if intra-specific competition is low, then females may forage on crops as frequently as males, perhaps even more because of their need to meet reproductive demands, as exemplified in a study of wild Sumatran orang-utans (*Pongo abelii*) in an agroforestry landscape (Campbell-Smith *et al.* 2011b).

Reduced fear of humans among primates can also exacerbate primate–people interactions and elicit negative shifts in people's perceptions of them. For example, one quarter of mountain gorillas in Uganda's Bwindi Impenetrable National Park are reported to visit farms and plantations neighbouring the park, including habituated and unhabituated groups (Asuma pers. comm. 2008). Bwindi gorillas apparently foraged on crops before habituation for tourism began. However, prior to the gazetting of the national park, gorillas only infrequently ventured outside the current park boundaries and were easily chased away. Unfortunately, loss of fear of humans due to habituation for tourist viewing may have since heightened the gorillas' assertiveness when foraging on crops (Hockings and Humle 2009; Macfie and Williamson 2010). Reduced fear of humans can indeed exacerbate the frequency of crop damage events, encourage primates' presence around human activity, and increase the frequency of incursions into human settlements, as well as their use of roads and paths (e.g. great apes: Hockings and Humle (2009) and Hockings *et al.* (2006)). The effect of habituation, combined with the exodus of people away from rural areas into urban conglomerates in countries such as Japan, is also accentuating these issues (Knight 2000; Watanabe and Muroyama 2005). The increasingly elderly population in rural Japan is less able to chase off animals feeding on their crops and implement prevention strategies. Japanese macaques (*M. fuscata*) and wild boars (*Sus scrofa*) apparently thrive particularly well in such areas since risk associated with foraging on crops is much reduced (Knight 2000).

Where primates are habituated to human presence and when primates perceive people as threatening, the risk of aggressive acts between the two can be heightened (Hockings and Humle 2009). For example, all aggressive events reported between

1995 and 2009 between habituated chimpanzees and people at Bossou, Guinea, were linked to some sort of human provocation (Hockings *et al.* 2010). These events may vary in the temporal interval between provocation and potential retaliation by chimpanzees, especially since chimpanzees demonstrate a capacity for episodic memory (the what, where, and when) (Martin-Ordas *et al.* 2010) and for facial recognition (Tomonaga 1999). However, more recently, as people at Bossou have become less tolerant of chimpanzee foraging on their crops, some acts of aggression of chimpanzees towards people appear to reflect displacement acts of aggression rather than acts of retaliation or 'revenge' per se. For example, chimpanzees that were chased away from a field using sling shots subsequently behaved belligerently towards 'innocent' children who were travelling along a nearby road and whom the chimpanzees encountered shortly after the event; a situation which was in this case managed by the research field assistants present nearby (Humle pers. obs.). McLennan and Hill (2013) noted that unhabituated chimpanzees at Bulindi in Uganda started stalking researchers and their field assistants soon after an outbreak of small-scale logging in local forest patches. Provocation and/or a feeling of insecurity are clear drivers of aggression, whether of primates towards people or vice versa. Indeed, risk perception may affect inter-specific aggression and/or intolerance, as the landscape shifts from being stable and secure to less predictable and more precarious for both wildlife and people. Koutstaal (2013) described anecdotal cases of how people's behaviour towards, and encounters with, wild chacma baboons (*Papio ursinus*) in the Cape Peninsular of South Africa underlie the animals' behaviour towards people, from curious and harmless to aggressive depending on people's response to their presence. Hurn (2011) further detailed how growing fear among tourists and residents promulgated aggressive management measures directed at baboons, from paintball guns to euthanasia of 'problem' individuals. Such cases exemplify how different primate populations, groups and/or individuals, including humans, may display different tipping points in tolerance capacity, and how perceptions influence people–primate interactions.

14.4.3 Predicting human intolerance and its consequences on primates

People's tolerance towards wildlife that damage crops is influenced by an array of ecological, socio-cultural, and economic factors (Naughton *et al.* 1999; Naughton-Treves 1997). Ecological factors such as (1) crop attributes, such as size, value, seasonality, and growth patterns; (2) patterns in crop loss and damage, for example the timing of crop-loss events relative to harvest, crop part damaged, circadian timing of damage event, and the extent and frequency of the damage; and (3) landscape attributes, such as proximity to forest edge, field size and habitat heterogeneity, and crop diversity, are known to influence the probability and the extent of crop damage and therefore people's tolerance of wildlife-related crop loss (Hill 2000; Naughton-Treves 1997; Hill, 1997). However, predicting tolerance is not always a straightforward affair. While some studies suggest that people are more likely to resent wildlife damage to staple crops, such as rice or cassava (Mascarenhas 1971), others indicate that farmers are less tolerant of damage to high-value cash crops (Hockings and McLennan 2012). Such differences may reflect variations in people's perceptions and expectations as influenced by their cultural and educational backgrounds, social status, gender, and their capital and labour investment, as well as their financial security, needs, and aspirations (Naughton-Treves 1997).

In some cases, intolerance may result in the deliberate killing, lethal control, retaliation, persecution, or retributive or defensive killing of individuals. This is illegal when endangered species are concerned, creating conflicts between farmers, government, and/or protected area authorities and conservationists. Meijaard *et al.* (2011) carried out social surveys across several hundred villages and amassed nearly 7,000 responses from across Borneo, Indonesia. Nearly a quarter of those who reported 'conflict' with orang-utans also reported personally killing an orang-utan, as opposed to 7% of respondents who reported no 'conflict'. However, there did not appear to be a significant relationship between killing rates and frequency of conflict. As previously discussed, retaliatory or retributive killings can affect the local survival of a species, its ranging behaviour, its behaviour towards humans, its social organization, its genetics, its ecology and the ecosystem, and may also exacerbate the risk of zoonotic disease transmission. Great apes are particularly vulnerable as a result of their life history (see Table 14.1); they reproduce slowly and are therefore vulnerable to demographic disturbances. Minimum age at first pregnancy for great apes is between 8 and 15 years and the inter-birth interval is typically between 4 and 9 years (Wich *et al.* 2004; Williamson *et al.* 2013). Demographic recovery can be particularly slow, especially since infant mortality can also be relatively high in the first year, for example as high as 20% among some chimpanzee communities (Hill *et al.* 2001).

Legal, lethal control of non-threatened primate species also occurs and may be promoted and managed by government authorities and/or non-governmental organizations (NGOs) locally. Such is the case in South Africa with chacma baboons, where adult males are on occasion trapped and euthanized because they are considered too dangerous, threatening, and intrusive (Hurn 2011). It was rumoured that the Department of Wildlife and National Parks of Malaysia culled nearly 100,000 macaques in 2012 alone (Vinod 2013). However, plans to selectively remove or cull individuals could impact the population's genetic health, especially if the effective size of the population concerned is small (Sukumar 1991).

Lethal control was adopted as a strategy to deal with invasive rhesus and patas monkeys in Puerto Rico (Engeman *et al.* 2010). These monkeys were introduced to coastal islets in secure breeding facilities in the 1930s for medical research purposes for a period of more than 40 years. Escapees joined the mainland of southwestern Puerto Rico and quickly adapted to foraging on crops to survive. The Department of Agriculture estimated the total economic losses by commercial farmers ranged between USD1.13 million to USD1.46 million per year between 2002 and 2006; this is an underestimate because it fails to account for the destruction of native wildlife, the threat of disease spread—that is herpes and hepatitis, and property damage. A major campaign aimed at mitigating the problem resulted in the culling of 800 monkeys (mostly patas) in 2008 (Lin 2010). Such management strategies are highly contentious, ignite severe criticism from the public and animal welfare groups, and are of unproven effectiveness. For example, Quirin and Dixon (2012) revealed in a study in western Ethiopia that a baboon cull that took place in 2004 had a variable effect on farmers' perceptions of current crop losses. Some perceived crop damage had declined, while others thought crop losses due to other species had significantly increased compared with pre-cull levels; this suggests either that culling was ineffective, at least based on people's perceptions, or perhaps that the baboons were not as important a cause of crop losses as initially perceived by local people. The latter highlights the importance of understanding the issue fully before implementing any kind mitigation strategy.

14.5 Prevention and mitigation strategies

14.5.1 Primate translocation and other mitigation strategies

Translocation of primates can sometimes be viewed as the most suitable strategy to manage negative interaction between primates and people, especially if weighed against options such as culling or euthanasia. However, such an option should be considered only when the situation *in situ* is unmanageable and the impact on both people and/or primates is deemed irreducible by any other means (Hockings and Humle 2009). Another challenge facing translocation is that very few data are available on the factors underlining translocation success (Guy *et al.* 2013). Translocations have to date mainly concerned invasive primates, situations where primates hold a special cultural or religious significance for people (Kavanagh and Caldecott 2013), or when the problem(s) or human-induced impact concerns an endangered species, such as in the case of orang-utans in parts of Borneo and Sumatra (Hockings and Humle 2009) or barbary macaques (*M. Sylvanus*) on Gibraltar (Fa and Lind 1996).

However, translocations are most often not well managed and planned. For example, in India, protests against government apathy towards reported issues with urban primate damage to property and harassment of people encouraged the authorities to resort to indiscriminate trapping and release of individuals in rural areas (Pirta *et al.* 1997). Problems quickly arose in these areas between translocated individuals and local people, who were unused to bold and threatening urban primates who were quick to pass on their bad habits to their more naïve rural conspecifics (T. Simlai pers. comm.). Such unplanned and mismanaged translocations could spatially disseminate 'conflict issues' and further strain people's religious and cultural values with respect to wildlife in other regions of the country (Priston and McLennan 2013).

Reports of exemplary translocations are unfortunately rare. One such example is that of three olive baboon (*Papio anubis*) troops in an area in Kenya where baboons were perceived as 'pests' and human encroachment was increasingly threatening their habitat (Strum 1994; Beck, Chapter 15, this volume). This translocation involved the capture in 1984 of 131 baboons who were relocated more than 200 km away from their natal area into a drier zone also harbouring wild conspecifics in the Laikipia Plateau. This initiative involved extensive pre- and post-release monitoring of translocated individuals, especially of the Pumphouse Gang (PHG) (Strum 2005). The PHG troop adapted well to its novel environment, demonstrating that translocation can in some cases be a successful strategy (Strum 2005, 2010).

Nevertheless, translocation or relocation is rarely a feasible option for many species, given that suitable habitats are often scarce and the process is ethically and logistically complicated, especially for primate species that are long lived and live in complex social groupings (e.g. Hockings and Humle 2009). In addition, releasing individuals into areas already populated by conspecifics could result in mortalities as a result of intra-specific aggression, especially among males (e.g. chimpanzees: Goosens *et al.* (2005); Humle *et al.* (2011)) or disease transmission if individuals are not appropriately quarantined and tested prior to release (Beck *et al.* 2007; Kavanagh and Caldecott 2013).

Very few studies have tested alternative mitigation techniques. Hill and Wallace (2012) experimented with different locally appropriate methods and developed techniques aimed at reducing crop damage by baboons, chimpanzees, vervets, and red-tail and blue monkeys around Budongo Forest Reserve, Uganda. They developed and trialled four categories of deterrents in collaboration with local farmers: barriers, alarms, repellents, and systematic guarding. Systematic guarding proved highly effective, as did net fencing with bells attached along its length that functioned as an alarm, signalling primate attempts to negotiate the fence. Impenetrable living Jatropha (*Jatropha* sp.) hedges, multistrand barbed wire fences combined with ocimum (*Camphor basil*) planted along the bottom of the fence, and rope fences coated with chilli paste were also effective at reducing primate crop damage. Barbed wire fences on their own showed mixed results, and simple ropes strung with bells were ineffective on their own. These measures varied in their costs and practical implementation, as barbed wire is expensive and a hedge cannot readily be moved around in a landscape characterized by shifting agriculture, although such an approach could be highly effective in protecting permanent gardens. This study also revealed that these measures led to wildlife shifting their attention to unprotected neighbouring farms, displacing the problem and highlighting the importance of implementing mitigation schemes simultaneously across all neighbouring farms. Persistent efforts could eventually lead to a significant decrease in primate crop damage events where the animals have adequate natural forage available, something that should be assessed *a priori*. A year later, farmers around the Budongo Forest Reserve were still using and maintaining most of the barriers and warning systems trialled in the original study, and some neighbouring farmers had also adopted similar methods (Hsiao *et al.* 2013).

Campbell-Smith *et al.* (2012) trialled noise deterrents and netting of trees to deter Sumatran orang-utans from raiding fruit orchards in an agroforestry landscape. Implementing these measures improved local farmers' attitudes towards orang-utans. Netting of trees and noise deterrents proved highly effective when comparing pre-trial and post-trial damage events and there was no difference in crop damage incidents between pre-trial and post-trial control farms. Although netting trees proved most effective, resulting in a significant increase in crop yield, farmers were no longer employing this technique within six months of the end of the study. The authors argued that this technique was more expensive and logistically complex to put in place, and therefore recommended the development of nets that were easier to install. Interestingly, farmers who trialled the tree nets were more likely to inflate crop losses than were control participants, perhaps because they had hopes of receiving compensation (Campbell-Smith *et al.* 2012).

Reducing primate aggression towards people can be promoted by changing people's behaviour towards primates and vice versa (Hockings and Humle 2009), enforcing rules and regulations at tourist sites (Macfie and Williamson 2010), and managing provisioning and waste disposal and recovery (Fuentes *et al.* 2007). In some cases, preventing surprise encounters between people and ground travelling apes, such as chimpanzees, by improving visibility on shared paths could reduce aggressive incidents (Hockings and Humle 2009).

14.5.2 Agricultural practices, land-use management and policy

Land-use development, whether across rural or urban areas, will necessarily impact primate resource and space requirements. Understanding the requirements of displaced and isolated populations of primates is therefore essential for land-use management and conservation planning (Hoffman and O'Riain 2012; Sha *et al.* 2009). However, not all changes to the landscape restrict animal movements. Bali has extremely high human densities of approximately 500 individuals per square kilometre and Balinese macaques (*M. fuscicularis*) occur throughout the island, aside from the capital city (Fuentes *et al.* 2005). Still, macaque groups are substantially or integrally food enhanced, that is their nutritional requirements are met from human provisioning or activity (Wheatley 1999). Incidental or voluntary provisioning occurs at temples or shrines where the majority of macaques occur, and where people commonly make food offerings. Lansing (1991) described how Bali, over the course of millennia, has become a mosaic of riparian forest corridors and forest islands outlined by culturally specific land-use patterns, such as wet-rice agriculture, irrigation systems, and temple complexes. Fuentes *et al.* (2005) claim this landscape has strongly benefitted macaque ranging, foraging, and male dispersal patterns, demonstrating their ability to adapt and maintain genetic viability (Fuentes *et al.* 2006). Landscape features are valuable in promoting individual dispersal but may also help to reduce encounter rates between people and macaques. Indeed, Sha *et al.* (2009) recommend that designing buffer zones around reserve areas could help minimize primate foraging on crops and overlap in land use between people and primates.

Wich *et al.* (2012) argue that conservation planning depends on understanding how the ranges of wild orang-utans and other threatened wildlife overlap with protected areas and commercial-scale or industrial concessions. The current challenge for orang-utan conservation lies in the fact that more than 50% of orang-utan distribution lies in either natural forests, where timber exploitation is permitted but land conversion is prohibited, or undeveloped oil-palm and tree-planting concessions, whose exploitation would conflict with national laws concerned with species protection. Wich *et al.* (2012) therefore suggest, for example, that efforts need to focus on improving yields in current oil-palm plantations and on expanding concessions in deforested areas. It is also urgent to align land-use management, conservation, and development policy with the valuation of ecosystem services of forested areas, such as water provision, flood control, carbon sequestration, and provisioning of natural resources supporting people's livelihoods (Garcia-Ulloa and Koh, Chapter 16, this volume).

Where primates can survive and thrive on natural resources and share the landscape with people, agricultural development should focus on maintaining natural resources, and preserve or promote connectivity between forest fragments, to ensure wildlife population viability and to help minimize the risk of crop damage or loss. Increasing connectivity to preserve and promote access to natural resources alone, however, will not necessarily prevent some species from preferring to forage on cultivars or planted trees (e.g. chimpanzee: Hockings *et al.* (2009)). One way of preventing crop loss or damage might be to switch land-use activities or promote zero- or low-risk crops rather than cultivation of high-risk crops (Hockings and McLennan 2012). Such strategies may not, however, always result in equal or greater economic benefit to farmers or landowners and may therefore be untenable. For example, to avoid damage to fruit and vegetable crops by invasive patas and rhesus monkeys on Puerto Rico, some landowners converted their land to pasture for livestock or the production of forage; however, this land-use shift resulted in significant economic losses for the landowners making this an unviable option for them (Engeman *et al.* 2010). Nevertheless, some crops could help balance both economic and conservation objectives. Hockings and Sousa (2012) demonstrated that cashew (*Anacardium occidentalis*) production across a forested–agricultural matrix around the Cantanhez National Park in Guinea-Bissau, West Africa, can benefit both wild chimpanzees and people, providing an example of co-utilization. While this tree species is of high economic value, it is also nutritionally beneficial to wild chimpanzees. Chimpanzees focus on the fleshy part of the fruit leaving behind the valuable casing, where the seed, that is the cashew nut, is located, for farmers to harvest. Although this crop species appears to meet both livelihood and conservation objectives, it must be noted that unmanaged expansion of cashew plantations, or any other low-conflict crop of high market value, could result in significant habitat loss for wild chimpanzees and other wildlife. Such risks highlight the necessity of promoting mixed approaches to rendering agricultural development compatible with wildlife conservation.

14.5.3 Increasing mutual tolerance

Sillero-Zuberi *et al.* (2007) proposed an 'impact reducing scheme' aimed at mitigating 'conflict' issues based on whether problems can be reduced effectively or not, leaving behind residual impact, that is impact that is currently irreducible (see Figure 14.2). According to this scheme, the only mechanism by which residual impact can be managed is by influencing people's tolerance capacity via different approaches including education or direct

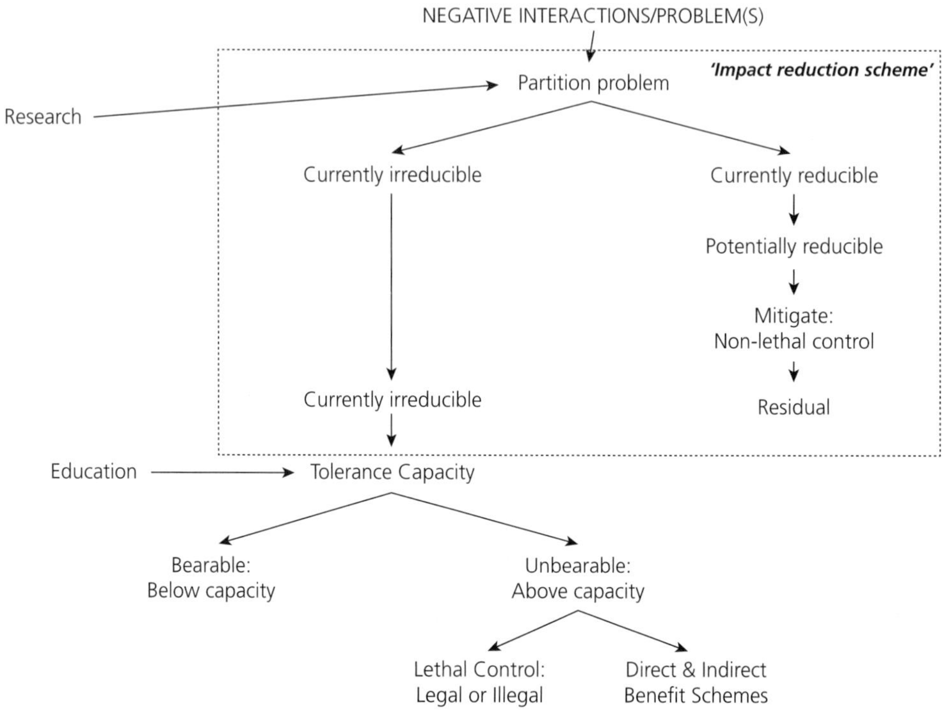

Figure 14.2 'Impact reduction scheme'. This figure is adapted from figure in 1st edition of *Topics in Conservation Biology* (2007)—chapter by Sillero-Zuberi *et al.*

and indirect benefit schemes. However, factors influencing people's willingness or capacity to tolerate sharing landscapes with wildlife are not well understood (Treves and Bruskotter 2014). It is often assumed that a lack of tolerance towards wildlife, including primates, is a consequence of people's concerns about economic losses. While this is often a focus of people's expressed concerns, it is not always the primary factor affecting people's tolerance towards their animal neighbours. For example, research by Marchini and Macdonald demonstrates that ranchers' expression of intent to kill jaguars is not necessarily a retaliatory response to livestock losses or even perceived threats to people, but is better explained by 'social norms', that is, ranchers were more likely to kill or threaten to kill a jaguar if that was the locally acceptable response (Marchini and Macdonald 2012). Recent research in Uganda has revealed that while people generally do not regard chimpanzees as a significant threat to their crops, the fact that people fear them is an important determinant of people's attitudes towards them, and perhaps their willingness to tolerate chimpanzee presence within a shared landscape (McLennan and Hill 2012) (see Table 14.1). The important point here is that, while most researchers concur that increasing people's willingness or capacity to tolerate wildlife, including primates, is key to improving relationships between people and wildlife, how improved tolerance is to be achieved is not necessarily straightforward or obvious. Should increasing tolerance not abate the problem, then it is expected that lethal control, whether legal or not, will ensue. Community outreach programmes aim to help rural villagers acquire or develop practical skills or new tools for defending their crops and livestock, for managing waste from its handling, collection, transportation to its disposal, for minimizing the risk of attack when faced with primates, and for switching from high to low 'conflict' crops (e.g. Hockings and Humle 2009). Under an optimistic scenario, such an approach could strengthen local capacity for 'conflict' prevention and resolution, and change people's perceptions, attitudes, and behaviour towards primates. It could also result in reduced risks for both people and primates, improvements in people's livelihoods, and a reduction in their vulnerability to negative interactions with primates. This could potentially help to promote people's commitment towards conservation and raise awareness of the essential role of primates in ecosystem functioning and their intrinsic and economic importance.

Direct compensation schemes designed to increase people's tolerance levels to damage caused by primates and to prevent retaliation are often funded by conservation organizations, although government compensation programmes also exist. However, at best, compensation can only address the symptoms of the problem and not its root causes (Bulte and Rondeau 2005; Hoare 2001), which often have less to do with the animals' behaviour than conflict and disagreement between different human groups (Dickman 2010; Madden and McQuinn 2014). The failure of most direct compensation schemes can be attributed to bureaucratic inadequacies, corruption, fraudulent claims, and the practical barriers that less literate farmers must overcome to submit a compensation claim (Hoare 2001). Such schemes are also difficult to manage logistically and are most often financially unsustainable (Dickman *et al.* 2011; Nyhus *et al.* 2003).

Indirect compensation programmes typically imply wildlife valuing schemes through their approach, often requiring people to stop using, or limit their use of, certain natural resources in exchange for compensatory incentives that might include support of local infrastructure through provision of schools, health centres, or access roads, for example, or the development of alternative income streams through agricultural or livestock initiatives or micro-business opportunities, such as Community-Based Natural Resource Management (CBNRM) programmes (Hockings and Humle 2009) or Integrated Conservation and Development (ICD) schemes (e.g. Blomley *et al.* 2010). But such benefit-sharing approaches are expensive and require sustained funding and logistical support (Hockings and Humle 2009). Lamarque *et al.* (2009) highlight several other issues related to indirect compensation schemes. Generated income is often insufficient to counter losses and damages, let alone to share these revenues with neighbouring communities. Compensation schemes also suffer from issues of ownership, participation, administrative arrangements, and disbursement of income; to be effective these need to be universally agreed upon before any venture is attempted. However, other schemes have been developed with potential to increase people's tolerance of primates and wildlife in general. Conservation through Public Health (CTPH) is one such example. This NGO based in Uganda runs programmes aimed at improving public, livestock, and wildlife health in order to minimize the risk of zoonosis around gorilla habitat. They also facilitate family planning and assist with local-level development and raising environmental awareness by providing people with Internet access and training them in using information and communication technology. A similar programme linking human and environmental health is ongoing in west Kalimantan, Indonesia, across 23 villages that surround the Gunung Palung National Park.[3] However, the effectiveness of these programmes in increasing tolerance towards endangered wildlife and primates in particular remains to be established.

Innovative compensation schemes also include insurance schemes whereby farmers pay a premium for coverage against a defined risk (Lamarque *et al.* 2009). Such schemes require an accurate assessment of the cause of crop damage or injury caused by wildlife. Because such schemes can operate on a more local scale, it is easier to verify reported cases. However, the scheme's efficacy can be improved if certain practices are imposed upon participants, for example if farmers are proactive in preventing crop foraging by wildlife by adopting deterrent techniques. Such schemes have to date primarily been applied to livestock depredation. It therefore remains to be ascertained whether such an approach could help with addressing issues between people and primates.

14.6 Conclusion: a matter of values?

This chapter highlights the complexities of interactions between people and primates and the different scales of interactions influencing co-existence, stressing the potentially significant impact of conflict among human stake-holders on the relationship between people and primates. Clearly some primate species are more able than others to accommodate themselves to changing landscapes and co-habitation with humans. Co-existence depends on a multitude of often temporally and spatially dynamic factors that affect socio-economic and cultural norms and values, as well as people's and primates' perceptions. Understanding animal–human interactions requires a multifaceted approach that brings together a detailed understanding of the context and perspectives of both the primates and people and relevant organizations and institutions, as well as the drivers of change of patterns of co-existence. Failing to grasp and address these could represent one of the biggest current threats to the long-term survival of many primate species across the globe. As demonstrated above, no single approach can mitigate or prevent all negative interactions between people and animals. Disentangling issues at stake on a case-by-case basis can help to inform grassroots schemes and policy across a wider landscape and to manage contexts demonstrating deteriorating relationships between people and primates.

[3] http://www.healthinharmony.org/asri/klinik-asri/.

Acknowledgements

We are grateful to Augustin Fuentes and Gail Campbell-Smith for the constructive comments they provided on an earlier draft of this chapter.

References

Ancrenaz, M., Marshall, A., Goossens, B., Van Schaik, C., Sugardjito, J., *et al.* (2008). *Pongo pygmaeus* [Online]. IUCN 2013. IUCN Red List of Threatened Species. Version 2013.2. Available at: www.iucnredlist.org [Accessed 22 March 2014].

Baker, J., Milner-Gulland, E. J., and Leader-Williams, N. (2012). Park gazettement and integrated conservation and development as factors in community conflict at Bwindi Impenetrable Forest, Uganda. *Conservation Biology* **26**: 160–170.

Baker, L. R., Tanimola, A. A., Olubode, O. S., and Garshelis, D. L. (2009). Distribution and abundance of sacred monkeys in Igboland, Southern Nigeria. *American Journal of Primatology* **71**: 574–586.

Baranga, D., Basuta, G. I., Teichroeb, J. A., and Chapman, C. A. (2012). Crop raiding patterns of solitary and social groups of red-tailed monkeys on cocoa pods in Uganda. *Tropical Conservation Science* **5**: 104–111.

Barua, M., Bhagwat, S. A., and Jadhav, S. (2013). The hidden dimensions of human–wildlife conflict: health impacts, opportunity and transaction costs. *Biological Conservation* **157**: 309–316.

Bearder, S. K., Nekaris, K. A. I., and Buzzell, C. A. (2002). *Dangers in the Night: Are some Nocturnal Primates Afraid of the Dark?* Cambridge: Cambridge University Press.

Beck, B., Walkup, K., Rodrigues, M., Unwin, S., Travis, D., and Stoinski, T. (2007). *Best Practice Guidelines for the Reintroduction of Great Apes.* Gland, Switzerland: SSC Primate Specialist Group of the World Conservation Union.

Berman, C. M., Li, J., Ogawa, H., Ionica, C., and Yin, H. (2007). Primate tourism, range restriction, and infant risk among macaca thibetana at Mt. Huangshan, China. *International Journal of Primatology* **28**: 1123–1141.

Blomley, T., Namara, A., Mcneilage, A., Franks, P., Rainer, H., *et al.* (2010). Assessing fifteen years of integrated conservation and development in south-western Uganda. In: IIED (Ed.), *Natural Resource Issues*. London: IIED.

Bulte, E. H. and Rondeau, D. (2005). Why compensating wildlife damages may be bad for conservation. *Journal of Wildlife Management* **69**: 14–19.

Butler, J. R. A. (2000). The economic costs of wildlife predation on livestock in Gokwe communal land, Zimbabwe. *African Journal of Ecology* **38**: 23–30.

Campbell-Smith, G., Simanjorang, H. V. P., Leader-Williams, N., and Linkie, M. (2010). Local attitudes and perceptions toward crop-raiding by orangutans (*Pongo abelii*) and other nonhuman primates in Northern Sumatra, Indonesia. *American Journal of Primatology* **72**: 866–876.

Campbell-Smith, G., Campbell-Smith, M., Singleton, I., and Linkie, M. (2011a). Apes in space: saving an imperilled orangutan population in Sumatra. *PLoS One* **6**.

Campbell-Smith, G., Campbell-Smith, M., Singleton, I., and Linkie, M. (2011b). Raiders of the lost bark: orangutan foraging strategies in a degraded landscape. *PLoS One* **6**.

Campbell-Smith, G., Sembiring, R., and Linkie, M. (2012). Evaluating the effectiveness of human-orangutan conflict mitigation strategies in Sumatra. *Journal of Applied Ecology* **49**: 367–375.

Carter, J., Ndiaye, S., Pruetz, J., and Mcgrew, W. C. (2003). Senegal. In: Kormos, R., Boesch, C. M. I. B., and Butynski, T. M. (Eds), *Status Survey and Conservation Action Plan: West African Chimpanzees.* Gland, Switzerland and Cambridge, UK: IUCN/SSC Primate Specialist Group.

Chapman, C. A. (1995). Primate seed dispersal: coevolution and conservation implications. *Evolutionary Anthropology* **4**: 74–82.

Chapman, C. A. and Onderdonk, D. A. (1998). Forests without primates: primate/plant codependency. *American Journal of Primatology* **45**: 127–141.

Chapman, C. A., Gillespie, T. R., and Goldberg, T. L. (2005). Primates and the ecology of their infectious diseases: how will anthropogenic change affect host-parasite interactions? *Evolutionary Anthropology* **14**: 134–144.

Chhangani, A. K. and Mohnot, S. M. (2004). Crop raid by Hanuman langur, *Semnopithecus entellus* in and around Aravallis, India and its management. *Primate Report* **29**: 35–47.

Cormier, L. (2006). A preliminary review of neotropical primates in the subsistence and symbolism of indigenous lowland South American peoples. *Ecological and Environmental Anthropology* **2**: 14–32.

Davis, J. T., Mengersen, K., Abram, N. K., Ancrenaz, M., Wells, J. A., *et al.* (2013). It's not just conflict that motivates killing of orangutans. *PLoS One* **8**.

De Freitas, C. H., Setz, E. Z. F., Araujo, A. R. B., and Gobbi, N. (2008). Agricultural crops in the diet of bearded capuchin monkeys, *Cebus libidinosus Spix* (Primates: Cebidae), in forest fragments in southeast Brazil. *Revista Brasileira De Zoologia* **25**: 32–39.

Dickman, A. J. (2010). Complexities of conflict: the importance of considering social factors for effectively resolving human–wildlife conflict. *Animal Conservation* **13**: 458–466.

Dickman, A. J., Macdonald, E. A., and Macdonald, D. W. (2011). A review of financial instruments to pay for predator conservation and encourage human-carnivore coexistence. *Proceedings of the National Academy of Sciences of the United States of America* **108**: 13937–13944.

Engeman, R. M., Laborde, J. E., Constantin, B. U., Shwiff, S. A., Hall, P., *et al.* (2010). The economic impacts to commercial farms from invasive monkeys in Puerto Rico. *Crop Protection* **29**: 401–405.

Fa, J. E. and Lind, R. (1996). Population management and viability of the Gibraltar Barbary macaques. In: Fa, J. E. and Lindburg, D. G. (Eds), *Evolution and Ecology of Macaque Societies.* Cambridge: Cambridge University Press.

Fa, J. E. and Brown, D. (2009). Impacts of hunting on mammals in African tropical moist forests: a review and synthesis. *Mammal Review* **39**: 231–264.

Fa, J. E., Peres, C. A., and Meeuwig, J. (2002). Bushmeat exploitation in tropical forests: an intercontinental comparison. *Conservation Biology* **16**: 232–237.

Forthman-Quick, D. L. (Ed.) (1986). *Activity Budgets and the Consumption of Human Food in Two Troops of Baboon,* Papio anubis, *at Gilgil, Kenya. Primate Ecology and Conservation.* Cambridge: Cambridge University Press.

Fuentes, A. (2010). Natural cultural encounters in Bali: monkeys, temples, tourists, and ethnoprimatology. *Cultural Anthropology* **25**: 600–624.

Fuentes, A. (2012). Ethnoprimatology and the anthropology of the human–primate interface. *Annual Review of Anthropology* **41**(41): 101–117.

Fuentes, A. and Hockings, K. J. (2010). The ethnoprimatological approach in primatology. *American Journal of Primatology* **72**: 841–847.

Fuentes, A., Lane, K. E., Johnson, N., Watanaskul, L., Rompis, A. L. T., *et al.* (2006). Assessing genetic structure in Balinese macaques and its implications for disease transmission. *American Journal of Physical Anthropology* **91**: S42.

Fuentes, A., Southern, M., and Suaryana, K. G. (2005). Monkey forests and human landscapes: is extensive sympatry sustainable for *Homo sapiens* and *Macaca fascicularis* in Bali? In: Patterson, J. and Wallis, J. (Eds), *Commensalism and Conflict: The Primate-Human Interface*. Norman: American Society of Primatology Publications.

Fuentes, A., Shaw, E., and Cortes, J. (2007). Qualitative assessment of macaque tourist sites in Padangtegal, Bali, Indonesia, and the Upper Rock Nature Reserve, Gibraltar. *International Journal of Primatology* **28**: 1143–1158.

Gillespie, T. R. and Chapman, C. A. (2008). Forest fragmentation, the decline of an endangered primate, and changes in host-parasite interactions relative to an unfragmented forest. *American Journal of Primatology* **70**: 222–230.

Goldenberg, D. (2013). *Hunting a Chimp on a Killing Spree* [Online]. New York: *The New York Times*. Available at: http://www.nytimes.com/2013/10/27/magazine/hunting-a-chimp-on-a-killing-spree.html?_r=0 [Accessed 22 March 2013].

Goossens, B., Setchell, J. M., Tchidongo, E., Dilambaka, E., Vidal, C., Ancrenaz, A., *et al.* (2005). Survival, interactions with conspecifics and reproduction in 37 chimpanzees released into the wild. *Biological Conservation* **123**: 461–475.

Gumert, M. D., Kluck, M., and Malaivijitnond, S. (2009). The physical characteristics and usage patterns of stone axe and pounding hammers used by long-tailed macaques in the Andaman sea region of Thailand. *American Journal of Primatology* **71**: 594–608.

Guy, A. J., Curnoe, D., and Banks, P. B. (2013). A survey of current mammal rehabilitation and release practices. *Biodiversity and Conservation* **22**: 825–837.

Higham, J. P., Warren, Y., Adanu, J., Umaru, B. N., Maclarnon, A. M., *et al.* (2009). Living on the edge: life-history of olive baboons at Gashaka-Gumti National Park, Nigeria. *American Journal of Primatology* **71**: 293–304.

Hill, C. M. (1997). Crop-raiding by wild vertebrates: the farmer's perspective in an agricultural community in western Uganda. *International Journal of Pest Management* **43**: 77–84.

Hill, C. M. (2000). Conflict of interest between people and baboons: crop raiding in Uganda. *International Journal of Primatology* **21**: 299–315.

Hill, C. M. (2002). Primate conservation and local communities—Ethical issues and debates. *American Anthropologist* **104**: 1184–1194.

Hill, C. M. and Webber, A. D. (2010). Perceptions of non-human primates in human–wildlife conflict scenarios. *American Journal of Primatology* **71**: 1–6.

Hill, C. M. and Wallace, G. E. (2012). Crop protection and conflict mitigation: reducing the costs of living alongside non-human primates. *Biodiversity and Conservation* **21**: 2569–2587.

Hill, D. A. (1999). Effects of provisioning on the social behaviour of Japanese and rhesus macaques: implications for socioecology. *Primates* **40**: 187–198.

Hill, K., Boesch, C., Goodall, J., Pusey, A., Williams, J., *et al.* (2001). Mortality rates among wild chimpanzees. *Journal of Human Evolution* **40**: 437–450.

Hiser, K. (2012). *Crop Raiding and Conflict: Farmers' Perceptions of Human–Wildlife Interactions in Hoima District, Uganda*. Department of Social Sciences. Oxford, UK: Oxford Brookes University.

Hoare, R. (2001). A decision support system (DSS) for managing human-elephant conflict situations in Africa. In: AFESG (Ed.), *IUCN/SSC African Elephant Specialist Group*. Nairobi: IUCN/SSC African Elephant Specialist Group.

Hockings, K. J. (2007). Human–Chimpanzee Coexistence at Bossou, The Republic of Guinea: A Chimpanzee Perspective. Ph.D., University of Stirling.

Hockings, K. J. and Humle, T. (2009). *Best Practice Guidelines for the Prevention and Mitigation of Conflict between Humans and Great Apes*. Gland, Switzerland: IUCN/SSC Primate Specialist Group (PSG).

Hockings, K. J. and Mclennan, M. R. (2012). From forest to farm: systematic review of cultivar feeding by chimpanzees—management implications for wildlife in anthropogenic landscapes. *PLoS One* **7**.

Hockings, K. J. and Sousa, C. (2012). Differential utilization of cashew-a low-conflict crop-by sympatric humans and chimpanzees. *Oryx* **46**: 375–381.

Hockings, K. J., Anderson, J. R., and Matsuzawa, T. (2006). Road crossing in chimpanzees: a risky business. *Current Biology* **16**: R668–R670.

Hockings, K. J., Anderson, J. R., and Matsuzawa, T. (2009). Use of wild and cultivated foods by chimpanzees at Bossou, Republic of Guinea: feeding dynamics in a human-influenced environment. *American Journal of Primatology* **71**: 636–646.

Hockings, K. J., Yamakoshi, G., Kabasawa, A., and Matsuzawa, T. (2010). Attacks on local persons by chimpanzees in Bossou, Republic of Guinea: long-term perspectives. *American Journal of Primatology* **72**: 887–896.

Hockings, K. J., Anderson, J. R., and Matsuzawa, T. (2012). Socioecological adaptations by chimpanzees, *Pan troglodytes verus*, inhabiting an anthropogenically impacted habitat. *Animal Behaviour* **83**: 801–810.

Hoffman, T. S., & O'Riain, M. J. (2012). Landscape requirements of a primate population in a human-dominated environment. *Frontiers in Zoology* **9**: 1–17. doi: 110.1186/1742-9994-9-1.

Hsiao, S. S., Ross, C. Hill, C. M., & Wallace, G. E. (2013). Crop-raiding deterrents around Budongo Forest Reserve: an evaluation through farmer actions and perceptions. *Oryx* **47**(4): 569–577. doi: 10.1017/s0030605312000853

Humle, T. and Kormos, R. (2011). Chimpanzees in Guinea and in West Africa. In: Matsuzawa, T., Humle, T., and Sugiyama, Y. (Eds), *Chimpanzees of Bossou and Nimba*. Tokyo: Springer-Verlag Tokyo.

Humle, T., Colin, C., Laurans, M., and Raballand, E. (2011). Group release of sanctuary chimpanzees (*Pan troglodytes*) in the Haut Niger National Park, Guinea, West Africa: ranging patterns and lessons so far. *International Journal of Primatology* **32**: 456–473.

Hurn, S. (2011). 'Like herding cats!' Managing conflict over wildlife heritage on South Africa's Cape Peninsula. *Ecological and Environmental Anthropology* **6**: 39–53.

Jones-Engel, L., May, C. C., Engel, G. A., Steinkraus, K. A., Schillaci, M. A., *et al.* (2008). Diverse contexts of zoonotic transmission of simian foamy viruses in Asia. *Emerging Infectious Diseases* **14**: 1200–1208.

Kamenya, S. (2002). *Human baby killed by Gombe chimpanzee* [Online]. Available at: http://mahale.main.jp/PAN/9_2/9(2)-06.html [Accessed 17 March 2014].

Kavanagh, M. and Caldecott, J. O. (2013). Strategic guidelines for the translocation of primates and other animals. *Raffles Bulletin of Zoology*, Supplement 29: 203–209.

Knight, J. (2000). *Natural Enemies: Human-Wildlife Conflict in Anthropological Perspective*. London: Routledge.

Kondgen, S., Kuhl, H., N'goran, P. K., Walsh, P. D., Schenk, S., *et al.* (2008). Pandemic human viruses cause decline of endangered great apes. *Current Biology* **18**: 260–264.

Kortlandt, A. (1986). The use of stone tools by wild-living chimpanzees and earliest hominids. *Journal of Human Evolution* **15**: 77–132.

Koutstaal, K. (2013). How Different Views on Baboon Agency Shape the Conservation Policy Making Dialogue in Cape Town, South Africa. Research Master in African Studies (RESMAAS), Leiden University.

Lamarque, F., Anderson, J., Fergusson, R., Lagrange, M., Osei-Owusu, Y., *et al.* (2009). Human-Wildlife Conflict in Africa. Food and Agriculture Organization of the United Nations.

Lambert, J. E. (2005). Competition, predation, and the evolutionary significance of the cercopithecine cheek pouch: the case of Cercopithecus and Lophocebus. *American Journal of Physical Anthropology* **126**: 183–192.

Lane, K. E., Lute, M., Rompis, A., Wandia, N., Arta Putra, H., *et al.* (2010). Pests, pestilence, and people: the long-tailed macaques and its role in the cultural complexities of Bali. In: Gursky-Doyen, S. and Supriatna, J. (Eds), *Indonesian Primates*. Berlin: Springer Verlag.

Lansing, S. J. (1991). *Priests and Programmers: Technologies and Power in the Engineered Landscape of Bali*. Princeton: Princeton University Press.

Lee, P. C. and Priston, N. E. C. (2005). Human attitudes to primates: perceptions of pests, conflict and consequences for conservation. In: Paterson, J. D. (Ed.), *Commensalism and Conflict: The Primate-Human Interface*. Norman, OK: American Society of Primatology.

Lin, D. (2010). *Officials kill 800 monkeys in Puerto Rico* [Online]. Available at: http://animalrights.about.com/b/2010/01/09/officials-kill-800-monkeys-in-puerto-rico.htm [Accessed 24 March 2014].

Lingomo, B. and Kimura, D. (2009). Taboo of eating bonobo among the Bongando people in the Wamba region, Democratic Republic of Congo. *African Study Monographs* **30**: 209–225.

Linkie, M., Dinata, Y., Nofrianto, A., and Leader-Williams, N. (2007). Patterns and perceptions of wildlife crop raiding in and around Kerinci Seblat National Park, Sumatra. *Animal Conservation* **10**: 127–135.

Loudon, J. E., Sauther, M. L., Fish, K. D., Hunter-Ishikawa, M., and Ibrahim, Y. J. (2006). One reserve, three primates: applying a holistic approach to understand the interconnections among ring-tailed lemurs (*Lemur catta*), Verreaux's sifaka (*Propithecus verreauxi*), and humans (*Homo sapiens*) at Beza Mahafaly Special Reserve, Madagascar. *Ecological and Environmental Anthropology* **2**: 54–74.

Macfie, E. J. and Williamson, E. A. (2010). *Best Practice Guidelines for Great Ape Tourism*. Gland, Switzerland: IUCN/SSC Primate Specialist Group (PSG).

Mackinnon, K. C. and Riley, E. P. (2010). Field primatology of today: current ethical issues. *American Journal of Primatology* **72**: 749–753.

Madden, F. and Mcquinn, B. (2014). Conservation's blind spot: the case for conflict transformation in wildlife conservation. *Biological Conservation* **178**: 97–106.

Malone, N. M., Fuentes, A., and White, F. J. (2010). Ethics commentary: subjects of knowledge and control in field primatology. *American Journal of Primatology* **72**: 779–784.

Marchini, S. and Macdonald, D. W. (2012). Predicting ranchers' intention to kill jaguars: case studies in Amazonia and Pantanal. *Biological Conservation* **147**: 213–221.

Martin-Ordas, G., Haun, D., Colmenares, F., and Call, J. (2010). Keeping track of time: evidence for episodic-like memory in great apes. *Animal Cognition* **13**: 331–340.

Mascarenhas, A. (1971). Agricultural vermin in Tanzania. In: Ominde, S. H. (Ed.), *Studies in East African Geography and Development*. London: Heinemann.

Mckinney, T. (2011). The effects of provisioning and crop-raiding on the diet and foraging activities of human-commensal white-faced capuchins (*Cebus capucinus*). *American Journal of Primatology* **73**: 439–448.

Mclennan, M. R. and Hill, C. M. (2012). Troublesome neighbours: changing attitudes towards chimpanzees (*Pan troglodytes*) in a human-dominated landscape in Uganda. *Journal for Nature Conservation* **20**: 219–227.

Mclennan, M. R. and Huffman, M. A. (2012). High frequency of leaf swallowing and its relationship to intestinal parasite expulsion in 'village' chimpanzees at Bulindi, Uganda. *American Journal of Primatology* **74**: 642–650.

Mclennan, M. R. and Hill, C. M. (2013). Ethical issues in the study and conservation of an African Great Ape in an unprotected, human-dominated landscape in Western Uganda. In: Macclancy, J. and Fuentes, A. (Eds.), *Ethics in the Field: Contemporary Challenges*. New York, Oxford: Berghahn Books.

Meijaard, E., Albar, G., Nardiyono, Rayadin, Y., Ancrenaz, M., and Spehar, S. (2010). Unexpected ecological resilience in Bornean orangutans and implications for pulp and paper plantation management. *PLoS One* **5**.

Meijaard, E., Buchori, D., Hadiprakarsa, Y., Utami-Atmoko, S. S., Nurcahyo, A., *et al.* (2011). Quantifying killing of orangutans and human-orangutan conflict in Kalimantan, Indonesia. *PLoS One* **6**.

Meijaard, E. and Nijman, V. (2000). The local extinction of the proboscis monkey Nasalis larvatus in Pulau Kaget Nature Reserve, Indonesia. *Oryx* **34**: 66–70.

Mittermeier, R. A. (1987). Effects of hunting on rain forest primates. In: Marsh, C. W. and Mittermeier, R. A. (Eds), *Primate Conservation in the Tropical Rain Forest*. New York: Alan R. Liss.

Morgan, D. and Sanz, C. (2007). *Best Practice Guidelines for Reducing the Impact of Commercial Logging on Great Apes in Western Equatorial Africa*. Gland, Switzerland: IUCN/SSC Primate Specialist Group.

Morgan, D., Sanz, C., Greer, D., Rayden, T., Maisels, F., *et al.* (2013). Great Apes and FSC: Implementing 'Ape Friendly' Practices in Central Africa's Logging Concessions. IUCN/SSC Primate Specialist Group, G., Switzerland Occasional Paper. Gland, Switzerland: IUCN/SSC Primate Specialist Group.

Muehlenbein, M. P., Martinez, L. A., Lemke, A. A., Ambu, L., Nathan, S., *et al.* (2010). Unhealthy travelers present challenges to sustainable primate ecotourism. *Travel Medicine and Infectious Disease* **8**: 169–175.

Naughton, L., Rose, R., and Treves, A. (1999). The social dimensions of human-elephant conflict in Africa: A literature review and case studies from Uganda and Cameroon. A Report to the African Elephant Specialist, Human-Elephant Task Conflict Task Force, of IUCN, Glands, Switzerland.

Naughton-Treves, L. (1997). Farming the forest edge: vulnerable places and people around Kibale National Park, Uganda. *Geographical Review* **87**: 27–46.

Naughton-Treves, L., Treves, A., Chapman, C., and Wrangham, R. (1998). Temporal patterns of crop-raiding by primates: linking food availability in croplands and adjacent forest. *Journal of Applied Ecology* **35**: 596–606.

Nyhus, P. J., Fisher, H., Osofsky, S., and Madden, F. (2003). Taking the bite out of wildlife damage: the challenges of wildlife compensation schemes. *Conservation Magazine* **4**: 37–40.

Oates, J. F. (1996). Habitat alteration, hunting and the conservation of folivorous primates in African forests. *Australian Journal of Ecology* **21**: 1–9.

Paterson, J. (Ed.) (2005). *Commensalism and Conflict: The Human Primate Interface*. Norman, OK: American Society of Primatology.

Peterson, J. V., Riley, E. P., and Putu Oka, N. (2015). Macaques and the ritual production of sacredness among Balinese Transmigrants in South Sulawesi, Indonesia. *American Anthropologist* **117**: 71–85.

Peterson, M. N., Birckhead, J. L., Leong, K., Peterson, M. J., and Peterson, T. R. (2010). Rearticulating the myth of human-wildlife conflict. *Conservation Letters* **3**: 74–82.

Pirta, R. S., Gadgil, M., and Kharshikar, A. V. (1997). Management of the rhesus monkey Macaca mulatta and Hanuman langur Presbytis entellus in Himachal Pradesh, India. *Biological Conservation* **79**: 97–106.

Poulsen, J. R., Clark, C. J., Mavah, G., and Elkan, P. W. (2009). Bushmeat supply and consumption in a tropical logging concession in Northern Congo. *Conservation Biology* **23**: 1597–1608.

Priston, N. E. C. and Mclennan, M. (2013). Managing humans, managing macaques: human–macaque conflict in Asia and Africa. In: Radhakrishna, S., Huffman, M. A., and Sinha, A. (Eds), *The Macaque Connection: Cooperation and Conflict between Humans and Macaques*. Springer.

Priston, N. E. C., Wyper, R. M., and Lee, P. C. (2012). Buton macaques (*Macaca ochreata brunnescens*): crops, conflict, and behavior on farms. *American Journal of Primatology* **74**: 29–36.

Quirin, C. and Dixon, A. (2012). Food security, politics and perceptions of wildlife damage in Western Ethiopia. *International Journal of Pest Management* **58**: 101–114.

Radhakrishna, S., Huffman, M. A., and Sinha, A. (Eds) (2013). *The Macaque Connection: Cooperation and Conflict between Humans and Macaques*. Springer.

Redpath, S. M., Young, J., Evely, A., Adams, W. M., Sutherland, W. J., *et al.* (2013). Understanding and managing conservation conflicts. *Trends in Ecology and Evolution* **28**: 100–109.

Remis, M. J. and Hardin, R. (2009). Transvalued species in an African forest. *Conservation Biology* **23**: 1588–1596.

Richard, A. F., Goldstein, S. J., and Dewar, R. E. (1989). Weed macaques—the evolutionary implications of macaque feeding ecology. *International Journal of Primatology* **10**: 569–597.

Riley, E. P. (2008). Ranging patterns and habitat use of Sulawesi Tonkean macaques (Macaca tonkeana) in a human-modified habitat. *American Journal of Primatology* **70**: 670–679.

Riley, E. P. (2010). The importance of human-macaque folklore for conservation in Lore Lindu National Park, Sulawesi, Indonesia. *Oryx* **44**: 235–240.

Riley, E. P. and Fuentes, A. (2011). Conserving social-ecological systems in Indonesia: human-nonhuman primate interconnections in Bali and Sulawesi. *American Journal of Primatology* **73**: 62–74.

Robins, J. G., Ancrenaz, M., Parker, J., Goossens, B., Ambu, L., *et al.* (2013). The release of northeast Bornean orangutans to Tabin Wildlife Reserve, Sabah, Malaysia. In: Pritpal, S. S. (Ed.), *Global Reintroduction Perspectives: Additional Case Studies from Around the Globe*. IUCN/SSC Reintroduction Specialist Group (RSG).

Russon, A. E. (2009). Orangutan rehabilitation and reintroduction. In: Wich, S. A., Utami Atmoko, S. S., Mitra Setia, T., and Van Schaik, C. P. (Eds), *Orangutans: Geographic Variation in Behavioral Ecology and Conservation*. Oxford, UK: Oxford University Press.

Saj, T., Sicotte, P., and Paterson, J. D. (1999). Influence of human food consumption on the time budget of vervets. *International Journal of Primatology* **20**: 977–994.

Saj, T. L., Mather, C., and Sicotte, P. (2006). Traditional taboos in biological conservation: the case of Colobus vellerosus at the Boabeng-Fiema Monkey Sanctuary, Central Ghana. *Social Science Information Sur Les Sciences Sociales* **45**: 285–310.

Sandbrook, C. and Adams, W. M. (2012). Accessing the impenetrable: the nature and distribution of tourism benefits at a Ugandan National Park. *Society and Natural Resources* **25**: 915–932.

Sapolsky, R. M. and Share, L. J. (2004). A Pacific culture among wild baboons: its emergence and transmission. *PLoS Biology* **2**: e106.

Schillaci, M. A., Engel, G. A., Fuentes, A., Rompis, A., Arta Putra, I., *et al.* (2010). The not-so-sacred monkeys of Bali: A radiographic study of human primate commensalism. In: Gursky-Doyen, S. and Supriatna, J. (Eds), *Indonesian Primates. Developments in Primatology: Progress and Prospects*. New York: Springer Science.

Schoener, T. W. (1974). Resource partitioning in ecological communities. *Science* **185**: 27–39.

Sha, J. C. M., Gumert, M. D., Lee, B., Fuentes, A., Rajathurai, S., *et al.* (2009). Status of the long-tailed macaque Macaca fascicularis in Singapore and implications for management. *Biodiversity and Conservation* **18**: 2909–2926.

Siex, K. S. and Struhsaker, T. T. (1999). Colobus monkeys and coconuts: a study of perceived human-wildlife conflicts. *Journal of Applied Ecology* **36**: 1009–1020.

Sillero-Zubiri, C., Sukumar, R., and Treves, A. (2007). Living with wildlife: the roots of conflict and the solutions. In: Macdonald, D. (Ed.), *Key Topics in Conservation Biology*. Oxford: Blackwell.

Sinha, A., Mukhopadhyay, K., Datta-Roy, A., and Ram, S. (2005). Ecology proposes, behaviour disposes: ecological variability in social organization and male behavioural strategies among wild bonnet macaques. *Current Science* **89**: 1166–1179.

Sitati, N. W., Walpole, M. J., Smith, R. J., and Leader-Williams, N. (2003). Predicting spatial aspects of human-elephant conflict. *Journal of Applied Ecology* **40**: 667–677.

Smith, K., Barrett, C. B., and Box, P. W. (2000). Participatory risk mapping for targeting research and assistance: with an example from east African pastoralists. *World Development* **28**: 1945–1959.

Sogbohossou, E. A., De Iongh, H. H., Sinsin, B., De Snoo, G. R., and Funston, P. J. (2011). Human-carnivore conflict around Pendjari Biosphere Reserve, northern Benin. *Oryx* **45**: 569–578.

Sponsel, L. E., Ruttanadakul, N., and Natadecha-Sponsel, P. (2002). Monkey business? The conservation implications of macaque ethnoprimatology in Southern Thailand. In: Fuentes, A. and Wolfe, L. D. (Eds), *Primates Face to Face: The Conservation Implications of Human-Nonhuman Primate Interconnections*. New York, NY: Cambridge University Press.

Strum, S. C. (1994). Prospects for management of primate pests. *Revue D Ecologie-La Terre Et La Vie* **49**: 295–306.

Strum, S. C. (2005). Measuring success in primate translocation: a baboon case study. *American Journal of Primatology* **65**: 117–140.

Strum, S. C. (2010). The development of primate raiding: implications for management and conservation. *International Journal of Primatology* **31**: 133–156.

Sugiyama, Y. and Fujita, S. (2011). The demography and reproductive parameters of Bossou Chimpanzees. In: Matsuzawa, T., Humle, T., and Sugiyama, Y. (Eds), *Chimpanzees of Bossou and Nimba*. Tokyo: Springer-Verlag Tokyo.

Sukumar, R. (1991). The management of large mammals in relation to male strategies and conflict with people. *Biological Conservation* **55**: 93–102.

Tappen, N. C. (1964). Primate studies in Sierra Leone. *Current Anthropology* **5**: 339–340.

Tomonaga, M. (1999). Inversion effect in perception of human faces in a chimpanzee (*Pan troglodytes*). *Primates* **40**: 417–438.

Treves, A. and Bruskotter, J. (2014). Tolerance for predatory wildlife. *Science* **344**: 475–476.

Vinod, G. (2013). *Exposed! Mass killing of macaques* [Online]. FMT News. Available http://www.freemalaysiatoday.com/category/nation/2013/03/26/exposed-mass-killing-of-macaques/ [Accessed 26 March 2013].

Wallace, G. E. and Hill, C. M. (2012). Crop damage by primates: quantifying the key parameters of crop-raiding events. *Plos One* **7**.

Watanabe, K. and Muroyama, Y. (2005). Recent expansion of the range of Japanese macaques, and associated management problems. In: Paterson, J. D. and Wallis, J. (Eds), *Commensalism and Conflict: The Human–Primate Interface*. American Society of Primatologists.

Wheatley, B. P. (1999). *The Sacred Monkeys of Bali*. Prospect Heights: Waveland Press.

Wich, S. A., Utami-Atmoko, S. S., Mitra Setia, T., Rijksen, H. D., Schürmann, C. L., *et al.* (2004). Life history of wild Sumatran orangutans (*Pongo abelii*). *Journal of Human Evolution* **47**: 385–398.

Wich, S. A., Gaveau, D., Abram, N., Ancrenaz, M., Baccini, A., *et al.* (2012). Understanding the impacts of land-use policies on a threatened species: is there a future for the Bornean Orang-utan? *Plos One* **7**.

Wilkie, D., Shaw, E., Rotberg, F., Morelli, G., and Auzel, P. (2000). Roads, development, and conservation in the Congo Basin. *Conservation Biology* **14**: 1614–1622.

Wilkie, D. S. and Carpenter, J. F. (1999). Bushmeat hunting in the Congo Basin: an assessment of impacts and options for mitigation. *Biodiversity and Conservation* **8**: 927–955.

Williamson, E. A., Maisels, F., and Groves, C. P. (2013). Hominidae. In: Mittermeier, R. A., Rylands, A. B., and Wilson, D. E. (Eds), *Handbook of the Mammals of the World*. Barcelona, Spain: Lynx Edicions.

Wilson, M. L., Hauser, M. D., and Wrangham, R. W. (2007). Chimpanzees (*Pan troglodytes*) modify grouping and vocal behaviour in response to location-specific risk. *Behaviour* **14**: 1621–1653.

Woodroffe, R., Thirgood, S., and Rabinowitz, A. (Eds) (2005). *People and Wildlife: Conflict or Coexistence?* Cambridge: Cambridge University Press.

Zhao, Q. K. (2005). Tibetan macaques, visitors, and local people at Mt. Emei: problems and countermeasures. In: Paterson, J. (Ed.), *Commensalism and Conflict: The Human–Primate Interface*. Norman, OK: American Society of Primatologists.

CHAPTER 15

The role of translocation in primate conservation

Benjamin B. Beck

Body size affects reintroduction in many ways, including the logistics of animal transportation (see section 15.5.4).
Photo copyright: Golden lion tamarin photograph by the author. Gorilla photograph by Tony.

15.1 Introduction and definitions

Humans 'translocate' non-human primates (and other animals and plants) by moving them from one place in the wild to another place in the wild, or from captivity to a place in the wild, for purposes of conservation and/or for other reasons. The International Union for the Conservation of Nature (IUCN) has recently published revised guidelines for translocations of plants and animals for conservation purposes (IUCN/SSC 2013). According to the guidelines and to Ewen *et al.* (2013), there are four types of translocation. 'Reinforcement' is the intentional movement and release of organisms to a site that is within the indigenous range of the species when conspecifics are already present. 'Reintroduction' is the intentional movement and release of organisms to a site that is within the indigenous range of the species when conspecifics are absent. 'Assisted Colonization' is the intentional movement and release of organisms to a site that is outside the indigenous range of the species to avoid extinction of the species. 'Ecological Replacement' is the intentional movement and release of organisms to a site that is outside the indigenous range of the species to restore a specific ecological function. The IUCN guidelines focus on conservation as the primary purpose for translocations. Translocations 'must be intended to yield a measurable conservation benefit at the levels of a population, species, or ecosystem, and not only provide benefit to translocated individuals' (IUCN/SSC 2013: VIII). There are also guidelines that are targeted specifically to translocation of non-human primates (Baker 2002). These too focus on conservation as the primary goal of translocation.

The 2013 IUCN guidelines recommend that the destination sites for translocations of living plants and animals: (1) be large enough to support a self-sustaining population of the species in question, or be connected to other areas where the species occurs; (2) contain sufficient resources to meet all of the needs of a self-sustaining population of the species; (3) have a climate that is appropriate for the species; (4) be securely protected and free of anthropomorphic threats to the species; (5) be part of a landscape that meets the needs of surrounding human

Beck, B.B., *The role of translocation in primate conservation*. In: *An Introduction to Primate Conservation*. Edited by: Serge A. Wich and Andrew J. Marshall, Oxford University Press (2016). © Oxford University Press.
DOI 10.1093/acprof:oso/9780198703389.003.0015

communities; and (6) have land-use regulations that allow translocation of the species in question.

15.2 Purposes of the chapter

The purposes of this chapter are to consider the range of reasons that motivate humans to translocate non-human primates (hereafter 'primates'), to examine the accomplishments and promise of translocation as a tool for primate conservation, and to explore the influence of some characteristics of primate biology on the potential success of translocations.

15.3 Reasons for translocation

There are many different reasons for translocating primates, and some translocations may be motivated by more than one reason.

15.3.1 Conservation-motivated translocations

The gravest threats to the continued survival of all primates are the conversion (some would say 'degradation' or 'destruction') and fragmentation of suitable native habitat by humans and hunting by humans (Oates 2013; Shumaker and Beck 2003; Strier 2007). Disease is also a major threat for some species (e.g. Vogel 2007). It has therefore been difficult to identify translocation destination sites that meet all of the *Guidelines'* requirements, particularly the requirement that the destination site be large enough to support a self-sustaining population of the species and the requirement that the site be protected and free of anthropomorphic threats. This suggests that, at least at the present time, translocation can play only a limited role in primate conservation. The creation of new protected areas with suitable primate habitat (Oates 1999), strengthening of wildlife protection laws and their enforcement, and creation of local economic incentives to replace hunting and logging are likely to be more important than translocation for primate conservation. However, the creation of new protected areas, increased law enforcement and wildlife protection, and economic and educational interventions that improve the livelihoods of people who share habitats with primates may create new opportunities for translocation. The identification of 'empty forests' in which primates have been extirpated by hunting will also create opportunities for translocation when hunting can be controlled or replaced by other livelihoods.

Some translocations have indeed been conducted for conservation purposes (e.g. golden lion tamarins, Kierulff *et al.* (2012)), and Section 15.4 of this chapter examines nine successful primate conservation translocations.

15.3.2 Economic, political, aesthetic, religious, scientific, and accidental translocations

Although there is presently only a limited role for translocation in primate conservation, thousands of non-human primates have been and continue to be translocated (Beck in prep.). Fewer than five individuals were translocated in some cases (e.g. three black lion tamarins, Valladares-Padua *et al.* (2000); one orphan chimpanzee, Treves and Naughton-Treves (1997)), but hundreds of primates were involved in other attempts (e.g. olive baboons, Strum (2005)). Some translocations were conducted at least in part for economic reasons (e.g. 'monkey farming' to provide crab-eating macaques and rhesus macaques for biomedical research laboratories: Kyes (1993), Taub and Mehlman (1989), respectively), political reasons (e.g. to keep Barbary macaques on Gibraltar as a symbol of the might of the British Empire, Candland and Bush (1995)), recreational/aesthetic reasons (e.g. common marmosets in the city of Rio de Janeiro, Konstant and Mittermeier (1982)), religious reasons (e.g. avoiding offense to Indian Hindus and Buddhists by discontinuing capture of rhesus monkeys for biomedical research, Carpenter (1940)), and/or scientific reasons (e.g. Japanese macaques into a large naturalistic enclosure in Texas, USA, Fedigan (1991)). Some translocations were accidental, resulting from escapes of captive animals (e.g. lion tamarins, Coimbra-Filho and Mittermeier (1978)).

There was 'no definite reason' for the release of golden langurs (*Trachypithecus geei*) in Tripura, India in 1988 (Gupta 2002). The managers of this translocation violated nearly every IUCN guideline. Not surprisingly, the translocation failed. Gupta is to be credited for a frank and useful account of how *not* to translocate primates.

15.3.3 Translocations intended to enhance animal welfare or as an alternative to death

The most commonly cited reason for primate translocations can generally be called 'animal welfare'. The benefits are intended to be for the released individuals, not for a species or ecosystem. Examples include releasing animals from sanctuaries or other captive sites (e.g. Bornean and Sumatran orang-utans in Indonesia, Warren and Swan (2002), Bornean orang-utans in Indonesia, Siregar *et al.* (2010)); non-lethal control of primates that raid crops, steal food, or harass people (e.g. rhesus monkeys in India, Imam *et al.* (2002), Southwick *et al.* (1998)); non-lethal removal of primates that are living as alien species outside of their historic range (e.g. golden-headed lion tamarins in the city of Niteroi, Rio de Janeiro state, Brazil, Kierulff (2012)); relocation of primates whose lives are in imminent danger (e.g. monkeys living in a forest that is threatened by rising water caused by dam construction in Surinam, Konstant and Mittermeier (1982)); and disposition of primates that become unwanted as pets, biomedical research animals, or attractions in bankrupt tourist operations (e.g. chimpanzees from a biomedical research laboratory in Liberia, Hannah and McGrew (1991); pet squirrel monkeys in Florida, Florida Fish and Game Commission[1] website).

Welfare translocations are considered by many professionals to be unwise because there is evidence that the welfare of the reintroduced individuals is not always enhanced, and because there is a potential for ecological disruption, the introduction of inappropriate genes, and disease transmission at the release sites, which may be within or outside of the historic range of the species (Guy *et al.* 2014). Beck *et al.* (2007), in guidelines for the reintroduction of great apes, recognize the importance and legitimacy of welfare-based translocations (at least of great apes) under specific, controlled conditions. These ape guidelines specify that translocation should 'not be conducted solely to dispose of surplus animals or to relieve overcrowding' (10). However, the IUCN translocation guidelines that pertain to all animals (see Section 15.1) consider welfare-related releases as 'being outside of the scope of the guidelines' (40), and as noted in Section 15.1, specify that translocations must be intended to yield a conservation benefit and not only provide benefit to the translocated individuals. Welfare translocations do not meet this requirement. The primate guidelines (see Section 15.1) likewise state that 'rescue/welfare is not considered a re-introduction [translocation] approach because the aim is motivated by goals other than conservation . . . ' (34).

Both sets of guidelines do urge that welfare translocations follow the *Guidelines'* procedural recommendations, but, given the reality that most primate translocations have been and probably will continue to be undertaken for welfare-related reasons, a set of guidelines targeted specifically to translocations that are intended to improve the welfare of individual primates would be beneficial. Welfare translocation guidelines might improve the quality of such operations and increase the probability that welfare would actually be enhanced. Welfare-related translocations with well-monitored outcomes, for example Rodriguez-Luna and Cortés-Ortiz (1994) and Strum (2005), can also provide scientific knowledge that might inform conservation-motivated translocations (Guy *et al.* 2014). Welfare-motivated translocations could also result in heightened public awareness of the plight of wildlife and thus enhance public support for conservation.

15.3.4 A successful welfare-motivated translocation of olive baboons

Strum (2002, 2005) translocated three groups of wild olive baboons (131 individuals) in Kenya in 1984. Olive baboons are common animals. The translocation did not enhance the probability of survival of the species nor restore a critical ecological function at the destination sites. These baboons had become crop- and garden-raiders and were entering peoples' homes to steal food. Farmers and soldiers began to shoot them. Strum and her associates had studied the ecology, behaviour, and reproduction

[1] http://myfwc.com/wildlifehabitats/nonnatives/mammals.

of many of these baboons for 14 years. She knew the identity and history of every individual. She decided to save these baboons by moving them to a more remote site that had fewer human inhabitants and less agricultural activity. This was a welfare translocation. It is unique and exemplary among all primate translocations for its meticulous planning and management and its degree of documentation.

The translocated baboons were medically screened before translocation to reduce the possibility of transfer of communicable disease. Females were released while males were held in cages at the release site to increase the probability that the baboons would remain near the destination site. The baboons were given some food supplementation immediately after release.

Strum was committed to a scientifically rigorous examination of the outcome of the translocation. She had documented the birth and death rates, mortality and survivorship, group sizes, body condition, and the occurrence of intestinal parasites of the baboon groups for years, and used this as a pre-release baseline for post-release comparisons. The groups were observed with comparable methodology for 15 years after they were translocated. Strum also studied a non-translocated group near the destination site as a 'control' for further comparison. Although there was short-term variation, the long-term similarities in birth rates, death rates, survivorship, mortality, group sizes, and causes of death before and after translocation and between translocated and non-translocated groups were compelling evidence for the success of the translocation. The translocated groups survived and were spared what was likely to have been total extirpation or at least violent deaths and a lifetime of conflict with humans.

15.3.5 An unsuccessful welfare-motivated translocation of gorillas

As part of an ongoing gorilla translocation programme in Gabon (see Section 15.4.8), the Aspinall Foundation released an intact group of nine captive-born and one wild-born zoo gorillas near the Mpassa River in 2014. This programme had previously been motivated by conservation, but the sponsor decided to release these ten zoo gorillas because he felt that it was morally wrong to keep wild animals in zoos, even in the zoos that are managed by the Foundation.[2] I would therefore conclude that the translocation of this cohort was primarily a welfare-motivated translocation. There is no evidence that the translocation of this cohort was intended to contribute to the establishment of a self-sustaining wild population, although these gorillas might have increased the size of the previously reintroduced population if they had survived. A large area of suitable forest habitat had been brought under increased protection in preparation for this and previous translocations.

The ten gorillas were first trained at the zoo and, after their arrival in Gabon, they were acclimated on a forested island at the release site, where they had access to natural foods. They received veterinary screening and support. After acclimation, they were allowed to leave the island and enter the forest. Shortly after this release, five of the ten gorillas were found dead and one disappeared, possibly due to conflict with a wild (previously reintroduced) gorilla(s).[3] The remaining four were returned to the island and thrive there, two years later. This is a wrenching example of the compromise in animal welfare that can be associated with translocation. Some would argue that the welfare of these ten zoo gorillas would have been better had they remained in captivity in a good zoo.

15.4 Case studies of successful translocation as a primate conservation tool

As noted earlier, the limited availability of suitable, protected habitat has limited the number of primate translocations that have been conducted. However, there have been nine clearly successful primate conservation translocations, that is they have arguably increased the probability of the establishment of a self-sustaining population of the species in question and/or restored critical ecological functions. In some cases, the translocation has helped to secure

[2] http://www.cbsnews.com/news/zoo-gorilla-family-freed-to-wild-60-minutes, 15 March 2015.

[3] http://www.KentOnline.co.uk, 6 September 2014.

and protect habitat that might otherwise not have been available for the conservation of animals and plants.

15.4.1 Ruffed lemurs

Thirteen captive-born black-and-white ruffed lemurs (*Varecia v. variegata*) were released in the Betampona Reserve in Madagascar between 1997 and 2001 to reinforce the reserve's diminished population of approximately 35 individuals (Britt *et al.* 2002, 2004 a, b). Betampona has a total area of 22 km^2, of which approximately half is intact rainforest. There are ten other lemur species, and a number of rare, endemic birds, mammals and plants that live in the reserve. A systematic site evaluation and choice of release sites preceded the translocation. The 13 individuals met rigorous selection criteria, among which was being mother reared. The lemurs all had pre-release experience in large forested enclosures. One of the released males and one of the females were integrated into the wild population and reproduced with wild mates. Three others survived at least a year. Five of the lemurs were killed by predators. The translocation was a component of a larger commitment by the Madagascar Faunal Group to the conservation of Betampona, which suggests that the conservation benefits of the translocation were greater than solely the reinforcement of the size and genetic diversity of the wild black-and-white ruffed lemur population.[4]

15.4.2 Golden lion tamarins

Golden lion tamarins (*Leontopithecus rosalia*, GLTs) were successfully translocated to a single watershed in the state of Rio de Janeiro, Brazil (Beck *et al.* 2002; Kierulff *et al.* 2002 a, b, 2012; Ruiz-Miranda *et al.* 2010). The destination sites were within the historic range of the species and were near but not fully connected with the range of a diminished, non-sustainable wild GLT population. Forty-two wild-born GLTs that had been rescued from isolated forest fragments that were slated for destruction, and 146 captive-born GLTs that had been born in North American and European zoos, were translocated during a 20-year period. The total wild population was estimated to exceed 3,000 in 2014 (J. Dietz, pers. comm.), a number that would be self-sustaining if all of the habitat that they occupy were protected and connected. The tamarins are independent of human support except that 'mini-translocations' of individuals or families between isolated forest fragments are still conducted to maintain genetic diversity. These actions will no longer be necessary when the habitat is completely interconnected. Connection of forest fragments is now a major goal of the programme.[5] The species has been downlisted from 'Critically Endangered' to 'Endangered'.

15.4.3 Golden-headed lion tamarins

Golden-headed lion tamarins (*Leontopithecus chyrsomelas*, GHLTs) are also being translocated, apparently successfully, from the Serra da Tiririca State Park and the Darcy Ribeiro Municipal Reserve within the city limits of Niteroi, in the state of Rio de Janeiro, Brazil. The founders of this population were released decades ago by a private collector. GHLTs are not native to this area of Brazil, and conservationists feared that they might hybridize with the wild golden lion tamarins (see Section 15.4.2) that live as close as 50 km away (Kierulff 2011, 2012). As of May 2015, 530 GHLTs had been captured, of which 293 (49 groups) had been translocated to destination sites in a forest that is privately owned by a paper pulp manufacturer in the state of Bahia, Brazil, within their historic range (C. Kierulff, pers. comm.). The translocated groups were quarantined at the Rio de Janeiro Primate Center before release. Some were radiocollared, and were monitored for three months to a year. There has been reproduction. One hundred and eighty individuals that were deemed unsuitable for release were sterilized and remain in captivity. There are still approximately 50 GHLTs living in the parks in Niteroi. These must be captured and then the area must be proven to be free of GHLTs for five years to completely eliminate the threat of hybridization with GLTs. This GHLT

[4] http://www.waza.org.

[5] http://www.savetheliontamarin.org.

population had an extraordinarily high density, most likely because they were being provisioned by people living near the parks. These human neighbours have been informed of the nature and reasons for the operation because the unauthorized capture of endangered animals is illegal in Brazil.

From the point of view of the GHLTs, this is a welfare translocation because it provides a non-lethal way to control animals that had become 'problems', that is, animals of an exotic species that were threatening to hybridize with a native endangered species. From the point of view of the GLTs, this was a conservation translocation. We are reminded that translocations can have multiple purposes.

15.4.4 Howler monkeys

A self-sustaining population of black howler monkeys (*Alouatta pigra*) was restored by the translocation of 62 wild-born individuals in 1992, 1993, and 1994 to a 518 km^2 protected forest in Belize, the Cockscomb Basin Wildlife Sanctuary (the word 'sanctuary' is usually used to describe a facility that rehabilitates animals in a captive setting, but here it means a 'reserve') (Horwich *et al.* 1993, 2002). Black howler monkeys had lived previously in the area before it had been declared a reserve and had been eradicated by disease, hurricane damage, and overhunting. Hunting had been controlled at the time of the translocation. Thirty-two of the translocated monkeys and 28 surviving offspring were found in a 9 km^2 study area within the reserve in 1997, and others were thought to be alive outside of the study area at that time. The current population is now estimated to be between 300 and 500 (R. Horwich, pers. comm.) and is free of human support. The ranging patterns of some of the translocated howler monkeys were compared systematically to those of wild, non-translocated groups; the translocated groups had established their home ranges approximately six months after they were released (Ostro *et al.* 1999).

15.4.5 Chimpanzees

Representatives of the Frankfurt Zoological Society released 17 chimpanzees (*Pan troglodytes*) on Rubondo Island in Lake Victoria in Tanzania between 1966 and 1969 (Borner 1985). All of the Rubondo chimpanzees had been born in the wild, captured, and kept in European zoos for three months to nine years. Rubondo, which became a national park in 1977, is 240 km^2 in area and has ample evergreen deciduous forest. No other apes lived on the island, but it was within the range of wild chimpanzees. There was little post-release monitoring, but the chimpanzees are known to have quickly begun to eat natural foods and build nests. At least two of the translocated chimpanzees, both females, were known to be alive in 1985. One or two of the males were shot by unknown parties because of aggressive behaviour towards humans. The chimpanzees continued to harass people and invade homes (with decreasing frequency) at least through 1985. Island-born infants were first observed in 1968, and by 1985 the population numbered about 20. Moscovice *et al.* (2007) estimated that there were between 27 and 35 chimpanzees on Rubondo in 2006. The number of chimpanzees that were translocated in this operation is very small compared to the total population of eastern chimpanzees. However, their translocation may have contributed to Rubondo being designated as a national park. This translocation may in part have been motivated by zoo officials who were seeking a humane way of disposing of some chimpanzees that were not adapting well to zoo life. However, the translocation was also intended to serve conservation, and it succeeded.

15.4.6 Chimpanzees

Project HELP (Habitat Ecologique et Liberté des Primates) released 37 rehabilitated chimpanzees from a sanctuary into the Conkouati Triangle in the Republic of Congo between 1996 and 2001 (Farmer and Jamert 2002; Farmer *et al.* 2010; Goossens *et al.* 2005). The Triangle has an area of 21 km^2 and is bounded by water, but there are natural connections through the canopy and over fallen trees to a neighbouring reserve. There was a small natural chimpanzee population at the release site (Tutin *et al.* 2001). As of early 2004, there had been 5 confirmed deaths, 9 disappearances, and there were 23 survivors (Goossens *et al.* 2005). As of 2009, 17 infants had been born (Farmer *et al.* 2010), some of which were offspring from matings between translocated and wild chimpanzees. Many of the losses were related to aggression between wild and translocated males, which is a common occurrence in chimpanzee translocations and mirrors the behaviour of males in wild populations (Goossens *et al.* 2005). The translocation did reinforce a wild population, and there have been reductions in poaching and deforestation in the release area (Tutin *et al.* 2001). Again, this translocation may have been motivated in part by the need to decrease a sanctuary population of rehabilitated chimpanzees humanely, but each of the cited articles stresses that it was primarily intended to enhance the probability of survival of chimpanzees in the area.

15.4.7 Chimpanzees

The Chimpanzee Conservation Centre released six male and six female rehabilitated wild-born chimpanzees into the Haut Niger National Park in 2008 (Humle *et al.* 2011). There was already a viable population of chimpanzees living in the park, but the purposes of this translocation were said to be reinforcement of the wild population, increasing its genetic diversity, and enhancing protection of the park. The translocation also served to reduce the population in the Centre's overcrowded sanctuary. Post-release monitoring has revealed high survivorship, birth of at least two infants, and contact with the wild population (Humle *et al.* 2011).

15.4.8 Gorillas

The Aspinall Foundation has translocated 43 rehabilitated wild-born lowland gorillas (*Gorilla g. gorilla*), 1 gorilla that had been born in an in-country sanctuary, and 7 gorillas that had been born in the Aspinall zoos in the UK (King *et al.* 2012, 2014). This does not include the 10 gorillas that were translocated in 2014 (see Section 15.3.5). The destination sites comprised more than 800 km^2 of intact lowland forests in the Batéké Plateau region of the Republic of the Congo (in the Lesio Louna Reserve) and in Gabon (in the Batéké Plateau National Park). Gorillas had lived in these areas but had been driven to extinction by overhunting in the twentieth century. Despite some mortality and removals because of gorilla–human conflict, the translocations seem to have resulted in self-sustaining populations at both sites. More than 30 infants have been born (King, pers. comm.). Hunting has declined at both sites because of protected area management projects that are associated with the translocation.

15.4.9 Orang-utans

The Sumatran Orangutan Conservation Programme has translocated more than 200 rehabilitated wild-born orphan Sumatran orang-utans (*Pongo abelii*) and two adults that had been born in the Perth Zoo in Australia. One destination site is at the edge of the Bukit Tigapuluh National Park in Jambi province and another in the Pinus Jantho Nature Reserve in Aceh province, both in Sumatra. Orang-utans had been absent in these areas since the late nineteenth century. Bukit Tigapuluh and surrounding buffer zones have about 350 km^2 of lowland forest (Trayford *et al.* 2010), and Jantho and surrounding areas have an additional 75 km^2. There has been post-release monitoring (five infants are known to have been born), and survival is estimated to be 70% (Riedler *et al.* 2010).[6] There continue to be alarming rates of habitat conversion in the area. However, the establishment of new subpopulations of *P. abelii* increases the probability of survival of this critically endangered species by creating new, potentially self-sustaining subpopulations, reinforcing numbers and most likely the genetic diversity of the species, and adding strength to arguments for increased protection in the area.

I have probably overlooked some other successful primate translocations whose purpose was at least in part to increase the probability of establishment of a self-sustaining population of the translocated species, that is conservation. There are other primate translocations that claim to be motivated by conservation, but the number of translocated animals is so small that the benefits to the species are negligible, and there is no evidence that significant tracts of habitat have been secured and protected or that critical ecological functions have been restored by the operation. However, even if the number of conservation-motivated primate translocation

[6] http://www.orangutan.org.au.

projects and the number of individual primates that were translocated primarily for the purpose of conservation were twice what I have reported here, they would still be a small fraction of the overall number of recorded primate translocations (Beck, in prep.).

15.4.10 Success

The most common definition of success of conservation-motivated reintroductions is 'establishment of a self-sustaining population' (Beck *et al.* 1994; Fischer and Lindenmeyer 2000; Griffith *et al.* 1989). The nine programmes described here have '*increased the probability* of establishment of a self-sustaining population', which is a less demanding definition of success. Only the golden lion tamarin, howler monkey, and Conkouati chimpanzee programmes have met the more demanding definition.

Beck *et al.* (1994) in a review of 145 reintroductions (translocations) of animals of all taxa found that only 11% could be determined to be successful by the more demanding criterion. Fischer and Lindenmeyer (2000) reported a success rate of 23% in a review of 116 animal translocations. At present there are no accurate numbers of attempted conservation-motivated primate translocations that would allow calculation of an overall success rate.

Each of these nine projects adhered closely to the IUCN guidelines during planning, execution, and follow-up monitoring. The lemurs, tamarins, and the Conkouati and Haut Niger chimpanzees were fitted with radiocollars to allow them to be tracked. Each of these translocation projects (with the possible exception of the golden-headed lion tamarin project, which is still in its early stages) can claim success in terms of high post-release survivorship and some post-release reproduction. Most of the losses have been documented (with the apparent exception of the Sumatran orang-utan operation, which involves individuals that are dispersed over vast areas of steep, swampy terrain) and the causes of many of the deaths have been identified. Many of the cases of dispersal of the translocated primates have been documented, and social dynamics have been described. Births and partial genealogies have been recorded. Conflicts between the translocated primates and humans have been documented. All of these accomplishments demonstrate adherence to reintroduction guidelines and are indicators of translocation success.

15.4.11 Animal welfare and conservation-motivated primate translocations

Harrington *et al.* (2013) reviewed a large sample of conservation-motivated translocations of all types of animals, and found that the welfare of the animals that were involved was rarely mentioned explicitly, and that measures to enhance welfare were included in only about two-thirds of the projects. However, welfare is explicitly mentioned and welfare-enhancing techniques were employed in the nine successful conservation-motivated primate translocation projects that are described in Section 15.4 and in many of the primate translocations that were conducted for other purposes. The managers of these nine primate translocations were extraordinarily and explicitly considerate of the health and wellbeing of the translocated primates and strove to maximize post-release survival. There was extensive pre-release preparation and veterinary support and, in the cases of the lemurs and golden lion tamarins, there was post-release supplemental feeding and veterinary support. One of the best indications of the concern of these managers for wellbeing is the decision in each of these projects (with the apparent exception of the Rubondo chimpanzee operation) to *not* introduce some individuals that would have been unlikely to survive after being translocated (e.g. Pearson *et al.* 2007). Perhaps this extraordinary mindfulness to animal welfare is unique to primate translocations. Animal welfare may not be a salient consideration in the economically motivated translocation of oysters.

These managers could confidently predict and were willing to accept that the translocated primates would be stressed by the change in their lives and by the challenges of becoming independent of human support. The managers were also able to predict that some of the animals would suffer and die. The managers did their best to minimize suffering and death, but they accepted the inevitability of both. Some question whether the suffering and death of individual animals is justified by the attempt to restore self-sustaining animal populations by means of translocation (e.g. Bekoff 2013). Of course, animals also sometimes suffer and die even as a result of welfare translocations that are ostensibly designed to improve their wellbeing and health. For example, of 87 confiscated Bornean gibbons (*Hylobates muelleri*) that were released from a rehabilitation centre in Sarawak between 1976 and 1988, fewer than 10 could be located in 1988 (Bennett 1992). Wild animals suffer and die as well. One of the criteria of a self-sustaining population is that it be sufficiently large and independent for natural selection to be able to operate, and natural selection often involves debilitating and stressful competition and injury and death.

Every responsible translocation manager seeks to maximize the wellbeing, health, and survival of translocated animals, and every responsible animal welfarist wishes to see self-sustaining populations of animals in the wild. No responsible person wants to see animals, especially primates, languish in substandard captive facilities, but few want to see them released into the wild without adequate precautions for the prevention of suffering and death.

15.5 The relationship of body size to primate translocation

Body size affects translocation in direct and indirect ways.

15.5.1 Body size and home range size

Body size, in combination with diet and niche breadth (see Section 15.6), is a major determinant of home range size (e.g. Furuichi *et al.* 1997). Larger primates tend to have larger home ranges than smaller primates with similar diets and niche breadth. For example, 100 unprovisioned wild chimpanzees require a home range of 25 to 100 km^2 (e.g. Wrangham 1986) but 100 unprovisioned golden lion tamarins can live in only 3 to 10 km^2 (Dietz *et al.* 1997). Therefore, chimpanzees (and other large primates) need a larger expanse of suitable habitat to be successfully translocated and, as noted earlier, suitable habitat is rarely available for translocation. Chimpanzees and tamarins can exist at higher densities, but only in seriously disturbed or disconnected fragments, on islands, and/or where there is provisioning. Such settings may be appropriate for welfare-based translocations but are usually not suitable for conservation translocations because they cannot support self-sustaining populations (Beck *et al.* 2007).

15.5.2 Body size and reproduction

Additionally, the body size of terrestrial mammals (including primates) is typically positively related to the period of gestation and the age at sexual maturity, that is, bigger primates have longer gestation periods and mature more slowly (Eisenberg (1981) for the general relationship, Inskipp (2005) for chimpanzees, French *et al.* (2002) for tamarins). The chimpanzee gestation period is approximately 35 weeks; the tamarin gestation period is approximately 18 weeks. Chimpanzee infants are dependent on their mothers until they are 5 years of age and reach sexual maturity at 10 to 12 years of age. Tamarins are dependent on their parents and siblings until they are approximately 4 months of age and reach sexual maturity at approximately 1.5 years of age. Because of their longer gestation periods and the slower developmental rate of their offspring, the inter-birth interval is longer in larger primates; females of larger primate species may give birth only once every 2 to 12 years. Smaller primates, such as marmosets and tamarins, may give birth every year or even twice a year.

Marmosets and tamarins, and some prosimians, have even greater reproductive output by having multiple offspring per birth event, that is they have litter sizes that are greater than one. Most larger primates typically have only one offspring per birth event.

The reproductive correlates of larger body size (longer gestation periods, later attainment of sexual maturity, smaller litter size) therefore result in slower post-release population growth for larger primates. Translocated populations of smaller primates will increase in size more quickly than translocated populations of larger primates. Growth in the size of the post-release population is often cited as a criterion of translocation success, at least when the establishment of a self-sustaining population is a goal, because primate (at least ape) populations generally require 250 to 500 individuals to be self-sustaining (Beck *et al.* 2007). A confident assessment of success of translocations of large primates may take decades (King *et al.* 2014), but post-release population growth will suggest success more quickly for smaller primates.

15.5.3 Body size and conflict with humans

Wild and translocated primates can come into conflict with humans living in communities near them. The problem is especially serious with great apes, and the IUCN has published a set of guidelines specifically on conflict between apes and humans (Hockings and Humle 2009). Crop-raiding is the most common form of conflict. A few individuals of a large primate species, such as chimpanzees, can destroy a farmer's crop or her domestic bee colony in a single day, and there is little that she can do to deter the raid. Larger numbers of medium-sized primates can cause equivalent damage and are equally difficult to deter. Macaques, baboons, and other cercopithecines even enter homes to steal food. The smaller prosimians, cebids, and callitrichids are less frequent crop-raiders and food-thieves, and they don't cause much loss when they do maraud because they don't need as much food.

Wild and translocated primates have also confronted, harassed, and even attacked people who live near them (e.g. gorillas, King *et al.* (2012)). Translocated apes have also attacked people who were conducting post-release monitoring (e.g. bonobos, D. Morel, pers. comm. (2011)). The problem is more noticeable and serious when large primates are involved in these aggressive interactions because their mass, strength, and dentition are more formidable, and they can inflict grievous damage in a short period of time. Medium and small primates appear to actually attack people rarely, and the attack would most likely be less severe if they did. Humans living near the destination sites of translocated primates are therefore more likely to be supportive if the translocation involves small rather than large primate species.

15.5.4 Body size and the logistics of translocation

Large primates, such as apes, need larger, heavier, and sturdier containers if they are to be moved. Eight people and a heavy truck (or helicopter) may be required to move one great ape to a translocation destination site. However, one person can carry a family of six or more tamarins in two lightweight containers for many kilometres (Beck and Rambaldi 2006).

15.5.5 Body size and hunting

Body size most likely also affects the probability that translocated primates will be hunted by people for food, because it is more efficient and profitable to kill a large primate. One shotgun shell produces more kilograms of bushmeat. Of course, this relationship is relative to the other species that are present. A spider monkey is smaller than a chimpanzee but larger than a sympatric squirrel monkey and is thus more likely to be hunted than the squirrel monkey.

15.5.6 Summary of the effects of body size on translocation success

Body size is a life-history variable that affects the ease and probable success of translocations. Large translocated primates require more extensive suitable destination habitats, are logistically more difficult to translocate, are more likely to be hunted by humans, and more likely to engage in conflict with humans that can have serious economic, even life-threatening, consequences. Large primate species reproduce more slowly than smaller primate species, and their post-release numbers will therefore increase more slowly.

15.6 The relationship of niche breadth to primate translocation

Niche breadth is another variable that affects the probability of success of a translocation. Primate generalists, that is, those that eat many types of foods, tolerate a broad range of ambient temperatures and humidity, can live in a variety of habitat types, and can use a variety of locomotor substrates, may be more likely than primate specialists to survive after being translocated to a new environment (Strum 2005). The success of translocations of chimpanzees, rhesus and crab-eating macaques, and baboons appears to support this conclusion. However, herbivorous birds and mammals in general are more likely to be successfully translocated than carnivores and omnivores (Griffith *et al.* 1989), and herbivory can be considered to be a dietary specialization. Folivory is an even more specialized form of herbivory. Nonetheless, primate folivores have been translocated successfully, perhaps because they eat leaves of many different plant species (e.g. howler monkeys, Santini (1985), Horwich *et al.* (1993, 2002); Zanzibar red colobus monkeys, Camperio Ciani *et al.* (2001), Struhsaker and Siex (1998)). More data are needed before we can conclude with confidence that translocations of primate generalists are more likely to be successful.

15.7 The relationship of being captive-bred to primate translocation

An important variable in the context of translocation of primates (and probably many other animals) is the place of birth of the animals to be translocated. It appears that primates born in captivity are less likely to adapt to life in the wild than primates that were born in the wild and then captured and held in captivity for some time before being translocated to another wild destination site.

This difference was hypothesized as early as 1978 (Coimbra-Filho and Mittermeier 1978) and was documented for animals in general in earlier meta-analyses (Beck *et al.* 1994; Griffth *et al.* 1989). Vogel *et al.* (2002) released captive- and wild-born squirrel monkeys (*Saimiri sciureus*) simultaneously in a single group and observed that the wild-borns adapted more successfully. Most of the captive-borns had to be recaptured and returned to captivity. Kierulff *et al.* (2002b) provide the most convincing case. They studied the post-release success of dozens of captive-born and wild-born golden lion tamarins, using comparable methodologies, except that the captive-borns received more post-release support than the wild-borns. Despite this difference, the wild-borns had higher post-release survival rates, reproduced more quickly, and became independent of all human support more quickly.

Captive-born primates of species that have highly specialized feeding patterns may be especially difficult to reintroduce. Primates that feed on embedded foods (Parker and Gibson 1977) are good examples. Embedded foods include nuts and fruits whose edible portion is embedded within a shell or tough rind; insect larvae that are embedded beneath tree bark, in cavities in tree branches, and in subterranean or arboreal nests; and insects and amphibians that live within curled leaves. Golden lion tamarins feed on embedded insects and amphibians with a behaviour that is known as 'micromanipulation', in which they insert their hands into cavities and crevices and probe tactilely for prey. Captive tamarins do micromanipulate, but after reintroduction they need to learn to identify sites that are likely to contain embedded prey. It appears that they learn more quickly if they have a skilled parent or sibling to observe. Reintroduced groups of captive-borns that lack a skilled model are more difficult to reintroduce (Beck *et al.* 2002). Orang-utans, chimpanzees, and capuchin monkeys use tools in a variety of ways to access embedded foods (Shumaker *et al.* 2011). Social learning appears to be critical in acquiring and perfecting these behaviours (e.g. Beck 2010; Custance *et al.* 2002). Captive-born individuals of these species that are reintroduced without skilled models are likely to be disadvantaged in acquiring these behaviours. Structured training by humans may not be effective for helping captive-born primates acquire these specialized feeding patterns (Beck *et al.* 2002).

Nonetheless, four of the nine successful programmes described in Section 15.4 involved some captive-bred primates. Captive-born primates can be translocated successfully but the effort required, time invested, mortality, and costs will be greater (Beck *et al.* 2002).

15.7.1 Zoos and conservation-motivated primate translocation

The primates that are currently held in North American, European, Australian, and Japanese zoos are now almost all captive-born. There is probably very little need (from a conservation perspective) to reintroduce or translocate captive-born primates at the present time. Suitable habitat is limited, and there are wild-born orphans and displaced individuals of virtually every primate species that are available to be translocated to any suitable, vacant habitat that is found. Zoos will therefore most likely play a limited role in conservation-motivated primate translocations in the foreseeable future. However, this could change suddenly and catastrophically. Disease, political instability, hunting, unchecked habitat conversion, and climate change are predicted to continue to reduce the numbers and distribution of wild primates. Sumatran orangutans are critically endangered and are predicted to become extinct within 100 years if hunting and economically motivated habitat conversion are not eliminated (Singleton *et al.* 2008). Many lemur species teeter at the brink of extinction because of agricultural expansion and illegal hunting and logging (Britt *et al.* 2002). We can't predict which primate species may be driven to extinction, and when, but it will happen. A remnant wild population or a back-up captive-born zoo population could be translocated for restoration of the species when (and if) the underlying causes of extinction are ameliorated. The combined contributions of countless translocation practitioners are now teaching us how to translocate captive-born and wild-born primates, and our successors may one day have to draw on this knowledge. We must work to continually revise a body of best practices.

15.8 Conclusions

To summarize the chapter, the key take-home lessons for conservationists are:

(1) There are many different reasons for translocating primates. Improving the welfare of individual primates and the conservation of primate species and ecosystems are the most common.
(2) There have been surprisingly few successful conservation-motivated primate translocations, but translocation may play a larger role in primate conservation in the future.
(3) Body size, niche breadth, and being born in captivity or in the wild have direct and indirect effects on the probability of success of primate translocations.

Acknowledgements

I am grateful to the editors and anonymous reviewers for helpful suggestions for improving this chapter, the Smithsonian National Zoo and the government of Brazil for giving me the opportunity to participate in the reintroduction of golden lion tamarins, and to the many colleagues who have assisted, informed, and supported my work. I am also mindful of the thousands of non-human primates that have endured the stresses and negative consequences of translocation.

References

Baker, L. R. (2002). IUCN/SSC Re-introduction Specialist Group: guidelines for nonhuman primate re-introductions. *Re-introduction News* **21**: 29–57.

Beck, B. B. (2010). Chimpanzee orphans: sanctuaries, reintroduction, and cognition. In Lonsdorf, E. V., Ross, S. R., and Matsuzawa, T. (Eds), *The Mind of the Chimpanzee*. pp. 332–346. Chicago: The University of Chicago Press.

Beck, B. B. and Rambaldi, D. M. (2006). Reintroduction of golden lion tamarins (*Leontopithecus r. rosalia*): implications for African great apes. Paper presented at the XXIst Congress of the International Primatological Society; Entebbe, Uganda.

Beck, B. B., Rapaport, L. G., Stanley Price, M. R., and Wilson, A. (1994). Reintroduction of captive-born animals. In: Olney, P. J. S., Mace, G. M., and Feistner, A. T. C. (Eds), *Creative Conservation*. pp. 265–286. London: Chapman and Hall.

Beck, B. B., Castro, M. I., Stoinski, T. S., and Ballou, J. D. (2002). The effects of prerelease environments and postrelease management on survivorship in reintroduced golden lion tamarins. In: Kleiman, D. G. and Rylands, A.B. (Eds), *Lion Tamarins: Biology and Conservation*. pp. 283–300. Washington, DC: Smithsonian Institution Press.

Beck, B. B., Walkup, K., Rodrigues, M., Unwin, S., Travis, D., *et al.* (2007). *Best Practice Guidelines for the Re-introduction of Great Apes*. Gland: IUCN/SSC Primate Specialist Group.

Bekoff, M. (2013). Thirteen Gold Monkeys: bringing conservation to the public. Blog, Animal Emotions. *Psychology Today*, 13 April.

Bennett, J. (1992). A glut of gibbons in Sarawak—is rehabilitation the answer? *Oryx* **26**: 157–164.

Borner, M. (1985). The rehabilitated chimpanzees of Rubondo Island. *Oryx* **19**: 151–154.

Britt, A., Welch, C., and Katz, A. (2002). The release of captive-bred black and white ruffed lemurs into the

Betampona Reserve, eastern Madagascar. *Re-introduction News* **21**: 18–20.

Britt, A., Welch, C., Katz, A., Iambana, B., Porton, I., *et al.* (2004a). The re-stocking of captive-bred ruffed lemurs (*Varecia variegata variegata*) into the Betampona Reserve, Madagascar. *Biodiversity and Conservation* **13**: 635–657.

Britt, A., Welch, C., and Katz, A. (2004b). Can small, isolated primate populations be effectively reinforced through the release of individuals from a captive population? *Biological Conservation* **115**: 319–327.

Camperio Ciani, A., Palentini, L., and Finotti, E. (2001). Survival of a small translocated *Procolobus kirkii* population on Pemba Island. *Animal Biodiversity and Conservation* **24**: 15–18.

Candland, D. K. and Bush, S. L. (1995). Primates and behavior. In: Gibbons, E. F., Durrant, B. S., and Demarest, J. (Eds), *Conservation of Endangered Species in Captivity*. pp. 521–551. Albany: State University of New York Press.

Carpenter, C. R. (1940). Rhesus monkeys for American laboratories. *Science* **92**: 284–286.

Coimbra-Filho, A. F. and Mittermeier, R. A. (1978). Reintroduction and translocation of lion tamarins: a realistic appraisal. In: Rothe, H., Wolters, H. J., and Hearn, J. P. (Eds), *Biology and Behaviour of Marmosets*. pp. 41–46. Gottingen: Eigenverlag Hartmut Rothe.

Custance, D. M., Whiten, A., and Fredman, T. (2002). Social learning and primate reintroduction. *International Journal of Primatology* **23**: 479–499.

Dietz, J. M., Peres, C. A., and Pinder, L. (1997). Foraging ecology and use of space in wild golden lion tamarins (*Leontopithecus rosalia*). *American Journal of Primatology* **41**: 289–305.

Eisenberg, J. F. (1981). *The Mammalian Radiations*. Chicago: University of Chicago Press.

Ewen, J. G., Armstrong, D. P., Parker, K. A., and Seddon, P. J. (2013). *Reintroduction Biology: Integrating Science and Management*. London: Wiley-Blackwell.

Farmer, K. H., Honig, N., Goossens, B., and Jamart, A. (2010). Habitat Ecologique et Liberté des Primates: re-introduction of central chimpanzees to the Conkouati-Douli National Park, Republic of Congo. In: Soorae, M. (Ed.), *Global Re-introduction Perspectives: 2010*. pp. 231–237. Abu Dhabi, UAE: IUCN/SSC Re-introduction Specialist Group.

Farmer, K. H. and Jamart, A. (2002). Habitat Ecologique et Liberté des Primates: a case study of chimpanzee re-introduction in the Republic of Congo. *Re-introduction News* **21**: 16–18.

Fedigan, L. M. (1991). History of the Arashiyama west Japanese macaques in Texas. In: Fedigan, L. M. and Asquith, P. J. (Eds), *The Monkeys of Arashiyama: Thirty-Five years of Research in Japan and the West*. pp. 140–154. Albany: State University of New York Press.

Fischer, J. and Lindenmeyer, D. B. (2000). An assessment of the published results of animal relocations. *Biological Conservation* **96**: 1–11.

French, J. A., de Vleeschouwer, K., Bales, K., and Heistermann, M. (2002). Lion tamarin reproductive biology. In: Kleiman D. G. and Rylands, A. B. (Eds), *Lion Tamarins: Biology and Conservation*. pp. 133–156. Washington, DC: Smithsonian Institution Press.

Furuichi, T., Inagaki, H., and Angove-Ovono, S. (1997). Population density of chimpanzees and gorillas in the Petit Loango Reserve, Gabon: employing a new method to distinguish between nests of the two species. *International Journal of Primatology* **18**: 1029–1046.

Goossens, B., Setchell, J. M., Tchidongo, E., Dilambaka, E. Vidal, C., *et al.* (2005). Survival, interactions with conspecifics and reproduction in 37 chimpanzees released into the wild. *Biological Conservation* **123**: 461–475.

Griffith, B., Scott, M., Carpenter, J. W., and Reed, C. (1989). Translocation as a species conservation tool: status and strategy. *Science* **24**: 477–480.

Gupta, A. K. (2002). Release of golden langurs in Tripura, India. *Re-introduction News* **21**: 26–28.

Guy, A. J., Curnoe, D., and Banks P. B. (2014). Welfare based primate rehabilitation as a potential conservation strategy: does it measure up? *Primates* **55**: 139–147.

Hannah, A. C. and McGrew, W. (1991). Rehabilitation of captive chimpanzees. In Box, H. O. (Ed.), *Primate Response to Environmental Change*. pp. 167–186. London: Chapman & Hall.

Harrington, L. A., Moehrenschlager, A., Merryl Gelling, A., Atkinson, R. P. D., Hughes, J., *et al.* (2013). Conflicting and complementary ethics of animal welfare considerations in reintroductions. *Conservation Biology* **27**: 486–500.

Hockings, K. and Humle, T. (2009). *Best Practice Guidelines for the Prevention and Mitigation of Conflict Between Humans and Great Apes*. Gland, Switzerland: IUCN/SSC Primate Specialist Group.

Horwich, R. H., Koontz, F., Saqui, E., Saqui, H., and Glander, K. (1993). A reintroduction program for the conservation of the black howler monkey in Belize. *Endangered Species Update* **10**(6): 1–6.

Horwich, R. H., Koontz, F., Saqui, E., Ostro, L., Silver, S., *et al.* (2002). Translocation of black howler monkeys in Belize. *Re-introduction News* **21**: 10–12.

Humle, T., Colin, C., Laurans, M., and Raballand, E. (2011). Group release of sanctuary chimpanzees (*Pan troglodytes*) in the Haut Niger National Park, Guinea, West Africa: ranging patterns and lessons so far. *International Journal of Primatology* **32**(3): 456–473.

Imam, E., Yaha, H. S. A., and Malik. I. (2002). A successful mass translocation of commensal rhesus monkeys *Macaca mulatta* in Vrindaban, India. *Oryx* **36**: 87–93.

Inskipp, T. (2005). Chimpanzee (*Pan troglodytes*). In: Caldecott, J. and Miles, L. (Eds), *World Atlas of Great Apes and Their Conservation*. pp. 53–82. Cambridge, UK: UNEP World Conservation Monitoring Centre.

IUCN/SSC (2013). *Guidelines for Reintroductions and Other Conservation Translocations. Version 1.0*. Gland, Switzerland: IUCN Species Survival Commission.

Kierulff, M. C. M. (2011). Invasive introduced golden-headed lion tamarins. *Tamarin Tales* **10**: 5–7.

Kierulff, M. C. M. (2012). The removal of golden-headed lion tamarin invaders. *Tamarin Tales* **11**: 2–3.

Kierulff, M. C. M., Beck, B. B., Kleiman, D. G., and Procopio, P. (2002a). Re-introduction and translocation as conservation tools for golden lion tamarins in Brazil. *Re-introduction News* **21**: 7–10.

Kierulff, M. C., Procopio de Oliveira, P., Beck, B. B., and Martins, A. (2002b). Reintroduction and translocation as conservation tools for golden lion tamarins. In: Kleiman D. G. and Rylands, A. B. (Eds), *Lion Tamarins: Biology and Conservation*. pp. 271–282. Washington, DC: Smithsonian Institution Press.

Kierulff, M. C. M., Ruiz-Miranda, C. R., Procópio de Oliveira, P., Beck, B. B., Martins, A., *et al.* (2012). The Golden lion tamarin *Leontopithecus rosalia*: a conservation success story. *International Zoo Yearbook* **46**: 36–45.

King, T., Chamberlan, C., and Courage, A. (2012). Assessing initial reintroduction success in long-lived primates by quantifying survival, reproduction, and dispersal parameters: western lowland gorillas (*Gorilla gorilla gorilla*) in Congo and Gabon. *International Journal of Primatology* **33**: 134–149.

King, T., Chamberlan, C., and Courage, A. (2014). Assessing reintroduction success in long-lived primates through population viability analysis: western lowland gorillas *Gorilla gorilla gorilla* in Central Africa. *Oryx* **48**: 294–303.

Konstant, W. R. and Mittermeier, R. A. (1982). Introduction, reintroduction and translocation of Neotropical primates: past experience and future possibilities. In: Olney, P. J. S. (Ed.), *International Zoo Yearbook, Volume 22*. pp. 69–77. London: Zoological Society of London.

Kyes, R. C. (1993). Survey of long-tailed macaques introduced onto Tinjil Island, Indonesia. *American Journal of Primatology* **31**: 77–83.

Moscovice, L. R., Issa, M. H., Petrzelkova, K. J., Keuler, N. S., Snowdon, C. T., *et al.* (2007). Fruit availability, chimpanzee diet, and grouping patterns on Rubondo Island, Tanzania. *American Journal of Primatology* **69**: 487–502.

Oates, J. F. (1999). *Myth and Reality in the Rain Forest*. Berkeley: University of California Press.

Oates, J. F. (2013). Primate conservation: unmet challenges and the role of the International Primatological Society. *International Journal of Primatology* **34**: 235–245.

Ostro, L. E. T., Silver, S. C., Koontz, F. W., Young, T. P., and Horwich, R. H. (1999). Ranging behavior of translocated and established groups of black howler monkeys *Alouatta pigra* in Belize, Central America. *Biological Conservation* **87**: 181–190.

Parker, S. T. and Gibson, K. R. (1977). Object manipulation, tool use, and sensorimotor intelligence as feeding adaptations in cebus monkeys and great apes. *Journal of Human Evolution* **6**: 623–641.

Pearson, L., Aczel, P., Mahé, S., Courage, A., and King, T. (2007). Gorilla Reintroduction to the Batéké Plateau National Park, Gabon: An Analysis of the Preparations and Initial Results with Reference to the IUCN Guidelines for the Re-introduction of Great Apes. Hythe, Kent, UK: The Aspinall Foundation.

Riedler, B., Millesi, E., and Pratje, P.H. (2010). Adaptation to forest life during the reintroduction process of immature *Pongo abelii*. *International Journal of Primatology* **31**: 647–663.

Rodriguez-Luna, E. and Cortés-Ortiz, L. (1994). Translocacion y seguimiento de un grupo de monos *Alouatta palliata* liberado en una isla (1988–1994). *Neotropical Primates* **2**: 1–5.

Ruiz-Miranda, C. R., Beck, B. B., Kleiman, D. G., Martins, A., Dietz, J. M., *et al.* (2010). Re-introduction and translocation of golden lion tamarins, Atlantic Coastal Forest, Brazil: the creation of a metapopulation. In: Soorae, P. S. (Ed.), *Global Re-introduction Perspectives: Additional Case-Studies from Around the Globe*. pp. 225–230. Abu Dhabi, UAE: IUCN/SSC Re-introduction Specialist Group.

Santini, M. E. L. (1985). Modificações temporais na dieta de *Alouatta caraya* (Primates, Cebidae), reintroduzido no Parque Nacional de Brasília. *A Primatologia no Brasil* **2**: 269–292.

Shumaker, R. W. and Beck, B. B. (2003). *Primates in Question*. Washington, DC: Smithsonian Institution Press.

Shumaker, R.W., Walkup, K. R., and Beck, B. B. (2011). *Animal Tool Behavior*. Baltimore: The Johns Hopkins University Press.

Singleton, I., Wich, S. A., and Griffiths, M. (2008). *Pongo abelii*. IUCN Red List of Endangered Species. Version 2014.3. www.iucnredlist.org.

Siregar, R. S. E., Farmer, K. H., Chivers, D., and Saragih, B. (2010). Re-introduction of Bornean orang-utans to Meratus protected forest, East Kalimantan, Indonesia. In: Soorae, P. S. (Ed.), *Global Re-introduction Perspectives: Additional Case-Studies from Around the Globe*. pp. 243–248. Abu Dhabi, UAE: IUCN/SSC Re-introduction Specialist Group.

Southwick, C. H., Malik, I., and Siddiqi, M. E. (1998). Translocations of rhesus monkeys in India: prospects and outcomes. *American Journal of Primatology* **45**: 209–210.

Strier, K. B. (2007). *Primate Behavioral Ecology*. Boston: Allyn and Bacon.

Struhsaker, T. T. and Siex, K. S. (1998). Translocation and introduction of the Zanzibar red colobus monkey: success and failure with an endangered island endemic. *Oryx* **32**: 277–284.

Strum, S. C. (2002). Translocation of three wild troops of baboons in Kenya. *Re-introduction News* **21**: 12–15.

Strum, S. C. (2005). Measuring success in primate translocations: a baboon case study. *American Journal of Primatology* **65**: 117–140.

Taub, D. M. and Mehlman, P. T. (1989). Development of the Morgan Island rhesus monkey colony. *Puerto Rico Health Sciences Journal* **8**: 159–169.

Trayford, H., Pratje, P., and Singleton, I. (2010). Re-introduction of the Sumatran orangutan in Sumatra, Indonesia. In: Soorae, P. S. (Ed.), *Global Re-introduction Perspectives: Additional Case-Studies from Around the Globe*. pp. 238–242. Abu Dhabi, UAE: IUCN/SSC Re-introduction Specialist Group.

Treves, A. and Naughton-Treves, L. (1997). Case study of a chimpanzee recovered from poachers and temporarily released with wild chimpanzees. *Primates* **38**: 315–324.

Tutin, C., Ancrenaz, M., Paredes, J., Vacher-Vallas, M., Vidal, C., *et al.* (2001). Conservation biology framework for the release of wild-born orphaned chimpanzees into the Conkouati Reserve, Congo. *Conservation Biology* **15**: 1247–1257.

Valladares-Padua, C., Martins, C. S., Wormell, D., and Setz, E. Z. (2000). Preliminary evaluation of the reintroduction of a mixed wild-captive group of black lion tamarins *Leontopithecus chrysopygus*. *Dodo* **36**: 30–38.

Vogel, G. (2007). Scientists say Ebola has pushed Western gorillas to the brink. *Science* **317**: 1484.

Vogel, I., Glöwing, B., Saint Pierre, I., Bayart, F., Contamin, H., *et al.* (2002). Squirrel monkey (*Saimiri sciureus*) rehabilitation in French Guiana: a case study. *Neotropical Primates* **10**: 147–149.

Warren, K. S. and Swan, R. A. (2002). Re-introduction of orang-utans in Indonesia. *Re-introduction News* **21**: 24–26.

Wrangham, R. W. (1986). Ecology and social relationships in two species of chimpanzee. In: Rubenstein D. L. and Wrangham, R. W. (Eds), *Ecological Aspects of Social Evolution*. pp. 352–378. Princeton: Princeton University Press.

CHAPTER 16

Payment for ecosystem services: the role of REDD + in primate conservation

John Garcia-Ulloa and Lian Pin Koh

An aerial view of a forest canopy in Sumatra. Tropical forests provide important climate regulation services and are key for the conservation of primates. Photo copyright: ConservationDrones.org.

16.1 Introduction

Increasing pressures from global human activities have compromised the biological heritage of our planet, including the future of primate species. The survival of this group is threatened by a combination of factors, such as habitat loss, fragmentation, habitat degradation, poaching, hunting, and climate change (see chapters in this volume). According to assessments by the International Union for Conservation of Nature (IUCN), the pressures resulting from these processes have put at risk 50% of primate species worldwide (IUCN 2014). This figure is likely an underestimate given a lack of information for at least another 30% of primate species (i.e. those considered as data deficient by the IUCN). The clearance, fragmentation, and degradation of tropical forests remain the main contributing factors in the disappearance of wild primate populations, as most species depend on these

Garcia-Ulloa, J. and Koh, L.P., *Payment for ecosystem services: the role of REDD + in primate conservation.* In: *An Introduction to Primate Conservation.* Edited by: Serge A. Wich and Andrew J. Marshall, Oxford University Press (2016). © Oxford University Press.
DOI 10.1093/acprof:oso/9780198703389.003.0016

ecosystems. These processes are largely driven by the large-scale expansion of agriculture, forestry, mining, and infrastructure projects, which are expected to increase as the human population grows.

The disappearance of tropical forests is not only detrimental to primates and biodiversity, but it also negatively affects the many benefits we receive from these ecosystems. Widespread deforestation results in disruptions of the carbon cycle, declines in soil fertility, and alteration of hydrological regimes (Foley *et al.* 2005). These impacts negatively affect global climate, compromise the primary productivity of the land, and result in significant social and economic consequences (Malhi *et al.* 2008).

Efforts to conserve tropical forests have increased in recent years (Rudel *et al.* 2009). Forest conservation initiatives have transitioned from an initial focus on protected areas to include a broader range of strategies, such as the preservation of habitats within agricultural and urban landscapes, environmentally sound land-use planning, and the use of economic tools to incentivize beneficial land-management practices (Rands *et al.* 2010). In this chapter, we discuss the concept of ecosystem services and Payment for Ecosystem Services (PES), within the context of the conservation of tropical forests as primate habitat. In particular, we focus on the largest PES initiative to date, Reducing Emissions from Deforestation and Forest Degradation (REDD +), which has emerged as a promising global mitigation strategy for climate change. We discuss key aspects of this initiative that are relevant to the realization of its potential benefits for conservation.

16.2 Ecosystem services

Various global initiatives have emerged to counteract environmental degradation caused by human activities. In 2005, more than 1,000 researchers, from 95 nations, published the Millennium Ecosystem Assessment (MEA). The MEA, following requests from governments received through four international conventions,[1] attempted for the first time to summarize the state and importance of all benefits that society receives from ecosystems. The report became a crucial step in designing a policy and action framework to adapt to and mitigate many of the problems related to environmental degradation, and also cemented the concept of ecosystem services.

Ecosystem services are all the benefits that society receives from ecosystems; in many cases these services are crucial for our livelihoods and quality of life (Perrings *et al.* 2010). The supply of clean air, potable water, food, and many other natural assets are probably the most obvious examples of these services. However, there are other less conspicuous ones, that either regulate (e.g. climate regulation) or support (e.g. nutrient cycling) other services, or that enrich our lives through the provision of non-material benefits (e.g. landscape aesthetics or outdoor recreation) (Figure 16.1).

Ecosystem services are inherently connected to ecosystem functioning and result from the intricate network interactions among the biotic and the abiotic components of an ecosystem (Cardinale *et al.* 2012). The contribution of species (i.e. the biotic component) is evident in certain services such as pollination or pest control, but less apparent in others such as climate regulation. Understanding the relationship between species and ecosystem services has been an intense area of ecological research in the last two decades, with many studies focusing on aspects such as how biodiversity and functional diversity affect ecosystem functioning and resilience (Kinzig 2009; Hooper *et al.* 2012).

In the case of primates, however, explicit references to their role in the maintenance and delivery of ecosystems services are scarce in the literature. Perhaps the most well-known example is their contribution to seed dispersal and forest regeneration, which have implications for net primary production, carbon sequestration, and timber supply (Box 16.1). Primates, however, provide other services (Marshall and Wich, Chapter 2, this volume). They are an important component in the diet of many forest-dwelling human communities, are key models in the study and control of many tropical diseases, and have a spiritual and cultural significance to many of us: after all they are our closest relatives.

The services that we obtain from ecosystems are in many cases crucial and not substitutable; maintaining

[1] The Convention on Biological Diversity, the United Nations Convention to Combat Desertification, the Ramsar Convention on Wetlands, and the Convention on Migratory Species.

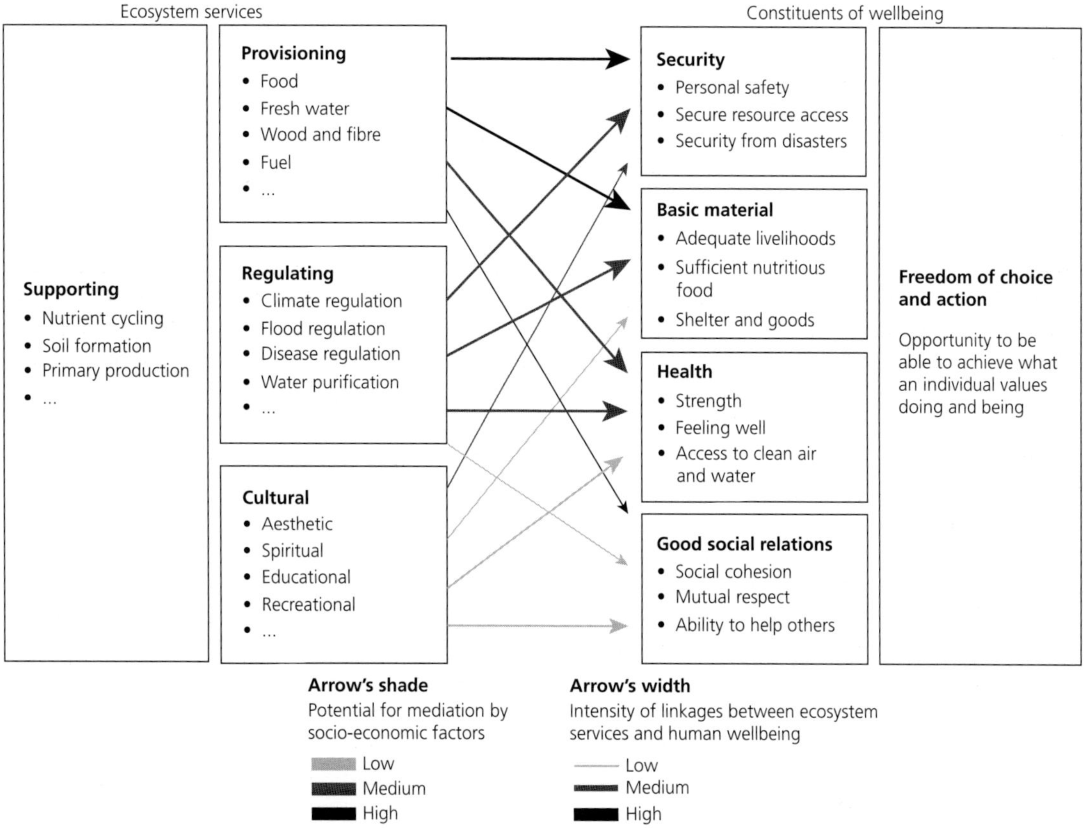

Figure 16.1 Conceptual framework showing the different types of ecosystem services and their connection with human wellbeing. (Adapted from the Millennium Ecosystem Assessment, http://www.millenniumassessment.org/en/GraphicResources.html.)

them is essential to humankind. The MEA (2005: v) states that, 'even though humans and society are buffered against environmental changes by culture and technology, we are fundamentally dependent on the flow of ecosystem services'. Paradoxically, however, the integrity and functioning of many of these services is at risk because of human activities.

The degradation of ecosystems, the loss of biodiversity, and the depletion of the services they provide to humanity, is rooted in the way society values different land uses (Ghazoul and Sheil, 2010) and the insufficient investment in their protection and maintenance. For instance, the investment in conservation of avian species currently covers only 12% of the total funds needed to protect all avian species threatened worldwide (McCarthy *et al.* 2012). Moreover, land and resource management often fails to recognize the interactions and trade-offs between different ecosystem services (Bryan 2013). For example, maximizing the short-term harvesting of timber in tropical forests can have negative impacts on forest regeneration and biodiversity conservation and drastically affect future timber supply (Putz *et al.* 2008; Burivalova *et al.* 2014). Thus, an integrative approach that takes into account the value of multiple and overlapping services, together with their interactions and trade-offs, is required in the management of ecosystems (Schowalter 2013).

The cost of environmental degradation and loss of ecosystem services have not been taken into account when making decisions on human activities over the land (Pirard 2012). This situation is, however, increasingly changing through the widespread promotion of economic instruments for

Box 16.1 Primate species are key to forest regeneration

The role of biodiversity in ecosystem functioning and in the delivery of ecosystem services has been the subject of intense ecological research in recent years. Evidence suggests that biodiversity is crucial for the maintenance of ecosystem functioning (Hooper *et al.* 2012). The role of primate species in seed dispersal is well recognized; primates species are highly mobile and can disperse seeds across larger areas (Figure 16.2). To understand the importance of primate-mediated seed-dispersal in forest functioning, Effiom *et al.* (2013) explored the effects in seed recruitment resulting from changes in mammal communities due to hunting. In their study, Effiom *et al.* compared paired forest sites with high and low hunting pressures in south-eastern Nigeria. They found that mammal composition had drastically changed in highly hunted areas. Large primates were rare, while rodent species were more abundant, even though both groups of species were affected by hunting. The community composition of mature trees was the same in both hunted and non-hunted sites, while the composition of seedling communities was significantly different. In non-hunted forests the seedlings of species dispersed by primates were dominant, whereas in hunted sites seedlings were mostly from abiotically dispersed species. Effiom *et al.* argued that this was a consequence of the combined effects of both reduced seed dispersal by primates and increased seed predation by rodents, as seed mortality increases with the proximity to a parent or conspecific tree because of the higher likelihood of being predated or attacked by pathogens. Furthermore, follow-up studies didn't find evidence that other groups, such as frugivore birds, could compensate the loss of seed dispersal in the absence of primates (Effiom *et al.* 2014). The results of these studies demonstrate that declines in medium and large primates can alter, to a degree, forest regeneration capacity. This can potentially affect ecosystems services, such as primary net production and the availability of timber and non-timber resources.

Figure 16.2 Proboscis monkey, *Nasalis larvatus*, feeding on a fruiting tree in the island of Borneo. Primate species are key seed-dispersal agents and play a significant role in the regeneration of tropical forests. Courtesy of mongabay.com/Rhett A. Butler.

environmental protection. One of such initiatives consists of the direct payment to landholders for the preservation and restoration of ecosystem services, and is referred to as 'payments for ecosystem services (PES)' (Gomez-Baggethun *et al.* 2010).

16.3 Payment for ecosystem services

PES are based on the beneficiary pays principle (i.e. rewarding a desired behaviour) rather than the polluter pays principle (i.e. punishing an unwanted behaviour). As such, it is a transaction between a supplier (i.e. landholders), who ensures the provision of an ecosystem service, and a group of buyers (i.e. beneficiaries), who have an interest in maintaining that service (Engel *et al.* 2008). The transaction is voluntary and conditional to the provision of the service, and can be performed through either direct agreements between landowners and beneficiaries (donor-funded programmes), voluntary markets (where demand is created by voluntary buyers), or compliance markets (where demand is created by regulatory frameworks) (Muradian *et al.* 2013).

Many countries have advanced greatly in the implementation of PES initiatives. In Costa Rica, for instance, a programme was established 20 years ago to pay landholders for conserving forested areas in their lands. More than 1 million ha of forests have been part of the programme since then, and the country's total forest area has increased from 20% in 1980s to 50% in 2010s (Sanchez-Azofeifa *et al.* 2007). The programme has been considered pivotal in the increase of forested area and it is internationally regarded as a success (Porras *et al.* 2013; Pirard 2012).

Because of the potential benefits for habitat protection and restoration, many conservationists consider payment for ecosystem services programmes as an opportunity to stream much needed resources towards the protection of threatened species, including primates (Rands *et al.* 2010). In fact, the potential role of PES has been highlighted in conservation planning of emblematic species such as orang-utans (Wich *et al.* 2011) and chimpanzees (IIED 2015). It can be argued, however, that the most promising and controversial PES initiative for conservation has emerged from the international negotiations on climate change: the 'Reducing Emissions from Deforestation and Forest Degradation (REDD+)' initiative, the largest PES programme to be implemented worldwide.

16.4 Reducing emissions from deforestation and forest degradation (REDD +)

REDD + is a climate change mitigation strategy designed to curb emissions originating from land-cover changes (Birdsey *et al.* 2013; Venter and Koh 2011), which is the third largest source of emissions after the energy and industry sectors (IPCC 2013). Between 1990 and 2010, up to 15% of the total CO_2 emissions worldwide originated from such changes (Houghton *et al.* 2012), the majority of which came from deforestation and the degradation of peatlands and soils in the tropics (Hooijer *et al.* 2010). The management and conversion of natural forests have thus taken a central role in our efforts to mitigate climate change, because of the carbon-related services that these ecosystems provide. Forests store an estimated 45% of all carbon present in terrestrial ecosystems (Bonan 2008), and remove a considerable proportion of CO_2 from the atmosphere every year (almost the equivalent to a third of current emissions) (Pan *et al.* 2011). The protection of carbon stocks and the carbon sequestration capacity of forests within a global climate change agreement is, thus, necessary.

REDD was originally conceived as a performance-based mechanism where forest-rich developing nations could receive monetary incentives, from international public and private donors, if their deforestation or forest degradation rates are lowered with respect to a baseline level. A second iteration of this scheme, which is represented by the added ' + ' in the acronym, additionally recognizes reforestation and sustainable forestry efforts as valid activities to enhance carbon sequestration (Venter and Koh 2011).

In practice, REDD + can be viewed as the first global-scale payment for an ecosystem services initiative (Corbera 2012), where landholders are remunerated for practices that protect and restore existing carbon stocks. Its mechanism is based in the issuance of carbon credits (i.e. certified carbon benefits), the subsequent sale of those credits in international markets, and the verification and monitoring of carbon benefits during the committed period. As such, the implementation of REDD + is a considerable challenge that requires: (i) broad international coordination between countries; (ii) significant institutional changes to enable the issuance, verification, and monitoring of carbon credits; (iii) a sufficient funding mechanism that secures long-term economic viability; and (iv) the development of appropriate methods to assess and verify potential benefits or impacts.

Although the implementation of REDD + has been under international negotiations within the United Nations Framework on Climate Change for a decade now, a full adoption of REDD + has yet to happen. So far, its implementation has only been done on an individual project basis, financed mainly by direct donor-funded programmes and more recently through voluntary carbon markets, with the latter trading in 2010 an all-time-high total value of USD124 million (Peters-Stanley *et al.* 2013). International negotiations, however, suggest that REDD + initiatives will transition from project-based mechanisms to national-scale policies and become pivotal in future land-use planning in forest-rich countries. In fact, organizations such as the United Nations Environmental Programme and United Nations Development Programme view these initiatives as potential catalysts for change towards a green economy.[2] Embarking on this low-carbon pathway, however, includes addressing the full range of risks and benefits that REDD + could provide (Ebeling and Yasue 2008).

16.5 REDD + and biodiversity

Forests not only play an important role in combatting climate change globally, but also provide other essential ecosystem services, such as the regulation of the hydrological cycle or the provision of food and materials to many forest-dwelling communities (Wunder 2001; Bonan 2008). Tropical forests are also key to the survival of many unique taxa, and their disappearance is one of the main causes of species losses worldwide (Ghazoul and Sheil 2010). In fact, the highest number of threatened species, for groups such as vertebrates, is found in tropical forests (Vié *et al.* 2009). Consequently, many conservationists have eagerly, but cautiously, followed the developments of REDD + negotiations, and hope that it becomes a useful tool for the conservation of threatened species (Harvey *et al.* 2010).

Global biodiversity losses and climate change are not only connected through deforestation, but are also closely interlinked and feedback on each other (Driscoll *et al.* 2012; Busch and Grantham 2013). On the one hand, the rapid nature of climate change suggests that many species, including forests primate species, will not be able to adapt and may face extinction (Colwell *et al.* 2008; Thomas *et al.* 2004; see Korstjens and Hillyer, Chapter 11, this volume). On the other hand, species losses can be detrimental to ecosystem processes and functions (Hooper *et al.* 2012), and therefore have significant repercussions in the provision of ecosystem services, including climate regulation (Bunker *et al.* 2005). In addition, biodiversity might well play an important role in the adaptation to climate change, by making ecosystems and agricultural systems more resilient (Chapin *et al.* 2000; Millar *et al.* 2007). Natural genetic diversity, for instance, can be crucial for the adaptation of our crops to a new climate (Takeda and Matsuoka 2008), while the presence of primate populations increases the regeneration capacity of forests (Box 16.1). Therefore, designing policies that tackle both the threats of biodiversity loss and climate change would improve the prospects for adapting successfully to the challenges of the coming decades.

Advocates of REDD + have lauded its potential to deliver not only carbon benefits, but also other multiple benefits for both environmental protection and socio-economic development (Venter and Koh 2011). REDD + detractors, on the other hand, remain sceptical and argue that these programmes may actually result in unintended impacts, such as communities' displacement, local restricted access to resources, unequal benefit sharing, speculative profit-seeking behaviours, and higher pressures to low-carbon ecosystems (Resosudarmo *et al.* 2014; Karsenty *et al.* 2014).

16.6 Realizing the potential of REDD + for biodiversity conservation

The delivery conservation benefits, and other multiple benefits, have been debated since the conception of REDD + initiatives (Venter *et al.* 2009). To date, however, REDD + only addresses biodiversity issues through qualitative safeguards (UNFCCC 2014). Such safeguards aim to ensure that no negative impacts on biodiversity arise from REDD + implementation, but fail to establish a mechanism to ensure the delivery of additional conservation benefits. Maximizing biodiversity conservation opportunities within REDD + activities will require an explicit integration within the technical and policy frameworks of REDD + programmes (Grainger *et al.* 2009; Gardner *et al.* 2012).

[2] Green economy is an initiative promoted by the United Nations Environmental Programme that seeks transition to an economy that results in improved human wellbeing and social equity, while significantly reducing environmental risks and ecological scarcities (UNEP 2011). It emphasizes reductions in carbon emissions and pollution, improvements in energy and resource efficiency, and minimal or no loss of biodiversity and ecosystem services. A green economy favours renewable energy and low-carbon and environmentally friendly economic development.

Although REDD + interventions will not be a 'silver-bullet' solution for the global biodiversity crisis (Harvey *et al.* 2010; Venter *et al.* 2013), they may have significant effects on biodiversity by boosting conservation actions in five ways: (i) the protection of suitable habitat by preventing forest conversion inside and outside the network of protected areas; (ii) the reforestation activities that restore habitat and re-establish populations (Ansell *et al.* 2011; Pichancourt *et al.* 2014); (iii) connection of isolated populations by establishing and protecting natural corridors, resulting in enhanced genetic flow and dispersal (Jantz *et al.* 2014); (iv) promotion of biodiversity friendly practices in the forestry and logging sectors, through the use of certification standards (Nepstad *et al.* 2013); and (v) the enhancement of funding and institutional support of conservation actions, using the institutional and legal arrangements constructed for REDD +.

Despite the synergies of REDD + policies and biodiversity conservation, many researchers and policy-makers question the feasibility and potential of REDD + to deliver real biodiversity benefits (Stickler *et al.* 2009), because the areas with high-carbon stocks that may be targetted by these policies are not necessarily the most important for biodiversity conservation (Paoli *et al.* 2010). In addition, is it possible that REDD + implementation will increase human pressure on non-carbon-rich ecosystems or on forests not protected by these policies? This process, known as leakage, could result in high biodiversity losses elsewhere (de Lima *et al.* 2013). Furthermore, afforestation and reforestation initiatives based on commercial plantations will unlikely deliver broad conservation benefits, since biodiversity levels in these systems are considerably lower than in natural forest (Pichancourt *et al.* 2014), although these interventions have helped in the recovery of local populations (Brown and Yoder 2015). Finally, it has also been argued that a focus on multiple benefits may fail to maximize carbon benefits, thus decreasing the potential of REDD + to mitigate climate change (McAfee 2012).

As proposed in the international climate change negotiations, qualitative safeguards will indeed play a vital role in avoiding impacts on biodiversity. However, maximizing biodiversity benefits will require a more proactive approach and a framework that takes into consideration biodiversity and carbon trade-offs (Phelps *et al.* 2012a). Thus, quantitative and spatial tools will be needed to prioritize conservation needs within REDD + policies. There are three key actions that can enable and enhance this process: (i) integration of biodiversity and carbon data; (ii) development of specific methods to assess and compare biodiversity benefits of land-development pathways; and (iii) integration of monitoring and verification of biodiversity impacts or benefits within the monitoring and verification activities for carbon accounting.

16.6.1 Integrating biodiversity and carbon data

The first step to maximize synergies between REDD + and biodiversity conservation is analysing the distribution patterns of carbon and biodiversity (Phelps *et al.* 2012b). At the global scale, carbon-rich and biodiversity-high areas overlap greatly in the tropics. However, this correspondence does not always occur at the regional or local scale, and varies for different species groups (Strassburg *et al.* 2010). In Indonesia, for instance, peatland forests hold the largest carbon stocks per unit area and are important areas for remaining orang-utan populations; however, these forests host lower biodiversity levels for most taxonomic groups than their counterparts on mineral soils (Paoli *et al.* 2010). Similarly, young secondary forests can sequester more carbon annually in the short term (Silver *et al.* 2000; Lewis *et al.* 2009), but primary forests have higher biodiversity levels (Gibson *et al.* 2011). This complex spatial pattern of biodiversity and carbon in tropical forests needs to be considered by policy-makers and practitioners in order to realize the conservation potential of REDD +. Various organizations have recently taken the first steps to integrate biodiversity and carbon data, including national assessments (e.g. Runsten *et al.* 2013) and taxon-specific assessments such as the mapping initiatives carried out for great apes (e.g. Wich *et al.* 2011; UN-REDD 2014).

One of the major obstacles to the analysis of carbon and biodiversity distributions is the different definitions of forest habitats used in various land-use and biodiversity datasets (Schmitt 2013). This is due to the use of different ecological classification systems, which vary greatly in their definitions of tree cover (Sasaki and Putz 2009). For instance, the habitat typology used by the IUCN Red List and species assessments is not always comparable with commonly used land-cover maps, such as the Globcover (Arino *et al.* 2008), or by the ecoregion classification developed by WWF (Olson *et al.* 2001). Integration of a broad range of vegetation, biodiversity, and carbon datasets will be necessary to conduct biodiversity and carbon-benefit analysis, especially in areas where regional or local biodiversity information is scarce or unavailable (Imai *et al.* 2014; Gardner *et al.* 2012). Thus, the development and agreement on a common conceptual framework and forest typology could facilitate the delivery of multiple benefits and minimize trade-offs between biodiversity and REDD + implementation (Schmitt 2013).

16.6.2 Assessing biodiversity value of specific REDD + activities

The delivery of multiple benefits from REDD + programmes will likely be budget constrained regardless of the financial mechanism used (e.g. voluntary vs compliance markets, national programmes vs individual project basis). Therefore, comprehensive assessments are needed in order to prioritize cost-effective interventions that deliver both biodiversity and carbon benefits (Kapos *et al.* 2012). Although ecologists and conservationists have developed a broad range of tools to identify species conservation priorities, there is a lack of an appropriate method to systematically quantify biodiversity outcomes of land-use change policies. Recent studies have taken first steps to overcome these methodological constraints, through the use of 'Species–Area Relationship' models and land-use scenario analysis (Koh and Ghazoul 2010; Garcia-Ulloa *et al.* 2012). These types of tools allow policy-makers and landholders to make better informed decisions and prioritize actions that could realize the potential multiple benefits of REDD + in their regions. In addition, identifying REDD + projects or interventions that have a real potential for biodiversity conservation could access larger financial resources through premium credits that reward the delivery of multiple benefits (Busch 2013; Dinerstein *et al.* 2013).

It is important to recognize, however, that biodiversity assessments should go beyond quantitative analyses of changes in species richness (He and Hubbell 2011). In reality, a complete picture of the benefits or impacts of REDD + projects for biodiversity should include many other aspects, such as genetic diversity or population dynamics. The development of methods to integrate these other aspects of biodiversity into land-use planning remains a research priority.

16.6.3 Verifying and monitoring biodiversity impacts and benefits from REDD + interventions

Understanding the synergies and trade-offs between biodiversity and carbon conservation will be crucial at the planning and implementing phases of REDD + activities (Phelps *et al.* 2012a). Monitoring, however, will play a vital role in verifying the benefits and impacts of these policies (Dickson and Kapos 2012). Monitoring efforts will need to (i) address impacts at various scales in order to facilitate the identification of unintended consequences such as degradation leakage[3] (Kapos *et al.* 2012), and (ii) be based on low-cost and readily accessible methods (Imai *et al.* 2014). In this sense, the increased use of camera-traps and drones in conservation is a step in the right direction. Drones, for instance, have been proposed as a cost-effective method to survey orang-utans through the count of nests in the canopy (Koh and Wich 2012), or as a key tool to improve land-use change monitoring efforts (Figure 16.3). Finally, the monitoring of biodiversity benefits and impacts should be integrated within the platforms that are being developed for the monitoring, verifying, and reporting of carbon benefits of REDD + initiatives (Gardner *et al.* 2012). Coupling both processes could reduce costs, facilitate analysis of policy outcomes, and steer policy implementation.

[3] Leakage occurs when environmental degradation in one area increases as a result of the implementation of environmental policies in a second area. For instance, policies directing expansion of oil-palm cultivation in agricultural land to avoid deforestation, may result in indirect land-use changes by displacing other agricultural activities towards forested areas (Wicke 2014).

Figure 16.3 New technologies can provide cost-efficient methods to monitor changes in land uses or wildlife populations. This entire landscape was mapped by an inexpensive and unmanned drone in a single 30 minute mission, covering about 200 hectares. This type of technology is able to detect illegal logging, agricultural encroachment, and deforestation. Courtesy of conservationdrones.org.

16.7 Final remarks

Payment for ecosystem services is having a significant impact on the way we carry out conservation. Conservation practitioners will need to consider how the arrival of these strategies can enhance conservation efforts, and must strive to profit from the technological, institutional, and policy architecture of PES programmes. REDD + initiatives, in particular, can play important roles in primate conservation by diverting considerable funding to projects that avoid habitat loss or promote restoration. The delivery of these benefits will depend, however, on whether policies explicitly integrate conservation as a policy objective. In the case of REDD +, biodiversity and carbon benefits can be simultaneously optimized when planning and management actively takes into account their synergies and trade-offs. There are three methodological aspects that would facilitate this. First, biodiversity and carbon databases should be integrated in a common framework with shared habitat typologies and definitions. Second, quantitative methods to model biodiversity outcomes in future land-use change scenarios could inform policy-makers on the potential impacts and benefits for conservation. Finally, monitoring impacts will be crucial to secure benefits and avoid negative outcomes through the continuous feedback on policy implementation.

Filling these methodological gaps should be a priority in ecology and conservation sciences. Primatologists, in particular, could play a pivotal role in the development of these methods, because of the close connection between primates and forests, the high vulnerability of primates to deforestation, and the cultural importance of primate species.

Acknowledgements

We thank the editors and reviewers of this book who provided valuable feedback. John Garcia-Ulloa is supported by the Mercator Foundation Switzerland. Lian Pin Koh is supported by the Australian Research Council.

References

Ansell, F. A., Edwards, D. P., and Hamer, K. C. (2011). Rehabilitation of logged rain forests: avifaunal composition, habitat structure, and implications for biodiversity-friendly REDD +. *Biotropica* **43**: 504–511.

Arino, O., Bicheron, P., Achard, F., Latham, J., Witt, R., *et al.* (2008). GLOBCOVER: the most detailed portrait of Earth. *Esa Bulletin-European Space Agency* **136**: 24–31.

Birdsey, R., Pan, Y., and Houghton, R. (2013). Sustainable landscapes in a world of change: tropical forests, land use and implementation of REDD plus: Part I Foreword. *Carbon Management* **4**: 465–468.

Bonan, G. B. (2008). Forests and climate change: forcings, feedbacks, and the climate benefits of forests. *Science* **320**: 1444–1449.

Brown, J. L. and Yoder, A. D. (2015). Shifting ranges and conservation challenges for lemurs in the face of climate change. *Ecology and Evolution* **5**: 1131–1142.

Bryan, B. A. (2013). Incentives, land use, and ecosystem services: synthesizing complex linkages. *Environmental Science and Policy* **27**: 124–134.

Bunker, D. E., DeClerck, F., Bradford, J. C., Colwell, R. K., Perfecto, I., *et al.* (2005). Species loss and aboveground carbon storage in a tropical forest. *Science* **310**: 1029–1031.

Burivalova, Z., Sekercioglu, C. H., and Koh, L. P. (2014). Thresholds of logging intensity to maintain tropical forest biodiversity. *Current Biology* **24**: 1893–1898.

Busch, J. (2013). Supplementing REDD + with biodiversity payments: the paradox of paying for multiple ecosystem services. *Land Economics* **89**: 655–675.

Busch, J. and Grantham, H. S. (2013). Parks versus payments: reconciling divergent policy responses to biodiversity loss and climate change from tropical deforestation. *Environmental Research Letters* **8**: 034028.

Cardinale, B. J., Duffy, J. E., Gonzalez, A., Hooper, D. U., Perrings, C., *et al.* (2012). Biodiversity loss and its impact on humanity. *Nature* **486**: 59–67.

Chapin, F. S., Zavaleta, E. S., Eviner, V. T., Naylor, R. L., Vitousek, P. M., *et al.* (2000). Consequences of changing biodiversity. *Nature* **405**: 234–242.

Colwell, R. K., Brehm, G., Cardelus, C. L., Gilman, A. C., and Longino, J. T. (2008). Global warming, elevational range shifts, and lowland biotic attrition in the wet tropics. *Science* **322**: 258–261.

Corbera, E. (2012). Problematizing REDD + as an experiment in payments for ecosystem services. *Current Opinion in Environmental Sustainability* **4**: 612–619.

de Lima, R. F., Olmos, F., Dallimer, M., Atkinson, P. W., and Barlow, J. (2013). Can REDD plus help the conservation of restricted-range island species? Insights from the Endemism hotspot of Sao Tome. *PLoS One* **8**: e74148.

Dickson, B. and Kapos, V. (2012). Biodiversity monitoring for REDD +. *Current Opinion in Environmental Sustainability* **4**: 717–725.

Dinerstein, E., Varma, K., Wikramanayake, E., Powell, G., Lumpkin, S., *et al.* (2013). Enhancing conservation, ecosystem services, and local livelihoods through a wildlife premium mechanism. *Conservation Biology* **27**: 14–23.

Driscoll, D. A., Felton, A., Gibbons, P., Felton, A. M., Munro, N. T., *et al.* (2012). Priorities in policy and management when existing biodiversity stressors interact with climate-change. *Climatic Change* **111**: 533–557.

Ebeling, J. and Yasue, M. (2008). Generating carbon finance through avoided deforestation and its potential to create climatic, conservation and human development benefits. *Philosophical Transactions of the Royal Society B—Biological Sciences* **363**: 1917–1924.

Effiom, E. O., Nunez-Iturri, G., Smith, H. G., Ottosson, U., and Olsson, O. (2013). Bushmeat hunting changes regeneration of African rainforests. *Proceedings of the Royal Society B—Biological Sciences* **280**: 20130246.

Effiom, E. O., Birkhofer, K., Smith, H. G., and Olsson, O. (2014). Changes of community composition at multiple trophic levels due to hunting in Nigerian tropical forests. *Ecography* **37**: 367–377.

Engel, S., Pagiola, S., and Wunder, S. (2008). Designing payments for environmental services in theory and practice: an overview of the issues. *Ecological Economics* **65**: 663–674.

Foley, J. A., DeFries, R., Asner, G. P., Barford, C., Bonan, G., *et al.* (2005). Global consequences of land use. *Science* **309**: 570–574.

Garcia-Ulloa, J., Sloan, S., Pacheco, P., Ghazoul, J., and Koh, L. P. (2012). Lowering environmental costs of oil-palm expansion in Colombia. *Conservation Letters* **5**: 366–375.

Gardner, T. A., Burgess, N. D., Aguilar-Amuchastegui, N., Barlow, J., Berenguer, E., *et al.* (2012). A framework for integrating biodiversity concerns into national REDD + programmes. *Biological Conservation* **154**: 61–71.

Ghazoul, J. and Sheil, D. (2010). *Tropical Rain Forest Ecology, Diversity, and Conservation*. Oxford: Oxford University Press.

Gibson, L., Lee, T. M., Koh, L. P., Brook, B. W., Gardner, T. A., *et al.* (2011). Primary forests are irreplaceable for sustaining tropical biodiversity. *Nature* **478**: 378–381.

Gomez-Baggethun, E., de Groot, R., Lomas, P. L., and Montes, C. (2010). The history of ecosystem services in economic theory and practice: from early notions to markets and payment schemes. *Ecological Economics* **69**: 1209–1218.

Grainger, A., Boucher, D. H., Frumhoff, P. C., Laurance, W. F., Lovejoy, T., *et al.* (2009). Biodiversity and REDD at Copenhagen. *Current Biology* **19**: R974–R976.

Harvey, C. A., Dickson, B., and Kormos, C. (2010). Opportunities for achieving biodiversity conservation through REDD. *Conservation Letters* **3**: 53–61.

He, F. L. and Hubbell, S. P. (2011). Species-area relationships always overestimate extinction rates from habitat loss. *Nature* **473**: 368–371.

Hooijer, A., Page, S., Canadell, J. G., Silvius, M., Kwadijk, J., *et al.* (2010). Current and future CO_2 emissions from drained peatlands in Southeast Asia. *Biogeosciences* **7**: 1505–1514.

Hooper, D. U., Adair, E. C., Cardinale, B. J., Byrnes, J. E. K., Hungate, B. A., *et al.* (2012). A global synthesis reveals biodiversity loss as a major driver of ecosystem change. *Nature* **486**: 105–U129.

Houghton, R. A., House, J. I., Pongratz, J., van der Werf, G. R., DeFries, R. S., *et al.* (2012). Carbon emissions from land use and land-cover change. *Biogeosciences* **9**: 5125–5142.

IIED (2015). Paying Local Communities for Ecosystem Services: The Chimpanzee Conservation Corridor. Available: http://www.iied.org/paying-local-communities-for-ecosystem-services-chimpanzee-conservation-corridor [Accessed 1 July 2015].

IPCC (2013). Summary for policymakers. In: Stocker, T. F., Qin, D., Plattner, G.-K., Tignor, M., Allen, S. K., Boschung, J., *et al.* (Eds), *Climate Change 2013: The Physical Science Basis. Contribution of Working Group I to the*

Fifth Assessment Report of the Intergovernmental Panel on Climate Change. Cambridge, United Kingdom and New York, USA: Intergovernmental Panel on Climate Change.

IUCN (2014). IUCN Red List of Threatened Animals. Available: http://www.iucnredlist.org [Accessed 1 December 2014].

Imai, N., Tanaka, A., Samejima, H., Sugau, J. B., Pereira, J. T., *et al.* (2014). Tree community composition as an indicator in biodiversity monitoring of REDD. *Forest Ecology and Management* **313**: 169–179.

Jantz, P., Goetz, S., and Laporte, N. (2014). Carbon stock corridors to mitigate climate change and promote biodiversity in the tropics. *Nature Climate Change* **4**: 138–142.

Kapos, V., Kurz, W. A., Gardner, T., Ferreira, J., Guariguata, M., *et al.* (2012). Impacts of forest and land management on biodiversity and carbon. In: Parrotta, J. A., Wildburger, C., and Mansourian, S. (Eds), *Understanding Relationships Between Biodiversity, Carbon, Forests and People: The Key to Achieving REDD + Objectives*. Vienna, Austria: International Union of Forest Research Organizations.

Karsenty, A., Vogel, A., and Castell, F. (2014). 'Carbon rights', REDD plus and payments for environmental services. *Environmental Science and Policy* **35**: 20–29.

Kinzig, A. P. (2009). VI Ecosystem Services. *The Princeton Guide to Ecology*. Princeton: Princeton University Press.

Koh, L. P. and Ghazoul, J. (2010). Spatially explicit scenario analysis for reconciling agricultural expansion, forest protection, and carbon conservation in Indonesia. *Proceedings of the National Academy of Sciences of the United States of America* **107**: 11140–11144.

Koh, L. P. and Wich, S. A. (2012). Dawn of drone ecology: low-cost autonomous aerial vehicles for conservation. *Tropical Conservation Science* **5**: 121–132.

Lewis, S. L., Lopez-Gonzalez, G., Sonke, B., Affum-Baffoe, K., Baker, T. R., *et al.* (2009). Increasing carbon storage in intact African tropical forests. *Nature* **457**: 1003–1006.

MEA (2005). *Ecosystems and Human Well-being: Synthesis*. Washington, DC: Millennium Ecosystem Assessment.

Malhi, Y., Roberts, J. T., Betts, R. A., Killeen, T. J., Li, W., *et al.* (2008). Climate change, deforestation, and the fate of the Amazon. *Science* **319**: 169–172.

McAfee, K. (2012). The contradictory logic of global ecosystem services markets. *Development and Change* **43**: 105–131.

McCarthy, D. P., Donald, P. F., Scharlemann, J. P. W., Buchanan, G. M., Balmford, A., *et al.* (2012). Financial costs of meeting global biodiversity conservation targets: current spending and unmet needs. *Science* **338**: 946–949.

Millar, C. I., Stephenson, N. L., and Stephens, S. L. (2007). Climate change and forests of the future: managing in the face of uncertainty. *Ecological Applications* **17**: 2145–2151.

Muradian, R., Arsel, M., Pellegrini, L., Adaman, F., Aguilar, B., *et al.* (2013). Payments for ecosystem services and the fatal attraction of win-win solutions. *Conservation Letters* **6**: 274–279.

Nepstad, D. C., Boyd, W., Stickler, C. M., Bezerra, T., and Azevedo, A. A. (2013). Responding to climate change and the global land crisis: REDD +, market transformation and low-emissions rural development. *Philosophical Transactions of the Royal Society B-Biological Sciences* **368**: 20120167.

Olson, D. M., Dinerstein, E., Wikramanayake, E. D., Burgess, N. D., Powell, G. V. N., *et al.* (2001). Terrestrial ecoregions of the world: a new map of life on Earth. *Bioscience* **51**: 933–938.

Pan, Y. D., Birdsey, R. A., Fang, J. Y., Houghton, R., Kauppi, P. E., *et al.* (2011). A large and persistent carbon sink in the world's forests. *Science* **333**: 988–993.

Paoli, G., Wells, P., Meijaard, E., Struebig, M., Marshall, A., *et al.* (2010). Biodiversity conservation in the REDD. *Carbon Balance and Management* **5**: 7.

Perrings, C., Naeem, S., Ahrestani, F., Bunker, D. E., Burkill, P., *et al.* (2010). Ecosystem Services for 2020. *Science* **330**: 323–324.

Peters-Stanley, M., Gonzalez, G., and Yin, D. (2013). *Covering New Ground: State of the Forest Carbon Markets 2013*. Washington, DC: Ecosystem Marketplace.

Phelps, J., Friess, D. A., and Webb, E. L. (2012a). Win-win REDD + approaches belie carbon-biodiversity trade-offs. *Biological Conservation* **154**: 53–60.

Phelps, J., Webb, E. L., and Adams, W. M. (2012b). Biodiversity co-benefits of policies to reduce forest-carbon emissions. *Nature Climate Change* **2**: 497–503.

Pichancourt, J.-B., Firn, J., Chades, I., and Martin, T. G. (2014). Growing biodiverse carbon-rich forests. *Global Change Biology* **20**: 382–393.

Pirard, R. (2012). Market-based instruments for biodiversity and ecosystem services: a lexicon. *Environmental Science and Policy* 19–20: 59–68.

Porras, I., Barton, D. N., Chacón-Cascante, A., and Miranda, M. (2013). *Learning from 20 Years of Payments for Ecosystem Services in Indonesia*. London: International Institute for Environment and Development.

Putz, F. E., Sist, P., Fredericksen, T., and Dykstra, D. (2008). Reduced-impact logging: challenges and opportunities. *Forest Ecology and Management* **256**: 1427–1433.

Rands, M. R. W., Adams, W. M., Bennun, L., Butchart, S. H. M., Clements, A., *et al.* (2010). Biodiversity conservation: challenges beyond 2010. *Science* **329**: 1298–1303.

Resosudarmo, I. A. P., Atmadja, S., Ekaputri, A. D., Intarini, D. Y., Indriatmoko, Y., *et al.* (2014). Does tenure security lead to REDD plus project effectiveness?

Reflections from five emerging sites in Indonesia. *World Development* **55**: 68–83.

Rudel, T. K., Defries, R., Asner, G. P., and Laurance, W. F. (2009). Changing drivers of deforestation and new opportunities for conservation. *Conservation Biology* **23**: 1396–1405.

Runsten, L., Ravilious, C., Kashindye, A., Giliba, R., Hailakwahi, V., *et al.* (2013). Using Spatial Information to Support Decisions on Safeguards and Multiple Benefits for REDD + in Tanzania. Prepared by UNEP-WCMC, Cambridge, UK; published by Ministry of Natural Resources & Tourism, Dar es Salaam, United Republic of Tanzania.

Sanchez-Azofeifa, G. A., Pfaff, A., Robalino, J. A., and Boomhower, J. P. (2007). Costa Rica's payment for environmental services program: intention, implementation, and impact. *Conservation Biology* **21**: 1165–1173.

Sasaki, N. and Putz, F. E. (2009). Critical need for new definitions of 'forest' and 'forest degradation' in global climate change agreements. *Conservation Letters* **2**: 226–232.

Schmitt, C. B. (2013). Global tropical forest types as support for the consideration of biodiversity under REDD. *Carbon Management* **4**: 501–517.

Schowalter, T. D. (2013). *Insects and Sustainability of Ecosystem Services*. Oxford: CRC Press.

Silver, W. L., Ostertag, R., and Lugo, A. E. (2000). The potential for carbon sequestration through reforestation of abandoned tropical agricultural and pasture lands. *Restoration Ecology* **8**: 394–407.

Stickler, C. M., Nepstad, D. C., Coe, M. T., McGrath, D. G., Rodrigues, H. O., *et al.* (2009). The potential ecological costs and cobenefits of REDD: a critical review and case study from the Amazon region. *Global Change Biology* **15**: 2803–2824.

Strassburg, B. B. N., Kelly, A., Balmford, A., Davies, R. G., Gibbs, H. K., *et al.* (2010). Global congruence of carbon storage and biodiversity in terrestrial ecosystems. *Conservation Letters* **3**: 98–105.

Takeda, S. and Matsuoka, M. (2008). Genetic approaches to crop improvement: responding to environmental and population changes. *Nature Reviews Genetics* **9**: 444–457.

Thomas, C. D., Cameron, A., Green, R. E., Bakkenes, M., Beaumont, L. J., *et al.* (2004). Extinction risk from climate change. *Nature* **427**: 145–148.

UN-REDD (2014). Forests of Hope: The UNREDD Programme and GRASP Collaborate to Conserve Great Ape Habitat. Available: https://unredd.wordpress.com/ [Accessed 25 May 2015].

UNEP (2011). *Towards a Green Economy: Pathways to Sustainable Development and Poverty Eradication*. Nairobi, Kenya: UNEP.

UNFCCC (2014). Decisions adopted by the Conference of the Parties—FCCC/CP/2013/10/Add.1. Report of the Conference of the Parties on its Nineteenth Session. United Nations Framework Convention on Climate Change.

Venter, O., Hovani, L., Bode, M., and Possingham, H. (2013). Acting optimally for biodiversity in a world obsessed with REDD. *Conservation Letters* **6**: 410–417.

Venter, O. and Koh, L. P. (2011). Reducing emissions from deforestation and forest degradation (REDD +): game changer or just another quick fix? *Annals of the New York Academy of Sciences* **1249**: 137–150.

Venter, O., Laurance, W. F., Iwamura, T., Wilson, K. A., Fuller, R. A., *et al.* (2009). Harnessing carbon payments to protect biodiversity. *Science* **326**: 1368–1368.

Vié, J.-C., Hilton-Taylor, C., and Stuart, S. N. (2009). *Wildlife in a Changing World: An Analysis of the 2008 IUCN Red List of Threatened Species*. Gland: IUCN.

Wich, S., Jenson, R. J., Refisch, J., and Nellemann, C. (2011). Orangutans and the Economics of Sustainable Forest Management in Sumatra. UNEP/GRASP/PanEco/YEL/ICRAF/GRID-Arendal.

Wicke, B. (2014). Palm oil as a case study of distal land connections. In: Seto, K. (Ed.), *Rethinking Global Land Use in an Urban Era*. pp. 163–180. MIT Press.

Wunder, S. (2001). Poverty alleviation and tropical forests—what scope for synergies? *World Development* **29**: 1817–1833.

CHAPTER 17

The role of evidence-based conservation in improving primate conservation

Sandra Tranquilli

Law enforcement guard in Law enforcement guard in Lyondji Community Bonobo Reserve, Democratic Republic of Congo. Photo copyright: Billy Dodson, African Wildlife Foundation.

17.1 Introduction

With every passing year, primates are falling under increasingly high anthropogenic pressure, making it crucial to develop effective conservation strategies to preserve the remaining populations (Tranquilli *et al.* 2012). In this chapter, I will discuss the evidence-based approach to conservation. The evidence-based approach uses scientific evidence to determine the effectiveness of conservation actions and informs conservation decision-making (Pullin and Knight 2001). Over the last decade, the assessment of management effectiveness has been increasingly recognized as crucially important to highlight deficiencies, to set priorities, to guide project managers, and to inform policy-makers, funding bodies, and conservationists (Hockings 2003).

In the first part of this chapter, I will examine the origins of evidence-based conservation (EBC), and how it has been used to objectively assess the success of conservation activities.

Tranquilli, S., *The role of evidence-based conservation in improving primate conservation* In: *An Introduction to Primate Conservation.* Edited by: Serge A. Wich and Andrew J. Marshall, Oxford University Press (2016). © Oxford University Press.
DOI 10.1093/acprof:oso/9780198703389.003.0017

Next, I will describe the methodological aspects of the approach and demonstrate how these are implemented to improve the effectiveness of primate conservation actions. This will be illustrated step by step using a hypothetical example of a threatened primate species. Then, I will provide examples of recent studies, one large-scale and one with local focus, which tested the effectiveness of interventions. Finally, I will outline factors that can improve the success of conservation measures, such as law enforcement, tourism, and research, before a primate population reaches the critical stage of decline.

17.2 The evidence-based approach to conservation

In order to effectively ensure the conservation of wildlife populations and natural resources, it is crucial that management practices should be informed by scientific evidence (Sutherland *et al.* 2004; Pullin and Knight 2009). Such evidence should be empirical and collected in accordance with scientific research methods, with results based on statistical analysis. The EBC approach uses the best available evidence to determine the effectiveness of conservation actions (Pullin and Knight 2001; Sutherland 2008a, b). The use of such an approach helps conservationists understand which conservation actions are required in the field, and how these should be implemented to ensure maximum effectiveness. The EBC is therefore an approach that aims to help practitioners to design better management plans (Matzek 2014).

17.2.1 The gap between scientists and practitioners

Both conservation science and practice seek solutions to conservation problems such as wildlife population decline or habitat loss. While a few reports have suggested that conservation practice has been considerably influenced by scientific knowledge over the last 20 years (Robinson 2006), a significant and problematic gap persists between the two fields (Knight *et al.* 2008).

This is a major problem because the gap or absence of communication between the two fields often leads to a lack of scientific understanding of the natural systems that are the focus of conservation efforts, as well as a failure to find the most effective strategies to solve problems.

There are several reasons for the relative lack of EBC. For example, the effectiveness of conservation activities is often only partially monitored, if at all (Smith *et al.* 2014). This may be due to the limited availability of funding, or to the inability of practitioners to spend the time that is necessary for monitoring changes in population size or structure. It may also be due to the fact that some consider the importance of gathering scientific evidence to support conservation strategies to be low priority (Oates 1999; Stem *et al.* 2005; Mascia *et al.* 2014). When conservation activities are monitored, their effectiveness is frequently not subjected to scientific analysis (Ray *et al.* 2005). Moreover, researchers sometimes pose questions that are considered to be irrelevant or useless by practitioners (e.g. Braunisch *et al.* 2012; Balme *et al.* 2014). As a result, few empirical data exist with which to assess the effectiveness of management practices (Pullin and Salasfky 2010).

Often, conservation practitioners are unfamiliar with scientific research related to their work and therefore are not aware of the most effective and appropriate actions to undertake to protect their target conservation zones (Dicks *et al.* 2014; Walsh *et al.* 2014). This may be a result of several factors, such as:

(i) A lack of communication between conservation scientists and practitioners (Knight *et al.* 2008; Arlettaz *et al.* 2010). This may result from the failure of scientists to clearly present their research findings to managers and to inform them of their recommendations. For example, researchers often fail to outline the management and policy implications of their research in the conclusions of their published papers, or their research may not be presented in a practitioner-friendly format (Meijaard and Sheil 2007; Arlettaz *et al.* 2010).

(ii) Limited access to the primary information (Pullin *et al.* 2004). Scientific journals in theory play a crucial role in disseminating information to the global scientific community. Conservation practitioners and the organizations with whom they work often cannot afford to access the scientific literature due to high journal subscription fees (Fuller *et al.* 2014; Gossa *et al.* 2014).

This is particularly the case for practitioners based in developing countries (Sunderland *et al.* 2009). Many peer-reviewed journals provide an option for papers to be open access, meaning that they are freely available. As a complicating factor, however, some of these journals require the researchers to pay in order for their articles to be open access. Academics are often unable to afford these payments without the support of their universities or funding bodies. To increase free access to journal articles for conservation practitioners and scientists in developing countries, primatologists and other researchers should consider submitting their scientific manuscripts to open-access journals, which also incurs a cost on the scientists. These and other factors mean that few scientific papers are read outside of academic circles (Pullin *et al.* 2004; Arlettaz *et al.* 2010). Access to scientific journals is further limited by the fact that practitioners sometimes lack the knowledge or training to read, understand, and/or interpret findings in scientific papers (de la Rosa 2000; Pullin *et al.* 2004). In addition, practitioners are often already operating with significant restraints on their time and resources. Reporting their own experiences in the scientific literature and devoting time to read the relevant work of others are additional time-consuming tasks. In the cases where practitioners do report their experiences, it is often not in a format that is useful to conservation scientists (Knight 2006; Pullin and Salafsky 2010). Some field managers may not recognize the benefits of exchanging their experiences, or they may not have the opportunity to do so, especially if they work in remote regions or for organizations that do not have a good system set up for the exchange of information (Sutherland 2008a, b; Matzek *et al.* 2014).

(iii) Publication bias. Research papers are more likely to be published if they report statistically significant findings (CEBC 2006). Negative results often remain unpublished. The same pattern has also been found in medical research, with significant results being three times more likely to be published than those experiments in which a medical treatment produced non-significant results (Crowley *et al.* 1990). Non-significant results can be particularly useful for conservation practitioners, as knowing what does not work (and under what conditions) can be at least as useful in selecting management interventions as knowing what does work (Martínez-Abraín 2013).

Other factors that can lead to a gap between science and conservation management are the lack of organizational capacity and resources (Pullin and Knight 2005), political and logistical factors influencing decisions (Roux *et al.* 2006), or poor will to implement management recommendations (Knight *et al.* 2008).

As a result of the barriers described above, the gap that exists between scientific research and conservation practice may lead to weak conservation management decisions driven primarily by personal or organizational experience, anecdotal evidence, or traditional management practices, rather than being based on effective methods supported by sound evidence (Pullin *et al.* 2004; Cook *et al.* 2010, 2013; Young and Van Aarde 2011). 'Experience-based' conservation methods may occasionally turn out to be beneficial, but are rarely as effective as practices that are based on robust scientific findings (Walsh *et al.* 2014). Failure to use validated methods may also result in an inefficient allocation of funds or ineffective interventions (Pullin and Knight 2001; McConnachie and Cowling 2013).

17.2.2 The development of the evidence-based conservation approach

Evidence-based conservation focuses on understanding problems and solutions using the best available scientific evidence, and integrating this with experience. This approach often involves a 'systematic review' (Dawes 2000) of all relevant research bearing on a specific conservation problem, which, crucially, implements statistical analyses. In particular, this process consists of the search for and the extraction, compilation, and analysis of quantitative and qualitative data.

Over the last two decades, the application of evidence-based methods and the dissemination of information relating to the effectiveness of particular conservation interventions in the form of recommendations or guidelines have led to dramatic shifts in practice within the fields of medicine and healthcare (e.g. Stevens and Milne 1997; Friedland 1998; Graham *et al.* 2011). Evidence-based medicine originally emerged as an approach to determine the most effective medical treatments from among a suite of alternatives and apply these consistently. This approach has revolutionized clinical practice (the 'effectiveness revolution') and has recently become a routine part of medical training (Stevens and Milne 1997). The field of conservation has been compared to those of medicine and public health (Pullin and Knight 2001; Fazey *et al.* 2004); in fact, both fields often require urgent decisions to be made (e.g. the implementation of actions combatting the decline of wildlife populations before they become locally extinct requires similar decision-making processes to be made as for the implementation of a vaccine to prevent an epidemic prior to its outbreak).

Pullin and Knight (2001) first introduced the EBC framework, which they derived from the similar approaches established in medicine and public health, calling for an 'effectiveness revolution in conservation'. This new approach in the field of nature conservation is rapidly gaining traction, and it complements the 'adaptive management' approach which 'uses management actions as experiments to provide data supporting, or failing to support, competing hypotheses when there is uncertainty regarding the response of ecological systems to management activities, to better meet management objectives over time' (Allen and Gunderson 2011: 1379).

EBC seeks to close the gap between scientists and practitioners. An important prerequisite for EBC is research on the effectiveness of conservation interventions, which requires the systematic collection and analysis of (i) accurate information on conservation interventions; (ii) the evaluation of such information; and (iii) dissemination of the conclusions and recommendations to inform and advise politicians, field managers, donors, and the general public. As such, it provides an objective way of measuring the effectiveness of conservation efforts on the ground at both the small- and large-scale levels (Segan *et al.* 2010).

17.3 Methodological approach

In order to assess the effectiveness of particular conservation interventions, it is usually best to follow a systematic approach. This involves building a framework that specifies the definition of the main questions, followed by data collection and analysis, eventually leading to the dissemination of conclusions and recommendations based on the evidence (Pullin and Stewart 2006). I describe these steps in detail here (see Figure 17.1 for the main steps), with each one followed by a hypothetical scenario.

17.3.1 Formulation of the questions

The first step in assessing the success of conservation efforts is to define the main research questions

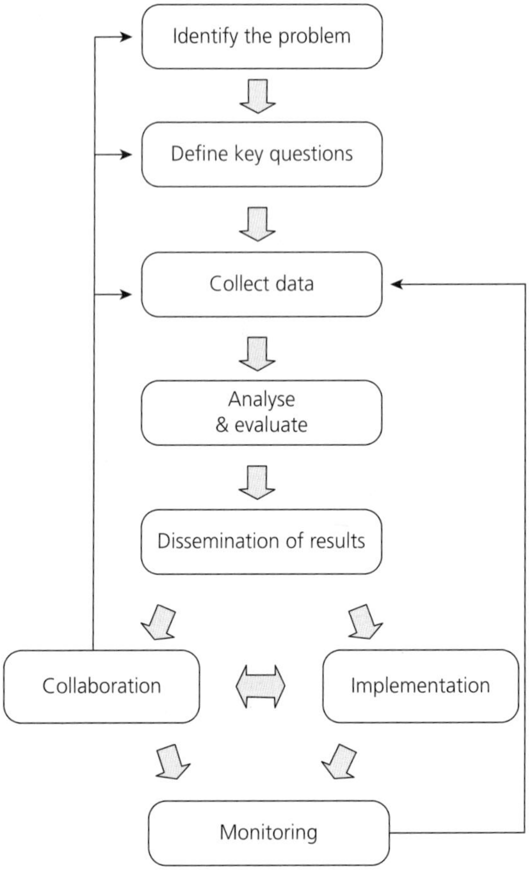

Figure 17.1 Scheme of the methodology used in the EBC approach.

of the study in terms of aim(s) and objective(s). Often, researchers address questions that are not useful in resolving conservation problems and thus are not relevant to practitioners (e.g. Braunisch *et al.* 2012; Cook *et al.* 2013). Prior to formulating their research questions, primatologists should have a clear idea of the threat(s) affecting their particular primate populations, as well as having come up with a definition of success for each particular conservation intervention and a means to measure this success. These should be based on needs and priorities that are relevant to practitioners and/or decision-makers and should be generated in collaboration with them (Pullin and Stewart 2006; Pullin and Knight 2009).

Examples of research questions may be:

- Is a primate population increasing, decreasing, or stable?
- What are the main threats to the population?
- Have any conservation interventions been implemented?
- If so, how successful were these interventions?

Key questions could also be formulated following the 'PICO' framework (Populations, Interventions, Control or Comparisons and Outcomes of interest) (Counsell 1997). This helps to formulate the main research question/s into a searchable query. Each element of the PICO framework provides an opportunity for the investigator to reformulate the questions so that they are better understood, as well as to access the study's feasibility and to guide the collection, analysis, and interpretation of the data (Counsell 1997; Huang *et al.* 2006). For example, for each element of a particular challenge in primate conservation we can associate a question: Population = what is the specific population of primates and what are the problems it faces?; Interventions = which major conservation intervention should be carried out?; Comparison = are there any alternative solutions to the intervention?; Outcome = what are we trying to measure or improve? In general, the identification of such questions allows specific hypotheses to be tested as well as generating literature search terms or keywords that permit systematic review.

As a hypothetical example, consider the following scenario, which has unfortunately become a common situation: empirical evidence (e.g. estimates of encounter rates or population density surveys) indicates that a primate population has declined over a period of several years or has gone extinct in an area. There may be several factors contributing to such a population decline, and the primary underlying cause may not always be obvious. If the population is in decline, the researcher should first define the possible mechanisms underlying the decline (e.g. was the decline caused by anthropogenic activity? Would a particular conservation intervention have helped? Was there insufficient funding for the protection of the population?).

17.3.2 Data collection

Systematic review

One potentially helpful analytical approach to the EBC is a systematic review of all previous studies. The goal is to collect the best background information available to aid in the development of the EBC framework. This is generally based on an exhaustive literature search using certain key words extracted from the research questions as search terms. Potential sources of information can be accessed through a global search of:

(a) Peer-reviewed scientific articles. These articles, which have been reviewed by multiple experts in the relevant fields, are accessible via research journals, most of which are available on the Internet in the form of electronic databases such as Google Scholar, Web of Science, or Scopus.

(b) The so-called 'grey literature' refers to reports that have not been published in peer-reviewed journals and, as a consequence, are often much more difficult to access. These include records that have been compiled by individual NGOs or GOs (Governmental Organizations) covering their *in-situ* work, for example, biodiversity surveys and management plans. The most effective way to gather such information is to make direct contact with the regional leaders of the organizations and/or meet them at their headquarters and ask permission to have access to their monthly and annual reports.

(Lists of peer-reviewed scientific articles and grey literature are also available on specific online databases (see Box 17.1)).

Once the initial search is complete and the relevant studies have been selected, it is of critical importance to assess the quality of the data and the methodology used, using a standardized and repeatable assessment method (CEBC 2006). Factors relevant to assessing a study's quality are its design, execution, and analysis as well as the validity of the results. Ideally, it is best to include peer-reviewed papers in the analysis. These methods must be critically appraised, in order to both prevent systematic errors or bias (Moher *et al.* 1995; Bilotta *et al.* 2014) to ensure that a consistent format is used, and to ensure that it will permit comparisons, either in one area over time, between different sites in the same country or region, or among different countries or regions altogether.

Searching for relevant information from other fields such as economics, sociology, and politics may provide valuable alternative perspectives on species conservation status (Sunderland *et al.* 2009).

Concerning the population decline in our hypothetical example, we could search the literature for potential explanatory variables for declines of primate populations. These might include, for example, the successes or failures of conservation efforts performed, threats to which the population is subject, or threats to which populations at other sites have been subjected under similar scenarios (i.e. development or road-building). We might use as search term the following: primate common name + latin name, primate common name + latin name + location, primate common name + latin name + location + conservation efforts, primate common name + latin name + location + hunting + deforestation + conservation efforts.

Conservation evidence synopses

Another method that can be used to collate pertinent evidence is the 'synopsis', a summary of the relevant scientific literature on a specific conservation topic (Walsh *et al.* 2014). This is a useful way to provide easy access to scientific information for conservation practitioners. In particular, such a convenient summary reduces the workload of busy practitioners, avoiding the need to search through all of the primary literature (Dicks *et al.* 2014). This approach is already well established in the field of medicine (British Medical Journal Group 2014). Recently it has been shown that conservation practitioners are more likely to make use of scientific evidence if they have good access to such a summary; the summary, therefore, helps conservation practitioners to improve their management decisions (Walsh *et al.* 2014). Several initiatives already exist, including the Conservation Evidence Synopses and Conservation for Environmental Evidence (see Box 17.1), both of which summarize and evaluate research findings in a clear and concise package.

A novel study

When relevant evidence is not available in the literature, scientists are obliged to conduct new research (e.g. conduct surveys to monitor the population status and/or threats facing a species, or collect information regarding any conservation intervention that was performed at the site). Consultation with experts (e.g. park managers in cases where primates occur inside protected areas) may provide additional help, in particular when the information required has not been published. Experts are defined by their qualifications and their long-term personal experience and knowledge of a study site, as well as their familiarity with the research that has been conducted there (Raymond *et al.* 2010; Burgman *et al.* 2011;Turnhout *et al.* 2012). They may help, for example, with the development and evaluation of projects, or provide information about model parameters (Martin *et al.* 2012; Sutherland *et al.* 2013).

For cases in which experts are involved, information can be collected using the 'expert elicitation approach', which includes the five steps described in Martin *et al.* (2012): (i) decide how the information will be used; (ii) determine what information to elicit; (iii) design the process of eliciting judgements from the expert; (iv) acquire the information from the expert; and finally (v) translate the information elicited so that it can be used in a model (this translation process is referred to as 'encoding').

Questionnaires can be developed using, for example, the Delphi Method, which is well established in the field of ecology (Eycott *et al.* 2011), and has recently been used extensively in conservation

Box 17.1 Web-based databases

The Centre for Evidence-Based Conservation (CEBC)[1]

The Centre for Evidence-Based Conservation (CEBC) provides support to those involved in conservation decision-making and management. The Centre is based at Bangor University, UK, and it produces systematic reviews both of the impact of human activities on natural resources and of the effectiveness of management and policy interventions and publishes the open access *Environmental Evidence Journal*. CEBC is supported by a wide range of environmental organizations and academic institutions and acts as the central headquarters of the Collaboration for Environmental Evidence (CEE).[2] CEE provides information on how to conduct systematic literature reviews and provides a comprehensive list of all such reviews that were conducted in the past and that are in progress. CEE also works at developing methodology for systematic reviews and map-making.

Conservation Evidence

The Conservation Evidence website[3] provides data aimed to support conservation decision-makers and to provide them with guidance on how to maintain and restore global biodiversity. It publishes an open-access online scientific journal (*Conservation Evidence*) containing primary studies on the effects of previous conservation interventions. It fills a publication gap that other journals tend not to publish in, such as small local studies or studies that fail to find positive significant effects of interventions. It does not require articles to have long introduction and discussion sections; therefore it is well suited for the publication of reports written by conservation practitioners. In addition, it serves to effectively summarize the evidence for or against the effectiveness of conservation actions and management decisions, using the scientific literature contained in the Conservation Evidence Synopses page of the website. All of this information is presented in the form of a searchable database. This site provides easy access for both scientists and conservationists searching for current knowledge about conservation projects and outcomes.

IUCN/SSC—A.P.E.S. portal

The IUCN/SSC–A.P.E.S. (Apes, Populations, Environment, and Surveys)[4] portal is an online initiative that collates information from conservation scientists and practitioners, as well as NGOs and GOs. It provides information on what is known about the population status and distribution of bonobos, chimpanzees, gorillas, gibbons, and orang-utans. It aims to help researchers and conservationists to better understand the threats faced by great apes as well as their conservation needs. This is done by providing interactive maps combined with information on the distribution of apes, along with information on their status, threats, and temporal trends. Finally, the website includes a regularly updated list of literature on all of the field studies on African great apes over the last 50 years, including survey datasets, guidelines, and descriptions of the conservation efforts implemented and surveys undertaken.

IUCN/SSC Primate Specialist Group

The International Union for Conservation of Nature (IUCN)/Species Survival Commissions (SSC) Primate Specialist Group[5] is composed of both scientists and conservation practitioners who work on primate conservation in tropical areas across Africa, Asia, and Latin America. This website provides information from reviews on the conservation status of threatened primates, with a list of publications available for downloading, including action plans, reports, scientific publications, and newsletters, all with the latest information on the status of endangered primates.

Miradi—Adaptive Management Software for Conservation projects

Miradi[6] is a portal available in five languages that guides practitioners through a step-by-step interview process designed to define their project's scope and produce conceptual models, manage their data, monitor the progress of their project, and prioritize threats and conservation actions in order to more effectively meet their goals. Moreover, it supports the development of long-term projects. First, the practitioner enters basic information about the project, including team composition and location. Next, the online tool helps to create conceptual diagrams of threats and their contributing factors, conservation strategies, and targets to assess the social, economic, and cultural contexts. A result chain then appears from the conceptual model to clarify and assess assumptions about the most appropriate strategies to undertake. Finally, it provides information on the success of particular conservation interventions for a number of projects.

[1] http://www.cebc.bangor.ac.uk/index.php.en?menu=0andamp;catid=0.

[2] http://www.environmentalevidence.org/.

[3] http://www.conservationevidence.com.

[4] http://apesportal.eva.mpg.de/.

[5] http://www.primate-sg.org/.

[6] http://www.Miradi.org.

management (e.g. McBride *et al.* 2012). This method involves asking experts to answer questions on a specific questionnaire over some set number of feedback rounds. After completing each round, experts are provided with an anonymous summary of the decisions of other experts from the previous round. This is done in order to encourage experts to revise their answers. After a certain number of rounds, a pre-specified level of agreement is reached (McBride *et al.* 2012).

There is, nevertheless, an ongoing controversy involving the use of such expert judgement (Ludwig *et al.* 2001; Kuhnert 2011). It is feared that expert judgement may be biased, which may thus lead to poor decision-making. However, substantial research has been conducted to overcome this type of problem (Kynn 2008; Burgman *et al.* 2011). For example, Burgman *et al.* (2011) outline three options: (i) the use of analytical tests to measure the knowledge of experts; (ii) training experts to improve their performance, bias, and overconfidence; and (iii) the use of structured elicitation procedures that encourages participation and involves cross-examination of the evidence, in order to deal with biases.

17.3.3 Evaluation of the results

The evaluation of the results is obviously an important step that aims to reveal whether or not particular conservation interventions were successes or failures. The evaluation also seeks to highlight the relative importance of other possible factors that lead to the decline of populations of primates or other wildlife species. The outcomes may also involve the prediction of the future status of a particular population (i.e. how it will be impacted) without the rapid implementation of an effective intervention.

Returning to our hypothetical example, at this point we have collected information on the primate population trend and its threats. The outcomes of the model show that deforestation has a highly significant negative effect the primate population in question, while law enforcement, on the other hand, has a positive effect on the survival of the population. In summary, the results suggest that our hypothetical primate population may quickly disappear if no conservation intervention is implemented.

17.3.4 Dissemination of the results

Following analysis of the data, and our prognosis regarding the status of a species and its habitat, threats encountered, and conservation interventions recommended, an important task remains: we must clarify how and to whom we want to disseminate these research findings. As discussed above, scientific results are often reported in the form of peer-reviewed articles and/or final reports. In order to aid both practitioners and decision-makers, these should be accompanied by recommendations that can more easily be translated into policy and practical action. Recommendations should be clear and pragmatic, as well as quantitative and explicit (Prendergast *et al.* 1999). Not all conservation practitioners and decision-makers have a scientific background, thus it is important to synthesize the findings in a format that is easy for them to understand (Pullin and Salafsky 2010).

The information we presented in our hypothetical example from a primate population would be relevant to both conservation practitioners and decision-makers. Some effective ways to disseminate the information would be through direct meetings, workshops, and popular science or conservation news articles. These should include recommendations on which conservation intervention programme is likely to be more effective for the long-term protection of the primate population.

In order to make research results more easily accessible, the information should be made available online (i.e. via open-access journals or web-based databases) and/or directly provided to conservation practitioners (i.e. at local, regional, or international workshops and conferences, and via newsletters).

Several organizations have recently established central web-based databases that provide easy-to-access information on research guidelines as well as instruction on how to assess the available scientific evidence relating to a particular topic. These are usually produced in collaboration with scientists, practitioners, and policy-makers (Sutherland *et al.* 2004; Box 17.1).

The significance of these databases is that they:

- consolidate data and information from different technical fields;
- store information on the past and current distribution of species, population densities, threats and conservation efforts, conservation projects, and outcomes in an open-access format so that anyone can query, use, and update them;
- provide open-access scientific journals;
- provide examples of which conservation actions are most appropriate to solve particular problems.

The dissemination of results may serve to help practitioners to: (1) adjust conservation plans if they fail to meet their predefined goals, thus improving the managers' performance at the local scale (adaptive management); (2) report their achievements to stake-holders and government (accountability); or (3) improve the efficiency of funding allocation among different conservation activities (resource allocation) (Hockings *et al.* 2006).

Results may be disseminated to the general public both at the local and national level by providing key messages via the media, publicity campaigns, and other public channels. Such efforts may change the attitudes, perceptions, and/or behaviour of the general public and may attract conservation funding from interested private donors (Johns 2005).

17.4 Evidence-based conservation of primates

Only rarely have primatological studies been conducted using an EBC approach. Recent studies have confirmed the importance of specific conservation efforts by measuring the effectiveness of interventions (e.g. Campbell *et al.* 2011; Tranquilli *et al.* 2012; N'Goran *et al.* 2012). In this section, I will discuss two case studies that can be used as evidence of conservation effectiveness as part the EBC process. I selected these particular studies, both conducted in sub-Saharan Africa, as they are the most recent ones to have examined the effectiveness of conservation practices, one on a small scale and the other on a large scale.

17.4.1 Long-term research sites as wildlife refugia: a study at a local scale

It has been suggested that long-term research sites are beneficial to the survival of vulnerable species in the surrounding areas (Wrangham and Ross 2008). This has never, however, been tested empirically at a local scale.

Campbell *et al.* (2011) conducted a study in the Taï National Park in Cote D'Ivôire, a 5,400 km^2 park that is the home of two long-term research projects, the Taï Chimpanzee Project and the Taï Monkey Project. These projects, which were established in 1979 and 1989, respectively, are based at adjacent field camps and sample overlapping areas.

Campbell's study collected spatial data on observations and signs (i.e. nests and faeces) of primates and duikers. The researchers walked line transects (75 transects, 1 km long and 1 km distant from each other) covering a study area of 200 km^2, focusing on population densities of three animal species: Diana monkeys (*Cercopithecus diana*), red colobus monkeys (*Proclobus badius*), and Maxwell's duikers (*Philatomba maxwelli*). These three species are traditionally the animals that are the most affected by poaching (Caspary *et al.* 2001). The study also analysed the spatial distribution of all poaching signs (i.e. camps, trails, empty cartridges, and snares) detected within the study area. The density of poaching signs was used as a proxy measure for hunting pressure.

The study used a generalized linear model (GLM) to test the effect of the long-term presence of a research area on animals (duikers, primates, and the three focused animal species listed above) and poaching sign encounter rates. The model variables included forest type, human density, distance to research area, and distance to the border of the national park.

The results demonstrated that proximity to the long-term research area had a significant positive impact on the presence of primate and duiker species, including all three of the abovementioned overharvested species. This was further supported by the cumulative weighted Akaike Information Criterion (AIC). These were calculated for all the predictor variables based on their contribution to the AIC relating to each possible combination of

GLMs. The variable that had the strongest importance was 'distance to research area'. In contrast, there was a significant decline in signs of poaching with proximity to the research area. This was clearly shown through interpolated maps from encounter rates of poaching signs and of red colobus groups observed (Figure 17.2).

This is the first study to demonstrate empirically that long-term research study sites may act as deterrents for poachers and serve as refugia for wildlife.

17.4.2 The assessment of conservation efforts in preventing African great ape extinction risk: a study at the continental scale

Over the past 20 to 30 years, human population growth has rapidly increased in many sub-Saharan African countries, with consequent increases in land use and other human activities that have caused a large amount of habitat loss and drastic declines in local wildlife populations (Smith *et al.* 2003; Struhsaker *et al.* 2005). All African great apes (bonobos, *Pan paniscus*; chimpanzees subspecies, *Pan troglodytes verus, P. t. troglodytes, P. t. schweinfurthii, P.t. vellerosus*; and gorillas, *Gorilla gorilla diehli, G. g. gorilla, G. beringei beringei, G. b. graueri*) are listed as Endangered or Critically Endangered on the IUCN Red List of Threatened Species (IUCN 2014). Their populations are rapidly declining and have been subject to local extinctions (see, e.g., Walsh *et al.* (2003); Campbell *et al.* (2008), Greengrass (2009); Gatti (2009); Plumptre *et al.* (2010)). In total, only 11% of the land within their habitat is regarded as protected (nationally and internationally), and it is in these areas where they have the greatest probability of long-term survival (Tranquilli 2013). It is therefore crucial to understand how effective protected areas (PAs) actually are at protecting ape populations.

A large-scale study conducted by Tranquilli *et al.* (2012) examined 109 PAs across 16 countries occurring within the range of African great apes. The study provided a continent-wide assessment of the relative significance of four different types of conservation efforts. These included: law enforcement (considered as primary conservation activity that directly protects the PA on the ground), tourism and research (both considered as secondary conservation activities, which actions are indirect but supportive), and finally the support (technical and/or financial) of national and international non-governmental organizations (NGOs).

The overall aim of the study, using data collected over a 20-year period, was to evaluate the effectiveness of these four conservation efforts at lowering the extinction risk of African great apes. Between 1990 and 2009, historical and current records of the presence and absence of great apes were collected from published and unpublished literature. Gaps in information for specific areas were filled by using responses to questionnaires from park managers and researchers who had worked in the PAs over extended periods of time. The study also considered other possible variables that may have influenced ape survival or extinction inside the PAs and within 50-km buffer zones around the areas. These variables included PA characteristics, such as size and years of protection since the establishment, as well as degraded area and socio-economic variables, such as gross domestic product per capita, armed conflicts, and human population density.

The study conducted two analyses. In the first analysis, all conservation efforts were considered as a single variable ('overall conservation effort') in a general linear mixed model (GLMMs), and the second analysis considered each individual conservation effort as a single variable in four separate GLMMs.

The results of the first analysis demonstrated that the likelihood of ape population survival increased significantly with the duration of the overall conservation effort. Specifically, the higher proportion of years with the presence of conservation efforts (i.e. the number of years with presence over the total of years considered) in a PA leads to a decreased probability of apes surviving (Figure 17.3). In addition, human population densities and national economic development were found to have a significant negative effect on ape population survival, although they were of smaller importance with respect to overall conservation efforts. The results of the second analysis revealed similar results to the first analysis. The proportion of years with presence of each conservation effort has a positive effect on ape persistence. Moreover, the analysis highlighted

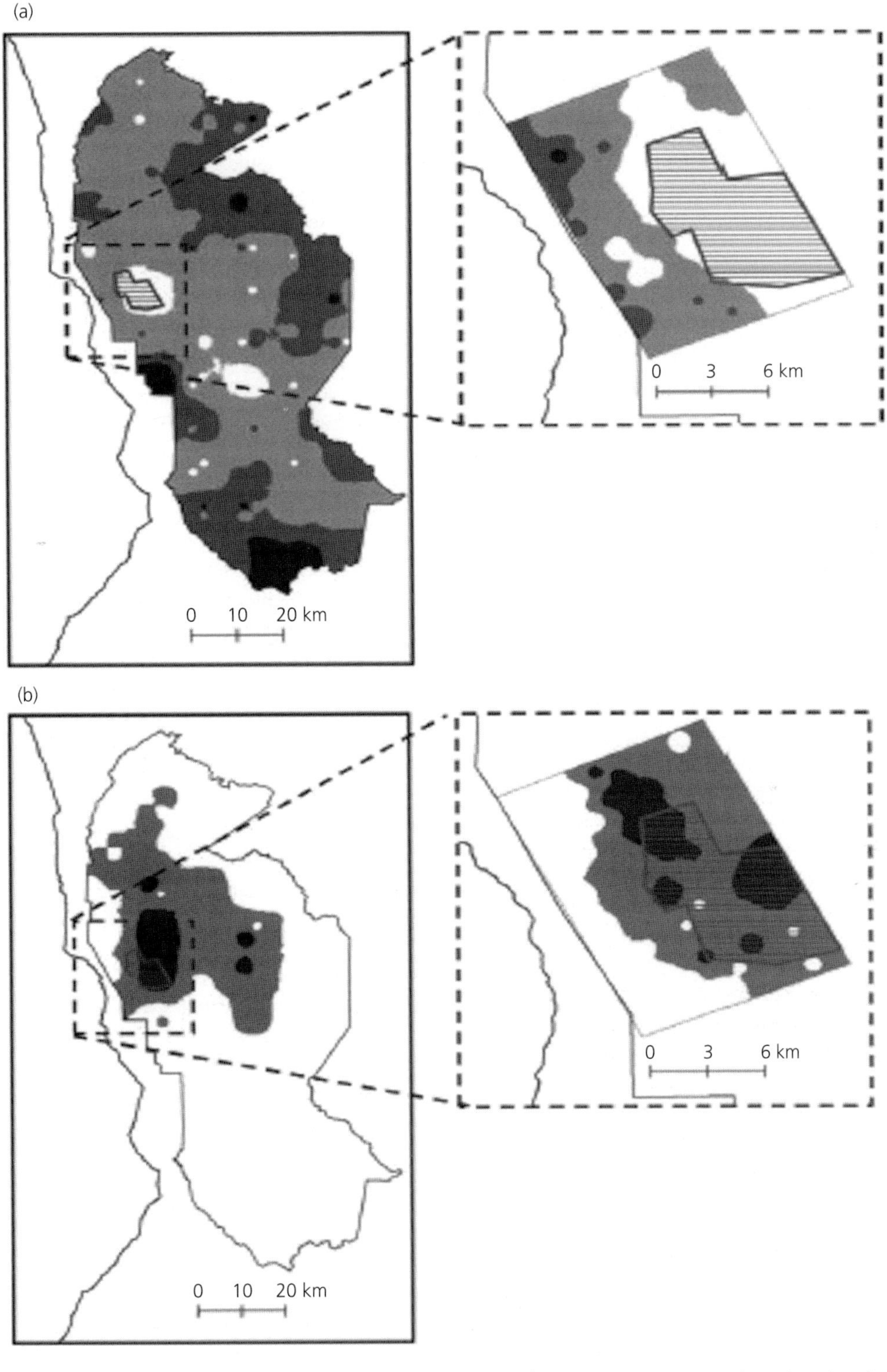

Figure 17.2 Interpolated maps of encounter rates of (a) poaching signs (sign km^{-1}) and (b) red colobus groups observed. On the right is represented data collected from the study of Campbell *et al.* (2011), on the left is represented data collected throughout the Tai National Park. Shaded regions represent different encounter rates: (a) white, 0–1; light grey, 2–4; dark grey, 5–7; black, 8–15; (b) white, 0–0.49; grey, 0.5–1.49; black, 1.50–3.00. Figure extracted from Campbell *et al.* (2011).

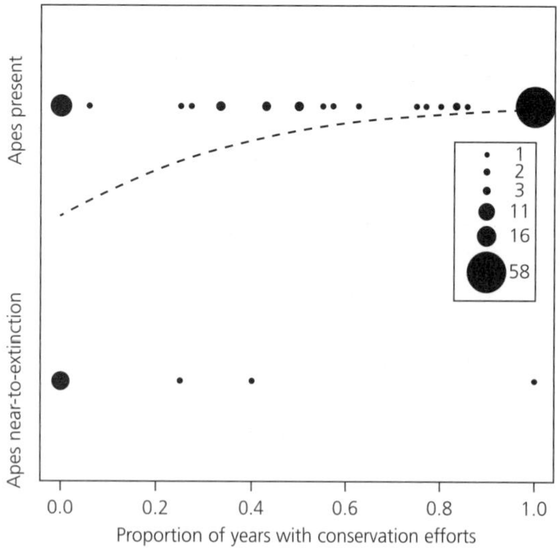

Figure 17.3 Great ape persistence/extinction in relation to the proportion of years with conservation efforts. The area of the circle represents the number of protected areas. The curve represents the probability of apes surviving as a function of the proportion of years with conservation efforts. Figure extract from Tranquilli *et al.* (2012).

which conservation efforts are the most effective. The results on the relative importance of the four different conservation efforts revealed the proportion of years with NGO involvement and the proportion of years with law enforcement guard presence to have the greatest impact on ape survival and more so than tourism and research station presence. Moreover, this was supported by comparing the AIC values of the four models.

In conclusion, this study confirms the importance of law enforcement as a primary conservation activity for the survival of primates and other wildlife survival. This is the first large-scale study to have assessed which conservation activities had the most influence on survival of an endangered species on the continent scale.

17.5 Improving the success of conservation actions

Once communication between scientists and conservation practitioners has been established, it is critical that both sides keep one another updated on the results of research, the progress made (or not made) in ongoing interventions, and problems that are faced on the ground. In this chapter, I have used the example of a species that is experiencing a population decline; it is crucial, however, that effective conservation measures are put in place if possible before a population reaches this stage. Within the domains of organizational capacity and available skills, the success of conservation activities (such as law enforcement, tourism, research programmes, see Tranquilli *et al.* (2012)) could be improved by several factors, for example long-term funding, long-term monitoring, the involvement of local communities, and stake-holder collaboration.

17.5.1 Long-term funding

The economic cost of implementing conservation plans to effectively conserve protected areas or to preserve a particular species of primate may vary considerably. Conservation funding can come from a number of sources, including national and international NGOs and GOs, private organizations, and research institutes. A number of studies have shown, however, that funding is often insufficient to provide adequate protection for wildlife and other natural resources, and to ensure effective management, especially in developing countries (Jachmann 2008; Waldron *et al.* 2013). This is partly because many conservation projects are funded only for a limited period of time (Oates 1999; Wilson *et al.* 2006). In addition, armed conflict and corruption can lead to a reduction in the financial resources available (e.g. Oates 1999; Price 2003; Struhsaker *et al.* 2005; Smith and Walpole 2005). As a result, conservation projects often find themselves stuck with insufficient staff and equipment or other basic necessities required for effective management. Such cases highlight the challenging problem of how to best allocate the limited available funding in order to effectively protect wildlife.

Funding constraints may affect not only on-site conservation activities, but also the monitoring programmes required to understand temporal trends in the conservation status of particular species (Balmford and Whitten 2003).

Increasing awareness of the challenges faced by conservationists and how these might be solved may help to gather the needed funding for the most effective conservation activities. In order to achieve this, it is important to maintain collaborations and the flow of information between local people and general public, conservation practitioners, policy-makers, and donors (Ferraro and Pattanayak 2006; Cook *et al.* 2013).

17.5.2 Long-term biodiversity monitoring

Long-term monitoring is a valuable tool for the assessment of temporal changes in the conservation status of a species, including degradation of its environment and the major threats it faces (Margoulis and Salafsky 1998; Elzinga *et al.* 2001; Tranquilli *et al.* 2014). Population monitoring has often accompanied long-term research studies of habituated primate groups, for example the chimpanzees (*Pan troglodytes schweinfurthii*) of the Gombe and Mahale National Parks (Nishida *et al.* 2003; Pusey *et al.* 2007), the Japanese macaques (*Macaca fuscata*) of Arashiyama (Fedigan and Asquith 1991), and the yellow baboons (*Papio cynocephalus*) of Amboseli National Park (Alberts and Altmann 2012). However, many projects or conservation plans do not include long-term monitoring programmes (Stokes *et al.* 2010). This may be because monitoring is not recognized as a priority or because there are insufficient funds to support it (Oates 1999). Additionally, in the case of primate research, populations of interest often occur in dense tropical forests where the visibility is very poor and where accessibility to the areas and logistics are difficult (Laurance *et al.* 2006). As a result of the rarity of primate sightings, it can take several months to collect the data required to cover an area adequately, which is often unfeasible when time and funds are limited. Consequently, there are often insufficient data to accurately evaluate the impact of conservation efforts in remote areas.

Monitoring schemes must employ repeated, empirical measures that directly address the goals specified in conservation plans, and can address the questions via measurements that can be taken. Standardized methods for collecting and analysing data are crucial to permit the assessment of spatial variation and temporal changes (Kühl *et al.* 2008; Campbell *et al.*, Chapter 6, this volume). Monitoring must be conducted on a sufficiently regular basis to permit early detection of population changes (Plumptre 2000) and to evaluate the effectiveness of conservation interventions (Sunderland *et al.* 2013). Rapid analysis of data and reporting of results ensures that a conservation management plan can be modified in the next round of sampling to suit changing conditions on the ground. This assures that the conservation management plan is flexible enough to be effective; the results of any modifications can be obtained during the next round of monitoring.

17.5.3 The involvement of local communities

The strong integration and involvement of local people and communities into conservation management is now recognized as essential for the success of biodiversity conservation (Roe *et al.* 2009; Waylen *et al.* 2010). Conservation efforts must take into account the legitimate needs of the local people, even though fulfilment of these needs may sometimes be at odds with conservation objectives. Communities often live in or around protected areas or in non-protected areas occupied by endangered primate species. Including local people in conservation decision-making processes may promote their cooperation in protecting the area from external intruders and help them to regulate their use of natural resources within these areas in a sustainable direction (Horowitz 1998; Pretty and Smith 2004). The sharing of information with local people about the conservation status of an area or species, through awareness programmes and involving locals in conservation activities—such as tourism, research, law enforcement, and other conservation programmes—provides conservation workers with several opportunities. It provides opportunities for locals to: (i) contribute to the long-term success of conservation actions (Kaltenborn *et al.* 2008); (ii) enhance their own understanding of the importance of natural resources (Wrangham and Ross 2008; Boissieere *et al.* 2009); and (iii) improve local and national economies (Wrangham and Ross 2008). In exchange, conservationists can benefit from the vast amount of traditional knowledge that locals often have about protected species and their environment, as well as in-country political and social realities.

17.5.4 Stake-holder collaboration

Continued collaboration between researchers, field managers, decision-makers, and conservation organizations is crucial for facilitating the regular exchange of information, although distinct perspectives, goals, and approaches held by each of these parties can potentially give rise to disagreement. Therefore it is important that the terms of collaborations are made explicitly clear at the beginning of a project. All participants, such as field managers, decision-makers, and conservation organizations, should be involved throughout the process. Collaborations can be initiated and maintained via email or directly through meetings (e.g. local workshops, or national and international conferences), with individuals working in different sectors being brought together when and where it is possible to exchange ideas and identify major challenges. Direct meetings are probably more effective then email contacts, as this allows both personal relationships and trust to be more easily developed. Multi-co-authorship of scientific papers and reports is of fundamental importance, given that the involvement of a broad range of partners adds substantial weight to the outcome and suggests broad support and commitment by all stake-holders. Media coverage may be important in order to provide further publicity for the park or the organization that supported the research, while reaching potential donors who may be interested in contributing to the protection of the area and species.

17.6 Conclusion

In this chapter I have presented an overview of the recent development of the evidence-based conservation approach. I have discussed the importance of continuous communication between conservation scientists and practitioners in facilitating the development of effective conservation plans, and have outlined the steps required to assess the impact of conservation actions. Specifically, I have highlighted the importance of following a systematic approach involving the formulation of key questions, the collection and analysis of primary information, and its subsequent dissemination. Primate conservation efforts can be improved by developing collaborations between researchers, managers, decision-makers, and local communities. In order to improve the effectiveness of conservation actions, long-term funding is required to sustain monitoring programmes, and these programmes need to be implemented on a regular basis to allow early detection of population changes. Because many primate conservation programmes still lack an evidence-based approach, there is an urgent need for its more widespread application in order to enhance our ability to assess conservation efficacy, and, ultimately, to promote effective conservation.

Acknowledgements

I wish to thank the editors for inviting me to contribute to this important book and for their comments that greatly improved the manuscript. I also thank Dr Thurston Cleveland Hicks, Jessica Walsh, and one anonymous reviewer for editing revisions of this manuscript.

References

Alberts, S. C. and Altmann, J. (2012). The Amboseli Baboon Research Project: 40 years of continuity and change. In: Kappeler O. and Watts, D. P. (Eds), *Long-Term Field Studies of Primates*. pp. 261–288. New York: Springer-Verlag.

Allen, C. R. and Gunderson, L. (2011). Pathology and failure in the design and implementation of adaptive management. *Journal of Environmental Management* **92**: 1379–1384.

Arlettaz, R., Schaub, M., Fournier, J., Reichlin, T. S., Sierro, A., *et al.* (2010). From publications to public actions: when conservation biologists bridge the gap between research and implementation. *Bioscience* **60**: 835–842.

Balmford, A. and Whitten, T. (2003). Who should pay for tropical conservation, and how could the costs be met? *Oryx* **37**: 238–250.

Bilotta, G. S., Milner, A. M., and Boyd, I. (2014). Quality assessment tools for evidence from environmental science. *Environmental Evidence* **3**: 14.

Boissière, M., Sheil, D., Basuki, I., Wan, M., and Le, H. (2009). Can engaging local people's interests reduce forest degradation in central Vietnam? *Biodiversity Conservation* **18**: 2743–2757.

Braunisch, V., Home, R., Pellet, J., and Arlettaz, R. (2012). Conservation science relevant to action: a research agenda identified and prioritized by practitioners. *Biological Conservation* **153**: 201–210.

British Medical Journal Group (2014). Learn, Teach and Practise Evidence-based Medicine. Clinical Evidence. Available at http://www.clinicalevidence.bmj.com/x/index.html [Accessed February 2014].

Burgman, M., Carr, A., Godden, L., Gregory, R., McBride, M., *et al.* (2011). Redefining expertise and improving ecological judgment. *Conservation Letters* **4**: 81–87.

Campbell, G., Kühl, H., N'Goran, K.P., and Boesch, C. (2008). Alarming decline of West African chimpanzees in Côte d'Ivoire. *Current Biology* **18**: 903–904.

Campbell, G., Kühl, H., Diarrassouba, A., N'Goran, P. K., and Boesch, C. (2011). Long-term research sites as *refugia* for threatened and over-harvested species. *Biology Letters* **7**: 723–726.

Caspary, H.U., Koné, I., Prouot, C., and De Pauw, M. (2001). *La chasse et la filière viande de brousse dans l'espace Taï, Côte d'Ivoire. Tropenbos Serie 2*. pp. 21. Côte d'Ivoire: Tropenbos.

CEBC (2006). Guidelines for Systematic Review in Conservation and Environmental Management, Centre for Evidence-Based Conservation Report. pp. 23. Birmingham: University of Birmingham.

Cook, C. N., Hockings, M., and Carter, R. W. (2010). Conservation in the dark? The information used to support management decisions. *Frontiers in Ecology and the Environment* **8**: 181–186.

Cook, C. N., Mascia, M. B., Schwartz, M. W., Possingham, H. P., and Fuller, R. A. (2013). Achieving conservation science that bridges the knowledge-action boundary. *Conservation Biology* **27**: 669–678.

Counsell, C. (1997). Formulating questions and locating primary studies for inclusion in systematic reviews. *Annals of Internal Medicine* **127**: 380–387.

Crowley, P., Chalmers, I. G., and Keirse, M. J. (1990). The effects of corticosteroid administration before preterm delivery: an overview of the evidence from controlled trials. *British Journal of Obstetrics and Gynaecology* **97**:11–25.

Dawes, M. (2000). *Evidence Based Practice. Health Service Journal Monographs*. p. 20. London: Emap Public Sector Management Publications.

De la Rosa, C. L. (2000). Improving Science Literacy and Conservation in Developing Countries. Available at http://www.actionbioscience.org/education [Accessed January 2014].

Dicks, L. V., Walsh, J. C., and Sutherland, W. J. (2014). Organizing evidence for environmental management decisions: a '4S' hierarchy. *Trends in Ecology and Evolution* **29**: 607–613.

Elzinga, C. L., Salzer, D. W., Willoughby, J. W., and Gibbs, J. (2001). *Monitoring Plant and Animal Populations*, p. 360. Abingdon: Blackwell Scientific Publications.

Eycott, A. E., Marzano, M., and Watts, K. (2011). Filling evidence gaps with expert opinion: the use of Delphi analysis in least-cost modelling of functional connectivity. *Landscape and Urban Planning* **13**: 400–409.

Fazey, I., Salisbury, J. G., and Lindenmayer, D. B. (2004). Can methods applied in medicine be use do summarize and disseminate conservation research? *Environmental Conservation* **31**: 190–198.

Fedigan, L. M. and Asquith, P. J. (1991). *The Monkeys of Arashyiama: Thirty-five Years of Research in Japan and the West*. p. 353. Albany: State University of New York Press.

Ferraro, P. J. and Pattanayak, S. K. (2006). Money for nothing? A call for empirical evaluation of biodiversity conservation investments. *PLoS Biology* **4**: e105.

Friedland, D.J. (1998). *Evidence-based Medicine: A Framework for Clinical Practice*. p. 263. Stamford: Appleton and Lange.

Fuller, R. A., Lee J. R., and Watson, J. E. M. (2014). Achieving open access to conservation science. *Conservation Biology* **28**: 1550–1557.

Gatti, S. (2009). *Status of Primate Populations in Protected Areas Targeted by the Community Forest Biodiversity Project*. pp. 1–42. Accra: WAPCA report Ghana.

Gossa, C., Fisher, M., and Milner-Gulland, E. J. (2014). The research-implementation gap: how practitioners and researchers from developing countries perceive the role of peer-reviewed literature in conservation science. *Oryx* **49**: 80–87.

Graham, R, Mancher, M., Wolman, D. M., Greenfield, S., and Steinberg, E. (2011). *Clinical Practice Guidelines We Can Trust*. p. 300. Washington: The National Academic Press.

Greengrass, E. J. (2009). Chimpanzees are close to extinction in southwest Nigeria. *Primate Conservation* **24**: 77–83.

Hockings, M. (2003). Systems for assessing the effectiveness of management in protected areas. *BioScience* **53**: 823–832.

Hockings, M., Stolton, S., Leverington, F., Dudley, N., and Courrau, J. (2006). *Evaluating Effectiveness: A Framework for Assessing Management Effectiveness of Protected Areas*. Gland: IUCN.

Horowitz, D. L. (1998). *Structure and Strategy in Ethnic Conflict*. p. 47. Washington: Annual World Bank Conference on Development Economics.

Huang, X., Lin, J., and Demner-Fushman, D. (2006). Evaluation of PICO as a Knowledge Representation for Clinical Questions. AMIA Annual Symposium Proceedings, 359–363.

IUCN (2014). *The IUCN Red List of Threatened Species*. Available at http://www.iucnredlist.org. [Accessed January 2014].

Jachmann, H. (2008). Monitoring law-enforcement performance in nine protected areas in Ghana. *Biological Conservation* **141**: 89–99.

Johns, D. (2005). The other connectivity: reaching beyond the choir. *Conservation Biology* **19**: 1681–1682.

Kaltenborn, B., Nyahongo, J., Kidegesho, J., and Haaland, H. (2008). Serengeti National Park and its neighbours—Do they interact? *Journal for Nature Conservation* **16**: 96–108.

Knight, A. (2006). Failing but learning: writing the wrongs after Redford and Taber. *Conservation Biology* **20**: 1312–1314.

Knight, A., Cowling, R. M., Rouget, M., Balmford, A., Lombard, A. T., *et al.* (2008). Knowing but not doing: selecting priority conservation areas and the research-implementation gap. *Conservation Biology* **22**: 610–617.

Kühl, J., Maisels, F., Ancrenaz, M., and Williamson, E. A. (2008). *Best Practice Guidelines for Survey and Monitoring of Great Ape Populations*. p. 32. Gland: IUCN SSC Primate Specialist Group (PSG).

Kuhnert, P. M. (2011). Four case studies in using expert opinion to inform priors. *Environmetrics* **22**: 662–674

Kynn, M. (2008). The 'heuristics and biases' bias in expert elicitation. *Journal of the Royal Statistical Society Series A* **171**: 239–264.

Laurance, W. F., Croes, B. M., Tchignoumba, L., Lahm, S. A., Alonso, A., *et al.* (2006). Impacts of roads and hunting on Central African rainforest mammals. *Conservation Biology* **20**: 1251–1261.

Ludwig, D., Mangel, M., and Haddad, B. (2001). Ecology, conservation, and public policy. *Annual Review of Ecology and Systematics* **32**: 481–517.

Margoulis, R. and Salafsky, N. (1998). *Measures of Success: A Systematic Approach to Designing, Managing and Monitoring Community-Oriented Conservation Projects*. p. 384. Washington: Island Press.

Martin, T. G., Burgman, M. A., Fidler, F., Kuhnert, P.M., Low-Choy, S., *et al.* (2012). Eliciting expert knowledge in conservation science. *Conservation Biology* **26**: 1–10

Martínez-Abraín, A. (2013). Why do ecologists aim to get positive results? Once again, negative results are necessary for better knowledge accumulation. *Animal Biodiversity and Conservation* **36**: 33–36.

Mascia, M. B., Pailler, S., Krithivasan, R., Roshchanka, V., Burns, D., *et al.* (2014). Protected area downgrading, downsizing, and degazettement (PADDD) in Africa, Asia, and Latin America and the Caribbean, 1900–2010. *Biological Conservation* **169**: 355–361.

Matzek, V., Covino, J., Funk, J. L., and Saunders, M. (2014). Closing the knowing-doing gap in invasive plant management: accessibility and interdisciplinary of scientific research. *Conservation Letters* **7**: 208–215.

McBride, M. F., Garnett, S. T., Szabo, J. K., Burbidge, A. H., Butchart, S. H. M., *et al.* (2012). Structured elicitation of expert judgment for threatened species assessment: a case study on a continental scale using email. *Methods in Ecology and Evolution* **3**: 906–920.

McConnachie, M. and Cowling, R. M. (2013). On the accuracy of conservation managers' beliefs and if they learn from evidence-based knowledge: a preliminary investigation. *Journal of Environmental Management* **128**: 7–14.

Meijaard, E. and Sheil, D. (2007). Is wildlife research useful for wildlife conservation in the tropics?: A review for Borneo with global implications. *Biodiversity and Conservation* **16**: 3053–3065.

Moher, D., Jadad, A. R., Nichol, G., Penman, M., Tugwell, P., *et al.* (1995). Assessing the quality of randomized controlled trials: an annotated bibliography of scales and checklists. *Controlled Clinical Trials* **16**: 62–73.

N'Goran P. K., Boesch C., Mundry R., Herbinger, I., Yapi, F. A., *et al.* (2012). Hunting, law enforcement, and African primate conservation. *Conservation Biology* **26**: 565–571.

Nishida, T., Corp, N., Hamai, M., Hasegawa, T., Hiraiwa-Hasegawa, M., *et al.* (2003). Demography, female life history, and reproductive profiles among the chimpanzees of Mahale. *American Journal of Primatology* **59**: 99–121.

Oates, J. F. (1999). *Myth and Reality in the Rain Forest*. p. 310. Berkeley: University of California Press.

Plumptre, A. J. (2000). Monitoring mammal populations with line transect techniques in African forests. *Journal of Applied Ecology* **37**: 356–368.

Plumptre, A. J., Rose, R., Nangendo, G., Williamson, E. A., Didier, K., *et al.* (2010). *Eastern Chimpanzee (Pan troglodytes schweinfurthii): Status Survey and Conservation Action Plan 2010–2020*. p. 48. Gland: IUCN.

Prendergast, J. R., Quinn, R. M., and Lawton, J. H. (1999). The gaps between theory and practice in selecting nature reserves. *Conservation Biology* **13**: 484–492.

Pretty, J. and Smith, D. (2004). Social capital in biodiversity conservation and management. *Conservation Biology* **18**: 631–638.

Price, S. V. (2003). *War and Tropical Forests: Conservation in Areas of Armed Conflicts*. p. 219. New York: Food Products Press.

Pullin, A. S. and Knight, T. M. (2001). Effectiveness in conservation practice: pointers from medicine and public health. *Conservation Biology* **15**: 50–54.

Pullin, A. S. and Knight, T. M. (2005). Assessing conservation management's evidence base: a survey of management-plan compilers in the United Kingdom and Australia. *Conservation Biology* **19**: 1989–1996.

Pullin, A. S. and Stewart, G. B. (2006). Guidelines for systematic review in conservation and environmental management. *Conservation Biology* **20**: 1647–1656.

Pullin, A. S. and Knight, T. M. (2009). Doing more good than harm—Building an evidence-base for conservation and environment management. *Biological Conservation* **142**: 931–934.

Pullin, A. S. and Salafsky, N. (2010). Save the whales? Save the rainforest? Save the data! *Conservation Biology* **24**: 915–917.

Pullin, A. S., Knight, T. M., Stone, D. A., and Charman, K. (2004). Do conservation managers use scientific evidence to support their decision-making? *Biological Conservation* **119**: 245–252.

Pusey, A. E., Pintea, L., Wilson, M. L., Kamenya, S., and Goodall, J. (2007). The contribution of long-term research at Gombe National Park to chimpanzee conservation. *Conservation Biology* **21**: 623–634.

Ray, J. C., Hunter, L. T. B., and Zigouris, J. (2005). *Setting Conservation and Research Priorities for Larger African Carnivores*. New York: Wildlife Conservation Society.

Raymond, C. M., Fazey, I., Reed, M. S., Stringer, L., Robinson, G. M., *et al.* (2010). Integrating local and scientific knowledge for environmental management. *Journal of Environmental Management* **91**: 1766–1777.

Robinson, J. G. (2006). Conservation biology and real-world conservation. *Conservation Biology* **20**: 658–669.

Roe, D., Nelson, F., and Sandbrook, C. (2009). Community Management of Natural Resources in Africa: Impacts, Experiences and Future Directions, Natural Resource Issues No. 18. p. 207. London: International Institute for Environment and Development.

Roux, D. J., Rogers, K. H., Biggs, H. C., Ashton, P. J., and Sergeant, A. (2006). Bridging the science-management divide: moving from unidirectional knowledge transfer to knowledge interfacing and sharing. *Ecology and Society* **11**: 4.

Segan, D. B., Bottrill, M. C., Baxter, P. W. J., and Possingham, H. P. (2010). Using conservation evidence to guide management. *Conservation Biology* **25**: 200–202.

Smith, R. K., Dicks, L. V., Mitchell, R., and Sutherland, W. J. (2014). Comparative effectiveness research: the missing link in conservation. *Conservation Evidence* **11**: 2–6.

Smith, R. J., Muir, R. D. J., Walpole, M. J., Balmford, A., and Leader-Williams, N. (2003). Governance and loss of biodiversity. *Nature* **426**: 67–70.

Smoth, R. J. and Walpole, M. J. (2005). Should conservationists pay more attention to corruption? *Oryx* **39**: 251–256.

Stem, C. Margoulis, R., Salafsky, N., and Brown, M. (2005). Monitoring and evaluation in conservation: a review of trends and approaches. *Conservation Biology* **2**: 295–309.

Stevens, A. and Milne, R. (1997). The effectiveness revolution and public health. In: Scally, G. (Ed.), *Progress in Public Health*. pp. 197–225. London: Royal Society of Medicine Press.

Stokes, E., Johnson, A., and Rao, M. (2010). *Monitoring Wildlife Populations for Management. Training Module 7 for the Network of Conservation Educators and Practitioners*. p. 32. Vientiane, Lao PDR: American Museum of Natural History and the Wildlife Conservation Society.

Struhsaker, T. T., Struhsaker P. J., and Siex, K. S. (2005). Conserving Africa's rain forests: problems in protected areas and possible solutions. *Biological Conservation* **123**: 45–54. [Accessed October 2013].

Sunderland, T., Sayer, J., and Hoang, M. (2013). *Evidence-Based Conservation. Lessons From the Lower Mekong*. p. 482. Stoodleigh: Florence Production Ltd.

Sunderland, T., Sunderland-Groves, J., Shanley, P., and Campbell, B. (2009). Bridging the gap: how can information access and exchange between conservation biologists and field practitioners be improved for better conservation outcomes? *Biotropica* **41**: 549–554.

Sutherland, W. J. (2008a). *The Conservation Handbook: Research, Management and Policy*. p. 296. Oxford: Blackwell Science.

Sutherland, T. (2008b). Evidence-based conservation. *Conservation Magazine*, July 29. Available at http://conservationmagazine.org/2008/07/evidence-based-conservation/.

Sutherland, W. J., Pullin, A. S., Dolman, P. M., and Knight, T. M. (2004). The need for evidence-based conservation. *Trends in Ecology and Evolution* **19**: 305–308.

Tranquilli, S. (2013). African Great Apes: Assessing Threats and Conservation Efforts. PhD thesis, University College London, London.

Tranquilli, S., Abedi-Lartey, M., Amsini, F., Arranz, L., Asamoah, A., *et al.* (2012). Lack of conservation effort rapidly increases African great ape extinction risk. *Conservation Letters* **5**: 48–55.

Tranquilli, S., Abedi-Lartey, M., Abernethy, K., Amsini, F., and Asamoah, A., *et al.* (2014). Protected areas in tropical Africa: assessing threats and conservation activities. *PLoS One* **9**: 1–21.

Turnhout, E., Bloomfield, B., Hulme, E., Vogel, J., and Wynne, B. (2012). Conservation policy: listen to the voices of experience. *Nature* **488**: 454–455.

Waldron, A., Mooers, A. O., Miller, D. C., Nibbelink, N., Redding, D.W., *et al.* (2013). Targeting global conservation funding to limit immediate biodiversity declines. *Proceedings of the National Academy of Science* **110**: 12144–12148.

Walsh, J. C., Dicks, L. V., and Sutherland, W. J. (2014). The effect of scientific evidence on conservation practitioners' management decisions. *Conservation Biology* **29**: 88–98.

Walsh, P. D., Abernethy, K. A., Bermejo, M., Beyers, R., De Wachter, P., *et al.* (2003). Catastrophic ape decline in western equatorial Africa. *Nature* **422**: 611–614.

Waylen, K. A., Fischer, A., McGowan, P. J. K., Thirgood, S.J., and Milner-Gulland, E. J. (2010). Effect of local cultural context on the success of community-based conservation interventions. *Conservation Biology* **24**: 1119–1129.

Wilson, K. A., McBride, M. F., Bode, M., and Possingham, H. P. (2006). Prioritizing global conservation efforts. *Nature* **440**: 337–340.

Wrangham, R. and Ross, E. (2008). *Science and Conservation in African Forests: the Benefit of Long-Term Research*. p. 254. Cambridge: Cambridge University Press.

Young, K. D. and Van Aarde, R. J. (2011). Science and elephant management decisions in South Africa. *Biological Conservation* **144**: 876–885.

CHAPTER 18

Some future directions for primate conservation research

Andrew J. Marshall and Serge A. Wich

Rhesus Macaque preparing to cross a road at Srisailam, Eastern Ghats, India. Photo copyright: Swapna Nelaballi.

18.1 Introduction

Primate researchers have made many positive contributes to conservation, including helping to form national parks (Wright 1992), founding or leading organizations to promote awareness and fund conservation activities (e.g. Jane Goodall's Roots and Shoots, Russell Mittermeier at Conservation International), raising the international profile of threatened primates (e.g. through work at the United Nations Environmental Programme, the International Union for the Conservation of Nature, and the Great Apes Survival Partnership), and contributing to the successful management of severely endangered taxa (e.g. mountain gorillas: Robbins *et al.* (2011); golden lion tamarins: Kleiman and Mallinson (1998)). The presence of primate researchers at field stations can facilitate law enforcement, provide alternative income to local communities, and promote awareness of the importance of biodiversity and its protection (Wrangham 2008; Campbell *et al.* 2008; Tranquilli *et al.* 2012; Laurance 2013). Involvement of students in research projects can help train the next generation of conservationists and natural resource managers (Blair *et al.* 2013). This training is particularly important in many habitat countries where capacity building is vital to ongoing conservation efforts and often an important activity at primate field sites. Primate field research documents the status of threatened species, assesses the effects of threats, examines interactions between non-human and human primates, and demonstrates the key ecological role many primates play in ecosystems (Cowlishaw and Dunbar 2000; Chapman and Peres 2001). Indeed, much basic primate research is framed in the context of primate conservation. Even studies that are not explicitly geared towards conservation (e.g. those focusing on social behaviour) can nevertheless have a positive effect by raising interest in primates among the broader public.

Marshall, A.J. and Wich, S.A., *Some future directions for primate conservation research.* In: *An Introduction to Primate Conservation.* Edited by: Serge A. Wich and Andrew J. Marshall, Oxford University Press (2016). © Oxford University Press.
DOI 10.1093/acprof:oso/9780198703389.003.0018

Despite these examples, both practitioners and academics have long questioned the extent to which conservation research on primates and other taxa has direct applicability for the protection of threatened populations and habitats (Harcourt 2000; Sheil 2001; Whitten *et al.* 2001; Terborgh 2004; Fazey *et al.* 2005). Indeed, beyond a handful of widely cited examples, it is harder than we might wish to find concrete examples of primate research that tangibly improves the conservation of particular primate populations at specific locations. Simply asserting that our research is beneficial for conservation or including discussion of the management implications of our work in academic papers does not ensure that our efforts will have a positive impact. Recently, attention has been paid to how research efforts can be made more directly relevant to conservation, producing a set of concrete recommendations for how research can contribute more effectively (Kareiva and Marvier 2012; Meijaard and Sheil 2007; Meijaard *et al.* 2012). A key message from this literature is that in order to enhance our contributions to conservation, researchers must keep pace with changing conditions, threats, and opportunities. This will entail both broadening the scope of our research and intensifying our efforts to answer specific applied questions of direct conservation relevance. We see potential in new approaches that acknowledge the conservation value of degraded lands, take a broad, landscape-level approach to land-use planning, and address in creative ways the often competing needs of primates and people. We are also convinced that success will require work with a wide range of stake-holders, including local communities and several entities—such as extractive industries (e.g. mining, forestry) and oil-palm companies involved in wholesale land conversion—that conservationists have historically viewed as adversaries.

In 2001, Chapman and Peres provided a comprehensive discussion of the role of scientists in promoting primate conservation (Chapman and Peres 2001). A decade and a half later, the core points they raised remain highly pertinent. The major threats to primates that they discussed persist (indeed, most have intensified), and while substantive progress has been made in several areas, much of the information they identified as necessary to understand key threats and evaluate proposed solutions remains unavailable for most species and ecosystems. This means that the need for rigorous primate conservation research is more urgent than ever. In this final chapter, we make several recommendations for how future research activity could make meaningful contributions to primate conservation. We have not attempted to provide a comprehensive list of the conservation topics in need of attention by primate researchers (c.f., Sutherland *et al.* 2009, 2013), and acknowledge that others would likely produce a rather different list of topics. Nevertheless, we hope that this subjective, and perhaps idiosyncratic, list will prove to be useful food for thought.

18.2 Fill gaps in taxonomic and geographic knowledge

Although primates are better studied than most other tropical taxa (see Marshall and Wich, Chapter 2, this volume), we lack even the most basic information about distribution, population status, and threats for many primate taxa. Published field studies on primates are not randomly distributed across phylogenetic space. For example, recent work published in the *International Journal of Primatology* is disproportionately skewed towards apes and away from strepsirrhines (Setchell 2012). A systematic examination of information available on the web about each primate taxon also demonstrated a heavy bias towards apes, and further showed that publically available information on primates did not correlate with extinction risk (Van Cleave 2012). This implies that for some taxa, basic field research or population surveys would make a real contribution (e.g. examples in Campbell *et al.*, Chapter 6, this volume). Indeed, without field research we may be ignorant of the fundamental taxonomic units of conservation (see Groves, Chapter 4, this volume); it is difficult to target conservation action towards a taxon that we do not know exists! Similarly, without good knowledge of a species' distribution and density, it is difficult to construct meaningful management plans (Campbell *et al.*, Chapter 6, this volume). Even if one eschews species as the fundamental units of conservation (Agapow *et al.* 2004; Mace 2004; Rylands and Mittermeier 2014), additional field research and surveys can provide valuable information about the distribution of threats and identify geographic areas or primate communities in need of conservation attention.

18.3 Make behavioural research more relevant to conservation

Primate behavioural research can make a greater contribution to conservation than it presently does. To date much behavioural work on primates has had limited relevance to applied conservation. While research that focuses on what are typically viewed as ecological topics (e.g. feeding ecology, seed dispersal, population density, demography) is often more obviously relevant to conservation than research on topics typically classified as behavioural (e.g. social interactions, mating behaviour, locomotion, endocrinology, communication), knowledge of individual behaviour is nevertheless indispensable. Behavioural studies can contribute to conservation and management by documenting the behavioural plasticity of species, examining their interactions with humans, informing reintroduction and captive breeding programmes, anticipating responses to habitat destruction and climate change, characterizing dispersal, and more (Swaisgood 2007; Sutherland *et al.* 2009; Caro and Sherman 2011). There are many good examples of primate behavioural research with conservation applicability, including work on crop-raiding, dispersal in fragmented landscapes, disease transmission, and behavioural responses to logging and hunting; more such work is needed.

18.4 Increase research in marginal habitats and outside protected areas

Much more research is needed on primates living outside protected areas, inhabiting marginal and degraded habitats, and ranging across complex, multi-use landscapes. Because primate habitats around the world are being fragmented and degraded, much of future primate conservation will take place in the context of suboptimal habitats and environments that are quite different from those to which many species are best adapted (Irwin, Chapter 7, this volume; Meijaard, Chapter 13, this volume; Chapman and Peres 2001). While there is increasing appreciation of this point, and greater attention is being paid to examining primates living in suboptimal ecological conditions (e.g. Irwin 2008; Arroyo-Rodríguez and Dias 2010; Campbell-Smith *et al.* 2011, Meijaard *et al.* 2010), most primate research is still being conducted inside protected areas and at sites comprising relatively high-quality habitats and that are relatively undisturbed. For instance, a recent survey of research published from all great ape range countries in Africa and Asia showed that research attention in protected areas is strongly biased towards large national parks containing great apes (Marshall *et al.* 2016). This bias may provide dangerously optimistic impressions of population vital rates (Marshall 2009) and result in crucial gaps in our knowledge of whether and how primates survive, adapt, and reproduce in lower quality and degraded environments. In cases where the majority of individuals live outside formally protected areas (e.g. orang-utans: Wich *et al.* (2012) and African great apes: Wich *et al.* (2014)), it also means that our research results may be representative of the minority of individuals of species that we seek to protect. A greater focus on primates outside protected areas, in disturbed and degraded habitats, and occupying mosaic habitats comprising a mixture of land covers and uses would render our research more directly applicable to the conservation and management of threatened primate populations.

18.5 Expand climate change research

Climate change research, which necessarily includes heavy use of species distribution modelling and climate projections, must also effectively incorporate biotic interactions, such as competition and resource availability (Blois *et al.* 2013; Kissling *et al.* 2012; Wisz *et al.* 2013) and the dispersal abilities of primates (Schloss *et al.* 2012). Most current studies of the effects of climate and land-cover change on primates and other mammals model habitat suitability based on basic ecological variables such as temperature, rainfall, altitude, or fairly course-grained indices of land cover (or a combination of these variables and time budgets, Korstjens *et al.* (2006); Lehmann *et al.* (2010); Struebig *et al.* (201)5; Korstjens and Hillyer, Chapter 11, this volume). While these variables are clearly important and can predict broad patterns of distribution and density (Korstjens and Dunbar 2007; Willems and Hill 2009: Wich *et al.* 2012), they are likely to be less successful on the local scale where differences in presence–absence and population density occur between areas that have very similar values for basic ecological variables. In such small-scale cases, measures such as disturbance or food availability are normally better predictors (Balcomb *et al.* 2000; Hanya *et al.* 2005; Wich *et al.* 2004; Marshall *et al.* 2006).

Moreover, the correlation between basic climatic variables (temperature, rainfall, and altitude) and the factors that directly influence the population density and distribution of most primate species (e.g. food availability) is likely to become weaker as global climate changes. This is because plant food distribution, for example, may change at a different rate than the basic ecological variables that influence it. For instance, the population density of white-bearded gibbons, *Hylobates albibarbis*, at Gunung Palung National Park, West Kalimantan, Indonesia, declines predictably with altitude (Marshall 2004, 2009), most likely due to decreases in the availability of figs, which are important fallback foods during periods of resource scarcity (Marshall and Leighton 2006; Marshall 2010). It would be dangerous to assume that the warming of higher elevation forests to conditions that are superficially comparable to present-day lowland forest types would mean that higher elevation forests would immediately provide high-quality habitat for gibbons, because the hemi-epiphytic figs (and their host trees) that are most important to gibbons may take hundreds of years to establish and grow at higher elevations (Leighton and Leighton 1983). This implies that the more rapidly climate changes, the less good simple ecological variables will be at predicting the distribution of primate species over the timescale of several decades at which most local conservation decisions are made. Field research and modelling that predict the rates and directions of changes in underlying ecological variables, such as the distribution and density of key lowland food plants (Lenoir and Svenning 2014), in response to climate change are greatly needed.

Most attempts to model shifts in species distribution resulting from climate change tacitly assume that species will be able to move to keep pace with changing climates. This assumption is unlikely to be true for primates, which appear to be surprisingly limited in their dispersal abilities (Beaudrot and Marshall 2011; Beaudrot *et al.* 2013, 2014). Indeed, a recent model of mammalian dispersal abilities suggests that primates will be one of the mammalian taxa least able to move to track changes in climate (Schloss *et al.* 2012). This suggests that studies of primate responses to climate change must include species-specific dispersal abilities, and that projections that fail to do so likely will provide dangerously optimistic estimates of their abilities to adapt to climate change.

18.6 Promote recognition of the value of ecosystem services provided by primates

Primates across the tropics perform a range of ecological functions that are critical to maintaining healthy, well-functioning ecosystems (e.g. pollination, seed dispersal, seed predation, folivory: Marshall and Wich, Chapter 2, this volume). There is mounting evidence that primates can be the sole providers of certain ecosystem services, and that the loss of primates (e.g. due to hunting) degrades ecosystem structure and function (Effiom *et al.* 2013; Nunez-Iturri *et al.* 2008). Despite their importance, the economic value of ecological services provided by primates is rarely estimated or incorporated into policy discussions regarding conservation and management. Widespread recognition of the important services provided by primates and explicit consideration of their economic value may provide useful justification for primate conservation (Wich and Marshall, Chapter 1, this volume; Garcia-Ulloa and Koh, Chapter 16, this volume). Primatologists could contribute immensely to this endeavour by conducting research that documents the ecological services provided by our study subjects and promoting awareness of the crucial role that primates play in maintaining ecosystem function.

18.7 Inform allocation of conservation funds

We must work to provide concrete information that will facilitate the process of setting priorities to determine how primate conservation funds should be allocated (Wilson *et al.* 2006; Brooks *et al.* 2006). Because funds and attention are limited (James *et al.* 1999; Balmford *et al.* 2003), trade-offs are inherent to conservation decision making. Although primates draw substantial public interest and conservation funding, resources are unlikely to be adequate to invest in actions that will save all primate populations, and probably not even all species, from extinction. This means that decisions will have to be made regarding wise allocation of funds towards specific conservation actions that prioritize some populations, sites, or species over others (Bottrill *et al.* 2008). Primate researchers can play a key role in this endeavour by providing the quantitative information necessary to conduct prioritizations and by contributing their expertise to more subjective discussions (Wilhere *et al.* 2012; Game *et al.* 2013).

We appreciate that discussing triage in the context of conservation is controversial (Parr *et al.* 2009; Jachowski and Kesler 2009), because it by definition results in decisions to support some actions over others and entails subjective value judgements (should prioritization of actions be based on species? phylogenetic diversity? ecosystem function? sustainable yield? benefits to people?). Nevertheless, it is important to recognize that all conservation plans are inherently prioritizations, whether we acknowledge the fact or not (Game *et al.* 2013). Therefore, the question is not whether or not we should engage in primate conservation triage, but rather whether we will do so in a transparent, clearly defined way that incorporates inherent uncertainties, costs, trade-offs, and risks of failure (Possingham *et al.* 2001; Regan *et al.* 2005; McDonald-Madden *et al.* 2008). We also note that funding conservation involves two components: (1) the societal and political decisions about how much funding to provide, and (2) the optimization decision about how to wisely allocate the available funds (Bottrill *et al.* 2009). Primate conservationists can make contributions in both areas, by advocating legislation and increased funding (e.g. Marshall *et al.* 1999) and engaging in decision-making processes that inform the allocation of conservation investment (e.g. Hannah *et al.* 1998; Whittaker 2006). In the case of the latter, primate conservation priority setting has to date generally been used to identify important locations for the conservation of specific taxa (e.g. Thorn *et al.* 2009; Davenport *et al.* 2014) or to direct attention towards specific species (e.g. Mittermeier *et al.* 2009) or populations (e.g. Caldecott and Miles 2005). Future recommendations are likely to be most useful for conservation managers if they are framed in the form of concrete suggestions for specific actions in specific places over specific timeframes, and available options are ranked in transparent ways that incorporate costs, benefits, risks, and uncertainty (Game *et al.* 2013).

18.8 Embrace interdisciplinarity

The research needed to answer some of the most pressing questions in primate conservation does not fit comfortably into traditional academic categories. As in other realms of conservation, much of the required research is interdisciplinary, necessitating close collaboration among diverse fields in the natural and social sciences (Daily and Ehrilch 1999; Mascia *et al.* 2003; Ostrom and Cox 2010; Reyers *et al.* 2010). Primatologists, who are often well versed in biology and ecology but affiliated with social science departments, are well placed to spearhead such collaborations. Interdisciplinary work can be challenging; things as simple as differences in terminology, dissimilar assumptions, and distinct perceptions about what constitutes evidence can complicate collaboration across disciplines (Brewer 1999; Golde and Gallagher 1999; Holt and Webb 2007). Nevertheless, there is now a substantial literature that provides concrete suggestions about how to overcome these challenges (e.g. Naiman 1999; Campbell 2005; Öberg 2009), and an increasing number of examples of successful interdisciplinary collaborations demonstrate that many teams have effectively negotiated them (e.g. Margles *et al.* 2010; Holt and Webb 2007; Rutherford *et al.* 2009). Other concerns once frequently expressed regarding interdisciplinary research seem to be diminishing. For instance, new journals dedicated to interdisciplinary research (e.g. *Ecosystems*, Turner and Carpenter (1999); *Ecology and Society*, Holling (1997)) and an increased appreciation of its importance by many editors (e.g. *Conservation Biology*, Holt and Webb (2007)) have alleviated reservations about limited outlets for publishing research that spans traditional fields. New initiatives at universities, foundations, and government agencies that fund collaboration across disciplines have eased concerns regarding limited financial resources for interdisciplinary work (Campbell 2005). The widespread acknowledgment of its crucial importance for conservation—and new incentives that promote it—suggests that primatologists could increase their contributions to conservation by embracing interdisciplinary research. For instance, there is likely real value in reading relevant literature outside our disciplines, collaborating with researchers in other fields, encouraging graduate students to engage in interdisciplinary work, and seriously entertaining worthy funding proposals and journal manuscripts that span multiple disciplines.

18.9 Acknowledge the value of applied work

Conservationists have long lamented the fact that some academic institutions do not value applied conservation research as highly as theoretical work, and fear this dissuades graduate students aspiring to academic positions or junior faculty seeking tenure from pursuing such research (Chapman and Peres 2001; Caro 2007; Caro and Sherman 2013). They have additionally noted that applied conservation research also often falls outside traditional sources of funding, making pursuit of such work difficult, especially for individuals at the start of their careers. Fortunately, this has changed considerably in recent years. Universities increasingly value conservation work, several new graduate programmes dedicated to primate conservation have emerged, and there are now multiple contexts for conducting conservation research outside academia (e.g. conservation organizations, government agencies). New funding sources for conservation research appear every year, and primatologists who add a conservation component to their research may well improve, not reduce, their chances of securing funding. Appreciation of the importance of applied conservation research is mounting, the need for it grows ever more urgent, and there are more opportunities than ever before to make positive contributions (Caro and Sherman 2011, 2013). All these things are likely to make applied conservation work a more attractive option for early career researchers. We certainly hope so, and encourage students to pursue such work and their advisors to support it.

18.10 Increase engagement outside academia

Conservation will only succeed with strong support from policy-makers and the public. Many academics involved in conservation actively work to build this support, and spend substantial time giving public lectures, writing articles for public media, helping with documentaries, and contributing their expertise to policy discussions with governments. Such activities may advance conservation goals and are a tangible way that academics can contribute to society at large, serving a public that (directly or indirectly) pays their salaries. Nevertheless, many academics perceive that their universities undervalue and provide little institutional support for their engagement outside academia. For example, in a recent survey, academics in the UK indicated that their public engagement is not well supported by their institutions, is not taken into consideration in career progression discussions, can be frowned upon by senior academics, and is not sufficiently valued by the UK Government Research Excellence Framework (Watermeyer 2015; Jump 2015). Academics may quite rationally respond to these disincentives by investing less in public engagement than they otherwise would. Rectifying this problem will require changes in both policy and perception at some universities. Students and faculty can help bring about these changes through more frequent, visible public engagement, by advocating changes in academic policy, by increased valuation of 'broader impacts' in hiring and promotion decisions, through support and encouragement of colleagues who invest in outreach work, and perhaps also through examination of their own attitudes and potential biases. Those who wish to engage outside academia could be supported by training that prepares them to communicate effectively with the public and contribute productively to policy discussions. Such training could help make us more effective advocates for the conservation of primates and their habitats.

Acknowledgements

We thank Lydia Beaudrot for helpful comments and discussion, and Swapna Nelaballi for providing the photograph for the chapter front page.

References

Agapow, P. M., Bininda-Emonds, O. R., Crandall, K. A., Gittleman, J. L., Mace, G. M., Marshall, J. C., *et al.* (2004). The impact of species concept on biodiversity studies. *The Quarterly Review of Biology* **79**(2): 161–179.

Arroyo-Rodríguez, V. and Dias, P. A. D. (2010). Effects of habitat fragmentation and disturbance on howler monkeys: a review. *American Journal of Primatology* **72**(1): 1–16.

Balcomb, S. R., Chapman, C. A., and Wrangham, R. W. (2000). Relationship between chimpanzee (Pan troglodytes) density and large, fleshy-fruit tree density: conservation implications. *American Journal of Primatology* **51**(3): 197–203.

Balmford, A., Gaston, K. J., Blyth, S., James, A., and Kapos, V. (2003). Global variation in terrestrial conservation costs, conservation benefits, and unmet conservation needs. *Proceedings of the National Academy of Sciences* **100**(3): 1046–1050.

Beaudrot, L., Kamilar, J., Marshall, A. J., and Reed, K. E. (2014). African primate assemblages exhibit a latitudinal gradient in dispersal limitation. *International Journal of Primatology* **35**: 1088–1104.

Beaudrot, L. H. and Marshall, A. J. (2011). Primate communities are structured more by dispersal limitation than by niches. *Journal of Animal Ecology* **80**: 332–341.

Beaudrot, L., Rejmánek, M., and Marshall, A. J. (2013). Dispersal modes affect tropical forest assembly across tropic levels. *Ecography* **36**: 984–993.

Blair, M. E., Bynum, N., and Sterling, E. J. (2013). Determining conservation status and contributing to *in situ* conservation action. In: Sterling, E. J., Bynum, N., and Blair, M. (Eds), *Primate Ecology and Conservation*. pp. 278–293. Oxford: Oxford University Press.

Blois, J. L., Zarnetske, P. L., Fitzpatrick, M. C., and Finnegan, S. (2013). Climate change and the past, present, and future of biotic interactions. *Science* **341**: 499–504.

Bottrill, M. C., Joseph, L. N., Carwardine, J., Bode, M., Cook, C., *et al.* (2008). Is conservation triage just smart decision making? *Trends in Ecology and Evolution* **23**: 649–654.

Bottrill, M. C., Joseph, L. N., Carwardine, J., Bode, M., Cook, C., *et al.* (2009). Finite conservation funds mean triage is unavoidable. *Trends in Ecology and Evolution* **24**(4): 183–184.

Brewer, G. D. (1999). The challenges of interdisciplinarity. *Policy Sciences* **32**(4): 327–337.

Brooks, T. M., Mittermeier, R. A., da Fonseca, G. A., Gerlach, J., Hoffmann, M., *et al.* (2006). Global biodiversity conservation priorities. *Science* **313**: 58–61.

Caldecott, J. O. and Miles, L. (Eds) (2005). *World Atlas of Great Apes and their Conservation*. California: University of California Press.

Campbell, L. M. (2005). Overcoming obstacles to interdisciplinary research. *Conservation Biology* **19**(2): 574–577.

Campbell, G., Kuehl, H., Diarrassouba, A., N'Goran, P. K., and Boesch, C. (2011). Long-term research sites as refugia for threatened and over-harvested species. *Biology Letters* **7**: 723–726.

Campbell, G., Kuehl, H., N'Goran Kouame, P., and Boesch, C. (2008). Alarming decline of West African chimpanzees in Cote d'Ivoire. *Current Biology* **18**: R903–R904.

Campbell-Smith, G., Campbell-Smith, M., Singleton, I., and Linkie, M. (2011). Apes in space: saving an imperilled orangutan population in Sumatra. *PLoS One* **6**(2): e17210. doi:10.1371/journal.pone.0017210.

Caro, T. (2007). Behavior and conservation: a bridge too far? *Trends in Ecology and Evolution* **22**(8): 394–400.

Caro, T. and Sherman, P. W. (2011). Endangered species and a threatened discipline: behavioural ecology. *Trends in Ecology and Evolution* **26**: 111–118.

Caro, T. and Sherman, P. W. (2013). Eighteen reasons animal behaviourists avoid involvement in conservation. *Animal Behaviour* **85**(2): 305–312.

Chapman, C. A. and Peres, C. A. (2001). Primate conservation in the new millennium: the role of scientists. *Evolutionary Anthropology* **10**: 16–33.

Cowlishaw, G. and Dunbar, R. I. (2000). *Primate Conservation Biology*. Chicago: University of Chicago Press.

Daily, G. C. and Ehrlich, P. R. (1999). Managing earth's ecosystems: an interdisciplinary challenge. *Ecosystems* **2**(4): 277–280.

Davenport, T. R., Nowak, K., and Perkin, A. (2014). Priority primate areas in Tanzania. *Oryx* **48**(01): 39–51.

Effiom, E. O., Birkhofer, K., Smith, H. G., and Olsson, O. (2013). Changes of community composition at multiple trophic levels due to hunting in Nigerian tropical forests. *Ecography* **37**: 367–377.

Fazey, I., Fischer, J., and Lindenmayer, D. B. (2005). What do conservation biologists publish? *Biological Conservation* **124**: 63–73.

Game, E. T., Kareiva, P., and Possingham, H. P. (2013). Six common mistakes in conservation priority setting. *Conservation Biology* **27**: 480–485.

Golde, C. M. and Gallagher, H. A. (1999). The challenges of conducting interdisciplinary research in traditional doctoral programs. *Ecosystems* **2**(4): 281–285.

Hannah, L., Rakotosamimanana, B., Ganzhorn, J., Mittermeier, R. A., Olivieri, S., *et al.* (1998). Participatory planning, scientific priorities, and landscape conservation in Madagascar. *Environmental Conservation* **25**(01): 30–36.

Hanya, G., Zamma, K., Hayaishi, S., Yoshihiro, S., Tsuriya, Y., *et al.* (2005). Comparisons of food availability and group density of Japanese macaques in primary, naturally regenerated, and plantation forests. *American Journal of Primatology* **66**(3): 245–262.

Harcourt, A. H. (2000). Conservation in practice. *Evolutionary Anthropology* **9**: 258–265.

Holling, C. S. (1997). The inaugural issue of *Conservation Ecology*. Conservation Ecology **1**(1): 1. http://www.consecol.org/vol1/iss1/art1/.

Holt, A. and Webb, T. (2007). Interdisciplinary research: leading ecologists down the route to sustainability? *Bulletin of the British Ecological Society* **38**(3): 2–13.

Irwin, M. T. (2008). Diademed sifaka (*Propithecus diadema*) ranging and habitat use in continuous and fragmented forest: higher density but lower viability in fragments? *Biotropica* **40**(2): 231–240.

Jachowski, D. and Kesler, D. (2009). Allowing extinction: are we ready to let species go? *Trends in Ecology and Evolution* **24**: 180.

James, A. N., Gaston, K. J., and Balmford, A. (1999). Balancing the Earth's accounts. *Nature* **401**(6751): 323–324.

Jump, P. (2015). Public engagement means 'sacrificing' academic career. *Times Higher Education*, 9 July 2015. https://www.timeshighereducation.com/news/public-engagement-means-sacrificing-academic-career.

Kareiva, P. and Marvier, M. (2012). What is conservation science? *BioScience* **62**: 962–969.

Kissling, W. D., Dormann, C. F., Groeneveld, J., Hickler, T., Kühn, I., *et al.* (2012). Towards novel approaches to modelling biotic interactions in multispecies assemblages at large spatial extents. *Journal of Biogeography* **39**: 2163–2178.

Kleiman, D. G. and Mallinson, J. J. (1998). Recovery and management committees for lion tamarins: partnerships in conservation planning and implementation. *Conservation Biology* **12**(1): 27–38.

Korstjens, A. H. and Dunbar, R. I. M. (2007). Time constraints limit group sizes and distribution in red and black-and-white colobus monkeys. *International Journal of Primatology* **28**: 555–575.

Korstjens, A. H., Verhoekx, I. L., and Dunbar, R. I. M. (2006). Time as a constraint on group size in spider monkeys. *Behavioral Ecology and Sociobiology* **60**: 683–694.

Laurance, W. F. 2013. Does research help to safeguard protected areas? *Trends in Ecology and Evolution* **28**: 261–266.

Lehmann, J., Korstjens, A. H., and Dunbar, R. I. (2010). Apes in a changing world—the effects of global warming on the behaviour and distribution of African apes. *Journal of Biogeography* **37**: 2217–2231.

Leighton, M. and Leighton, D. (1983). Vertebrate responses to fruiting seasonality within a Bornean rain forest. In: Sutton, S. L., Whitmore, T. C., and Chadwick, A. C. (Eds), *Tropical Rain Forest: Ecology and Management*. pp. 181–196. Boston: Blackwell Scientific.

Lenoir, J. and Svenning, J.-C. (2014). Climate-related range shifts—a global multidimensional synthesis and new research directions. *Ecography* **37**: 1–14.

Mace, G. M. (2004). The role of taxonomy in species conservation. *Philosophical Transactions of the Royal Society B: Biological Sciences* **359**(1444): 711–719.

Margles, S. W., Peterson, R. B., Ervin, J., and Kaplin, B. A. (2010). Conservation without borders: building communication and action across disciplinary boundaries for effective conservation. *Environmental Management* **45**(1): 1–4.

Marshall, A. J. (2004). Population Ecology of Gibbons and Leaf Monkeys across a Gradient of Bornean Forest Types. Ph.D. Dissertation, Harvard University.

Marshall, A. J. (2009). Are montane forests demographic sinks for Bornean white-bearded gibbons? *Biotropica* **41**: 257–267.

Marshall, A. J. (2010). Effect of habitat quality on primate populations in Kalimantan: gibbons and leaf monkeys as case studies. In: Supriatna, J. and Gursky, S. L. (Eds), *Indonesian Primates*. pp. 157–177. New York: Springer Science + Business Media.

Marshall, A. J. and Leighton, M. (2006). How does food availability limit the population density of white-bearded gibbons? In: Hohmann, G., Robbins, M., and Boesch C. (Eds), *Feeding Ecology of the Apes and Other Primates*. pp. 311–333. Cambridge, UK: Cambridge University Press.

Marshall, A. J., Jones, J. H., and Wrangham, R. W. (1999). The Plight of the Apes: a global survey of great ape populations. Congressional briefing to U.S. House of Representatives in support of Great Ape Conservation Act, H.R. 4320.

Marshall, A.J., Meijaard, E., Van Cleave, E., and Sheil, D. 2016. Charisma counts: the presence of great apes affects the allocation of research effort in the paleotropics. *Frontiers in Ecology and the Environment* **14**: 13–19.

Marshall, A. J., Nardiyono, L. M. Engström, B. Pamungkas, J. Palapa, E., *et al.* (2006). The blowgun is mightier than the chainsaw in determining population density of Bornean orangutans (*Pongo pygmaeus morio*) in the forests of East Kalimantan. *Biological Conservation* **129**: 566–578.

Mascia, M. B., Brosius, J. P., Dobson, T. A., Forbes, B. C., Horowitz, L., *et al.* (2003). Conservation and the social sciences. *Conservation Biology* **17**(3): 649–650.

McDonald-Madden, E., Baxter, P. W., and Possingham, H. P. (2008). Making robust decisions for conservation with restricted money and knowledge. *Journal of Applied Ecology* **45**: 1630–1638.

Meijaard, E. and Sheil, D. (2007). Is wildlife research useful for wildlife conservation in the tropics? A review for Borneo with global implications. *Biodiversity Conservation* **16**: 3053–3065.

Meijaard, E., Albar, G., Nardiyono, Rayadin, Y., Ancrenaz, M., *et al.* (2010). Unexpected ecological resilience in Bornean orangutans and implications for pulp and paper plantation management. *PLoS One* **5**(9): e12813.

Meijaard, E., Wich, S. A., Ancrenaz, M., and Marshall, A. J. (2012). Not by science alone: why orangutan conservationists must think outside the box. *Annals of the New York Academy of Sciences* **1249**: 29–44.

Mittermeier, R. A., Wallis, J., Rylands, A. B., Ganzhorn, J. U., Oates, J. F., *et al.* (2009). Primates in peril: the world's 25 most endangered primates 2008–2010. *Primate Conservation* **24**: 1–57.

Naiman, R. J. (1999). A perspective on interdisciplinary science. *Ecosystems* **2**(4): 292–295.

Nunez-Iturri, G., Olsson, O., and Howe, H. F. (2008). Hunting reduces recruitment of primate-dispersed trees in Amazonian Peru. *Biological Conservation* **141**: 1536–1546.

Öberg, G. (2009). Facilitating interdisciplinary work: using quality assessment to create common ground. *Higher Education* **57**(4): 405–415.

Ostrom, E. and Cox, M. (2010). Moving beyond panaceas: a multi-tiered diagnostic approach for social-ecological analysis. *Environmental Conservation* **37**: 451–463.

Parr, M. J., Bennun, L., Boucher, T., Brooks, T., Chutas, C. A., *et al.* (2009). Why we should aim for zero extinction. *Trends in Ecology and Evolution* **24**(4): 181.

Possingham, H. P., Andelman, S. J., Noon, B. R., Trombulak, S., and Pulliam, H. R. (2001). Making smart conservation decisions. In: Soulé, M. E. and Orians, G. H. (Eds), *Conservation Biology: Research Priorities for the Next Decade*. pp. 225–244. Washington, DC: Island Press.

Regan, H. M., Ben-Haim, Y., Langford, B., Wilson, W. G., Lundberg, P., *et al.* (2005). Robust decision-making under severe uncertainty for conservation management. *Ecological Applications* **15**(4): 1471–1477.

Reyers, B., Roux, D. J., and O'Farrell, P. J. (2010). Can ecosystem services lead ecology on a transdisciplinary pathway? Environ. *Conservation* **37**: 501–511.

Robbins, M. M., Gray, M., Fawcett, K. A., Nutter, F. B., Uwingeli, P., *et al.* (2011). Extreme conservation leads to recovery of the Virunga mountain gorillas. *PLoS One* **6**: e19788.

Rutherford, M. B., Gibeau, M. L., Clark, S. G., and Chamberlain, E. C. (2009). Interdisciplinary problem solving workshops for grizzly bear conservation in Banff National Park, Canada. *Policy Sciences* **42**(2): 163–187.

Rylands, A. B. and Mittermeier, R. A. (2014). Primate taxonomy: species and conservation. *Evolutionary Anthropology* **23**: 8–10.

Schloss, C. A., Nuñez, T. A., and Lawler, J. J. (2012). Dispersal will limit ability of mammals to track climate change in the Western Hemisphere. *Proceedings of the National Academy of Sciences* **109**(22): 8606–8611.

Setchell, J. M. (2012). On editing the *International Journal of Primatology*. *International Journal of Primatology* **33**: 1–9.

Sheil, D. (2001). Conservation and biodiversity monitoring in the tropics: realities, priorities, and distractions. *Conservation Biology* **15**: 1179–1182

Struebig, M. J., Wilting, A., Gaveau, D. L. A., Meijaard, E., Smith, R. J. *et al.* (2015). Targeted conservation to safeguard a biodiversity hotspot from climate and landcover change. *Current Biology* **25:** 372–378.

Sutherland, W. J., Adams, W. M., Aronson, R. B., Aveling, R., Blackburn, T. M., *et al.* (2009). One hundred questions of importance to the conservation of global biological diversity. *Conservation Biology* **23**: 557–567.

Sutherland, W. J., Freckleton, R. P., Godfray, H. C. J., Beissinger, S. R., Benton, T., *et al.* (2013). Identification of 100 fundamental ecological questions. *Journal of Ecology* **101**: 58–67.

Swaisgood, R. R. (2007). Current status and future directions of applied behavioral research for animal welfare and conservation. *Applied Animal Behaviour Science* **102**(3): 139–162.

Terborgh, J. (2004). Reflections of a scientist on the World Parks Congress. *Conservation Biology* **18**: 619–620

Thorn, J. S., Nijman, V., Smith, D., and Nekaris, K. A. I. (2009). Ecological niche modelling as a technique for assessing threats and setting conservation priorities for Asian slow lorises (Primates: Nycticebus). *Diversity and Distributions* **15**: 289–298.

Tranquilli, S., Abedi-Lartey, M., Amsini, F., Arranz, L., Asamoah, A., *et al.* (2012). Lack of conservation effort rapidly increases African great ape extinction risk. *Conservation Letters* **5**: 48–55.

Turner, M. G. and Carpenter, S. R. (1999). Tips and traps in interdisciplinary research. *Ecosystems* **2**(4): 275–276.

Van Cleave, E. (2012). Can Google Search Results Inform Primate Conservation Decisions? B.Sc. Honors thesis, Department of Anthropology, University of California, Davis.

Watermeyer, R. (2015). Lost in the 'third space': the impact of public engagement in higher education on academic identity, research practice and career progression. *European Journal of Higher Education* **5**: 331–347.

Whittaker, D. J. (2006). A conservation action plan for the Mentawai primates. *Primate Conservation* **20**: 95–105.

Whitten, T., Holmes, D., and MacKinnon, K. (2001). Conservation biology: a displacement behavior for academia? *Conservation Biology* **15**: 1–3.

Wich, S., Buij, R., and Van Schaik, C. (2004). Determinants of orangutan density in the dryland forests of the Leuser Ecosystem. *Primates* **45**(3): 177–182.

Wich, S. A, D. Gaveau, N. Abram, M. Ancrenaz, A. Baccini, S., *et al.* (2012). Understanding the impacts of land-use policies on a threatened species: is there a future for the Bornean orang-utan? *PLoS One* **7**(11): e49142.

Wich, S. A., Garcia-Ulloa, J., Hjalmar, S., Kühl, T., Humle, J., *et al.* (2014). Will oil palm's homecoming spell doom for Africa's great apes? *Current Biology* **24**(14): 1659–1663.

Wilhere, G. F., Maguire, L. A., Scott, J. M., Rachlow, J. L., Goble, D. D., *et al.* (2012). Conflation of values and science: response to Noss et al. *Conservation Biology* **26**: 943–944.

Willems, E. P. and Hill, R. A. (2009). A critical assessment of two species distribution models: a case study of the vervet monkey (Cercopithecus aethiops). *Journal of Biogeography* **36**: 2300–2312.

Wilson, K. A., McBride, M. F., Bode, M., and Possingham, H. P. (2006). Prioritizing global conservation efforts. *Nature* **440**: 337–340.

Wisz, M. S., Pottier, J., Kissling, W. D., Pellissier, L., Lenoir, J., *et al.* (2013). The role of biotic interactions in shaping distributions and realised assemblages of species: implications for species distribution modelling. *Biological Reviews* **88**: 15–30.

Wrangham, R. (2008). Why the link between long-term research and conservation is a case worth making. In: Wrangham, W.R. and Ross, E. (Eds), *Science and Conservation in African Forests*. pp. 1–8. Cambridge: Cambridge University Press.

Wright, P. C. (1992). Primate ecology, rainforest conservation, and economic development: building a national park in Madagascar. *Evolutionary Anthropology* **1**: 25–33.

Index

References to figures are given in italic type. References to tables are given in bold type.